# Student Solution Manual

## to accompany the 3rd edition of

# Vector Calculus, Linear Algebra, and Differential Forms: A Unified Approach

John Hamal Hubbard      Barbara Burke Hubbard

Cornell University
University of Provence

Matrix Editions   Ithaca, NY 14850   MatrixEditions.com

Copyright 2007 by Matrix Editions
214 University Ave. Ithaca, NY 14850
www.MatrixEditions.com

Printed in the United States of America

10 9 8 7 6 5 4 3 2 1

ISBN: 978-0-9715766-4-3

# Contents

# Note to Students

This manual gives solutions to all odd-numbered exercises in the third edition of *Vector Calculus, Linear Algebra, and Differential Forms: A Unified Approach*, (copyright 2007, ISBN 9780971576636). That is the version with the triangle on the cover, not the pale yellow cover (2nd edition) or dark blue (first edition).

You should not attempt to use this manual with the first or second edition of the textbook, as many exercise numbers have been changed (indeed, many exercises in the third edition are new).

Please think about the exercises before looking at the solutions! You will get more out of them that way. Some exercises are straightforward; their solutions are correspondingly straightforward. Others are "entertaining".

You may be tempted go straight to the solution manual to look for problems similar to those assigned as your homework, without reading the text. This is a mistake. You may learn some of the simpler material that way, but if you don't read the text, soon you will find you cannot understand the solutions. The textbook contains a lot of challenging material; you really have to come to terms with the underlying ideas and definitions.

In this manual, we do not number every equation. When we do, we use (1), (2), ... , starting again with (1) in a new solution.

While completing this manual, we discovered several typos in the statements of exercises. We mention them in the relevant solution, with the comment that "in the first printing of the text ... ". As of this writing, there is only one printing of the textbook. The printing number is indicated on the copyright page; below the words "Printed in the United States of America" are the decreasing numbers 10 9 8 7 6 5 4 3 2 1. The right-most number is the number of the printing; when a second printing is issued, the 1 at the end is dropped.

Errata for this manual and for the textbook will be posted at

www.matrixeditions.com

If you think you have found errors either in the text or in this manual, please email us at

hubbard@matrixeditions.com

or

jhh8@cornell.edu

so that we can make the errata lists as complete as possible.

John H. Hubbard and Barbara Burke Hubbard

Solution 0.2.1: The negation in part a is false; the original statement is true: $5 = 4+1$, $13 = 9+4$, $17 = 16 + 1$, $29 = 25 + 4, \ldots$, $97 = 81 + 16, \ldots$.

If you divide a whole number (prime or not) by 4 and get a remainder of 3, then it is never the sum of two squares: 3 is not, 7 is not, 11 is not, etc. You may well be able to prove this; it isn't all that hard. But the original statement about primes is pretty tricky.

The negation in part b is is false also, and the original statement is true: it is the definition of the mapping $x \mapsto x^2$ being continuous. But in part c, the negation is true, and the original statement is false. Indeed, if you take $\epsilon = 1$ and any $\delta > 0$ and set $x = \frac{1}{\delta}$, $y = \frac{1}{\delta} + \frac{\delta}{2}$, then

$$|y^2 - x^2| = |y - x||y + x|$$
$$= \frac{\delta}{2}\left(\frac{2}{\delta} + \frac{\delta}{2}\right) > 1.$$

The original statement says that the function $x \mapsto x^2$ is *uniformly continuous*, and it isn't.

**0.2.1 a.** The negation of the statement is: There exists a prime number such that if you divide it by 4 you have a remainder of 1, but which is not the sum of two squares.

**b.** The negation is: There exist $x \in \mathbb{R}$ and $\epsilon > 0$ such that for all $\delta > 0$ there exists $y \in \mathbb{R}$ with $|y - x| < \delta$ and $|y^2 - x^2| \geq \epsilon$.

**c.** The negation is: There exists $\epsilon > 0$ such that for all $\delta > 0$, there exist $x, y \in \mathbb{R}$ with $|x - y| < \delta$ and $|y^2 - x^2| \geq \epsilon$.

**0.3.1 a.** $(A * B) * (A * B)$      **b.** $(A * A) * (B * B)$      **c.** $A * A$

**0.4.1 a.** No: Many people have more than one aunt, and some have none.

**b.** No, $\frac{1}{0}$ is not defined.

**c.** No.

**0.4.3** Here are some examples:

**a.** "DNA of" from people (excluding clones and identical twins) to DNA patterns.

**b.** $f(x) = x$.

**0.4.5** Here are some examples:

**a.** "Eldest daughter of" from fathers to children.

**b.** $\arctan(x) : \mathbb{R} \to \mathbb{R}$.

**0.4.7** The following are well defined: $g \circ f : A \to C$, $h \circ k : C \to C$, $k \circ g : B \to A$, $k \circ h : A \to A$, and $f \circ k : C \to B$. The others are not, unless some of the sets are subsets of the others. For example, $f \circ g$ is not because the codomain of $g$ is $C$, which is not the domain of $f$, unless $C \subset A$.

**0.4.9 a.** $f\big(g(h(3))\big) = f\big(g(-1)\big) = f(-3) = 8$

**b.** $f\big(g(h(1))\big) = f\big(g(-2)\big) = f(-5) = 25$

**0.4.11** If the argument of the square root is nonnegative, the square root can be evaluated, so the open first and the third quadrants are in the natural domain. The $x$-axis is not (since $y = 0$ there), but the $y$-axis with the origin removed is in the natural domain, since $x/y = 0$ there.

**0.4.13** The function is defined for $\{ x \in \mathbb{R} \mid -1 \leq x < 0, \text{ or } 0 < x \}$. It is also defined for the negative odd integers.

**0.5.1** Without loss of generality we may assume that the polynomial is of odd degree $d$ and that the coefficient of $x^d$ is 1. Write the polynomial

$$x^d + a_{d-1}x^{d-1} + \cdots + a_0.$$

1

Let $A = |a_0| + \cdots + |a_{d-1}| + 1$. Then

$$\left| a_{d-1}A^{d-1} + a_{d-2}A^{d-2} + \cdots + a_0 \right| \le (A-1)A^{d-1},$$

and similarly,

$$\left| a_{d-1}(-A)^{d-1} + a_{d-2}(-A)^{d-2} + \cdots + a_0 \right| \le (A-1)A^{d-1}.$$

Therefore,

$$p(A) \ge A^d - (A-1)A^{d-1} > 0 \quad \text{and} \quad p(-A) \ge (-A)^d + (A-1)A^{d-1} < 0.$$

By the intermediate value theorem, there must exist $c$ with $|c| < A$ such that $p(c) = 0$.

**0.5.3**    Suppose $f : [a,b] \to [a,b]$ is continuous. Then the function $g(x) = x - f(x)$ satisfies the hypotheses of the intermediate value theorem (theorem 0.5.9): $g(a) \le 0$ and $g(b) \ge 0$. So there must exist $x \in [a,b]$ such that $g(x) = 0$, i.e., $f(x) = x$.

**0.5.5** *First solution*

Let $\sum_{k=1}^{\infty} b_k$ be a rearrangement of $\sum_{k=1}^{\infty} a_k$, which means that there exists a bijective map $\alpha : \mathbb{N} \to \mathbb{N}$ such that $b_{\alpha(k)} = a_k$.

Since the series $\sum_{k=1}^{\infty} a_k$ is absolutely convergent, we know it is convergent. Let $A = \sum_{k=1}^{\infty} a_k$. We must show that

We give two solutions to exercise 0.5.5.

$$(\forall \epsilon > 0)(\exists M) \left( m \ge M \implies \left| \sum_{k=1}^{m} b_k - A \right| < \epsilon \right).$$

First choose $N$ such that $\sum_{N+1}^{\infty} |a_k| < \epsilon/2$; in that case,

$$(n > N) \implies \left| \sum_{k=1}^{n} a_k - A \right| \le \sum_{k=n}^{\infty} |a_k| \le \sum_{k=N+1}^{\infty} |a_k| < \epsilon/2.$$

Now set

$$M = \max\{\alpha(1), \ldots, \alpha(N)\},$$

i.e., $M$ is so large that all of $a_1, \ldots, a_N$ appear among $b_1, \ldots, b_M$. Then

$$m > M \implies \left| \sum_{k=1}^{m} b_k - A \right| \le \left| \sum_{k=1}^{N} a_k - A \right| + \sum_{k=N+1}^{\infty} |a_k| < \epsilon.$$

*Second solution*

The series $\sum_{k=1}^{\infty} a_k + |a_k|$ is a convergent series of positive numbers, so its sum is the sup of all finite sums of terms.

If $\sum_{k=1}^{\infty} b_k$ is a rearrangement of the same series, then $\sum_{k=1}^{\infty} b_k + |b_k|$ is also a rearrangement of $\sum_{k=1}^{\infty} a_k + |a_k|$.

But the finite sums of terms of the series $\sum_{k=1}^{\infty} a_k + |a_k|$ and $\sum_{k=1}^{\infty} b_k + |b_k|$ are exactly the same set of numbers, so these two series converge to the same limit. The same argument says that

$$\sum_{k=1}^{\infty} |a_k| = \sum_{k=1}^{\infty} |b_k|.$$

The first equality is justified because both series

$$\sum_{k=1}^{\infty}(b_k+|b_k|) \quad \text{and} \quad \sum_{k=1}^{\infty}|b_k|$$

converge, so their difference does also.

Solution 0.6.1, part a: There are infinitely many ways of solving this problem, and many seem just as natural as the one we propose.

Solution 0.6.3: The same argument holds if one uses the function $g$, with derivative

$$g'(x)=\frac{\pi}{2(1+\cos^2\frac{\pi x}{2})}.$$

Thus

$$\sum_{k=1}^{\infty}b_k=\sum_{k=1}^{\infty}(b_k+|b_k|)-\sum_{k=1}^{\infty}|b_k|=\sum_{k=1}^{\infty}(a_k+|a_k|)-\sum_{k=1}^{\infty}|a_k|=\sum_{k=1}^{\infty}a_k.$$

**0.6.1**  **a.** Begin by listing the integers between $-1$ and $1$, then list the numbers between $-2$ and $2$ that can be written with denominators $\leq 2$ and which haven't already been listed, then list the rationals between $-3$ and $3$ that can be written with denominators $\leq 3$ and haven't already been listed, etc. This will eventually list all rationals. Here is the beginning of the list:

$$-1,0,1,-2,-\frac{3}{2},-\frac{1}{2},\frac{1}{2},\frac{3}{2},2,-3,-\frac{8}{3},-\frac{5}{2},-\frac{7}{3},-\frac{5}{3},-\frac{4}{3},-\frac{2}{3},\frac{1}{3},\frac{1}{3},\frac{2}{3},\cdots$$

**b.** Just as before, list first the finite decimals in $[-1,1]$ with at most one digit after the decimal point (there are 21 of them), then the ones in $[-2,2]$ with at most two digits after the decimal, and which haven't been listed earlier (there are 380 of these), etc.

**0.6.3** It is easy to write a bijective map $(-1,1)\to\mathbb{R}$. For instance,

$$f(x)=\frac{x}{1-x^2} \quad \text{or} \quad g(x)=\tan\frac{\pi x}{2}.$$

The derivative of the first mapping is

$$f'(x)=\frac{(1-x^2)+2x^2}{(1-x^2)^2}=\frac{1+x^2}{(1-x^2)^2}>0$$

so the mapping is monotone increasing on $(-1,1)$, hence injective. Since it is continuous on $(-1,1)$ and

$$\lim_{x\searrow-1}f(x)=-\infty \quad \text{and} \quad \lim_{x\nearrow+1}f(x)=+\infty,$$

it is surjective by the intermediate value theorem.

**0.6.5**  **a.** To the right, if $a\in A$ is an element of a chain, then it can be continued:

$$a,f(a),g\big(f(a)\big),f\big(g(f(a))\big),\ldots,$$

and this is clearly the only possible continuation.

Similarly, if $b$ is an element of a chain, then the chain can be continued

$$b,g(b),f\big(g(b)\big),g\big(f(g(b))\big),\ldots,$$

and again this is the only possible continuation.

Let us set $A_1=g(B)$ and $B_1=f(A)$. Let $f_1:A_1\to B$ be the unique map such that $f_1(g(b))=b$ for all $b\in B$, and let $g_1:B_1\to A$ be the unique map such that $g_1(f(a))=a$ for all $a\in A$. To the left, if $a\in A_1$ is an element of a chain, we can extend to $f_1(a),a$. Then if $f(a_1)\in B_1$, we can extend one further, to $g_1(f_1(a))$, and if $g_1(f_1(a))\in A_1$, we can extend further to

$$f_1\big(g_1(f_1(a))\big),\ g_1\big(f_1(a)\big),\ f_1(a),a.$$

This can either continue forever or at some point we will run into an element of $A$ that is not in $A_1$ or into an element of $B$ that is not in $B_1$; the chain necessarily ends there.

b. As we saw, any element of $A$ or $B$ is part of a unique infinite chain to the right and of a unique finite or infinite chain to the left.

**Part c:**

(1) can be uniquely continued forever

(2) can be continued to an element of $A$ that is not in the image of $g$

(3) can be continued to an element of $B$ that is not in the image of $f$

. c. Since every element of $A$ and of $B$ is an element of a unique chain, and since this chain satisfies either (1), (2), or (3) and these are exclusive, the mapping $h$ is well defined.

If $h(a_1) = h(a_2)$ and $a_1$ belongs to a chain of type 1 or 2, then so does $f(a_1)$, hence so does $f(a_2)$, hence so does $a_2$, and then $h(a_1) = h(a_2)$ implies $a_1 = a_2$, since $f$ is injective.

Now suppose that $a_1$ belongs to a chain of type 3. Then $a_1$ is not the first element of the list, so $a_1 \in A_1$, and $h(a_1) = f_1(a_1)$ is well defined. The element $h(a_2)$ is a part of the same chain, hence also of type 3, and $h(a_2) = f_1(a_2)$. But then $a_1 = g(f_1(a_1)) = g(f_1(a_2)) = a_2$. So $h$ is injective. Now any element $b \in B$ belongs to a maximal chain. If this chain is of type 1 or 2, then $b$ is not the first element of the chain, so $b \in B_1$ and $h(g_1(b)) = f(g_1(b)) = b$, so $b$ is in the image. If $b$ is in a chain of type 3, then $b = f_1(g(b)) = h(g(b))$, so again $b$ is in the image of $h$, proving that $h$ is surjective.

d. This is the only entertaining part of the problem. In this case, there is only one chain of type 1, the chain of all 0's. There are only two chains of type 2, which are

$$+1 \xrightarrow{f} \frac{1}{2} \xrightarrow{g} \frac{1}{2} \xrightarrow{f} \frac{1}{4} \xrightarrow{g} \frac{1}{4} \xrightarrow{f} \frac{1}{8} \cdots$$

$$-1 \xrightarrow{f} -\frac{1}{2} \xrightarrow{g} -\frac{1}{2} \xrightarrow{f} -\frac{1}{4} \xrightarrow{g} -\frac{1}{4} \xrightarrow{f} -\frac{1}{8} \cdots$$

All other chains are of type 3, and end to the left with a number in $(1/2, 1)$ or in $(-1, -1/2)$, which are the points in $B - f(A)$. Such a sequence might be

$$\frac{3}{4} \xrightarrow{g} \frac{3}{4} \xrightarrow{f} \frac{3}{8} \xrightarrow{g} \frac{3}{8} \xrightarrow{f} \frac{3}{16} \cdots$$

The existence of this function $h$ is *Bernstein's theorem*.

Following our definition of $h$, we see that

$$h(x) = \begin{cases} 0 & \text{if } x = 0 \\ x/2 & \text{if } x = \pm 1/2^k \text{ for some } k \geq 0 \\ x & \text{if } x \neq \pm 1/2^k \text{ for some } k \geq 0. \end{cases}$$

**Solution 0.6.7:** Two sets $A$ and $B$ have the same cardinality if there exists an invertible mapping $A \to B$. Here, finding such an invertible map would be difficult, so instead we use Bernstein's theorem, which says that if there is an injective map $A \to B$ and an injective map $B \to A$, then there is a invertible map $A \to B$.

**0.6.7 a.** There is an obvious injective map $[0, 1) \to [0, 1) \times [0, 1)$, given simply by $g : x \mapsto (x, 0)$.

We need to construct an injective map in the opposite direction; that is a lot harder. Take a point in $(x, y) \in [0, 1) \times [0, 1)$, and write both coordinates as decimals:

$$x = .a_1 a_2 a_3 \ldots \quad \text{and} \quad y = .b_1 b_2 b_3 \ldots ;$$

Remember that
injective = one to one
surjective = onto
bijective = invertible.

Bernstein's theorem is the object of exercise 0.6.5.

The map $(f, \mathrm{id}) : \mathbb{R} \times \mathbb{R} \to \mathbb{R}$ maps the first $\mathbb{R}$ to $\mathbb{R} \times \mathbb{R}$ and the second $\mathbb{R}$ to itself.

Solution 0.7.3, part b: If you had trouble with this, note that it follows from part a of proposition 0.7.5 (geometrical representation of multiplication of complex numbers) that

$$\left| \frac{1}{z} \right| = \frac{1}{|z|},$$

since $zz^{-1} = 1$, which has length 1. The modulus (absolute value) of $3 + 4i$ is $\sqrt{25} = 5$, so the modulus of $(3 + 4i)^{-1}$ is $1/5$. It follows from part b of proposition 0.7.5 that the polar angle of $z^{-1}$ is minus the polar angle of $z$, since the two angles sum to 0, which is the polar angle of the product $zz^{-1} = 1$. The polar angle of $3 + 4i$ is $\arccos(3/5)$.

if either number can be written in two ways, one ending in 0's and the other in 9's, use the one ending in 0's.

Now consider the number $f(x, y) = .a_1 b_1 a_2 b_2 a_3 b_3 \ldots$. The mapping $f$ is injective: using the even and the odd digits of $f(x, y)$ allows you to reconstruct $x$ and $y$. The only problem you might have is if $f(x, y)$ could be written in two different ways, but as constructed $f(x, y)$ will never end in all 9's, so this doesn't happen.

Note that this mapping is not surjective; for instance, $.191919\ldots$ is not in the image. But Bernstein's theorem guarantees that since there are injective maps both ways, there is a bijection between $[0, 1)$ and $[0, 1) \times [0, 1)$.

b. The proof of part a works also to construct a bijective mapping $(0, 1) \to (0, 1) \times (0, 1)$. But it is easy to construct a bijective mapping $(0, 1) \to \mathbb{R}$, for instance $x \mapsto \cot(\pi x)$. If we compose these mappings, we find bijective maps

$$\mathbb{R} \to (0, 1) \to (0, 1) \times (0, 1) \to \mathbb{R} \times \mathbb{R}.$$

c. We can use part b repeatedly to construct bijective maps

$$\mathbb{R} \xrightarrow{f} \mathbb{R} \times \mathbb{R} \xrightarrow{(f, \mathrm{id})} (\mathbb{R} \times \mathbb{R}) \times \mathbb{R} = \mathbb{R} \times \mathbb{R}^2 \xrightarrow{(f, \mathrm{id})} (\mathbb{R} \times \mathbb{R}) \times \mathbb{R}^2$$

$$= \mathbb{R} \times \mathbb{R}^3 \xrightarrow{(f, \mathrm{id})} (\mathbb{R} \times \mathbb{R}) \times \mathbb{R}^3 = \mathbb{R} \times \mathbb{R}^4 \ldots$$

Continuing this way, it is easy to get a bijective map $\mathbb{R} \to \mathbb{R}^n$.

**0.6.9** It is not possible. For suppose it were, and consider as in equation 0.6.2 the decimal made up of the entries on the diagonal. If this number is rational, then the digits are eventually periodic, and if you do anything systematic to these digits, such as changing all digits that are not 7 to 7's, and changing 7's to 5's, the sequence of digits obtained will still be eventually periodic, so it will represent a rational number. This rational number must appear someplace in the sequence, but it doesn't, since it has a different $k$th digit than the $k$th number for every $k$.

The only weakness in this argument is that it might be a number that can be written in two different ways, but it isn't, since it has only 5's and 7's as digits.

**0.7.1** "Modulus of $z$," "absolute value of $z$," and $|z|$ are synonyms. "Real part of $z$" is the same as $\operatorname{Re} z = a$. "Imaginary part of $z$" is the same as $\operatorname{Im} z = b$. The "complex conjugate of $z$" is the same as $\bar{z}$.

**0.7.3** a. The absolute value of $2 + 4i$ is $|2 + 4i| = 2\sqrt{5}$. The argument (polar angle) of $2 + 4i$ is $\arccos 1/\sqrt{5}$, which you could also write as $\arctan 2$.

b. The absolute value of $(3 + 4i)^{-1}$ is $1/5$. The argument (polar angle) is $-\arccos(3/5)$.

c. The absolute value of $(1 + i)^5$ is $4\sqrt{2}$. The argument is $5\pi/4$. (The complex number $1 + i$ has absolute value $\sqrt{2}$ and polar angle $\pi/4$. De Moivre's formula says how to compute these for $(1 + i)^5$.)

d. The absolute value of $1 + 4i$ is $\sqrt{17}$; the argument is $\arccos 1/\sqrt{17}$.

**0.7.5**  Parts 1–4 are immediate. For part 5, we find

$$(z_1 z_2)z_3 = \big((x_1 x_2 - y_1 y_2) + i(y_1 x_2 + x_1 y_2)\big)(x_3 + iy_3)$$
$$= (x_1 x_2 x_3 - y_1 y_2 x_3 - y_1 x_2 y_3 - x_1 y_2 y_3)$$
$$+ i(x_1 x_2 y_3 - y_1 y_2 y_3 + y_1 x_2 x_3 + x_1 y_2 x_3),$$

which is equal to

$$z_1(z_2 z_3) = (x_1 + iy_1)\big((x_2 x_3 - y_2 y_3) + i(y_2 x_3 + x_2 y_3)\big)$$
$$= (x_1 x_2 x_3 - x_1 y_2 y_3 - y_1 y_2 x_3 - y_1 x_2 y_3)$$
$$+ i(y_1 x_2 x_3 - y_1 y_2 y_3 + x_1 y_2 x_3 + x_1 x_2 y_3).$$

Parts 6 and 7 are immediate. For part 8, multiply out:

$$(a + ib)\left(\frac{a}{a^2 + b^2} - i\frac{b}{a^2 + b^2}\right) = \frac{a^2}{a^2 + b^2} + \frac{b^2}{a^2 + b^2} + i\left(\frac{ab}{a^2 + b^2} - \frac{ab}{a^2 + b^2}\right)$$
$$= 1 + i0 = 1.$$

Part 9 is also a matter of multiplying out:

$$z_1(z_2 + z_3) = (x_1 + iy_1)\big((x_2 + iy_2) + (x_3 + iy_3)\big)$$
$$= (x_1 + iy_1)\big((x_2 + x_3) + i(y_2 + y_3)\big)$$
$$= x_1(x_2 + x_3) - y_1(y_2 + y_3) + i\big(y_1(x_2 + x_3) + x_1(y_2 + y_3)\big)$$
$$= x_1 x_2 - y_1 y_2 + i(y_1 x_2 + x_1 y_2) + x_1 x_3 - y_1 y_3 + i(y_1 x_3 + x_1 y_3)$$
$$= z_1 z_2 + z_1 z_3.$$

Solution 0.7.7: Remember that the set of points such that the sum of their distances to two points is constant, is an ellipse, with foci at those points.

**0.7.7 a.** The equation $|z - u| + |z - v| = c$ represents an ellipse with foci at $u$ and $v$, at least if $c > |u - v|$. If $c = |u - v|$ it is the degenerate ellipse consisting of just the segment $[u, v]$, and if $c < |u - v|$ it is empty, by the triangle inequality, which asserts that if there is a $z$ satisfying the equality, then

$$c < |u - v| \le |u - z| + |z - v| = c.$$

b. Set $z = x + iy$; the inequality $|z| < 1 - \operatorname{Re} z$ becomes

$$\sqrt{x^2 + y^2} < 1 - x,$$

corresponding to a region bounded by the curve of equation

$$\sqrt{x^2 + y^2} = 1 - x.$$

We should worry whether the squaring introduced parasitic points, where $-\sqrt{x^2 + y^2} < 1 - x$, but this is not the case, since $1 - x$ is positive throughout the region.

If we square this equation, we will get the curve of equation

$$x^2 + y^2 = 1 - 2x + x^2, \quad \text{i.e.,} \quad x = \frac{1}{2}(1 - y^2),$$

which is a parabola lying on its side. The original inequality corresponds to the inside of the parabola.

**0.7.9** a. The quadratic formula gives $x = \dfrac{-i \pm \sqrt{-1-8}}{2}$, so the solutions are $x = i$ and $x = -2i$.

b. In this case, the quadratic formula gives

$$x^2 = \frac{-1 \pm \sqrt{1-8}}{2} = \frac{-1 \pm i\sqrt{7}}{2}.$$

Each of these numbers has two square roots, which we still need to find.

One way, probably the best, is to use the polar form; this gives

$$x^2 = r(\cos\theta \pm i\sin\theta),$$

where

$$r = \frac{\sqrt{1+7}}{2} = \sqrt{2}, \quad \theta = \pm\arccos{-\frac{1}{2\sqrt{2}}} \approx 1.2094\ldots \text{ radians.}$$

Thus the four roots are

$$\pm\sqrt[4]{2}(\cos\theta/2 + i\sin\theta/2) \quad \text{and} \quad \pm\sqrt[4]{2}(\cos\theta/2 - i\sin\theta/2).$$

c. Multiplying the first equation through by $(1+i)$ and the second by $i$ gives

$$i(1+i)x - (2+i)(1+i)y = 3(1+i)$$
$$i(1+i)x - \qquad\qquad\;\; y = 4i,$$

which gives

$$-(2+i)(1+i)y + y = 3 - i, \quad \text{i.e.,} \quad y = i + \frac{1}{3}.$$

Substituting this value for $y$ then gives $x = \frac{7}{3} - \frac{8}{3}i$.

**0.7.11** a. These are the vertical line $x = 1$ and the circle centered at the origin of radius 3.

b. Use $Z = X + iY$ as the variable in the codomain. Then

$$(1+iy)^2 = 1 - y^2 + 2iy = X + iY$$

gives $1 - X = y^2 = Y^2/4$. Thus the image of the line is the curve of equation $X = 1 - Y^2/4$, which is a parabola with horizontal axis.

The image of the circle is another circle, centered at the origin, of radius 9, i.e., the curve of equation $X^2 + Y^2 = 81$.

c. This time use $Z = X + iY$ as the variable in the domain. Then the inverse image of the line $= \operatorname{Re} z = 1$ is the curve of equation

$$\operatorname{Re}(X + iY)^2 = X^2 - Y^2 = 1,$$

which is a hyperbola. The inverse image of the curve of equation $|z| = 3$ is the curve of equation $|Z^2| = |Z|^2 = 3$, i.e., $|Z| = \sqrt{3}$, the circle of radius $\sqrt{3}$ centered at the origin.

**0.7.13** a. The cube roots of 1 are

$$1, \quad \cos\frac{2\pi}{3} + i\sin\frac{2\pi}{3} = -\frac{1}{2} + \frac{i\sqrt{3}}{2}, \quad \cos\frac{4\pi}{3} + i\sin\frac{4\pi}{3} = -\frac{1}{2} - \frac{i\sqrt{3}}{2}.$$

b. The fourth roots of 1 are $1$, $i$, $-1$, $-i$.

c. The sixth roots of 1 are

$$1, \quad -1, \quad \frac{1}{2}+i\frac{\sqrt{3}}{2}, \quad \frac{1}{2}-i\frac{\sqrt{3}}{2}, \quad -\frac{1}{2}+i\frac{\sqrt{3}}{2}, \quad -\frac{1}{2}-i\frac{\sqrt{3}}{2}$$

**0.7.15**  a. The fifth roots of 1 are

$$\cos 2\pi k/5 + i \sin 2\pi k/5, \quad \text{for } k = 0, 1, 2, 3, 4.$$

The point of the question is to find these numbers in some more manageable form. One possible approach is to set $\theta = 2\pi/5$, and to observe that $\cos 4\theta = \cos \theta$. If you set $x = \cos \theta$, this leads to the equation

$$2(2x^2 - 1)^2 - 1 = x \quad \text{i.e.,} \quad 8x^4 - 8x^2 - x + 1 = 0.$$

This still isn't too manageable, until you start asking what other angles satisfy $\cos 4\theta = \cos \theta$. Of course $\theta = 0$ does, meaning that $x = 1$ is one root of our equation. But $\theta = 2\pi/3$ does also, meaning that $-1/2$ is also a root. Thus we can divide:

$$\frac{8x^4 - 8x^2 - x + 1}{(x-1)(2x+1)} = 4x^2 + 2x - 1,$$

and $\cos 2\pi/5$ is the positive root of that quadratic equation, i.e.,

$$\cos \frac{2\pi}{5} = \frac{\sqrt{5}-1}{4}, \text{ which gives } \sin \frac{2\pi}{5} = \frac{\sqrt{10+2\sqrt{5}}}{4}.$$

The fifth roots of 1 are now

$$1, \quad \frac{\sqrt{5}-1}{4} \pm i\frac{\sqrt{10+2\sqrt{5}}}{4}, \quad -\frac{\sqrt{5}+1}{4} \pm i\frac{\sqrt{10-2\sqrt{5}}}{4}.$$

b. It is straightforward to draw a line segment of length $(\sqrt{5}-1)/4$: construct a rectangle with sides 1 and 2, so the diagonal has length $\sqrt{5}$. Then subtract 1 and divide twice by 2, as shown in the figure below.

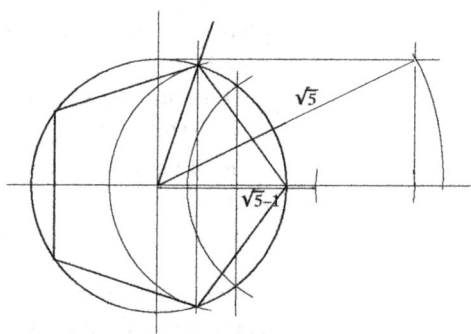

FIGURE FOR SOLUTION 0.7.15.

**1.1.1**  a. $\begin{bmatrix} 1 \\ 3 \end{bmatrix} + \begin{bmatrix} 2 \\ 1 \end{bmatrix} = \begin{bmatrix} 3 \\ 4 \end{bmatrix}$  b. $2\begin{bmatrix} 2 \\ 4 \end{bmatrix} = \begin{bmatrix} 4 \\ 8 \end{bmatrix}$

c. $\begin{bmatrix} 1 \\ 3 \end{bmatrix} - \begin{bmatrix} 2 \\ 1 \end{bmatrix} = \begin{bmatrix} 1-2 \\ 3-1 \end{bmatrix} = \begin{bmatrix} -1 \\ 2 \end{bmatrix}$  d. $\begin{bmatrix} 3 \\ 2 \end{bmatrix} + \vec{e}_1 = \begin{bmatrix} 3 \\ 2 \end{bmatrix} + \begin{bmatrix} 1 \\ 0 \end{bmatrix} = \begin{bmatrix} 4 \\ 2 \end{bmatrix}$

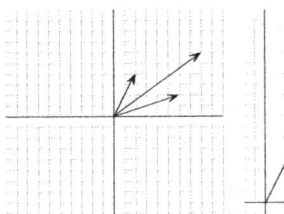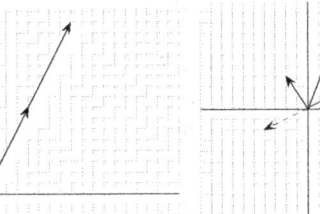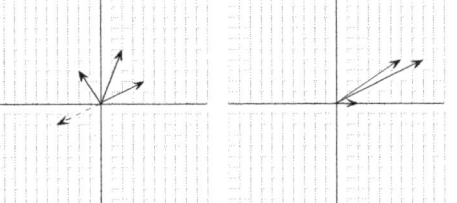

FIGURE FOR SOLUTION 1.1.1. From left: (a), (b), (c), and (d).

**1.1.3**  a. $\vec{v} \in \mathbb{R}^3$   b. $L \subset \mathbb{R}^2$   c. $C \subset \mathbb{R}^3$   d. $\mathbf{x} \in \mathbb{C}^2$,
e. $B_0 \subset B_1 \subset B_2, \ldots$

**1.1.5**  a. $\sum_{i=1}^n \vec{e}_i$   b. $\sum_{i=1}^n i\vec{e}_i$   c. $\sum_{i=3}^n i\vec{e}_i$

**1.1.7**  The vector field in part a points straight up everywhere. Its length depends only on how far you are from the $z$-axis, and it gets longer and longer the further you get from the $z$-axis; it vanishes on the $z$-axis. The vector field in part b is simply rotation in the $(x, y)$-plane, like (f) in exercise 1.1.6. But the $z$-component is down when $z > 0$ and up when $z < 0$. The vector field in part c spirals out in the $(x, y)$-plane, like (h) in exercise 1.1.6. Again, the $z$-component is down when $z > 0$ and up when $z < 0$.

a. 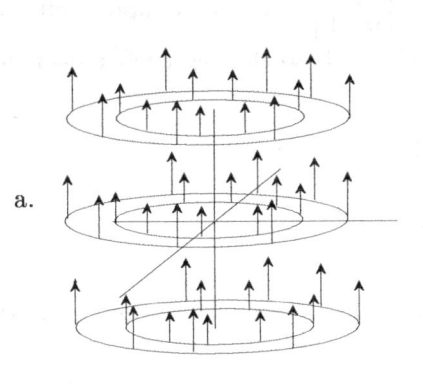  b. 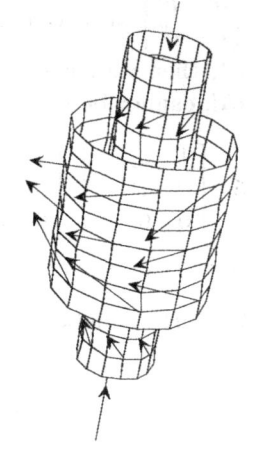  c.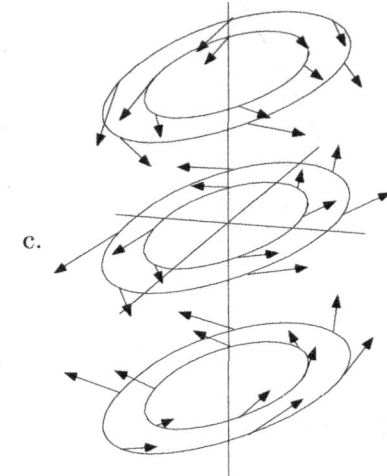

**1.2.1**  i. $2 \times 3$   ii. $2 \times 2$   iii. $3 \times 2$   iv. $3 \times 4$   v. $3 \times 3$

b. The matrices i and v can be multiplied on the right by the matrices iii, iv, v; the matrices ii and iii on the right by the matrices i and ii.

**1.2.3** a. $\begin{bmatrix} 5 \\ 2 \end{bmatrix}$     b. $\begin{bmatrix} 6 & 16 & 2 \end{bmatrix}$

**1.2.5** a. True: $(AB)^\top = B^\top A^\top = B^\top A$

b. True: $(A^\top B)^\top = B^\top (A^\top)^\top = B^\top A = B^\top A^\top$

c. False: $(A^\top B)^\top = B^\top (A^\top)^\top = B^\top A \neq BA$

d. False: $(AB)^\top = B^\top A^\top \neq A^\top B^\top$

**1.2.7** The matrices a and d have no transposes here. The matrices b and f are transposes of each other. The matrices c and e are transposes of each other.

**1.2.9** a. $A^\top = \begin{bmatrix} 1 & 1 \\ 0 & 0 \end{bmatrix}$, $B^\top = \begin{bmatrix} 1 & 2 \\ 0 & 1 \\ 1 & 0 \end{bmatrix}$   b. $(AB)^\top = B^\top A^\top = \begin{bmatrix} 1 & 1 \\ 0 & 0 \\ 1 & 1 \end{bmatrix}$

c. $(AB)^\top = \begin{bmatrix} 1 & 0 & 1 \\ 1 & 0 & 1 \end{bmatrix}^\top = \begin{bmatrix} 1 & 1 \\ 0 & 0 \\ 1 & 1 \end{bmatrix}$

d. The matrix multiplication $A^\top B^\top$ is impossible.

**1.2.11** The expressions b, c, d, f, g, and i make no sense.

**1.2.13** The trivial case is when $a = b = c = d = 0$; then obviously $ad - bc = 0$ and the matrix is not invertible. Let us suppose $d \neq 0$. (If we suppose that any other entry is nonzero, the proof would work the same way.) If $ad = bc$, then the first row is a multiple of the second: we can write $a = \frac{b}{d}c$ and $b = \frac{b}{d}d$, so the matrix is $A = \begin{bmatrix} \frac{b}{d}c & \frac{b}{d}d \\ c & d \end{bmatrix}$.

To show that $A$ is not invertible, we need to show that there is no matrix $B = \begin{bmatrix} a' & b' \\ c' & d' \end{bmatrix}$ such that $AB = \begin{bmatrix} 1 & 0 \\ 0 & 1 \end{bmatrix}$. But if the upper left corner of $AB$ is 1, then we have $\frac{b}{d}(a'c + c'd) = 1$, so the lower left corner, which is $a'c + c'd$, cannot be 0.

**1.2.15** $\begin{bmatrix} 1 & a & b \\ 0 & 1 & c \\ 0 & 0 & 1 \end{bmatrix} \begin{bmatrix} 1 & x & y \\ 0 & 1 & z \\ 0 & 0 & 1 \end{bmatrix} = \begin{bmatrix} 1 & a+x & az+b+y \\ 0 & 1 & c+z \\ 0 & 0 & 1 \end{bmatrix}$

So $x = -a$, $z = -c$, and $y = ac - b$.

**1.2.17** With the labeling shown in the margin, the adjacency matrices are

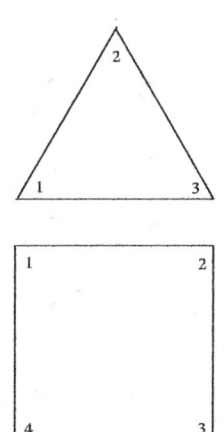

Labeling for solution 1.2.17

a.   $A_T = \begin{bmatrix} 0 & 1 & 1 \\ 1 & 0 & 1 \\ 1 & 1 & 0 \end{bmatrix}$     $A_S = \begin{bmatrix} 0 & 1 & 0 & 1 \\ 1 & 0 & 1 & 0 \\ 0 & 1 & 0 & 1 \\ 1 & 0 & 1 & 0 \end{bmatrix}$

b.    $A_T^2 = \begin{bmatrix} 2 & 1 & 1 \\ 1 & 2 & 1 \\ 1 & 1 & 2 \end{bmatrix}$    $A_T^3 = \begin{bmatrix} 2 & 3 & 3 \\ 3 & 2 & 3 \\ 3 & 3 & 2 \end{bmatrix}$    $A_T^4 = \begin{bmatrix} 6 & 5 & 5 \\ 5 & 6 & 5 \\ 5 & 5 & 6 \end{bmatrix}$

$$A_T^5 = \begin{bmatrix} 10 & 11 & 11 \\ 11 & 10 & 11 \\ 11 & 11 & 10 \end{bmatrix}$$

$$A_S^2 = \begin{bmatrix} 2 & 0 & 2 & 0 \\ 0 & 2 & 0 & 2 \\ 2 & 0 & 2 & 0 \\ 0 & 2 & 0 & 2 \end{bmatrix} \qquad A_S^3 = \begin{bmatrix} 0 & 4 & 0 & 4 \\ 4 & 0 & 4 & 0 \\ 0 & 4 & 0 & 4 \\ 4 & 0 & 4 & 0 \end{bmatrix} \qquad A_S^4 = \begin{bmatrix} 8 & 0 & 8 & 0 \\ 0 & 8 & 0 & 8 \\ 8 & 0 & 8 & 0 \\ 0 & 8 & 0 & 8 \end{bmatrix}$$

$$A_S^5 = \begin{bmatrix} 0 & 16 & 0 & 16 \\ 16 & 0 & 16 & 0 \\ 0 & 16 & 0 & 16 \\ 16 & 0 & 16 & 0 \end{bmatrix}$$

The diagonal entries of $A^n$ are the number of walks we can take of length $n$ that take us back to our starting point.

c. In a triangle, by symmetry there are only two different numbers: the number $a_n$ of walks of length $n$ from a vertex to itself, and the number $b_n$ of walks of length $n$ from a vertex to a different vertex. The recurrence relation relating these is

$$a_{n+1} = 2b_n \quad \text{and} \quad b_{n+1} = a_n + b_n.$$

These reflect that to walk from a vertex $V_1$ to itself in time $n+1$, at time $n$ we must be at either $V_2$ or $V_3$, but to walk from a vertex $V_1$ to a different vertex $V_2$ in time $n+1$, at time $n$ we must be either at $V_1$ or at $V_3$. If $|a_n - b_n| = 1$, then $a_{n+1} - b_{n+1} = |2b_2 - (a_n + b_n)| = |b_n - a_n| = 1$.

d. Color two opposite vertices of the square black and the other two white. Every move takes you from a vertex to a vertex of the opposite color. Thus if you start at time 0 on black, you will be on black at all even times, and on white at all odd times, and there will be no walks of odd length from a vertex to itself.

e. Suppose such a coloring in black and white exists; then every walk goes from black to white to black to white ..., in particular the $(B, B)$ and the $(W, W)$ entries of $A^n$ are 0 for all odd $n$, and the $(B, W)$ and $(W, B)$ entries are 0 for all even $n$. Moreover, since the graph is connected, for any pair of vertices there is a walk of some length $m$ joining them, and then the corresponding entry is nonzero for $m$, $m + 2$, $m + 4, \ldots$ since you can go from the point of departure to the point of arrival in time $m$, and then bounce back and forth between this vertex and one of its neighbors.

Conversely, suppose the entries of $A^n$ are zero or nonzero as described, and look at the top line of $A^n$, where $n$ is chosen sufficiently large so that any entry that is ever nonzero is nonzero for $A^{n-1}$ or $A^n$. The entries correspond to pairs of vertices $(V_1, V_i)$; color in white the vertices $V_i$ for which the $(1, i)$ entry of $A^n$ is zero, and in black those for which the $(1, i)$

entry of $A^{n+1}$ is zero. By hypothesis, we have colored all the vertices. It remains to show that adjacent vertices have different colors. Take a path of length $m$ from $V_1$ to $V_i$. If $V_j$ is adjacent to $V_i$, then there certainly exists a path of length $m+1$ from $V_1$ to $V_j$, namely the previous path, extended by one to go from $V_i$ to $V_j$. Thus $V_i$ and $V_j$ have opposite colors.

**1.2.19**

a. $B = \begin{bmatrix} 1 & 1 & 0 & 1 & 0 & 1 & 0 & 0 \\ 1 & 1 & 1 & 0 & 0 & 0 & 1 & 0 \\ 0 & 1 & 1 & 1 & 0 & 0 & 0 & 1 \\ 1 & 0 & 1 & 1 & 1 & 0 & 0 & 0 \\ 0 & 0 & 0 & 1 & 1 & 1 & 0 & 1 \\ 1 & 0 & 0 & 0 & 1 & 1 & 1 & 0 \\ 0 & 1 & 0 & 0 & 0 & 1 & 1 & 1 \\ 0 & 0 & 1 & 0 & 1 & 0 & 1 & 1 \end{bmatrix}$

b. $B^2 = \begin{bmatrix} 4 & 2 & 2 & 2 & 2 & 2 & 2 & 0 \\ 2 & 4 & 2 & 2 & 0 & 2 & 2 & 2 \\ 2 & 2 & 4 & 2 & 2 & 0 & 2 & 2 \\ 2 & 2 & 2 & 4 & 2 & 2 & 0 & 2 \\ 2 & 0 & 2 & 2 & 4 & 2 & 2 & 2 \\ 2 & 2 & 0 & 2 & 2 & 4 & 2 & 2 \\ 2 & 2 & 2 & 0 & 2 & 2 & 4 & 2 \\ 0 & 2 & 2 & 2 & 2 & 2 & 2 & 4 \end{bmatrix}$    $B^3 = \begin{bmatrix} 10 & 10 & 6 & 10 & 6 & 10 & 6 & 6 \\ 10 & 10 & 10 & 6 & 6 & 6 & 10 & 6 \\ 6 & 10 & 10 & 10 & 6 & 6 & 6 & 10 \\ 10 & 6 & 10 & 10 & 10 & 6 & 6 & 6 \\ 6 & 6 & 6 & 10 & 10 & 10 & 6 & 10 \\ 10 & 6 & 6 & 6 & 10 & 10 & 10 & 6 \\ 6 & 10 & 6 & 6 & 6 & 10 & 10 & 10 \\ 6 & 6 & 10 & 6 & 10 & 6 & 10 & 10 \end{bmatrix}$

Solution 1.2.19, part b: There are three ways to go from a vertex $V_i$ to an adjacent vertex $V_j$ and back again in three steps: go, return, stay; go, stay, return; and stay, go, return. Each vertex $V_i$ has three adjacent vertices that can play the role of $V_j$. That brings us to nine. The couch potato itinerary "stay, stay, stay" brings us to 10.

The diagonal entries of $B^3$ correspond to the fact that there are exactly 10 loops of length 3 going from any given vertex back to itself.

**1.2.21**  a. The proof is the same as with unoriented walks (proposition 1.2.23): first we state that if $B_n$ is the $n \times n$ matrix whose $i,j$th entry is the number of walks of length $n$ from $V_i$ to $V_j$, then $B_1 = A^1 = A$ for the same reasons as in the proof of proposition 1.2.23. Here again if we assume the proposition true for $n$, we have:

$$(B_{n+1})_{i,j} = \sum_{k=1}^{n} (B_n)_{i,k} (B_1)_{k,j} = \sum_{k=1}^{n} (A^n)_{i,k} A_{k,j} = (A^{n+1})_{i,j}.$$

So $A^n = B_n$ for all n.

Solution 1.2.21, part b: The first column of an adjacency corresponds to vertex 1, the second to vertex 2, and so on, and the same for the rows. If the matrix is upper triangle, for example,

$\begin{bmatrix} 1 & 1 & 1 & 0 \\ 0 & 0 & 1 & 0 \\ 0 & 0 & 1 & 1 \\ 0 & 0 & 0 & 1 \end{bmatrix}$, you can go from

vertex 1 to vertex 2, but not from 2 to 1; from 2 to 3, but not from 3 to 2, and so on: once you have gone from a lower-numbered vertex to a higher-numbered vertex, there is no returning.

b. If the adjacency matrix is upper triangular, then you can only go from a lower number vertex to a higher number vertex; if it is lower triangular, you can only go from a higher number vertex to a lower number vertex. If it is diagonal, you can never go from any vertex to any other.

**1.2.23**  a. Let $A = \begin{bmatrix} a & 1 & 0 \\ b & 0 & 1 \end{bmatrix}$    and    let $B = \begin{bmatrix} 0 & 0 \\ 1 & 0 \\ 0 & 1 \end{bmatrix}$. Then $AB = I$.

b. Whatever matrix one multiplies $B$ by on the right, the top left corner of the resultant matrix will always be 0 when we need it to be 1. So the matrix $B$ has no right inverse.

c. With $A$ and $B$ as in part a, write $I^\top = I = AB = (AB)^\top = B^\top A^\top$. So $A^\top$ is a right inverse for $B^\top$, so $B^\top$ has infinitely many right inverses.

**1.3.1 a.** Every linear transformation $T : \mathbb{R}^4 \to \mathbb{R}^2$ is given by a $2 \times 4$ matrix. For example, $A = \begin{bmatrix} 1 & 0 & 1 & 2 \\ 3 & 2 & 1 & 7 \end{bmatrix}$ is a linear transformation $T : \mathbb{R}^4 \to \mathbb{R}^2$.

**b.** Any row matrix 3 wide will do, for example, $[1, -1, 2]$; such a matrix takes a vector in $\mathbb{R}^3$ and gives a number.

**1.3.3** A. $\mathbb{R}^4 \to \mathbb{R}^3$        B. $\mathbb{R}^2 \to \mathbb{R}^5$        C. $\mathbb{R}^4 \to \mathbb{R}^2$        D. $\mathbb{R}^4 \to \mathbb{R}$.

**1.3.5** Multiply the original matrix by the vector $\vec{v} = \begin{bmatrix} .25 \\ .025 \\ .025 \\ .025 \\ .025 \\ .025 \\ .025 \\ .025 \\ .025 \\ .025 \\ .025 \\ .5 \end{bmatrix}$, putting $\vec{v}$ on the right.

**1.3.7** It is enough to know what $T$ gives when evaluated on the three standard basis vectors $\begin{bmatrix} 1 \\ 0 \\ 0 \end{bmatrix}, \begin{bmatrix} 0 \\ 1 \\ 0 \end{bmatrix}, \begin{bmatrix} 0 \\ 0 \\ 1 \end{bmatrix}$. The matrix of $T$ is $\begin{bmatrix} 3 & -1 & 0 \\ 1 & 1 & 2 \\ 2 & 3 & 1 \\ 1 & 0 & 1 \end{bmatrix}$.

**1.3.9** No, $T$ is not linear. If it were, the matrix would be $[T] = \begin{bmatrix} 2 & 1 & 1 \\ 1 & 2 & 0 \\ 1 & 1 & 1 \end{bmatrix}$, but $[T] \begin{bmatrix} 2 \\ -1 \\ 4 \end{bmatrix} = \begin{bmatrix} 7 \\ 0 \\ 5 \end{bmatrix}$, which contradicts the definition of the transformation.

**1.3.11** The rotation matrix is $\begin{bmatrix} \cos\theta & \sin\theta \\ -\sin\theta & \cos\theta \end{bmatrix}$. This transformation takes $\vec{e}_1$ to $\begin{bmatrix} \cos\theta \\ -\sin\theta \end{bmatrix}$, which is thus the first column of the matrix, by theorem 1.3.4; it takes $\vec{e}_2$ to $\begin{bmatrix} \cos(90° - \theta) = \sin\theta \\ \sin(90° - \theta) = \cos\theta \end{bmatrix}$, which is the second column.

One could also write this mapping as the rotation matrix of example 1.3.9, applied to $-\theta$:

$$\begin{bmatrix} \cos(-\theta) & -\sin(-\theta) \\ \sin(-\theta) & \cos(-\theta) \end{bmatrix}.$$

---

On one exam at Cornell University, the first question was "What is a linear transformation $T : \mathbb{R}^n \to \mathbb{R}^m$?" Several students gave answers like "A function taking some vector $\vec{v} \in \mathbb{R}^n$ and producing some vector $\vec{w} \in \mathbb{R}^m$."

This is wrong! The student who gave this answer may have been thinking of matrix multiplication. But a mapping from $\mathbb{R}^n$ to $\mathbb{R}^m$ need not be given by matrix multiplication: consider the mapping $\begin{bmatrix} x \\ y \end{bmatrix} \mapsto \begin{bmatrix} x^2 \\ y^2 \end{bmatrix}$. In any case, defining a linear transformation by saying that it is given by matrix multiplication begs the question, because it does not explain why matrix multiplication is linear.

The correct answer was given by another student in the class:

"A linear transformation

$$T : \mathbb{R}^n \to \mathbb{R}^m$$

is a mapping $\mathbb{R}^n \to \mathbb{R}^m$ such that for all $\vec{a}, \vec{b} \in \mathbb{R}^n$,

$$T(\vec{a} + \vec{b}) = T(\vec{a}) + T(\vec{b}),$$

and for all $\vec{a} \in \mathbb{R}^n$ and all scalar $r$,

$$T(r\vec{a}) = rT(\vec{a})."$$

Linearity and the approximation of nonlinear mappings by linear mappings are key motifs of this book. You must know the definition, which gives you a foolproof way to check whether or not a given mapping is linear.

**1.3.13** The expressions a, e, f, and j are not well-defined compositions. For the others:

    b. $C \circ B : \mathbb{R}^m \to \mathbb{R}^n$ (domain $\mathbb{R}^m$, codomain $\mathbb{R}^n$)    c. $A \circ C : \mathbb{R}^k \to \mathbb{R}^m$

    d. $B \circ A \circ C : \mathbb{R}^k \to \mathbb{R}^k$       g. $B \circ A : \mathbb{R}^n \to \mathbb{R}^k$

    h. $A \circ C \circ B : \mathbb{R}^m \to \mathbb{R}^m$    i. $C \circ B \circ A : \mathbb{R}^n \to \mathbb{R}^n$

**1.3.15** We need to show that $A(\vec{v} + \vec{w}) = A\vec{v} + A\vec{w}$ and that $A(c\vec{v}) = cA\vec{v}$. By definition 1.2.4,

Solution 1.3.15: With the dot product, introduced in the next section, the solution is simpler: Every entry in the vector $A\vec{v}$ is the dot product of $\vec{v}$ with one of the rows of $A$. The dot product is linear with respect to both of its arguments, so the mapping $A\vec{v}$ is also linear with respect to $\vec{v}$.

$$(A\vec{v})_i = \sum_{k=1}^{n} a_{i,k} v_k, \quad (A\vec{w})_i = \sum_{k=1}^{n} a_{i,k} w_k, \text{ and}$$

$$\left(A(\vec{v} + \vec{w})\right)_i = \sum_{k=1}^{n} a_{i,k}(v + w)_k = \sum_{k=1}^{n} a_{i,k}(v_k + w_k)$$

$$= \sum_{k=1}^{n} a_{i,k} v_k + \sum_{k=1}^{n} a_{i,k} w_k = (A\vec{v})_i + (A\vec{w})_i.$$

Similarly, $\left(A(c\vec{v})\right)_i = \sum_{k=1}^{n} a_{i,k}(cv)_k = \sum_{k=1}^{n} a_{i,k} cv_k = c\sum_{k=1}^{n} a_{i,k} v_k = c(A\vec{v})_i.$

**1.3.17**

$$\begin{bmatrix} \cos(2\theta) & \sin(2\theta) \\ \sin(2\theta) & -\cos(2\theta) \end{bmatrix}^2 = \begin{bmatrix} \cos(2\theta)^2 + \sin(2\theta)^2 & \cos(2\theta)\sin(2\theta) - \sin(2\theta)\cos(2\theta) \\ \sin(2\theta)\cos(2\theta) - \cos(2\theta)\sin(2\theta) & \sin(2\theta)^2 + \cos(2\theta)^2 \end{bmatrix}$$

$$\begin{bmatrix} \cos(2\theta) & \sin(2\theta) \\ \sin(2\theta) & -\cos(2\theta) \end{bmatrix}^2 = \begin{bmatrix} 1 & 0 \\ 0 & 1 \end{bmatrix} = I$$

**1.3.19** By commutativity of matrix addition, $\frac{AB + BA}{2} = \frac{BA + AB}{2}$, so the Jordan product is commutative. By non-commutativity of matrix multiplication:

$$\frac{\frac{AB+BA}{2}C + C\frac{AB+BA}{2}}{2} = \frac{ABC + BAC + CAB + CBA}{4}$$

$$\neq \frac{ABC + ACB + BCA + CBA}{4} = \frac{A\frac{BC+CB}{2} + \frac{BC+CB}{2}A}{2}$$

so the Jordan product is not associative.

**1.3.21** The number 0 is in the set, since $\text{Re}(0) = 0$. If $a, b$ are in the set, then $a+b$ is also in the set, since $\text{Re}(a+b) = \text{Re}(a) + \text{Re}(b) = 0$. If $a$ is in the set and $c$ is a real number, then $ca$ is in the set, since $\text{Re}(ca) = c\text{Re}(a) = 0$. So the set is a subspace of $\mathbb{C}$. The subspace is a line in $\mathbb{C}$ with a polar angle $\theta$ such that $\theta + \varphi = k\pi/2$, where $\varphi$ is the polar angle of $w$ and $k$ is an odd integer.

**1.4.1** a. Numbers:   $\vec{v} \cdot \vec{w}$, $|\vec{v}|$, $|A|$, and $\det A$. (If $A$ consists of a single row, then $A\vec{v}$ is also a number.)

Vectors:   $\vec{v} \times \vec{w}$ and $A\vec{v}$ (unless $A$ consists of a single row).

So far we have defined only determinants of $2 \times 2$ and $3 \times 3$ matrices; in section 4.8 we will define the determinant in general.

b. In the expression $\vec{v} \times \vec{w}$, the vectors must each have three entries. In the expression $\det A$, the matrix $A$ must be square.

**1.4.3** To normalize a vector, divide it by its length. This gives:

a. $\dfrac{1}{\sqrt{17}} \begin{bmatrix} 0 \\ 1 \\ 4 \end{bmatrix}$     b. $\dfrac{1}{\sqrt{58}} \begin{bmatrix} -3 \\ 7 \end{bmatrix}$     c. $\dfrac{1}{\sqrt{31}} \begin{bmatrix} \sqrt{2} \\ -2 \\ -5 \end{bmatrix}$

**1.4.5**  (a) $\cos(\theta) = \dfrac{1}{1 \times \sqrt{3}} \left( \begin{bmatrix} 1 \\ 0 \\ 0 \end{bmatrix} \cdot \begin{bmatrix} 1 \\ 1 \\ 1 \end{bmatrix} \right) = \dfrac{1}{\sqrt{3}}$, so

$$\theta = \arccos\left(\dfrac{1}{\sqrt{3}}\right) \approx .95532.$$

b. $\cos(\theta) = 0$, so $\theta = \pi/2$ .

**1.4.7**  a. $\det = 1$; the inverse is $\begin{bmatrix} 0 & 1 \\ -1 & 2 \end{bmatrix}$

b. $\det = 0$; no inverse

c. $\det = ad$; if $a, d \neq 0$, the inverse is $\dfrac{1}{ad}\begin{bmatrix} d & -b \\ 0 & a \end{bmatrix}$.

d. $\det = 0$; no inverse

**1.4.9**  a. $\begin{bmatrix} -6yz \\ 3xz \\ 5xy \end{bmatrix}$     b. $\begin{bmatrix} 6 \\ 7 \\ -4 \end{bmatrix}$     c. $\begin{bmatrix} -2 \\ -22 \\ 3 \end{bmatrix}$

**1.4.11** a. True by theorem 1.4.5, because $\vec{w} = -2\vec{v}$.

Solution 1.4.11, part c: Our answer depended on the vectors chosen. In general,

$$\det[\vec{a}, \vec{b}, \vec{c}] = -\det[\vec{a}, \vec{c}, \vec{b}];$$

the result here is true only because both sides are 0, since $\vec{v}$, $\vec{w}$ are linearly dependent.

b. False; $\vec{u} \cdot (\vec{v} \times \vec{w})$ is a number; $|\vec{u}|(\vec{v} \times \vec{w})$ is a number times a vector, i.e., a vector.

c. True: since $\vec{w} = -2\vec{v}$, we have $\det[\vec{u}, \vec{v}, \vec{w}] = 0$ and $\det[\vec{u}, \vec{w}, \vec{v}] = 0$.

d. False, since $\vec{u}$ is not necessarily (in fact almost surely not) a multiple of $\vec{w}$; the correct statement is $|\vec{u} \cdot \vec{w}| \leq |\vec{u}||\vec{w}|$.

e. True.      f. True.

**1.4.13** $\begin{bmatrix} xa \\ xb \\ xc \end{bmatrix} \times \begin{bmatrix} a \\ b \\ c \end{bmatrix} = \begin{bmatrix} xbc - xbc \\ -(xac - xac) \\ xab - xab \end{bmatrix} = \begin{bmatrix} 0 \\ 0 \\ 0 \end{bmatrix}$.

Solutions 1.4.13 and 1.4.15: Of course, the vectors must be in $\mathbb{R}^3$ for the cross product to be defined.

**1.4.15** $\begin{bmatrix} a \\ b \\ c \end{bmatrix} \times \begin{bmatrix} d \\ e \\ f \end{bmatrix} = \begin{bmatrix} bf - ce \\ cd - af \\ ae - bd \end{bmatrix} = - \begin{bmatrix} d \\ e \\ f \end{bmatrix} \times \begin{bmatrix} a \\ b \\ c \end{bmatrix} = - \begin{bmatrix} ec - bf \\ af - cd \\ bd - ae \end{bmatrix}$

**1.4.17**  a. It is the line of equation $\begin{bmatrix} x \\ y \end{bmatrix} \cdot \begin{bmatrix} 2 \\ -1 \end{bmatrix} = 2x - y = 0$.

b. It is the line of equation $\begin{bmatrix} x-2 \\ y-3 \end{bmatrix} \cdot \begin{bmatrix} 2 \\ -4 \end{bmatrix} = 2x - 4 - 4y + 12 = 0$, which you can rewrite as $2x - 4y + 8 = 0$.

**1.4.19  a.** The length of $\vec{v}_n$ is $|\vec{v}_n| = \sqrt{1 + \cdots + 1} = \sqrt{n}$.

**b.** The angle is $\arccos \dfrac{1}{\sqrt{n}}$, which tends to 0 as $n \to \infty$.

Solution 1.4.19, part b: We find it surprising that the diagonal vector $\vec{v}_n$ is almost orthogonal to all the standard basis vectors when $n$ is large.

**1.4.21  a.** $|A| = \sqrt{1 + 1 + 4} = \sqrt{6}; \qquad |B| = \sqrt{5}; \qquad |\vec{c}| = \sqrt{10}$

**b.** $|AB| = \left| \begin{bmatrix} 2 & 0 \\ 2 & 2 \end{bmatrix} \right| = \sqrt{12} \le \sqrt{30} = |A||B|$

$|A\vec{c}| = \sqrt{50} \le |A||\vec{c}| = \sqrt{60}; \qquad |B\vec{c}| = \sqrt{13} \le |B||\vec{c}| = \sqrt{50}$

**1.4.23  a.** The length is $|\vec{w}_n| = \sqrt{1 + 4 + \cdots + n^2} = \sqrt{\dfrac{n^3}{3} + \dfrac{n^2}{2} + \dfrac{n}{6}}$.

Solution 1.4.23, part a: This uses the fact that

$$1^2 + 2^2 + \cdots + n^2 = \frac{n^3}{3} + \frac{n^2}{2} + \frac{n}{6}.$$

**b.** The angle $\alpha_{n,k}$ is $\arccos \dfrac{k}{\sqrt{\frac{n^3}{3} + \frac{n^2}{2} + \frac{n}{6}}}$.

**c.** In all three cases, the limit is $\pi/2$. Clearly $\lim_{n \to \infty} \alpha_{n,k} = \pi/2$, since the cosine tends to 0.

The limit of $\alpha_{n,n}$ is $\lim_{n \to \infty} \arccos 0 = \pi/2$.

The limit of $\alpha_{n,\lfloor n/2 \rfloor}$ is also $\pi/2$, since it is the arccos of

$$\frac{\lfloor n/2 \rfloor}{\sqrt{\frac{n^3}{3} + \frac{n^2}{2} + \frac{n}{6}}}, \qquad \text{which tends to 0 as } n \to \infty.$$

**1.4.25**

$$\left( \sqrt{x_1^2 + x_2^2} \sqrt{y_1^2 + y_2^2} \right)^2 - \left( x_1 y_1 + x_2 y_2 \right)^2 = (x_2 y_1)^2 + (x_1 y_2)^2 - 2x_1 x_2 y_1 y_2$$

$$= (x_1 y_2 - x_2 y_1)^2 \ge 0,$$

so $(x_1 y_1 + x_2 y_2)^2 \le \left( \sqrt{x_1^2 + x_2^2} \sqrt{y_1^2 + y_2^2} \right)^2$, so

$$|x_1 y_1 + x_2 y_2| \le \sqrt{x_1^2 + x_2^2} \sqrt{y_1^2 + y_2^2}.$$

**1.4.27  a.** To show that the transformation is linear, we need to show that

$$T_{\vec{a}}(\vec{v} + \vec{w}) = T_{\vec{a}}(\vec{v}) + T_{\vec{a}}(\vec{w}) \quad \text{and} \quad \alpha T_{\vec{a}}(\vec{v}) = T_{\vec{a}}(\alpha \vec{v}).$$

For the first,

$$T_{\vec{a}}(\vec{v} + \vec{w}) = \vec{v} + \vec{w} - 2(\vec{a} \cdot (\vec{v} + \vec{w}))\vec{a} = \vec{v} + \vec{w} - 2(\vec{a} \cdot \vec{v} + \vec{a} \cdot \vec{w})\vec{a} = T_{\vec{a}}(\vec{v}) + T_{\vec{a}}(\vec{w}).$$

For the second,

$$\alpha T_{\vec{a}}(\vec{v}) = \alpha \vec{v} - 2\alpha(\vec{a} \cdot \vec{v})\vec{a} = \alpha \vec{v} - 2(\vec{a} \cdot \alpha \vec{v})\vec{a} = T_{\vec{a}}(\alpha \vec{v}).$$

**b.** We have $T_{\vec{a}}(\vec{a}) = -\vec{a}$, since $\vec{a} \cdot \vec{a} = a^2 + b^2 + c^2 = 1$:

$$T_{\vec{a}}(\vec{a}) = \vec{a} - 2(\vec{a} \cdot \vec{a})\vec{a} = \vec{a} - 2\vec{a} = -\vec{a}.$$

The transformation $T_{\vec{\mathbf{a}}}$ is the 3-dimensional version of the transformation shown in figure 1.3.4.

If $\vec{\mathbf{v}}$ is orthogonal to $\vec{\mathbf{a}}$, then $T_{\vec{\mathbf{a}}}(\vec{\mathbf{v}}) = \vec{\mathbf{v}}$, since in that case $\vec{\mathbf{a}} \cdot \vec{\mathbf{v}} = 0$. Thus $T_{\vec{\mathbf{a}}}$ is reflection in the plane that goes through the origin and is perpendicular to $\vec{\mathbf{a}}$.

c. The matrix of $T_{\vec{\mathbf{a}}}$ is

$$M = [T_{\vec{\mathbf{a}}}(\vec{\mathbf{e}}_1), T_{\vec{\mathbf{a}}}(\vec{\mathbf{e}}_2), T_{\vec{\mathbf{a}}}(\vec{\mathbf{e}}_3)] = \begin{bmatrix} 1 - 2a^2 & -2ab & -2ac \\ -2ab & 1 - 2b^2 & -2bc \\ -2ac & -2bc & 1 - 2c^2 \end{bmatrix}$$

Squaring the matrix gives the $3 \times 3$ identity matrix: if you reflect a vector, then reflect it again, you are back to where you started.

Solution 1.5.1: Note that part c is the same as part a.

**1.5.1** a. The set $\{\, x \in \mathbb{R} \mid 0 < x \le 1 \,\}$ is neither open nor closed: the point 1 is in the set, but $1 + \epsilon$ is not for every $\epsilon$, showing it isn't open, and 0 is not but $0 + \epsilon$ is for every $\epsilon > 0$, showing that the complement is also not open, so the set is not closed.

b. open         c. neither         d. closed         e. closed         f. neither

g. Both. That the empty set $\phi$ is closed is obvious. Showing that it is open is an "eleven-legged alligator" argument (see section 0.2). Is it true that for every point of the empty set, there exists $\epsilon > 0$ such that ... ? Yes, because there are no points in the empty set.

**1.5.3** a. Suppose $A_i, i \in I$ is some collection (probably infinite) of open sets. If $\mathbf{x} \in \bigcup_{i \in I} A_i$, then $\mathbf{x} \in A_j$ for some $j$, and since $A_j$ is open, there exists $\epsilon > 0$ such that $B_\epsilon(\mathbf{x}) \subset A_j$. But then $B_\epsilon(\mathbf{x}) \subset \bigcup_{i \in I} A_i$.

Solution 1.5.3, part b: There is a smallest $\epsilon_i$, because there are finitely many of them, and it is positive. If there were infinitely many, then there would be a greatest lower bound, but it could be 0.

Part c: In fact, every closed set is a countable intersection of open sets.

b. If $A_1, \ldots, A_j$ are open and $\mathbf{x} \in \cap_{i=1}^k A_i$, then there exist $\epsilon_1, \ldots, \epsilon_k > 0$ such that $B_{\epsilon_i}(\mathbf{x}) \subset A_i$, for $i = 1, \ldots, k$. Set $\epsilon$ to be the smallest of $\epsilon_1, \ldots, \epsilon_k$. Then $B_\epsilon(\mathbf{x}) \subset B_{\epsilon_i}(\mathbf{x}) \subset A_i$.

c. The infinite intersection of open sets $(-1/n, 1/n)$, for $n = 1, 2, \ldots$, is not open; as $n \to \infty$, $-1/n \to 0$ and $1/n \to 0$; the set $\{0\}$ is not open.

**1.5.5** a. This set is open. Indeed, if you choose $\begin{pmatrix} x \\ y \end{pmatrix}$ in your set, then $1 < \sqrt{x^2 + y^2} < \sqrt{2}$. Set

$$r = \min\left\{ \sqrt{x^2 + y^2} - 1, \sqrt{2} - \sqrt{x^2 + y^2} \right\} > 0.$$

Then the ball of radius $r$ around $\begin{pmatrix} x \\ y \end{pmatrix}$ is contained in the set, since if $\begin{pmatrix} u \\ v \end{pmatrix}$ is in that ball, then, by the triangle inequality,

Equation (1) uses the familiar form of the triangle inequality: if $\mathbf{a} = \mathbf{b} + \mathbf{c}$, then

$$|\mathbf{a}| \le |\mathbf{b}| + |\mathbf{c}|.$$

Equation (2) uses the variant

$$|\mathbf{a}| \ge \big||\mathbf{b}| - |\mathbf{c}|\big|.$$

$$\left\| \begin{bmatrix} u \\ v \end{bmatrix} \right\| \le \left\| \begin{bmatrix} u - x \\ v - y \end{bmatrix} \right\| + \left\| \begin{bmatrix} x \\ y \end{bmatrix} \right\| < r + \left\| \begin{bmatrix} x \\ y \end{bmatrix} \right\| \le \sqrt{2} \qquad (1)$$

$$\left\| \begin{bmatrix} u \\ v \end{bmatrix} \right\| \ge \left\| \begin{bmatrix} x \\ y \end{bmatrix} \right\| - \left\| \begin{bmatrix} u - x \\ v - y \end{bmatrix} \right\| > \left\| \begin{bmatrix} x \\ y \end{bmatrix} \right\| - r \ge 1. \qquad (2)$$

s

b. The locus $xy \ne 0$ is also open. It is the complement of the two axes, so that if $\begin{pmatrix} x \\ y \end{pmatrix}$ is in the set, then $r = \min\{|x|, |y|\} > 0$, and the ball $B$ of radius $r$ around $\begin{pmatrix} x \\ y \end{pmatrix}$ is contained in the set. Indeed, if $\begin{pmatrix} u \\ v \end{pmatrix}$ is $B$, then $|u| = |x + u - x| > |x| - |u - x| > |x| - r \geq 0$, so $u$ is not 0, and neither is $v$, by the same argument.

Many students have found exercise 1.5.5 difficult, even though they also thought it was obvious, but didn't know how to say it. If this applies to you, you should check carefully where we used the triangle inequality, and how. Almost everything concerning inequalities requires the triangle inequality.

c.    This time our set is the $x$-axis, and it is closed. We will use the criterion that a set is closed if the limit of a convergent sequence of elements of the set is in the set (proposition 1.5.17). If $\begin{pmatrix} x_n \\ y_n \end{pmatrix}$ is a sequence in the set, and converges to $\begin{pmatrix} x_0 \\ y_0 \end{pmatrix}$, then all $y_n = 0$, so $y_0 = \lim_{n \to \infty} y_n = 0$, and the limit is also in the set.

d. The rational numbers are neither open nor closed. Any rational number $x$ is the limit of the numbers $x + \sqrt{2}/n$, which are all irrational, so the rationals aren't closed. Any irrational number is the limit of the finite decimals used to write it, which are all rational, so the irrationals aren't closed either.

**1.5.7**  a. The natural domain is $\mathbb{R}^2$ minus the union of the two axes; it is open.

b. The natural domain is that part of $\mathbb{R}^2$ where $x^2 > y$ (i.e., the area "inside" the parabola of equation $y = x^2$). It is open since its "fence" $x^2$ belongs to its neighbor.

c.    The natural domain of $\ln \ln x$ is $\{x | x > 1\}$, since we must have $\ln x > 0$. This domain is open.

d. The natural domain of $\arcsin$ is $[-1, 1]$. Thus the natural domain of $\arcsin \frac{3}{x^2 + y^2}$ is $\mathbb{R}^2$ minus the open disc $x^2 + y^2 < 3$. Since this domain is the complement of an open disc, it is closed and not open.

e. The natural domain is all of $\mathbb{R}^2$, which is open.

f.    The natural domain is $\mathbb{R}^3$ minus the union of the three coordinate planes of equation $x = 0$, $y = 0$, $z = 0$; it is open.

**1.5.9**  For any $n > 0$ we have $\left| \sum_{i=1}^{n} \mathbf{x}_i \right| \leq \sum_{i=1}^{n} |\mathbf{x}_i|$ by the triangle inequality (theorem 1.4.9). Because $\sum_{i=1}^{\infty} \mathbf{x}_i$ converges, $\sum_{i=1}^{n} \mathbf{x}_i$ converges as $n \to \infty$. So :

$$\left| \sum_{i=1}^{\infty} \mathbf{x}_i \right| \leq \sum_{i=1}^{\infty} |\mathbf{x}_i|.$$

**1.5.11**  a. First let us see that

$$\left( (\forall \epsilon > 0)(\exists N)(n > N) \implies |\mathbf{a}_n - \mathbf{a}| < \varphi(\epsilon) \right) \implies \left( \mathbf{a}_n \text{ converges to } \mathbf{a} \right).$$

Choose $\eta > 0$. Since $\lim_{t \to 0} \varphi(t) = 0$, there exists $\delta > 0$ such that when $0 < t \leq \delta$ we have $\varphi(t) < \eta$. Our hypothesis guarantees that there exists $N$ such that when $n > N$, then $|\mathbf{a}_n - \mathbf{a}| \leq \varphi(\delta) = \eta$.

Now for the converse:

$$\left( \mathbf{a}_n \text{ converges to } \mathbf{a} \right) \implies \left( (\forall \epsilon > 0)(\exists N)(n > N) \implies |\mathbf{a}_n - \mathbf{a}| < \varphi(\epsilon) \right).$$

For any $\epsilon > 0$, we also have $\varphi(\epsilon) > 0$, so there exists $N$ such that

$$n > N \implies |\mathbf{a}_n - \mathbf{a}| < \varphi(\epsilon).$$

b. The analogous statement for limits of functions is:

Let $\varphi : [0, \infty) \to [0, \infty)$ be a function such that $\lim_{t \to 0} \varphi(t) = 0$. Let $U \subset \mathbb{R}^n$, $f : U \to \mathbb{R}^m$, and $\mathbf{x}_0 \in \overline{U}$. Then $\lim_{\mathbf{x} \to \mathbf{x}_0} f(\mathbf{x}) = \mathbf{a}$ if and only if for every $\epsilon > 0$ there exists $\delta > 0$ such that when $\mathbf{x} \in U$ and $|\mathbf{x} - \mathbf{x}_0| < \delta$, we have $|f(\mathbf{x}) - \mathbf{a}| < \varphi(\epsilon)$.

**1.5.13** Choose a point $\mathbf{a} \in \mathbb{R}^n - C$. Suppose that the ball $B_{1/n}(\mathbf{a})$ of radius $1/n$ around $\mathbf{a}$ satisfies $B_{1/n}(\mathbf{a}) \cap C \neq \phi$ for every $n$. Choose $\mathbf{a}_n$ in $B_{1/n}(\mathbf{a}) \cap C$. Then the sequence $(\mathbf{a}_n)$ converges to $\mathbf{a}$, since for any $\epsilon > 0$ we can find $N$ such that for $n > N$ we have $1/n < \epsilon$, so for $n > N$ we have $|\mathbf{a} - \mathbf{a}_n| < 1/n < \epsilon$. Then our hypothesis implies that $\mathbf{a} \in C$, a contradiction. Thus there exists $N$ such that $B_{1/N}(\mathbf{a}) \cap C = \phi$. This shows that $\mathbb{R}^n - C$ is open, so $C$ is closed.

Solution 1.5.15: Here, $x$ plays the role of the alligators in section 0.2, and $x \geq 0$ satisfying $|-2 - x| < \delta$ plays the role of eleven-legged alligators; the conclusion $|\sqrt{x} - 5| < \epsilon$ (i.e., that 5 is the limit) is the conclusion "are orange with blue spots" and the conclusion $|\sqrt{x} - 3| < \epsilon$ (i.e., that 3 is the limit) is the conclusion "are black with white stripes."

**1.5.15** Both statements are true. To show that the first is true, we say: for every $\epsilon > 0$, there exists $\delta > 0$ such that for all $x$ satisfying $x \geq 0$ and $|-2 - x| < \delta$, then $|\sqrt{x} - 5| < \epsilon$. For any $\epsilon > 0$, choose $\delta = 1$. Then there is no $x \geq 0$ satisfying $|-2 - x| < \delta$. So for those nonexistent $x$ satisfying $|-2 - x| < 1$, it is true that $|\sqrt{x} - 5| < \epsilon$. By the same argument the second statement is true.

**1.5.17** Set $\mathbf{a} = \lim_{m \to \infty} \mathbf{a}_m$, $\mathbf{b} = \lim_{m \to \infty} \mathbf{b}_m$  and  $c = \lim_{m \to \infty} c_m$.

1. Choose $\epsilon > 0$ and find $M_1$ and $M_2$ such that if $m \geq M_1$ then we have $|\mathbf{a}_m - \mathbf{a}| \leq \epsilon/2$, and if $m \geq M_2$ then $|\mathbf{b}_m - \mathbf{b}| \leq \epsilon/2$. Set

$$M = \max(M_1, M_2).$$

If $m > M$, we have

$$|\mathbf{a}_m + \mathbf{b}_m - \mathbf{a} - \mathbf{b}| \leq |\mathbf{a}_m - \mathbf{a}| + |\mathbf{b}_m - \mathbf{b}| \leq \frac{\epsilon}{2} + \frac{\epsilon}{2} = \epsilon.$$

So the sequence $(\mathbf{a}_m + \mathbf{b}_m)$ converges to $\mathbf{a} + \mathbf{b}$.

2. Choose $\epsilon > 0$. Find $M_1$ such that if

$$m \geq M_1, \quad \text{then} \quad |\mathbf{a}_m - \mathbf{a}| \leq \frac{1}{2} \inf\left(\frac{\epsilon}{|c|}, \epsilon\right).$$

The inf is there to guard against the possibility that $|c| = 0$. In particular, if $m \geq M_1$, then $|\mathbf{a}_m| \leq |\mathbf{a}| + \epsilon$. Next find $M_2$ such that if

$$m \geq M_2, \quad \text{then} \quad |c_m - c| \leq \frac{\epsilon}{2(|\mathbf{a}| + \epsilon)}.$$

If $M = \max(M_1, M_2)$ and $m \geq M$, then

$$|c_m \mathbf{a}_m - c\mathbf{a}| = |c(\mathbf{a}_m - \mathbf{a}) + (c_m - c)\mathbf{a}_m|$$

$$\leq |c(\mathbf{a}_m - \mathbf{a})| + |(c_m - c)\mathbf{a}_m| \leq \frac{\epsilon}{2} + \frac{\epsilon}{2} = \epsilon,$$

so the sequence $(c_m \mathbf{a}_m)$ converges and the limit is $c\mathbf{a}$.

3.   We can either repeat the argument above, or use parts 1 and 2 as follows:

$$\lim_{m\to\infty} \mathbf{a}_m \cdot \mathbf{b}_m = \lim_{m\to\infty} \sum_{i=1}^{n} a_{m,i} b_{m,i} = \sum_{i=1}^{n} \lim_{m\to\infty} (a_{m,i} b_{m,i})$$

$$= \sum_{i=1}^{n} \left( \lim_{m\to\infty} a_{m,i} \right) \left( \lim_{m\to\infty} b_{m,i} \right) = \sum_{i=1}^{n} a_i b_i = \mathbf{a} \cdot \mathbf{b}.$$

4. Find $C$ such that $|\mathbf{a}_m| \leq C$ for all $m$; saying that $\mathbf{a}_m$ is bounded means exactly that such a $C$ exists. Choose $\epsilon > 0$, and find $M$ such that when $m > M$, then $|c_m| < \epsilon/C$ (this is possible since the $c_m$ converge to 0). Then when $m > M$ we have

$$|c_m \mathbf{a}_m| = |c_m||\mathbf{a}_m| \leq \frac{\epsilon}{C} C = \epsilon.$$

**1.5.19**   a. Suppose $I - A$ is invertible, and write

$$I - A + C = I - A + C(I-A)^{-1}(I-A) = \big(I + C(I-A)^{-1}\big)(I-A),$$

so

$$(I - A + C)^{-1} = (I-A)^{-1}\Big(I + C(I-A)^{-1}\Big)^{-1}$$

$$= (I-A)^{-1}\Big(\underbrace{I - (C(I-A)^{-1}) + (C(I-A)^{-1})^2 - (C(I-A)^{-1})^3 + \cdots}_{\text{geometric series}}\Big)$$

so long as the series is convergent. By proposition 1.5.37, this will happen if

$$|C(I-A)^{-1}| < 1, \quad \text{in particular if} \quad |C| < \frac{1}{|(I-A)^{-1}|}.$$

Thus every point of $U$ is the center of a ball contained in $U$.

For the second part of the question, the matrices

$$C_n = \begin{bmatrix} 1 - 1/n & 0 \\ 0 & 1-1/n \end{bmatrix}, \quad n = 1, 2, \ldots$$

converge to $I$, and $C_n$ is in $U$, since $I - C_n = \begin{bmatrix} 1/n & 0 \\ 0 & 1/n \end{bmatrix}$ is invertible.

b. Simply factor: $(A+I)(A-I) = A^2 + A - A - I = A^2 - I$, so

$$(A^2 - I)(A-I)^{-1} = (A+I)(A-I)(A-I)^{-1} = A + I,$$

which converges to $2I$ as $A \to I$.

c.   Showing that $V$ is open is very much like showing that $U$ is open (part a). Suppose $B - A$ is invertible, and write

$$B - A + C = (I + C(B-A)^{-1})(B-A),$$

so

$$(B - A + C)^{-1} = (B-A)^{-1}\big(I + C(B-A)^{-1}\big)^{-1}$$

$$= (B-A)^{-1}\Big(I - (C(B-A)^{-1}) + (C(B-A)^{-1})^2 - (C(B-A)^{-1})^3 + \cdots\Big),$$

so long as the series is convergent. This will happen if

$$|C(B-A)^{-1}| < 1, \text{ in particular, if } |C| < \frac{1}{|(B-A)^{-1}|}.$$

Thus every point of $V$ is the center of a ball contained in $V$. Again, the matrices

$$\begin{bmatrix} 1+1/n & 0 \\ 0 & -1+1/n \end{bmatrix}, \quad n = 1, 2, \dots$$

do the trick.

d. This time the limit does not exist. Note that you cannot factor $A^2 - B^2 = (A+B)(A-B)$ if $A$ and $B$ do not commute.

First set

$$A_n = \begin{bmatrix} 1/n+1 & 1/n \\ 0 & -1+1/n \end{bmatrix}.$$

Part d: You may wonder how we came by the matrices $A_n$; we observed that

$$B\begin{bmatrix} 0 & 1 \\ 0 & 0 \end{bmatrix} = \begin{bmatrix} 0 & 1 \\ 0 & 0 \end{bmatrix}$$

$$\begin{bmatrix} 0 & 1 \\ 0 & 0 \end{bmatrix}B = \begin{bmatrix} 0 & -1 \\ 0 & 0 \end{bmatrix},$$

so these matrices do not commute.

Then

$$A_n^2 - B^2 = \begin{bmatrix} 2/n+1/n^2 & 2/n^2 \\ 0 & -2/n+1/n^2 \end{bmatrix} \quad \text{and} \quad (A-B)^{-1} = \begin{bmatrix} n & -n \\ 0 & n \end{bmatrix}.$$

Thus we find

$$(A_n^2 - B^2)(A_n - B)^{-1} = \begin{bmatrix} 2+1/n & -2+1/n \\ 0 & -2+1/n \end{bmatrix} \rightarrow \begin{bmatrix} 2 & -2 \\ 0 & -2 \end{bmatrix}$$

as $n \rightarrow \infty$.

Do the same computation with $A'_n = \begin{bmatrix} 1/n+1 & 0 \\ 0 & -1+1/n \end{bmatrix}$. This time we find

$$(A'^2_n - B^2)(A'_n - B)^{-1} = \begin{bmatrix} 2+1/n & 0 \\ 0 & -2+1/n \end{bmatrix} \rightarrow \begin{bmatrix} 2 & 0 \\ 0 & -2 \end{bmatrix} = 2B$$

as $n \rightarrow \infty$.

Since both sequence $A_n$ and $A'_n$ converge to $B$, this shows that there is no limit.

**1.5.21** a. This is a quotient of continuous functions where the denominator does not vanish at $\begin{pmatrix} 0 \\ 0 \end{pmatrix}$, so it is continuous at the origin.

b. Again, there is no problem: this is the square root of a continuous function, at a point where the function is 1, so it is continuous at the origin.

c. If we approach the origin along the x-axis, $f = 1$, and if we approach the origin along the y-axis, $f = |y|^{\frac{2}{3}}$ goes to 0, so $f$ is not continuous at the origin. There is no way of choosing a value of $f\begin{pmatrix} 0 \\ 0 \end{pmatrix}$ that will make $f$ continuous at the origin.

d. When $0 < x^2 + 2y^2 < 1$,

$$0 > (x^2+y^2)\ln(x^2+2y^2) \geq (x^2+y^2)\ln\big(2(x^2+y^2)\big)$$
$$= (x^2+y^2)\ln(x^2+y^2) + (x^2+y^2)\ln 2.$$

Part d uses the following statement from one-variable calculus:

$$\lim_{u \to 0} u \ln|u| = 0.$$

This can be proved by applying l'Hôpital's rule to $\dfrac{\ln|u|}{1/u}$.

The term $(x^2 + y^2) \ln(x^2 + y^2)$ tends to 0 using equation in the margin. The second term, $(x^2 + y^2) \ln 2$, obviously tends to 0. So if we choose $f\begin{pmatrix} 0 \\ 0 \end{pmatrix} = 0$, then $f$ is continuous.

e. The function is not continuous near the origin. Since $\ln 0$ is undefined, the diagonal $x + y = 0$ is not part of the function's domain of definition. However, the function is defined at points arbitrarily close to that line, e.g., the point $\begin{pmatrix} x \\ -x + e^{-1/x^3} \end{pmatrix}$. At this point we have

$$\left( x^2 + \left( -x + e^{-1/x^3} \right)^2 \right) \ln \left| x - x + e^{-1/x^3} \right| \geq x^2 \left| \frac{1}{x^3} \right| = \frac{1}{|x|},$$

which tends to infinity as $x$ tends to 0. But if we approach the origin along the $x$-axis (for instance), the function is $x^2 \ln|x|$, which tends to 0 as $x$ tends to 0.

**1.5.23** a. To say that $\lim_{B \to A} (A - B)^{-1}(A^2 - B^2)$ exists means that there is a matrix $C$ such that for all $\epsilon > 0$, there exists $\delta > 0$, such that when $|B - A| < \delta$ and $B - A$ is invertible, then

$$\left| (B - A)^{-1}(B^2 - A^2) - C \right| < \epsilon.$$

b. We will show that the limit exists, and is $\begin{bmatrix} 2 & 0 \\ 0 & 2 \end{bmatrix} = 2I$. Write $B = I + H$, with $H$ invertible, and choose $\epsilon > 0$. We need to show that there exists $\delta > 0$ such that if $|H| < \delta$, then

$$\left| (I + H - I)^{-1}(I + H)^2 - I^2 - 2I \right| < \epsilon. \tag{2}$$

Indeed,

$$\left| (I + H - I)^{-1}(I + H)^2 - I^2 - 2I \right| = \left| H^{-1}(I^2 + IH + HI + H^2 - I^2) - 2I \right|$$
$$= \left| H^{-1}(2H + H^2) - 2I \right| = |H|.$$

So if you set $\delta = \epsilon$, and $|H| \leq \delta$, then equation (2) is satisfied.

c. We will show that the limit does not exist. In this case, we find

$$(A + H - A)^{-1}(A + H)^2 - A^2 = H^{-1}(I^2 + AH + HA + H^2 - I^2)$$
$$= H^{-1}(AH + HA + H^2) = A + H^{-1}AH + H^2.$$

If the limit exists, it must be $2A$: choose $H = \epsilon I$ so that $H^{-1} = \epsilon^{-1} I$; then

$$A + H^{-1}AH + H^2 = 2A + \epsilon I$$

is close to $2A$.

But if you choose $H = \epsilon \begin{bmatrix} 1 & 0 \\ 0 & -1 \end{bmatrix}$, you will find that

$$H^{-1}AH = \begin{bmatrix} 1/\epsilon & 0 \\ 0 & -1/\epsilon \end{bmatrix} \begin{bmatrix} 0 & 1 \\ 1 & 0 \end{bmatrix} \begin{bmatrix} \epsilon & 0 \\ 0 & \epsilon \end{bmatrix} = \begin{bmatrix} 0 & -1 \\ -1 & 0 \end{bmatrix} = -A.$$

So with this $H$ we have

$$A + H^{-1}AH + H^2 = A - A + \epsilon H,$$

which is close to the zero matrix.

**1.6.1** Let $B$ be a set contained in a ball of radius $R$ centered at a point $\mathbf{a}$. Then it is also contained in a ball of radius $R + |\mathbf{a}|$ centered at the origin; thus it is bounded.

**1.6.3** The polynomial $p(z) = 1 + x^2 y^2$ has no roots because 1 plus something positive cannot be 0. This does not contradict the fundamental theorem of algebra because although $p$ is a polynomial in the real variables $x$ and $y$, it is not a polynomial in the complex variable $z$: it is a polynomial in $z$ and $\bar{z}$. It is possible to write $p(z) = 1 + x^2 y^2$ in terms of $z$ and $\bar{z}$. You can use

$$x = \frac{z + \bar{z}}{2} \quad \text{and} \quad y = \frac{z - \bar{z}}{2i},$$

and find

$$p(z) = 1 + \frac{z^4 - 2|z|^4 + \bar{z}^4}{-16} \tag{1}$$

but you simply cannot get rid of the $\bar{z}$.

**1.6.5**  a. Suppose $|z| > 3$. Then

$$|z|^6 - |q(z)| \geq |z|^6 - (4|z|^4 + |z| + 2) \geq |z|^6 - (4|z|^4 + |z|^4 + 2|z|^4)$$
$$= |z|^4(|z|^2 - 7) \geq (9 - 7) \cdot 3^4 = 162.$$

How did we come by the number 3? We started the computation, until we got to the expression $|z|^2 - 7$, which we needed to be positive. The number 3 works, and 2 does not; 2.7 works too.

b. Since $p(0) = 2$, but when $|z| > 3$ we have $|p(z)| \geq |z|^6 - |q(z)| \geq 162$, the minimum value of $|p|$ on the disc of radius $R_1 = 3$ around the origin must be the absolute minimum value of $|p|$. Notice there must be a point at which this minimum value is achieved, since $|p|$ is a continuous function on the closed and bounded set $|z| \leq 3$ of $\mathbb{C}$.

**1.6.7**  Consider the function $g(x) = f(x) - mx$. This is a continuous function on the closed and bounded set $[a, b]$, so it has a minimum at some point $c \in [a, b]$. Let us see that $c \neq a$ and $c \neq b$. Since $g'(a) = f'(a) - m < 0$, we have

$$\lim_{h \to 0} \frac{g(a + h) - g(a)}{h} < 0.$$

Let us spell this out: for every $\epsilon > 0$, there exists $\delta > 0$ such that $0 < |h| < \delta$ implies

$$\left| \frac{g(a + h) - g(a)}{h} - g'(a) \right| < \epsilon.$$

Solution 1.6.7: Although our function $g$ is differentiable on a neighborhood of $a$ and $b$, we cannot apply proposition 1.6.11 if the minimum occurs at one of those points, since $c$ would not be a maximum on a neighborhood of the point.

Choose $\epsilon = |g'(a)|/2$, and find a corresponding $\delta > 0$, and set $h = \delta/2$. Then the inequality

$$\left| \frac{g(a + h) - g(a)}{h} - g'(a) \right| < \frac{|g'(a)|}{2}$$

implies that

$$\frac{g(a+h) - g(a)}{h} < \frac{g'(a)}{2} < 0$$

and since $h > 0$ we have $g(a+h) < g(a)$, so $a$ is not the minimum of $g$.

Similarly, $b$ is not the minimum:

$$\lim_{h \to 0} \frac{g(b+h) - g(b)}{h} = g'(b) - m > 0.$$

Express this again in terms of $\epsilon$'s and $\delta$'s, choose $\epsilon = g'(b)/2$, and set $h = -\delta/2$. As above, we have

$$\frac{g(b+h) - g(b)}{h} > \frac{g'(b)}{2} > 0,$$

and since $h < 0$, this implies $g(b+h) < g(b)$.

So $c \in (a, b)$, and in particular $c$ in a minimum on $(a, b)$, so $g'(c) = f'(c) - m = 0$ by proposition 1.6.11.

**1.6.9** "Between 0 and $a$" means that if you plot $a$ as a point in $\mathbb{R}^2$ in the usual way (real part of $a$ on the $x$-axis, imaginary part on the $y$-axis), then $a + bu^j$ lies on the line segment connecting the origin and the point $a$. For this to happen, $bu^j$ must point in the opposite direction as $a$, and we must have $|bu^j| < |a|$. Write

$$a = r_1(\cos\omega_1 + i\sin\omega_1)$$
$$b = r_2(\cos\omega_2 + i\sin\omega_2)$$
$$u = p(\cos\theta + i\sin\theta).$$

Then

$$a + bu^j = r_1(\cos\omega_1 + i\sin\omega_1) + r_2 p^j \big(\cos(\omega_2 + j\theta) + i\sin(\omega_2 + j\theta)\big).$$

Then $bu^j$ will point in the opposite direction from $a$ if

$$\omega_2 + j\theta = \omega_1 + \pi + 2k\pi \text{ for some } k, \text{ i.e., } \theta = \frac{1}{j}(\omega_1 - \omega_2 + \pi + 2k\pi),$$

and we find $j$ distinct such angles by taking $k = 0, 1, \ldots, j - 1$.

The condition $|bu^j| < |a|$ becomes $r_2 p^j < r_1$, so we can take

$$0 < p < (r_1/r_2)^{1/j} \stackrel{\text{def}}{=} p_0.$$

**1.6.11** Set $p(x) = x^k + a_{k-1}x^{k-1} + \cdots + a_1 x + a_0$ with $k$ odd. Choose

$$C = \sup\{1, |a_{k-1}|, \ldots, |a_0|\}$$

and set $A = kC + 1$. Then if $x \leq -A$ we have

$$p(x) = x^k + a_{k-1}x^{k-1} + \cdots + a_1 x + a_0$$
$$\leq (-A)^k + CA^{k-1} + \cdots + C \leq -A^k + kCA^{k-1}$$
$$= A^{k-1}(kC - A) = -A^{k-1} \leq 0.$$

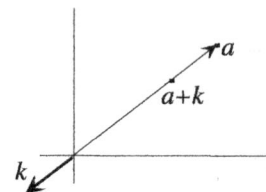

FIGURE FOR SOLUTION 1.6.9
A first error to avoid is writing "$a + bu^j$ is between 0 and $a$" as

"$0 < a + bu^j < a$."

Remember that $a$, $b$, and $u$ are complex numbers so that writing that sort of inequality doesn't make sense. If we set $k = bu^j$ to simplify notation, then $a + k$ is between 0 and $a$ if $a - (a+k) = k$ is on the same line as $a$ and points in the opposite direction, with $|k| < |a|$.

The proof given essentially reproves proposition 0.7.7. If you want to use that proposition instead, you could say:

If $a + bu^j$ is between 0 and $a$, then there exists $\rho$ with $0 < \rho < 1$ such that

$$a + bu^j = \rho a, \text{ i.e., } u^j = \frac{(\rho - 1)a}{b}.$$

This equation has $j$ solutions by proposition 0.7.7, and

$$|u| = (1 - \rho)|a/b| < |a/b|,$$

so we can take $p_0 = |a/b|^{1/j}$.

Similarly, if $x \geq A$ we have

$$p(x) = x^k + a_{k-1}x^{k-1} + \cdots + a_1 x + a_0$$
$$\geq (A)^k - CA^{k-1} - \cdots - C \geq A^k - kCA^{k-1}$$
$$= A^{k-1}(A - kC) = A^{k-1} \geq 0.$$

Since $p : [-A, A] \to \mathbb{R}$ is a continuous function (corollary 1.5.30) and we have $p(-A) \leq 0$ and $p(A) \geq 0$, then by the intermediate value theorem there exists $x_0 \in [-A, A]$ such that $p(x_0) = 0$.

**1.7.1**  a. $f(a) = 0$, $f'(a) = \cos(a) = 1$, so the tangent is $g(x) = x$.

b. $f(a) = \frac{1}{2}$, $f'(a) = -\sin(a) = -\frac{\sqrt{3}}{2}$, so the tangent is

$$g(x) = -\frac{\sqrt{3}}{2}\left(x - \frac{\pi}{3}\right) + \frac{1}{2}.$$

c. $f(a) = 1$, $f'(a) = -\sin(a) = 0$, so the tangent is $g(x) = 1$.

d. $f(a) = 2$, $f'(a) = -\frac{1}{a^2} = -4$, so the tangent is

$$g(x) = -4(x - 1/2) + 2 = -4x + 4.$$

**1.7.3**  a. $f'(x) = \left(3\sin^2(x^2 + \cos x)\right)\left(\cos(x^2 + \cos x)\right)\left(2x - \sin x\right)$

b. $f'(x) = \left(2\cos((x + \sin x)^2)\right)\left(-\sin((x + \sin x)^2)\right)\left(2(x + \sin x)\right)\left(1 + \cos x\right)$

c. $f'(x) = \left((\cos x)^5 + \sin x\right)\left(4(\cos x)^3\right)(-\sin(x)) = (\cos x)^5 - 4(\sin x)^2(\cos x)^3$

d. $f'(x) = 3(x + \sin^4 x)^2(1 + 4\sin^3 x \cos x)$

e. $f'(x) = \dfrac{\sin^3 x(\cos x^2 * 2x)}{2 + \sin(x)} + \dfrac{\sin x^2(3\sin^2 x \cos x)}{2 + \sin(x)} - \dfrac{(\sin x^2 \sin^3 x)(\cos x)}{\left(2 + \sin(x)\right)^2}$

f. $f'(x) = \cos\left(\dfrac{x^3}{\sin x^2}\right)\left(\dfrac{3x^2}{\sin x^2} - \dfrac{(x^3)(\cos x^2 * 2x)}{(\sin x^2)^2}\right)$

**1.7.5**  a. Compute the partial derivatives:

$$D_1 f\left(\begin{matrix} x \\ y \end{matrix}\right) = \frac{x}{\sqrt{x^2 + y}} \quad \text{and} \quad D_2 f\left(\begin{matrix} x \\ y \end{matrix}\right) = \frac{1}{2\sqrt{x^2 + y}}.$$

This gives

$$D_1 f\left(\begin{matrix} 2 \\ 1 \end{matrix}\right) = \frac{2}{\sqrt{2^2 + 1}} = \frac{2}{\sqrt{5}} \quad \text{and} \quad D_2 f\left(\begin{matrix} 2 \\ 1 \end{matrix}\right) = \frac{1}{2\sqrt{2^2 + 1}} = \frac{1}{2\sqrt{5}}.$$

At the point $\left(\begin{matrix} 1 \\ -2 \end{matrix}\right)$, we have $x^2 + y < 0$, so the function is not defined there, and neither are the partial derivatives.

b. Similarly, $D_1 f\left(\begin{matrix} x \\ y \end{matrix}\right) = 2xy$ and $D_2 f\left(\begin{matrix} x \\ y \end{matrix}\right) = x^2 + 4y^3$. This gives

$$D_1 f\left(\begin{matrix} 2 \\ 1 \end{matrix}\right) = 4 \quad \text{and} \quad D_2 f\left(\begin{matrix} 2 \\ 1 \end{matrix}\right) = 4 + 4 = 8;$$

$$D_1 f \begin{pmatrix} 1 \\ -2 \end{pmatrix} = -4 \quad \text{and} \quad D_2 f \begin{pmatrix} 1 \\ -2 \end{pmatrix} = 1 + 4 \cdot (-8) = -31.$$

c. Compute

$$D_1 f \begin{pmatrix} x \\ y \end{pmatrix} = -y \sin xy$$

$$D_2 f \begin{pmatrix} x \\ y \end{pmatrix} = -x \sin xy + \cos y - y \sin y.$$

This gives

$$D_1 f \begin{pmatrix} 2 \\ 1 \end{pmatrix} = -\sin 2 \quad \text{and} \quad D_2 f \begin{pmatrix} 2 \\ 1 \end{pmatrix} = -2 \sin 2 + \cos 1 - \sin 1$$

$$D_1 f \begin{pmatrix} 1 \\ -2 \end{pmatrix} = -2 \sin 2 \quad \text{and} \quad D_2 f \begin{pmatrix} 1 \\ -2 \end{pmatrix} = \sin 2 + \cos 2 - 2 \sin 2 = \cos 2 - \sin 2$$

d. Since

$$D_1 f \begin{pmatrix} x \\ y \end{pmatrix} = \frac{xy^2 + 2y^4}{2(x + y^2)^{3/2}} \quad \text{and} \quad D_2 f \begin{pmatrix} x \\ y \end{pmatrix} = \frac{2x^2 y + xy^3}{(x + y^2)^{3/2}},$$

we have

$$D_1 f \begin{pmatrix} 2 \\ 1 \end{pmatrix} = \frac{4}{2\sqrt{27}} \quad \text{and} \quad D_2 f \begin{pmatrix} 2 \\ 1 \end{pmatrix} = \frac{10}{\sqrt{27}};$$

$$D_1 f \begin{pmatrix} 1 \\ -2 \end{pmatrix} = \frac{36}{10\sqrt{5}} \quad \text{and} \quad D_2 f \begin{pmatrix} 1 \\ -2 \end{pmatrix} = -\frac{12}{5\sqrt{5}}.$$

**1.7.7**  Just pile up the partial derivative vectors side by side:

$$\text{a.} \quad \left[ \mathbf{D}\vec{f}\begin{pmatrix} x \\ y \end{pmatrix} \right] = \begin{bmatrix} -\sin x & 0 \\ 2xy & x^2 + 2y \\ 2x \cos(x^2 - y) & -\cos(x^2 - y) \end{bmatrix}$$

$$\text{b.} \quad \left[ \mathbf{D}\vec{f}\begin{pmatrix} x \\ y \end{pmatrix} \right] = \begin{bmatrix} \frac{x}{\sqrt{x^2+y^2}} & \frac{y}{\sqrt{x^2+y^2}} \\ y & x \\ 2y \sin xy \, \cos xy & 2x \sin xy \, \cos xy \end{bmatrix}.$$

**1.7.9**  a. The derivative is an $m \times n$ matrix

b. a $1 \times 3$ matrix (line matrix)

c. a $4 \times 1$ matrix (vector 4 high)

**1.7.11**

a. $\left[ y \cos(xy), \ x \cos(xy) \right]$   b. $\left[ 2x e^{x^2 + y^3}, \ 3y^2 e^{x^2 + y^3} \right]$

c. $\begin{bmatrix} y & x \\ 1 & 1 \end{bmatrix}$        d. $\begin{bmatrix} \cos \theta & -r \sin \theta \\ \sin \theta & r \cos \theta \end{bmatrix}$

**1.7.13**  For the first part, $|x|$ and $mx$ are continuous functions, hence so is $f(0 + h) - f(0) - mh = |h| - mh.$

For the second, we have

$$\frac{|h| - mh}{h} = \frac{-h - mh}{h} = -1 - m \quad \text{when } h < 0$$

$$\frac{|h| - mh}{h} = \frac{h - mh}{h} = 1 - m \quad \text{when } h > 0.$$

The difference between these values is always 2, and cannot be made small by taking $h$ small.

**1.7.15** a. There exists a linear transformation $[\mathbf{D}F(A)]$ such that

Solution 1.7.15, part a: The absolute value in the numerator is optional (but not in the denominator: you cannot divide by matrices).

Since $H$ is an $n \times m$ matrix, the $[0]$ in $\lim_{H \to [0]}$ is the $n \times m$ matrix with all entries 0.

$$\lim_{H \to [0]} \frac{|F(A + H) - F(A) - [\mathbf{D}F(A)]H|}{|H|} = 0.$$

b. The derivative is $[\mathbf{D}F(A)]H = AH^\top + HA^\top$. We found this by looking for linear terms in $H$ of the difference

$$F(A + H) - F(A) = (A + H)(A + H)^\top - AA^\top$$
$$= (A + H)(A^\top + H^\top) - AA^\top$$
$$= AH^\top + HA^\top + HH^\top;$$

see remark 1.7.6. The linear terms $AH^\top + HA^\top$ are the derivative. Indeed,

$$\lim_{H \to [0]} \frac{|(A + H)(A + H)^\top - AA^\top - AH^\top - HA^\top|}{|H|}$$

$$= \lim_{H \to [0]} \frac{|HH^\top|}{|H|} \leq \lim_{H \to [0]} \frac{|H||H^\top|}{|H|} = \lim_{H \to [0]} |H| = 0.$$

**1.7.17** The derivative of the squaring function is given by

$$[\mathbf{D}S(A)]H = AH + HA;$$

Solution 1.7.17: This is sort of a miracle; the expressions should not be equal, they should differ by terms in $\epsilon^2$. The reason why they are exactly equal here is that

$$\begin{bmatrix} 0 & 0 \\ \epsilon & 0 \end{bmatrix}^2 = \begin{bmatrix} 0 & 0 \\ 0 & 0 \end{bmatrix}.$$

substituting $A = \begin{bmatrix} 1 & 1 \\ 0 & 1 \end{bmatrix}$ and $H = \begin{bmatrix} 0 & 0 \\ \epsilon & 0 \end{bmatrix}$ gives

$$\begin{bmatrix} 1 & 1 \\ 0 & 1 \end{bmatrix}\begin{bmatrix} 0 & 0 \\ \epsilon & 0 \end{bmatrix} + \begin{bmatrix} 0 & 0 \\ \epsilon & 0 \end{bmatrix}\begin{bmatrix} 1 & 1 \\ 0 & 1 \end{bmatrix} = \begin{bmatrix} \epsilon & 0 \\ \epsilon & 0 \end{bmatrix} + \begin{bmatrix} 0 & 0 \\ \epsilon & \epsilon \end{bmatrix} = \begin{bmatrix} \epsilon & 0 \\ 2\epsilon & \epsilon \end{bmatrix}.$$

Computing $(A + H)^2 - A^2$ gives the same result;

$$(A + H)^2 - A^2 = \begin{bmatrix} 1 + \epsilon & 2 \\ 2\epsilon & 1 + \epsilon \end{bmatrix} - \begin{bmatrix} 1 & 2 \\ 0 & 1 \end{bmatrix} = \begin{bmatrix} \epsilon & 0 \\ 2\epsilon & \epsilon \end{bmatrix}.$$

**1.7.19** Since $\lim_{\vec{\mathbf{h}} \to \mathbf{0}} \frac{|\vec{\mathbf{h}}|\vec{\mathbf{h}}}{|\vec{\mathbf{h}}|} = \mathbf{0}$, the derivative exists at the origin and is the 0 linear transformation, represented by the $n \times n$ matrix with all entries 0.

**1.7.21** We will work directly from the definition of the derivative:

$$\det(I + H) - \det(I) - (h_{1,1} + h_{2,2})$$
$$= (1 + h_{1,1})(1 + h_{2,2}) - h_{1,2}h_{2,1} - 1 - (h_{1,1} + h_{2,2})$$
$$= h_{1,1}h_{2,2} - h_{1,2}h_{2,1}.$$

Each $h_{i,j}$ satisfies $|h_{i,j}| \le |H|$, so we have

$$\frac{|\det(I+H) - \det(I) - (h_{1,1} + h_{2,2})|}{|H|} \le \frac{|h_{1,1}h_{2,2} - h_{1,2}h_{2,1}|}{|H|} \le \frac{2|H|^2}{|H|} = 2|H|.$$

Thus

$$\lim_{H \to 0} \frac{|\det(I+H) - \det(I) - (h_{1,1} + h_{2,2})|}{|H|} \le \lim_{H \to 0} 2|H| = 0.$$

**1.8.1** Three make sense:

  c. $\mathbf{g} \circ \mathbf{f} : \mathbb{R}^2 \to \mathbb{R}^2$; the derivative is a $2 \times 2$ matrix

  d. $\mathbf{f} \circ \mathbf{g} : \mathbb{R}^3 \to \mathbb{R}^3$; the derivative is a $3 \times 3$ matrix

  e. $f \circ \mathbf{f} : \mathbb{R}^2 \to \mathbb{R}$; the derivative is a $1 \times 2$ matrix

**1.8.3** Yes: We have a composition of sine, the exponential function, and the function $\begin{pmatrix} x \\ y \end{pmatrix} \mapsto xy$, all of which are differentiable everywhere.

**1.8.5** One must also show that $fg$ is differentiable, working from the definition of the derivative.

**1.8.7** Since $[\mathbf{D}f(x)] = [\, x_2 \ \ x_1 + x_3 \ \ x_2 + x_4 \ \cdots \ x_{n-2} + x_n \ \ x_{n-1} \,]$, we have

$$[\mathbf{D}\big(f(\gamma(t))\big)] = [\, t^2 \ \ t + t^3 \ \ t^2 + t^4 \ \cdots \ t^{n-2} + t^n \ \ \ t^{n-1} \,]. \text{ In addition,}$$

$[\mathbf{D}\gamma(t)] = \begin{bmatrix} 1 \\ 2t \\ \vdots \\ nt^{n-1} \end{bmatrix}$. So the derivative of the function $t \to f(\gamma(t))$ is

$$\underbrace{[\mathbf{D}(f \circ \gamma)(t)]}_{\substack{\text{deriv. of comp. at } t}} = \underbrace{[\mathbf{D}f(\gamma(t))]}_{\substack{\text{deriv. of } f \\ \text{at } \gamma(t)}} \underbrace{[\mathbf{D}\gamma(t)]}_{\substack{\text{deriv. of } \gamma \\ \text{at } t}} = t^2 + \left( \sum_{i=2}^{n-1} it^{i-1}(t^{i-1} + t^{i+1}) \right) + nt^{2(n-1)}$$

Solution 1.8.9: This isn't really a good problem to test knowledge of the chain rule, because it is easiest to solve it without ever invoking the chain rule (at least in several variables).

**1.8.9** Clearly

$$D_1 f \begin{pmatrix} x \\ y \end{pmatrix} = 2xy\varphi'(x^2 - y^2); \qquad D_2 f \begin{pmatrix} x \\ y \end{pmatrix} = -2y^2\varphi'(x^2 - y^2) + \varphi(x^2 - y^2).$$

Thus

$$\frac{1}{x} D_1 f \begin{pmatrix} x \\ y \end{pmatrix} + \frac{1}{y} D_2 f \begin{pmatrix} x \\ y \end{pmatrix} = 2y\varphi'(x^2 - y^2) - 2y\varphi'(x^2 - y^2) + \frac{1}{y}\varphi(x^2 - y^2)$$

$$= \frac{1}{y^2} f \begin{pmatrix} x \\ y \end{pmatrix}.$$

To use the chain rule, write $f = k \circ \mathbf{h} \circ \mathbf{g}$, where

$$\mathbf{g}\begin{pmatrix} x \\ y \end{pmatrix} = \begin{pmatrix} x^2 - y^2 \\ y \end{pmatrix}, \qquad \mathbf{h}\begin{pmatrix} u \\ v \end{pmatrix} = \begin{pmatrix} \varphi(u) \\ v \end{pmatrix}, \qquad k\begin{pmatrix} s \\ t \end{pmatrix} = st.$$

This leads to

$$\left[ \mathbf{D}f\begin{pmatrix} x \\ y \end{pmatrix} \right] = [t, s] \begin{bmatrix} \varphi'(u) & 0 \\ 0 & 1 \end{bmatrix} \begin{bmatrix} 2x & -2y \\ 0 & 1 \end{bmatrix} = [2xt\varphi'(u), \ -2yt\varphi'(u) + s].$$

Insert the values of the variables; you find

$$D_1f\begin{pmatrix} x \\ y \end{pmatrix} = 2xy\,\varphi'(x^2-y^2) \text{ and } D_2f\begin{pmatrix} x \\ y \end{pmatrix} = -2y^2\varphi'(x^2-y^2)+\varphi(x^2-y^2).$$

Now continue as above.

**1.8.11** Using the chain rule in one variable,

$$D_1f = \varphi'\left(\frac{x+y}{x-y}\right)\left(\frac{1(x-y)-1(x+y)}{(x-y)^2}\right) = \varphi'\left(\frac{x+y}{x-y}\right)\left(\frac{-2y}{(x-y)^2}\right)$$

and

$$D_2f = \varphi'\left(\frac{x+y}{x-y}\right)\left(\frac{1(x-y)-(-1)(x+y)}{(x-y)^2}\right) = \varphi'\left(\frac{x+y}{x-y}\right)\left(\frac{2x}{(x-y)^2}\right)$$

so

$$xD_1f + yD_2f = \varphi'\left(\frac{x+y}{x-y}\right)\left(\frac{-2xy}{(x-y)^2}\right) + \varphi'\left(\frac{x+y}{x-y}\right)\left(\frac{2yx}{(x-y)^2}\right) = 0$$

**1.8.13** Call $S(A) = A^2 + A$ and $T(A) = A^{-1}$. We have $F = T \circ S$, so

$$[\mathbf{D}F(A)]H = [\mathbf{D}T(A^2+A)][\mathbf{D}S(A)]H$$
$$= [\mathbf{D}T(A^2+A)](AH+HA+H)$$
$$= -(A^2+A)^{-1}(AH+HA+H)(A^2+A)^{-1}.$$

It isn't really possible to simplify this much.

**1.9.1** Except at $\begin{pmatrix} 0 \\ 0 \end{pmatrix}$, the partial derivatives of $f$ are given by

$$D_1f\begin{pmatrix} x \\ y \end{pmatrix} = \frac{2x^5 + 4x^3y^2 - 2xy^4}{(x^2+y^2)^2} \quad \text{and} \quad D_2f\begin{pmatrix} x \\ y \end{pmatrix} = \frac{4x^2y^3 - x^4y + 2y^5}{(x^2+y^2)^2}.$$

At the origin, they are both given by

$$D_1f\begin{pmatrix} 0 \\ 0 \end{pmatrix} = D_2f\begin{pmatrix} 0 \\ 0 \end{pmatrix} = \lim_{h \to 0} \frac{1}{h}\left(\frac{h^4}{h^2}\right) = 0.$$

Thus there are partial derivatives everywhere, and we need to check that they are continuous. The only problem is at the origin. One easy way to show this is to remember that

$$|x| \le \sqrt{x^2+y^2} \quad \text{and} \quad |y| \le \sqrt{x^2+y^2}.$$

Then both partial derivatives satisfy

$$\left|D_if\begin{pmatrix} x \\ y \end{pmatrix}\right| \le 8\frac{(x^2+y^2)^{5/2}}{(x^2+y^2)^2} = \sqrt{x^2+y^2}.$$

Thus the limit of both partials at the origin is 0, so the partials are continuous and $f$ is differentiable everywhere.

**1.9.3** a. It means that there is a line matrix $[a, \ b]$ such that

$$\lim_{\vec{h} \to 0} \frac{\sin\left(\dfrac{h_1^2 h_2^2}{h_1^2 - h_2^2}\right) - ah_1 - bh_2}{(h_1^2 + h_2^2)^{1/2}} = 0.$$

b. Since $f$ vanishes identically on both axes, both partials exist and are 0 at the origin. In fact, $D_1 f$ vanishes on the $x$-axis and $D_2 f$ vanishes on the $y$-axis.

c.   We know that if $f$ is differentiable at the origin, then its partial derivatives exist at the origin and are the numbers $a, b$ of part a. Thus for $f$ to be differentiable at the origin, we must have

$$\lim_{\vec{h} \to 0} \frac{\sin\left(\dfrac{h_1^2 h_2^2}{h_1^2 - h_2^2}\right)}{(h_1^2 + h_2^2)^{1/2}} = 0,$$

and this is indeed the case, since

$$|\sin(h_1^2 h_2^2)| \leq |h_1^2 h_2^2| < |h_1|^2 + |h_2|^2 \quad \text{when } |h_1|, \ |h_2| < 1.$$

**1.1** a. Not a subspace: $\vec{\mathbf{0}}$ is not on the line.

b. Not a subspace: $\vec{\mathbf{0}}$ is not on the line.

c. A subspace.

**1.3** If $A$ and $B$ are upper triangular matrices, then if $i > j$ we know that $a_{i,j} = b_{i,j} = 0$. Using the definition of matrix multiplication:

$$c_{i,j} = \sum_{k=1}^{n} a_{i,k} b_{k,j},$$

we see that if $i > j$, then in the summation either $a_{i,k} = 0$ or $b_{k,j} = 0$, so $c_{i,j} = \sum_{k=1}^{n} 0 = 0$. If for all $i > j$ we have $c_{i,j} = 0$, then $C$ is upper triangular, so if $A$ and $B$ are upper triangular, $AB$ is also upper triangular.

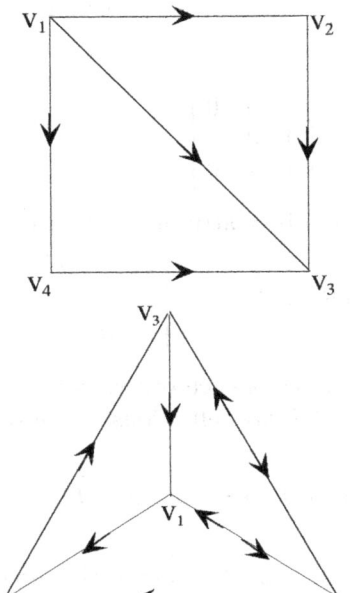

Labeling of vertices for solution 1.5, parts b and c.

**1.5** a. Labeling the vertices in the direction of the arrows: $\begin{bmatrix} 0 & 1 & 0 \\ 0 & 0 & 1 \\ 1 & 0 & 0 \end{bmatrix}$

b. With the labels shown in the figure: $\begin{bmatrix} 0 & 1 & 1 & 1 \\ 0 & 0 & 1 & 0 \\ 0 & 0 & 0 & 0 \\ 0 & 0 & 1 & 0 \end{bmatrix}$

c. Again, with the labeling as shown in the figure: $\begin{bmatrix} 0 & 1 & 0 & 1 \\ 0 & 0 & 1 & 0 \\ 1 & 0 & 0 & 1 \\ 1 & 1 & 1 & 0 \end{bmatrix}$

Solution 1.7: This uses theorem 1.3.4; the $i$th column of $[T]$ is $T\vec{e}_i$.

**1.7** a. Yes there is; its matrix is $[T] = \begin{bmatrix} 1 & 0 & 0 & 2 \\ 0 & 1 & 1 & -1 \\ 0 & 2 & 1 & -1 \end{bmatrix}$. We computed this

matrix by determining what combinations of the four input vectors give the four standard basis vectors in $\mathbb{R}^4$. For example, if we call the input vectors $\vec{v}_1, \vec{v}_2, \vec{v}_3, \vec{v}_4$, then $\vec{e}_4 = \vec{v}_3 - \vec{v}_2$, so the fourth column of the matrix is

$T\vec{e}_4 = T\vec{v}_3 - T\vec{v}_2 = \begin{bmatrix} 2 \\ -1 \\ -1 \end{bmatrix}$. Similarly, $\vec{e}_1 = \vec{v}_4 - \vec{v}_3 + \vec{v}_2$ so

$$T(\vec{e}_1) = T\vec{v}_4 - T\vec{v}_3 + T\vec{v}_2 = \begin{bmatrix} 1 \\ 0 \\ 0 \end{bmatrix}.$$

It is easy to confirm that this matrix does indeed satisfy the four equations of the exercise.

b. No, it is not linear: we have $[T] \begin{bmatrix} 1 \\ 1 \\ 1 \\ 1 \end{bmatrix} = \begin{bmatrix} 3 \\ 1 \\ 2 \end{bmatrix}$, not $\begin{bmatrix} 0 \\ 3 \\ 2 \end{bmatrix}$. Another way to see this is to say that if $S$ were linear, then by part a we would have

$$S\vec{e}_1 = \begin{bmatrix} 1 \\ 0 \\ 0 \end{bmatrix}, \quad S\vec{e}_2 = \begin{bmatrix} 0 \\ 1 \\ 2 \end{bmatrix}, \quad S\vec{e}_3 = \begin{bmatrix} 0 \\ 1 \\ 1 \end{bmatrix}, \quad S\vec{e}_4 = \begin{bmatrix} 2 \\ -1 \\ -1 \end{bmatrix},$$

which by linearity should give $S \begin{bmatrix} 1 \\ 1 \\ 1 \\ 1 \end{bmatrix} = S(\vec{e}_1 + \vec{e}_2 + \vec{e}_3 + \vec{e}_4) = \begin{bmatrix} 3 \\ 1 \\ 2 \end{bmatrix}$.

**1.9**  a. The matrices of $S$ and $T$ are

$$[S] = \begin{bmatrix} 0 & 1 & 0 \\ 1 & 0 & 0 \\ 0 & 0 & 1 \end{bmatrix} \quad \text{and} \quad [T] = \begin{bmatrix} 1 & 0 & 0 \\ 0 & 0 & 1 \\ 0 & 1 & 0 \end{bmatrix}.$$

b. The matrices of the compositions are given by matrix multiplication:

$$[S \circ T] = [S][T] = \begin{bmatrix} 0 & 0 & 1 \\ 1 & 0 & 0 \\ 0 & 1 & 0 \end{bmatrix} \quad \text{and} \quad [T \circ S] = [T][S] = \begin{bmatrix} 0 & 1 & 0 \\ 0 & 0 & 1 \\ 1 & 0 & 0 \end{bmatrix}.$$

c. The matrices $[S \circ T]$ and $[T \circ S]$ are inverses of each other: you can either compute it out, or note that since $S$ and $T$ are reflections, we have $S \circ S = T \circ T = I$, so

$$T \circ (S \circ S) \circ T = T \circ T = I \quad \text{and} \quad S \circ (T \circ T) \circ S = S \circ S = I.$$

d.    They are the rotations by $2\pi/3$ and $-2\pi/3$ around the line $x = y = z$, counterclockwise if you look from a point of this line with positive coordinates towards the origin.

**1.11**  Below we denote by $|\overrightarrow{\text{side}}|$ the length of the side.

a. Because the side and the diagonal define a right triangle,

$$\text{angle between } \overrightarrow{\text{side}} \text{ and } \overrightarrow{\text{diagonal}} = \arccos\left( \frac{|\overrightarrow{\text{side}}|}{|\overrightarrow{\text{diagonal}}|} \right)$$

$$\theta_x = \arccos\left( \frac{a}{\sqrt{a^2 + b^2 + c^2}} \right) \qquad \theta_y = \arccos\left( \frac{b}{\sqrt{a^2 + b^2 + c^2}} \right)$$

$$\theta_z = \arccos\left( \frac{c}{\sqrt{a^2 + b^2 + c^2}} \right)$$

b. Volume (parallelepiped) $= abc = \text{area(base)} \times \text{height}$, but height $= \text{length(diagonal)} \times \sin(\text{angle of diagonal with face})$, so

$$\text{angle} = \arcsin\left( \frac{abc}{\text{area(base)}\sqrt{a^2 + b^2 + c^2}} \right)$$

$$\theta_{x-y} = \arcsin\left( \frac{abc}{ab\sqrt{a^2 + b^2 + c^2}} \right) = \arcsin\left( \frac{c}{\sqrt{a^2 + b^2 + c^2}} \right)$$

$$\theta_{x-z} = \arcsin\left(\frac{abc}{ac\sqrt{a^2+b^2+c^2}}\right) = \arcsin\left(\frac{b}{\sqrt{a^2+b^2+c^2}}\right)\theta_{y-z}$$

$$= \arcsin\left(\frac{abc}{bc\sqrt{a^2+b^2+c^2}}\right) = \arcsin\left(\frac{a}{\sqrt{a^2+b^2+c^2}}\right)$$

**1.13** a. The normalized vectors are:

$$\text{i.}\quad \frac{1}{\sqrt{14}}\begin{bmatrix} 2 \\ 1 \\ 3 \end{bmatrix}, \qquad \text{ii.}\quad \frac{1}{\sqrt{13}}\begin{bmatrix} -2 \\ 3 \end{bmatrix}, \qquad \text{iii.}\quad \frac{1}{\sqrt{7}}\begin{bmatrix} \sqrt{3} \\ 0 \\ 2 \end{bmatrix}.$$

b. The angle $\theta$ satisfies $\cos\theta = \dfrac{2\sqrt{3}+6}{7\sqrt{2}}$, i.e., $\theta = \arccos\dfrac{2\sqrt{3}+6}{7\sqrt{2}}$.

Solution 1.15: Notation for sequences can be ambiguous. A sequence $x_1, x_2, \ldots$ may be written $x_i$, but $x_i$ can also refer to some collection of the elements of the sequence. The correct way to describe the sequence in part a is: Let $i \mapsto \mathbf{x}_i$ be a sequence in $\cap_i C_i$, i.e, a map $\mathbb{N} \to \cap_i C_i$.

**1.15** a. Let $C_i$, $i \in I$ be some collection of closed subsets of $\mathbb{R}^n$. We will use proposition 1.5.17 to show that their intersection is closed. Indeed, let $\mathbf{x}_1, \mathbf{x}_2, \ldots$ be a convergent sequence in $\cap_{i\in I} C_i$, converging in $\mathbb{R}^n$ to some $\mathbf{x}_0$. Then the sequence $\mathbf{x}_i$ belongs to each $C_i$, and since the $C_i$ are closed, we have $\mathbf{x}_0 \in C_i$ for each $i \in I$. Therefore $\mathbf{x}_0 \in \cap_{i\in I} C_i$.

b. Again, we will use proposition 1.5.17. Let $C_1, \ldots, C_m$ be a finite collection of closed subsets of $\mathbb{R}^m$. Suppose that $\mathbf{x}_1, \mathbf{x}_2, \ldots$ is a convergent sequence in the union $\cup_{i=1}^m C_i$, converging in $\mathbb{R}^n$ to some $\mathbf{x}_0$. Then infinitely many of the entries of the sequence must be elements of a single $C_k$; these form a subsequence, which still converges to $\mathbf{x}_0$ by proposition 1.5.19. Hence $\mathbf{x}_0$ is an element of $C_k$, hence also an element of $\cup_{i=1}^m C_i$. It follows from proposition 1.5.17 that the union is closed.

c. The union of the closed sets $[0, (n-1)/n]$, $n = 2, 3, 4, \ldots$ is the nonclosed set $[0,1)$.

**1.17** a. The derivative of the function $zy^2$ at $\mathbf{p} = \begin{pmatrix} 1 \\ 1 \\ 1 \end{pmatrix}$ is $[0\ 2\ 1]$, so the directional derivatives at $\mathbf{p}$ in the directions $\vec{e}_1$, $\vec{e}_2$, $\vec{e}_3$, $\vec{v}_1$, $\vec{v}_2$ are

$$[0\ 2\ 1]\vec{e}_1 = 0, \qquad [0\ 2\ 1]\vec{e}_2 = 2, \qquad [0\ 2\ 1]\vec{e}_3 = 1, \qquad [0\ 2\ 1]\vec{v}_1 = \sqrt{2}/2$$
$$[0\ 2\ 1]\vec{v}_2 = 3\sqrt{2}/2.$$

So the function $zy^2$ increases most slowly in the direction $\vec{e}_1$.

b. The derivative of the function $2x^2 - y^2$ at $\mathbf{p}$ is $[4\ -2\ 0]$, giving the directional derivatives at $\mathbf{p}$

$$[4\ -2\ 0]\vec{e}_1 = 4, \qquad [4\ -2\ 0]\vec{e}_2 = -2, \qquad [4\ -2\ 0]\vec{e}_3 = 0$$
$$[4\ -2\ 0]\vec{v}_1 = 2\sqrt{2}, \qquad [4\ -2\ 0]\vec{v}_2 = -\sqrt{2}.$$

So to make $2x^2 - y^2$ increase as much as possible, also choose direction $\vec{e}_1$.

3. To make $2x^2 - y^2$ decrease as much as possible, choose direction $\vec{e}_2$.

**1.19** a. This uses a trick:

$$\frac{x+y}{x^2-y^2} = \frac{x+y}{(x+y)(x-y)} = \frac{1}{x-y};$$

there is no limit as $\begin{pmatrix} x \\ y \end{pmatrix} \rightarrow \begin{pmatrix} 0 \\ 0 \end{pmatrix}$, since when $\begin{pmatrix} x \\ y \end{pmatrix}$ is close to the origin, $x - y$ is also small (perhaps 0), so the quotient is big (or undefined).

b. Again the limit does not exist: on the line $y = kx$, this function is

$$\frac{x^4(1+k^2)^2}{x(1+k)} = x^3 \frac{(1+k^2)^2}{1+k}.$$

Choose $\epsilon > 0$, and set $1 + k = \epsilon^4$ and $x = \epsilon$; the function becomes

$$\frac{\epsilon^3}{\epsilon^4}(1+k^2) > \frac{1}{\epsilon}.$$

Thus there are points near the origin where the function is arbitrarily large. But there are also points where the function is arbitrarily close to 0, taking $x = y = \epsilon$.

c. Since $\lim_{u \to 0} u \ln |u| = 0$, the limit is 0.

d. Let us look at the function on the line $y = \epsilon$, for some $\epsilon > 0$. The function becomes

$$(x^2 + \epsilon^2)(\ln|x| + \ln \epsilon) = x^2 \ln|x| + \epsilon^2 \ln|x| + x^2 \ln \epsilon + \epsilon^2 \ln \epsilon.$$

When $|x|$ is small, the first, third, and fourth terms are small. But the second is not. If $x = e^{-1/\epsilon^3}$, for instance, the second term is

$$-\frac{\epsilon^2}{\epsilon^3} = -\frac{1}{\epsilon},$$

which will become arbitrarily large as $\epsilon \to 0$. Thus along the curve given by $x = e^{-1/\epsilon^3}$ and $y = \epsilon$, the function tends to $-\infty$ as $\epsilon \to 0$. But along the curve $x = y = \epsilon$, the function tends to 0 as $\epsilon \to 0$, so it has no limit.

The statement

$$\lim_{u \to 0} u \ln |u| = 0,$$

can be proved using l'Hôpital's rule applied to $\dfrac{\ln|u|}{1/u}$. We used this already in solution 1.5.21.

**1.21**  Let $\pi_n$ be $\pi$ to $n$ places (e.g., $\pi_2 = 3.14$). Then $\pi - \pi_n < 10^{-n}$. We can do the same with $e$. So:

$$\left| \mathbf{a}_n - \begin{bmatrix} \pi \\ e \end{bmatrix} \right| < \sqrt{2 \cdot 10^{2(-n)}} = (\sqrt{2})10^{-n} < 10^{-n+1}.$$

So the best we can say in general is that to get $\left| \mathbf{a}_n - \begin{bmatrix} \pi \\ e \end{bmatrix} \right| < 10^{-m}$, we need $n = m + 1$. When $m = 3$, we can get away with $n = 3$, since $(.59\ldots)^2 + (.28\ldots)^2 < 1$. But when $m = 4$, we really need $n = 5$, since $(.92\ldots)^2 + (.81\ldots)^2 > 1$.

Solution 1.21: In the first printing of the text, we gave a wrong value for $e$. It is $e = 2.71828\ldots$.

For $m = 3$, we have

$$\left| \mathbf{a}_3 - \begin{bmatrix} \pi \\ e \end{bmatrix} \right| \approx \left| \begin{bmatrix} .00059 \\ .00028 \end{bmatrix} \right|$$

$$\approx \sqrt{(.0006)^2 + (.0003)^2}$$

$$= 10^{-3}\sqrt{.36 + .09} < 10^{-3}.$$

**1.23**  Let $a_n z^n$ be the term of highest degree. Let $c > 0$ be the real number such that $a_n c^n = p(c) - a_n c^n$. Then for any $R > c$ we know that $p(z) \neq 0$ for any $|z| \geq R$. (The term in $z^{10}$ is nonzero and the other terms together cannot match it.) We know by the fundamental theorem of algebra that $p(z)$ must have at least one root (and we know by corollary 1.6.14 that it has exactly ten). Therefore we know that $p(z)$ has a root for $|z| < R$ if $R > c$.

**1.25**  We have $\sqrt{x^2} = |x|$, so

$$\lim_{h \to 0} \frac{1}{h}(f(h) - f(0)) = \lim_{h \to 0} \frac{|h|}{h},$$

which does not exist (since it is $\pm 1$ depending on whether $h$ is positive or negative).

For $\sqrt[3]{x^2}$, we have

$$\lim_{h \to 0} \frac{1}{h}(f(h) - f(0)) = \lim_{h \to 0} \frac{h^{2/3}}{h} = \lim_{h \to 0} \frac{1}{h^{1/3}},$$

which tends to $\pm\infty$.

Of course, $\sqrt{x^4} = x^2$ is differentiable.

**1.27**  a. Both partials exist at $\begin{pmatrix} 0 \\ 0 \end{pmatrix}$ and are 0, but the function is not differentiable. Indeed, if it were, the derivative would necessarily be the Jacobian matrix, i.e., the 0 matrix, and we would have

$$\lim_{\vec{\mathbf{h}} \to \mathbf{0}} \frac{f(\vec{\mathbf{h}}) - f(\mathbf{0}) - [0,0]\vec{\mathbf{h}}}{|\vec{\mathbf{h}}|} = 0.$$

But writing the definition out leads to

$$\lim_{\vec{\mathbf{h}} \to \mathbf{0}} \frac{h_1^2 h_2}{(h_1^2 + h_2^2)\sqrt{h_1^2 + h_2^2}} = 0,$$

which isn't true: for instance, if you set $h_1 = h_2 = t$, the expression above becomes

$$\frac{t^3}{2\sqrt{2}\,|t|^3}, \quad \text{which does not become small as } t \to 0.$$

Solution 1.27, part b: If $f$ were differentiable, then by the chain rule the composition would be differentiable, and it isn't.

b.  This function is not differentiable. If you set $g(t) = \begin{pmatrix} t \\ t \end{pmatrix}$, then $(f \circ g)(t) = 2|t|$ is not differentiable at $t = 0$, but $g$ is differentiable at $t = 0$, so $f$ is not differentiable at the origin, which is $g(0)$.

c. Here are two proofs of part c:

*First proof* This isn't even continuous at the origin, although both partials exist there and are 0. But if you set $x = t, y = t$, then

$$\frac{\sin(xy)}{x^2 + y^2} = \frac{\sin t^2}{2t^2} \to \frac{1}{2}, \quad \text{as } t \to 0.$$

For $x$ small, $\sin x \approx x$; see equation 3.4.6 for the Taylor polynomial of $\sin(x)$.

*Second proof* This function is not differentiable; although both partial derivatives exist at the origin, the function itself is not continuous at the origin. For example, along the diagonal,

$$\lim_{t \to 0} \frac{\sin t^2}{2t^2} = \frac{1}{2},$$

but the limit along the antidiagonal is $-1/2$:

$$\lim_{t \to 0} \frac{\sin(-t^2)}{2t^2} = -\frac{1}{2}.$$

**1.29**  a. By the chain rule this is

$$[\mathbf{D}f(t)] = \left[ -\frac{1}{t + \sin t}, \; \frac{1}{t^2 + \sin(t^2)} \right] \begin{bmatrix} 1 \\ 2t \end{bmatrix} = \frac{2t}{t^2 + \sin(t^2)} - \frac{1}{t + \sin t}.$$

b. We defined $f$ for $t > 1$, but we will analyze it for all values of $t$. The function $f$ is increasing for all $t > 0$ and decreasing for all $t < 0$. First let us see that $f$ increases for $t > 0$, i.e., that the derivative is strictly positive for all $t > 0$. To see this, put the derivative on a common denominator:

$$[\mathbf{D}f(t)] = \frac{2t^2 + 2t\sin t - t^2 - \sin(t^2)}{(t^2 + \sin(t^2))(t + \sin t)}. \tag{1}$$

.The function $f$ tends to $-\infty$ as $t$ tends to 0. To see this, note that if $s$ is small, $\sin s$ is very close to $s$ (see the margin note for solution 1.27). So

$$\int_t^{t^2} \frac{ds}{s + \sin s} \approx \int_t^{t^2} \frac{ds}{2s}$$
$$= \frac{1}{2}\left(\ln|t^2| - \ln|t|\right)$$
$$= \frac{1}{2}(2\ln|t| - \ln|t|$$
$$= \frac{\ln|t|}{2}.$$

For $x > 0$, we have $x > \sin x$ (try graphing the functions $x$ and $\sin x$), so for $t > 0$, the denominator is strictly positive. We need to show that the numerator is also strictly positive, i.e.,

$$t^2 + 2t\sin t - \sin(t^2) > 0. \tag{2}$$

For $-\pi < t \le \pi$, this is true: for $-\pi < t \le \pi$, we know that $t$ and $\sin t$ are both positive or both negative, so $2t\sin t > 0$, and we have $t^2 > \sin(t^2)$ for $t \ne 0$. (Do not worry about $t^2 < t$ for small $t$; if you set $x = t^2$, the formula $x > \sin x$ still applies for $x > 0$.)

For $t > \pi$, equation (2) is also true: in that case,

$$t^2 + 2t\sin t - \sin(t^2) \ge t(t + 2\sin t) - 1 \ge t(\pi - 2) - 1 > \pi - 1.$$

To show that the function is decreasing for $t < 0$, we must show that the derivative is negative. For $t < 0$, the numerator is still strictly positive, by the argument above. But the denominator is negative: $t^2 + \sin(t^2)$ is positive, but $t + \sin t$ is negative.

**1.31** Set $f(A) = A^{-1}$ and $g(A) = AA^\top + A^\top A$. Then $F = f \circ g$, and we wish to compute

$$[\mathbf{D}F(A)]H = [\mathbf{D}f \circ g(A)]H = [\mathbf{D}f(g(A))][\mathbf{D}g(A)]H$$
$$= [\mathbf{D}f(AA^\top + A^\top A)] \underbrace{[\mathbf{D}g(A)]H}_{\substack{\text{new increment} \\ \text{for } \mathbf{D}f}}. \tag{1}$$

Solution 1.31: Remember (see example 1.7.18) that we cannot treat equation (1) as matrix multiplication. To treat the derivatives of $f$ and $g$ as matrices, we would have to identify $\mathrm{Mat}\,(n,n)$ with $\mathbb{R}^{n^2}$; the derivatives would be $n^2 \times n^2$ matrices. Instead we think of the derivatives as linear transformations.

The linear terms in $H$ of

$$g(A + H) - g(A) = (A + H)(A + H)^\top + (A + H)^\top(A + H) - AA^\top - A^\top A$$

are $AH^\top + HA^\top + A^\top H + H^\top A$; this is $[\mathbf{D}g(A)]H$, which is the new increment for $\mathbf{D}f$.

We know from proposition 1.7.19 that $[\mathbf{D}f(A)]H = -A^{-1}HA^{-1}$, which we will rewrite as

$$[\mathbf{D}f(B)]K = -B^{-1}KB^{-1} \tag{2}$$

to avoid confusion. We substitute $AH^\top + HA^\top + A^\top H + H^\top A$ for the increment $K$ in equation (2) and $g(A) = AA^\top + A^\top A$ for $B$. This gives

$$[\mathbf{D}F(A)]H = [\mathbf{D}f(AA^\top + A^\top A)][\mathbf{D}g(A)]H$$
$$= \underbrace{(-AA^\top + A^\top A)^{-1}}_{-B^{-1}} \underbrace{(AH^\top + HA^\top + A^\top H + H^\top A)}_{K} \underbrace{(AA^\top + A^\top A)^{-1}}_{B^{-1}}.$$

There is no obvious way to simplify this expression.

**1.33** a. Except at the origin, all partials exist by theorem 1.7.9. At the origin, all partials exist and are 0, since the function vanishes identically on all three coordinate axes.

b. By theorem 1.8.1, the function is differentiable everywhere except at the origin. At the origin, the function is not differentiable. In fact, it isn't even continuous: the limit

$$\lim_{h \to 0} f \begin{pmatrix} t \\ t \\ t \end{pmatrix} = \lim_{h \to 0} \frac{h^3}{3h^4} = \lim_{h \to 0} \frac{1}{3h} \quad \text{does not exist.}$$

**1.35** Compute

$$A^2 = \left( \begin{bmatrix} a & b \\ c & d \end{bmatrix} \right)^2 = \begin{bmatrix} a^2 + bc & b(a+d) \\ c(a+d) & bc + d^2 \end{bmatrix}$$

**Solution 1.35, part b:** The locus of equation $A^2 = I$ is given by four equations in four unknowns, so you might expect it to be a union of finitely many points.

This is not the case; the locus is essentially a surface: you can chose $a$ arbitrarily and set $d = -a$; then the point $\begin{pmatrix} b \\ c \end{pmatrix}$ belongs to a hyperbola.

The three choices $a = -1$ and 1 are special: in that case one of $b$ and $c$ must be 0, i.e., the point $\begin{pmatrix} b \\ c \end{pmatrix}$ belongs to the union of the axes, which is a degenerate hyperbola.

There are two additional points of the locus that sit by themselves, $A = \pm I$. That is why we said the locus is "essentially" a surface: it is a surface plus those two points.

You may wonder about the signs of the $\sin \omega$ terms. Once the telescope is level, it is pointing in the direction of the $x$-axis. You, the astronomer rotating the telescope, are at the negative $x$ end of the telescope. If you rotate it counterclockwise, as seen by you, the matrix is as we say. On the other hand, we are not absolutely sure that the problem is unambiguous as stated.

a. If we want $A^2 = \begin{bmatrix} 0 & 0 \\ 0 & 0 \end{bmatrix}$ we need to have $a^2 = d^2 = -bc$ $(a = \pm d)$ and either $b = c = 0$ or $a = -d$. If $a = d = 0$, then either $b = 0$ or $c = 0$. If $a = d \neq 0$, then $b = c = 0$, so $a = d = 0$, so all entries of $A$ are 0. If $a = -d \neq 0$, then we have $c = \frac{-a^2}{b}$.

b. If $A^2 = I$, then $a^2 + bc = d^2 + bc = 1$, and either $b = c = 0$ or $a = -d$. If $a = d = 0$, then $bc = 1$. If $a = d \neq 0$, then $b = c = 0$, and hence $A = \pm I$. Finally, if $a = -d \neq 0$, then $bc = 1 - a^2$

c. If we want $A^2 = -I$, then we have almost the same. If $a = d = 0$, then $b = \frac{-1}{c}$. If $b = c = 0$, then $a, d = \pm i$. If either $b$ or $c$ is nonzero, then we have $a = -d$, $c = \frac{-1-a^2}{b}$.

**1.37** a. This is a case where it is much easier to think of first rotating the telescope so that it is in the $(x, z)$-plane, then changing the elevation, then rotating back. This leads to the following product of matrices:

$$\begin{bmatrix} \cos\theta_0 & -\sin\theta_0 & 0 \\ \sin\theta_0 & \cos\theta_0 & 0 \\ 0 & 0 & 1 \end{bmatrix} \begin{bmatrix} \cos\varphi & 0 & -\sin\varphi \\ 0 & 1 & 0 \\ \sin\varphi & 0 & \cos\varphi \end{bmatrix} \begin{bmatrix} \cos\theta_0 & \sin\theta_0 & 0 \\ -\sin\theta_0 & \cos\theta_0 & 0 \\ 0 & 0 & 1 \end{bmatrix}$$

$$= \begin{bmatrix} \cos^2\theta_0 \cos\varphi - \sin^2\theta_0 & \cos\theta_0 \sin\theta_0(\cos\varphi - 1) & -\sin\varphi\cos\theta_0 \\ \cos\theta_0 \sin\theta_0(\cos\varphi - 1) & \cos\varphi\sin^2\theta_0 + \cos^2\theta_0 & -\sin\theta_0 \sin\varphi \\ \sin\varphi \cos\theta_0 & \sin\varphi\sin\theta_0 & \cos\varphi \end{bmatrix}$$

b. It is best to think of first rotating the telescope into the $(x, z)$-plane, then rotating it until it is horizontal (or vertical), then rotating it on its own axis, and then rotating it back (in two steps). This leads to the following product of matrices:

$$\begin{bmatrix} \cos\theta_0 & -\sin\theta_0 & 0 \\ \sin\theta_0 & \cos\theta_0 & 0 \\ 0 & 0 & 1 \end{bmatrix} \begin{bmatrix} \cos\varphi_0 & 0 & -\sin\varphi_0 \\ 0 & 1 & 0 \\ \sin\varphi_0 & 0 & \cos\varphi_0 \end{bmatrix} \begin{bmatrix} 1 & 0 & 0 \\ 0 & \cos\omega & \sin\omega \\ 0 & -\sin\omega & \cos\omega \end{bmatrix}$$

$$\begin{bmatrix} \cos\varphi_0 & 0 & \sin\varphi_0 \\ 0 & 1 & 0 \\ -\sin\varphi_0 & 0 & \cos\varphi_0 \end{bmatrix} \begin{bmatrix} \cos\theta_0 & \sin\theta_0 & 0 \\ -\sin\theta_0 & \cos\theta_0 & 0 \\ 0 & 0 & 1 \end{bmatrix}.$$

**2.1.1** a.
$\begin{bmatrix} 3 & 1 & -4 \\ 0 & 2 & 1 \\ 1 & -3 & 0 \end{bmatrix} \begin{bmatrix} x \\ y \\ z \\ \hline 0 \\ 4 \\ 1 \end{bmatrix}$
b. $\begin{bmatrix} 3 & 1 & -4 & 0 \\ 0 & 2 & 1 & 4 \\ 1 & -3 & 0 & 1 \end{bmatrix}$
c. $\begin{bmatrix} 1 & -7 & 2 & 1 \\ 1 & -3 & 0 & 2 \\ 2 & -2 & 0 & -1 \end{bmatrix}$

**2.1.3**

a. $\begin{bmatrix} 1 & 2 & 3 \\ 4 & 5 & 6 \end{bmatrix} \rightarrow \begin{bmatrix} 1 & 0 & -1 \\ 0 & 1 & 2 \end{bmatrix}$
b. $\begin{bmatrix} 1 & -1 & 1 \\ -1 & 0 & 2 \\ -1 & 1 & 1 \end{bmatrix} \rightarrow \begin{bmatrix} 1 & 0 & 0 \\ 0 & 1 & 0 \\ 0 & 0 & 1 \end{bmatrix}$

c. $\begin{bmatrix} 1 & 2 & 3 & 5 \\ 2 & 3 & 0 & -1 \\ 0 & 1 & 2 & 3 \end{bmatrix} \rightarrow \begin{bmatrix} 1 & 0 & 0 & 1 \\ 0 & 1 & 0 & -1 \\ 0 & 0 & 1 & 2 \end{bmatrix}$

d. $\begin{bmatrix} 1 & 3 & -1 & 4 \\ 1 & 2 & 1 & 2 \\ 3 & 7 & 1 & 9 \end{bmatrix} \rightarrow \begin{bmatrix} 1 & 0 & 5 & -2 \\ 0 & 1 & -2 & 2 \\ 0 & 0 & 0 & 1 \end{bmatrix}$

e. $\begin{bmatrix} 1 & 1 & 1 & 1 \\ 2 & -3 & 3 & 3 \\ 1 & -4 & 2 & 2 \end{bmatrix} \rightarrow \begin{bmatrix} 1 & 0 & 6/5 & 6/5 \\ 0 & 1 & -1/5 & -1/5 \\ 0 & 0 & 0 & 0 \end{bmatrix}$

**2.1.5** You can undo "multiplying row $i$ by $m \neq 0$" by "multiplying row $i$ by $1/m$" (which is possible because $m \neq 0$; see definition 2.1.1).

You can undo "adding row $i$ to row $j$" by "subtracting row $i$ from row $j$," i.e., "adding $(-\text{row } i)$ to row $j$".

You can undo "switching row $i$ and row $j$" by "switching row $i$ and row $j$" again.

**2.1.7** Switching rows 2 and 3 of the matrix $\begin{bmatrix} 1 & 0 & 0 & 2 \\ 0 & 0 & 1 & -1 \\ 0 & 1 & 0 & 1 \end{bmatrix}$ brings it to echelon form, giving $\begin{bmatrix} 1 & 0 & 0 & 2 \\ 0 & 1 & 0 & 1 \\ 0 & 0 & 1 & -1 \end{bmatrix}$.

The matrix $\begin{bmatrix} 1 & 1 & 0 & 1 \\ 0 & 0 & 2 & 0 \\ 0 & 0 & 0 & 1 \end{bmatrix}$ can be brought to echelon form by multiplying row 2 by 1/2, giving $\begin{bmatrix} 1 & 1 & 0 & 1 \\ 0 & 0 & 1 & 0 \\ 0 & 0 & 0 & 1 \end{bmatrix}$.

The matrix $\begin{bmatrix} 0 & 0 & 0 \\ 1 & 0 & 0 \\ 0 & 1 & 0 \end{bmatrix}$ can be brought to echelon form by switching first the first and second rows, then the second and third rows:

$$\begin{bmatrix} 0 & 0 & 0 \\ 1 & 0 & 0 \\ 0 & 1 & 0 \end{bmatrix} \rightarrow \begin{bmatrix} 1 & 0 & 0 \\ 0 & 0 & 0 \\ 0 & 1 & 0 \end{bmatrix} \rightarrow \begin{bmatrix} 1 & 0 & 0 \\ 0 & 1 & 0 \\ 0 & 0 & 0 \end{bmatrix}.$$

The matrix $\begin{bmatrix} 0 & 1 & 0 & 3 & 0 & -3 \\ 0 & 0 & -1 & 1 & 1 & 1 \\ 0 & 0 & 0 & 0 & 1 & 2 \end{bmatrix}$ can be brought to echelon form

by multiplying row 2 through by $-1$, then adding row 3 to row 2:

$$\begin{bmatrix} 0 & 1 & 0 & 3 & 0 & -3 \\ 0 & 0 & -1 & 1 & 1 & 1 \\ 0 & 0 & 0 & 0 & 1 & 2 \end{bmatrix} \rightarrow \begin{bmatrix} 0 & 1 & 0 & 3 & 0 & -3 \\ 0 & 0 & 1 & -1 & -1 & -1 \\ 0 & 0 & 0 & 0 & 1 & 2 \end{bmatrix} \rightarrow \begin{bmatrix} 0 & 1 & 0 & 3 & 0 & -3 \\ 0 & 0 & 1 & -1 & 0 & 1 \\ 0 & 0 & 0 & 0 & 1 & 2 \end{bmatrix}.$$

**2.1.9** The first problem occurs when you subtract $2 \cdot 10^{10}$ from 1 to get from the second to the third matrix of equation 2.1.12 (second row, second entry). The 1 is "invisible" if computing only to 10 significant digits, and disappears in the subtraction: $1 - 20000000000 = -19999999999$, which to 10 significant digits is $-20000000000$. Another "invisible" 1 is found in the second row, third entry.

Solution 2.1.9: This is the main danger in numerical analysis: adding (or subtracting) numbers of very different sizes loses precision.

**2.2.1**  a.  The augmented matrix $[A, \vec{\mathbf{b}}]$ corresponds to

$$2x + y + 3z = 1$$
$$x - y = 1$$
$$x + y + 2z = 1.$$

Since $[A, \vec{\mathbf{b}}]$ row reduces to

$$\begin{bmatrix} \underline{1} & 0 & 1 & 0 \\ 0 & \underline{1} & 1 & 0 \\ 0 & 0 & 0 & \underline{1} \end{bmatrix},$$

$x$ and $y$ are pivotal unknowns, and $z$ is a nonpivotal unknown.

b.  If we list first the variable $y$, then $z$, then $x$, the system of equations becomes

$$\begin{aligned} y &+ 3z &+ 2x &= 1 \\ -y & & + x &= 1 \\ y &+ 2z &+ x &= 1. \end{aligned}$$

The corresponding matrix is

$$\begin{bmatrix} 1 & 3 & 2 & 1 \\ -1 & 0 & 1 & 1 \\ 1 & 2 & 1 & 1 \end{bmatrix}, \text{ which row reduces to } \begin{bmatrix} 1 & 0 & -1 & 0 \\ 0 & 1 & 1 & 0 \\ 0 & 0 & 0 & 1 \end{bmatrix}.$$

This time $y$ and $z$ are the pivotal variables, and $x$ is the nonpivotal variable.

**2.2.3**  a.   Call the equations $A$, $B$, $C$, $D$.   Adding $A$ and $B$ gives $2x + 4y - 2w = 0$; comparing this with $C$ gives $-2w = z - 5w + v$, so

$$3w = z + v. \tag{1}$$

Comparing $C$ and $2D$ gives $15w = 5z + 3v$, which is compatible with equation (1) only if $v = 0$. So equation (1) gives $3w = z$.

Substituting 0 for $v$ and $3w$ for $z$ in each of the four equations gives $z + 2y - w = 0$.

b.  Since you can choose arbitrarily the value of $y$ and $w$, and they determine the values of the other variables, the family of solutions depends on two parameters.

**2.2.5**  a. This system has a solution for every value of $a$. If you row reduce $\begin{bmatrix} a & 1 & 0 & 2 \\ 0 & a & 1 & 3 \end{bmatrix}$ you may seem to get

$$\begin{bmatrix} 1 & 0 & -1/a^2 & -\left(2/a + 3/a^2\right) \\ 0 & 1 & 1/a & 3/a \end{bmatrix},$$

which seems to indicate that there is a solution for any value of $a$ except $a = 0$. However, obviously the system has a solution if $a = 0$; in that case, $y = 2$ and $z = 3$. The problem with the above row reduction is that if $a = 0$, then $a$ can't be used for a pivotal 1. If $a = 0$, the matrix row reduces to $\begin{bmatrix} 0 & 1 & 0 & 2 \\ 0 & 0 & 1 & 3 \end{bmatrix}$.

b.  We have two equations in three unknowns; there is no unique solution.

**2.2.7**  We can perform row operations to bring $\begin{bmatrix} 1 & 1 & 2 & 1 \\ 1 & -1 & a & b \\ 2 & 0 & -b & 0 \end{bmatrix}$ to

$$\begin{bmatrix} 1 & 0 & (2+a)/2 & (1+b)/2 \\ 0 & 1 & (2-a)/2 & (1-b)/2 \\ 0 & 0 & 2+a+b & 1+b \end{bmatrix}.$$

a.   There are then two possibilities.  If $a + b + 2 \neq 0$, the first three columns row reduce to the identity, and the system of equations has the unique solution

$$x = \frac{b(b+1)}{2+a+b}, \quad y = \frac{-b^2 - 3b + 2a}{2+a+b}, \quad x = \frac{1+b}{2+a+b}.$$

If $a + b + 2 = 0$, then there are two possibilities to consider: either $b + 1 = 0$ or $b + 1 \neq 0$. If $b + 1 = 0$, so that $a = b = -1$, the matrix row reduces to

$$\begin{bmatrix} 1 & 0 & 1/2 & 0 \\ 0 & 1 & 3/2 & 1 \\ 0 & 0 & 0 & 0 \end{bmatrix}.$$

In this case there are infinitely many solutions: the only nonpivotal variable is $z$, so we can choose its value arbitrarily; the others are $x = -z/2$ and

$y = 1 - (3z)/2$. If $a + b + 2 = 0$ and $b + 1 \neq 0$, then there is a pivotal 1 in the last column, and there are no solutions.

b. The first case, where $a + b + 2 \neq 0$, corresponds to an open subset of the $(a, b)$-plane. The second case, where $a = b = -1$, corresponds to a closed set. The third is neither open nor closed.

**2.2.9**  Row reducing

$$\begin{bmatrix} 1 & -1 & -1 & -3 & 1 & 1 \\ 1 & 1 & -5 & -1 & 7 & 2 \\ -1 & 2 & 2 & 2 & 1 & 0 \\ -2 & 5 & -4 & 9 & 7 & \beta \end{bmatrix} \text{ gives } \begin{bmatrix} 1 & 0 & 0 & -4 & 3 & 2 \\ 0 & 1 & 0 & -1/3 & 7/3 & 5/6 \\ 0 & 0 & 1 & -2/3 & -1/3 & 1/6 \\ 0 & 0 & 0 & 0 & 0 & \beta + 1/2 \end{bmatrix}.$$

There are then two possibilities: either $\beta \neq -1/2$ or $\beta = -1/2$. If $\beta \neq -1/2$, there will be a pivotal 1 in the last column (once we have divided by $\beta + 1/2$), so there is no solution. But if $\beta = -1/2$, there are infinitely many solutions: $x_4$ and $x_5$ are nonpivotal, so their values can be chosen arbitrarily, and then the values of $x_1, x_2$ and $x_3$ are given by

$$x_1 = 2 + 4x_4 - 3x_5$$
$$x_2 = 5/6 + x_4/3 - 7x_5/3$$
$$x_3 = 1/6 + 2x_4/3 + x_5/3.$$

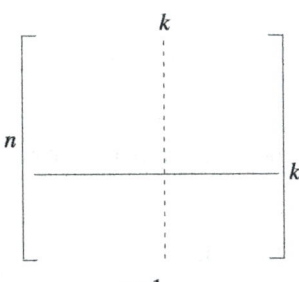

FIGURE FOR SOLUTION 2.2.11, part b: This $n \times (n + 1)$ matrix represents a system $A\vec{x} = \vec{b}$ of $n$ equations in $n$ unknowns. By the time we are ready to obtain a pivotal 1 at the intersection of the $k$th column (dotted) and $k$th row, all the entries on the $k$th row to the left of the $k$th column are 0, so we only need to place a 1 in position $k, k$ and then justify that act by dividing all the entries on the $k$th row to the right of $k$th column by the $(k, k)$ entry. There are $n + 1 - k$ such entries.

If the $(k, k)$ entry is 0, we go down the $k$th column until we find a nonzero entry. In computing the total number of computations, we are assuming the worse case scenario, where all entries of the $k$th column are nonzero.

**2.2.11**  a.  $R(1) = 1 + 1/2 - 1/2 = 1$,    $R(2) = 8 + 2 - 1 = 9$.

If we have one equation in one unknown, we need to perform one division. If we have two equations in two unknowns, we need two divisions to get a pivotal 1 in the first row (the 1 is free), followed by two multiplications and two additions to get a 0 in the first element of the second row (the 0 is free). One more division, multiplication and addition get us a pivotal 1 in the second row and a 0 for the second element of the first row, for a total of nine.

b. As illustrated by the figure in the margin, we need $n + 1 - k$ divisions to obtain a pivotal 1 in the column $k$. To obtain a 0 in another entry of column $k$ requires $n + 1 - k$ multiplications and $n + 1 - k$ additions. We need to do this for $n - 1$ entries of column $k$. So our total is

$$(n + 1 - k) + 2(n - 1)(n + 1 - k) = (2n - 1)(n - k + 1).$$

c.  For $n = 1$, we have $(2 - 1)(1 - 1 + 1) = 1 = 1^3 + \frac{1^2}{2} - \frac{1}{2}$, so the relationship is true for $n = 1$. If the relation is true for $n$, then

$$\sum_{k=1}^{n+1} \Big(2(n+1) - 1\Big)\Big((n+1) - k + 1\Big) = \sum_{k=1}^{n+1}(2n+1)(n - k + 2)$$

$$= 2n + 1 + \sum_{k=1}^{n}\Big((2n - 1)(n - k + 1) + (4n - 2k + 3)\Big)$$

$$= 3n^2 + 4n + 1 + \sum_{k=1}^{n}(2n - 1)(n - k + 1)$$

$$= 3n^2 + 4n + 1 + n^3 + \frac{n^2}{2} - \frac{n}{2} = (n+1)^3 + \frac{(n+1)^2}{2} - \frac{n+1}{2}.$$

So by recursion, the relation is true for all $n \geq 1$.

d.
$$Q(1) = \frac{2}{3} + \frac{3}{2} - \frac{7}{6} = 1,$$
$$Q(2) = \frac{2}{3}8 + \frac{3}{2}4 - \frac{7}{6}2 = 9,$$
$$Q(3) = \frac{2}{3}27 + \frac{3}{2}9 - \frac{7}{6}9 = 28.$$

The function $R(n) - Q(n)$ is the cubic polynomial $\frac{1}{3}n^3 - n^2 + \frac{2}{3}n$, with a root at $n = 2$. Its derivative, $n^2 - 2n + \frac{2}{3}$, has roots at $1 \pm \sqrt{1/3}$; it is strictly positive for $n \geq 2$. So the function $R(n) - Q(n)$ is increasing as a function of $n$ for $n \geq 2$, and hence is strictly positive for $n \geq 3$.

e. For partial row reduction for a single column, the operations needed are like those for full row reduction (part b) except that we are just putting zeros below the diagonal, so we can replace $n - 1$ in the total for full row reduction by $n - k$, to get

$$(n + 1 - k) + 2(n - k)(n + 1 - k) = (n - k + 1)(2n - 2k + 1)$$

total operations (divisions, multiplications, and additions).

f.   Denote by $P(n)$ the total computations needed for partial row reduction. By part e, we have

$$P(n) = \sum_{k=1}^{n} (n - k + 1)(2n - 2k + 1).$$

Let

$$P_1(n) = \frac{2}{3}n^3 + \frac{1}{2}n^2 - \frac{1}{6}n.$$

We will show by induction that $P = P_1$.

Clearly, $P(1) = P_1(1) = 1$. If $P(n) = P_1(n)$, we get:

$$P(n+1) = \sum_{k=1}^{n+1} (n - k + 2)(2n - 2k + 3) = 1 + \sum_{k=1}^{n} (n - k + 2)(2n - 2k + 3)$$

$$= 1 + \sum_{k=1}^{n} \Big( (n - k + 1) + 1 \Big)\Big( (2n - 2k + 1) + 2 \Big)$$

In line 4, we get the next-to-last term using

$$\sum_{k=1}^{n} k = \frac{n(n+1)}{2}.$$

$$= 1 + \underbrace{\sum_{k=1}^{n} (n - k + 1)(2n - 2k + 1)}_{P(n)} + \sum_{k=1}^{n} (4n - 4k + 5)$$

$$= 1 + \underbrace{\frac{2}{3}n^3 + \frac{1}{2}n^2 - \frac{1}{6}n}_{P_1(n) \text{ by inductive hypothesis}} + 4n^2 - 4\frac{n^2 + n}{2} + 5n$$

$$= \frac{2}{3}(n+1)^3 + \frac{1}{2}(n+1)^2 - \frac{1}{6}(n+1) = P_1(n+1).$$

So the relation is true for all $n \geq 1$.

g. We need $n - k$ multiplications and $n - k$ additions for the row $k$, so the total number of operations for back substitution is $B(n) = n^2 - n$.

h. So the total number of operations for $n$ equations in $n$ unknowns is

$$Q(n) = P(n) + B(n) = \frac{2}{3}n^3 + \frac{3}{2}n^2 - \frac{7}{6}n \quad \text{for all } n \geq 1.$$

**2.3.1** The inverse of $A$ is $A^{-1} = \begin{bmatrix} 3 & -1 & -4 \\ 1 & -1 & -1 \\ -2 & 1 & 3 \end{bmatrix}$. Now compute

$$\begin{bmatrix} 3 & -1 & -4 \\ 1 & -1 & -1 \\ -2 & 1 & 3 \end{bmatrix} \begin{bmatrix} 1 & 2 & 0 \\ 1 & 0 & 1 \\ 1 & 1 & 1 \end{bmatrix} = \begin{bmatrix} -2 & 2 & -5 \\ -1 & 1 & -2 \\ 2 & -1 & 4 \end{bmatrix}.$$

The columns of the product are the solutions to the three systems we were trying to solve.

**2.3.3** a. Let $A$ be an $n \times m$ matrix. Let us first see that saying that $A$ is invertible is the same as saying that the equation $A\vec{x} = \vec{b}$ has a unique solution for every $\vec{b} \in \mathbb{R}^n$. Our definition of invertible is that $A$ is invertible if there exists $B$ such that $AB = I_n$ and $BA = I_m$. If you multiply through $A\vec{x} = \vec{b}$ from the left by $B$, you find

$$\vec{x} = BA\vec{x} = B\vec{b},$$

indicating that $B\vec{b}$ is the only possible solution. But is it a solution? Yes: $A(B\vec{b}) = (AB)\vec{b} = \vec{b}$.

Now apply theorem 2.2.1 to see when the system of $m$ equations in $m$ variables $A\vec{x} = \vec{b}$ has a unique solution for every $\vec{b} \in \mathbb{R}^n$. The matrix $A$ cannot have any nonpivotal columns, so $A$ cannot have more columns than rows, i.e., we must have $n \leq m$. But if $n < m$, then $\tilde{A}$ will definitely have a row of 0's, so there will be $\vec{b}$'s for which $A\vec{x} = \vec{b}$ has no solutions. Thus $n = m$.

b. For instance,

$$\begin{bmatrix} 1 & 0 & 0 \\ 0 & 1 & 0 \end{bmatrix} \begin{bmatrix} 1 & 0 \\ 0 & 1 \\ 0 & 0 \end{bmatrix} = \begin{bmatrix} 1 & 0 \\ 0 & 1 \end{bmatrix}, \text{ but } \begin{bmatrix} 1 & 0 \\ 0 & 1 \\ 0 & 0 \end{bmatrix} \begin{bmatrix} 1 & 0 & 0 \\ 0 & 1 & 0 \end{bmatrix} = \begin{bmatrix} 1 & 0 & 0 \\ 0 & 1 & 0 \\ 0 & 0 & 0 \end{bmatrix}.$$

**2.3.5** a. Since $A = \begin{bmatrix} 3 & -1 & 3 & 1 \\ 2 & 1 & -2 & 1 \\ 1 & 1 & 1 & 1 \end{bmatrix}$ row reduces to $\begin{bmatrix} 1 & 0 & 0 & 3/8 \\ 0 & 1 & 0 & 1/2 \\ 0 & 0 & 1 & 1/8 \end{bmatrix}$,

the solution is $x = 3/8, y = 1/2, z = 1/8$.

b. Since $\begin{bmatrix} 3 & -1 & 3 & 1 & 0 & 0 \\ 2 & 1 & -2 & 0 & 1 & 0 \\ 1 & 1 & 1 & 0 & 0 & 1 \end{bmatrix}$ row reduces to

$$\begin{bmatrix} 1 & 0 & 0 & 3/16 & 1/4 & -1/16 \\ 0 & 1 & 0 & -1/4 & 0 & 3/4 \\ 0 & 0 & 1 & 1/16 & -1/4 & 5/16 \end{bmatrix}, \text{ we have}$$

$$A^{-1} = \begin{bmatrix} 3/16 & 1/4 & -1/16 \\ -1/4 & 0 & 3/4 \\ 1/16 & -1/4 & 5/16 \end{bmatrix} \quad \text{and} \quad \begin{bmatrix} 3/16 & 1/4 & -1/16 \\ -1/4 & 0 & 3/4 \\ 1/16 & -1/4 & 5/16 \end{bmatrix} \begin{bmatrix} 1 \\ 1 \\ 1 \end{bmatrix} = \begin{bmatrix} 3/8 \\ 1/2 \\ 1/8 \end{bmatrix}.$$

**2.3.7** It just so happens that $A = A^{-1}$:

$$\begin{bmatrix} 1 & -6 & 3 \\ 2 & -7 & 3 \\ 4 & -12 & 5 \end{bmatrix}^2 = \begin{bmatrix} 1 & 0 & 0 \\ 0 & 1 & 0 \\ 0 & 0 & 1 \end{bmatrix}. \quad \text{So by proposition 2.3.1, the solution is}$$

$$\vec{x} = A^{-1} \begin{bmatrix} 5 \\ 7 \\ 11 \end{bmatrix} = A \begin{bmatrix} 5 \\ 7 \\ 11 \end{bmatrix} = \begin{bmatrix} -4 \\ -6 \\ -9 \end{bmatrix}.$$

**2.3.9**  a.  The products will be

$$\begin{bmatrix} -2 & 3 & -14 \\ 0 & 2 & 3 \\ 1 & 0 & 4 \end{bmatrix} \quad \text{3 times the third row is subtracted from the first}$$

$$\begin{bmatrix} 1 & 3 & -2 \\ 0 & 4 & 6 \\ 1 & 0 & 4 \end{bmatrix} \quad \text{the second row is multiplied by 2}$$

$$\begin{bmatrix} 1 & 3 & -2 \\ 1 & 0 & 4 \\ 0 & 2 & 3 \end{bmatrix} \quad \text{the second and third rows are switched.}$$

b.

$$\begin{bmatrix} 1 & 3 & -2 \\ 0 & 2 & 3 \\ 1 & 0 & 4 \end{bmatrix} \qquad \begin{bmatrix} 1 & 3 & -2 \\ 0 & 2 & 3 \\ 1 & 0 & 4 \end{bmatrix} \qquad \begin{bmatrix} 1 & 3 & -2 \\ 0 & 2 & 3 \\ 1 & 0 & 4 \end{bmatrix}$$

$$\begin{bmatrix} 1 & 0 & -3 \\ 0 & 1 & 0 \\ 0 & 0 & 1 \end{bmatrix} \begin{bmatrix} -2 & 3 & -14 \\ 0 & 2 & 3 \\ 1 & 0 & 4 \end{bmatrix}, \quad \begin{bmatrix} 1 & 0 & 0 \\ 0 & 2 & 0 \\ 0 & 0 & 1 \end{bmatrix} \begin{bmatrix} 1 & 3 & -2 \\ 0 & 4 & 6 \\ 1 & 0 & 4 \end{bmatrix}, \quad \begin{bmatrix} 1 & 0 & 0 \\ 0 & 0 & 1 \\ 0 & 1 & 0 \end{bmatrix} \begin{bmatrix} 1 & 3 & -2 \\ 1 & 0 & 4 \\ 0 & 2 & 3 \end{bmatrix}.$$

**2.3.11** Let $A$ be an $n \times m$ matrix. Then

$AE_1(i, x)$ has the same columns as $A$, except the $i$th, which is multiplied by $x$.

$AE_2(i, j, x)$ has the same columns as $A$ except the $j$th, which is the sum of the $j$th column of $A$ (contributed by the 1 in the $(j, j)$th position), and $x$ times the $i$th column (contributed by the $x$ in the $(i, j)$th position).

$AE_3(i, j)$ has the same columns as $A$, except for the $i$th and $j$th, which are switched.

**2.3.13** Here is one way to show this. Denote by $a$ the $i$th row and by $b$ the $j$th row of our matrix. Assume we wish to switch the $i$th and the $j$th rows. Then multiplication on the left by $E_2(i, j, 1)$ turns the $i$th row into $a + b$. Multiplication on the left by $E_2(j, i, -1)$ then by $E_1(j, -1)$ turns the

$j$th row into $a$. Finally, we multiply on the left by $E_2(i, j, -1)$ to subtract $a$ from the $i$th row, making that row $b$. So we can switch rows by multiplying with only the first two types of elementary matrices.

Here is a different explanation of the same argument: Compute the product

$$\begin{bmatrix} 1 & -1 \\ 0 & 1 \end{bmatrix} \begin{bmatrix} 1 & 0 \\ 0 & -1 \end{bmatrix} \begin{bmatrix} 1 & 0 \\ -1 & 1 \end{bmatrix} \begin{bmatrix} 1 & 1 \\ 0 & 1 \end{bmatrix} = \begin{bmatrix} 0 & 1 \\ 1 & 0 \end{bmatrix}.$$

This certainly shows that the $2 \times 2$ elementary matrix $E_3(1, 2)$ can be written as a product of elementary matrices of type 1 and 2.

More generally,

$$E_2(i, j, -1)E_1(j, -1)E_2(j, i, -1)E_2(i, j, 1) = E_3(i, j).$$

**2.4.1** The only way you can write

$$\begin{bmatrix} 0 \\ 0 \\ \vdots \\ 0 \end{bmatrix} = a_1 \begin{bmatrix} 1 \\ 0 \\ \vdots \\ 0 \end{bmatrix} + \cdots + a_k \begin{bmatrix} 0 \\ \vdots \\ 0 \\ 1 \end{bmatrix} = \begin{bmatrix} a_1 \\ a_2 \\ \vdots \\ a_k \end{bmatrix},$$

is if $a_1 = a_2 = \cdots = a_k = 0$.

**2.4.3** To make the basis orthonormal, each vector needs to be normalized to give it length 1. This is done by dividing each vector by its length (see equation 1.4.6). So the orthonormal basis is $\begin{bmatrix} 1/\sqrt{2} \\ 1/\sqrt{2} \end{bmatrix}, \begin{bmatrix} 1/\sqrt{2} \\ -1/\sqrt{2} \end{bmatrix}$. These vectors form a basis of $\mathbb{R}^2$ because they are two linearly independent vectors in $\mathbb{R}^2$; they are orthogonal because

Solution 2.4.3: The vectors in equation (1) are orthogonal, by corollary 1.4.8; they are orthonormal because they are orthogonal and each vector has length 1.

$$\begin{bmatrix} 1/\sqrt{2} \\ 1/\sqrt{2} \end{bmatrix} \cdot \begin{bmatrix} 1/\sqrt{2} \\ -1/\sqrt{2} \end{bmatrix} = 0. \tag{1}$$

**2.4.5** To show that span $(\vec{v}_1, \ldots, \vec{v}_k)$ is a subspace of $\mathbb{R}^n$, we need to show that it is closed under addition and under multiplication by scalars. This follows from the computations

Solution 2.4.5: Recall that only a very few, very special subsets of $\mathbb{R}^n$ are subspaces of $\mathbb{R}^n$; see definition 1.1.5. Roughly, a subspace is a flat subset that goes through the origin: to be closed under multiplication, a subspace must contain the zero vector, so that $0 \cdot \vec{v} = \vec{0}$.

$$c(a_1\vec{v}_1 + \cdots + a_k\vec{v}_k) = ca_1\vec{v}_1 + \cdots + ca_k\vec{v}_k;$$
$$(a_1\vec{v}_1 + \cdots + a_k\vec{v}_k) + (b_1\vec{v}_1 + \cdots + b_k\vec{v}_k)$$
$$= (a_1 + b_1)\vec{v}_1 + \cdots + (a_k + b_k)\vec{v}_k.$$

To see that it is the smallest subspace that contains the $\vec{v}_i$, note that any subspace that contains the $\vec{v}_i$ must contain their linear combinations, hence the smallest such subspace is span $(\vec{v}_1, \ldots, \vec{v}_k)$.

**2.4.7** Let $A$ be an $n \times n$ matrix. The product $A^\top A$ is then

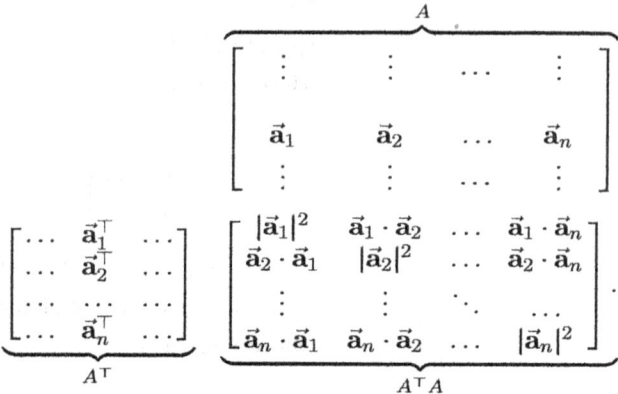

An *orthogonal* $n \times n$ matrix is a matrix whose columns form an *orthonormal* basis of $\mathbb{R}^n$.

The diagonal entries are given by the length squared of the columns of $A$, since $\vec{\mathbf{a}}_i^\top \vec{\mathbf{a}}_i = \vec{\mathbf{a}}_i \cdot \vec{\mathbf{a}}_i = |\vec{\mathbf{a}}_i|^2$. All other entries are dot products of two different columns of $A$. If $A^\top A = I$, so that all entries not on the diagonal are 0, while those on the diagonal are 1, then the columns of $A$ are orthogonal and have length 1. Thus they form an orthonormal basis of $\mathbb{R}^n$, and $A$ is said to be orthogonal.

Similarly, if $A$ is orthogonal, then the length of each column vector is 1, so that $A^\top A$ has 1's on the diagonal, and the dot product of two non-identical columns is 0, giving 0 for all other entries of $A^\top A$.

**2.4.9** To see that condition 2 implies condition 3, first note that $2 \implies 3$ is logically equivalent to (not 3) $\implies$ (not 2). Now suppose $\{\vec{\mathbf{v}}_1, \ldots, \vec{\mathbf{v}}_k\}$ is a *linearly dependent* set spanning $V$, so by definition 2.4.10, there exists a nontrivial solution to

$$a_1 \vec{\mathbf{v}}_1 + \cdots + a_k \vec{\mathbf{v}}_k = \mathbf{0}.$$

Without loss of generality, we may assume that $a_k$ is nonzero (if it isn't, renumber the vectors so that $a_k$ is nonzero). Using the above relation, we can solve for $\vec{\mathbf{v}}_k$ in terms of the other vectors:

$$\vec{\mathbf{v}}_k = -\left( \frac{a_1}{a_k} \vec{\mathbf{v}}_1 + \cdots + \frac{a_{k-1}}{a_k} \vec{\mathbf{v}}_{k-1} \right).$$

This implies that $\{\vec{\mathbf{v}}_1, \ldots, \vec{\mathbf{v}}_k\}$ cannot be a minimal spanning set, because if we were to drop $\vec{\mathbf{v}}_k$ we could still form all the linear combinations as before. So (not 3) $\implies$ (not 2), and we are finished.

To show that $3 \implies 1$:

The vectors $\vec{\mathbf{v}}_1, \ldots, \vec{\mathbf{v}}_k$ span $V$, so for any vector $\vec{\mathbf{w}} \in V$, there exist some numbers $a_1, \ldots, a_n$ such that

$$a_1 \vec{\mathbf{v}}_1 + \cdots + a_n \vec{\mathbf{v}}_n = \vec{\mathbf{w}}.$$

Thus, if we add this vector to $\vec{\mathbf{v}}_1, \ldots, \vec{\mathbf{v}}_k$, we will have a linearly dependent set because

$$a_1 \vec{\mathbf{v}}_1 + \cdots + a_n \vec{\mathbf{v}}_n - \vec{\mathbf{w}} = 0$$

is a nontrivial linear combination of the vectors that equals $\mathbf{0}$. Since $\vec{\mathbf{w}}$ can be any vector in $V$, the set $\{\vec{\mathbf{v}}_1, \ldots, \vec{\mathbf{v}}_k\}$ is a maximal linearly independent set.

**2.4.11** a. For any $n$, we have $n+1$ linear equations for the $n+1$ unknowns $a_{0,n}$, $a_{1,n}$, ..., $a_{n,n}$, which say

$$a_{0,n}\left(\frac{0}{n}\right)^k + a_{1,n}\left(\frac{1}{n}\right)^k + a_{2,n}\left(\frac{2}{n}\right)^k + \cdots + a_{n,n}\left(\frac{n}{n}\right)^k = \int_0^1 x^k\, dx = \frac{1}{k+1},$$

one for each $k = 0, 1, \ldots, n$.

These systems of linear equations are:

- When $n = 1$

$$a_{0,1}1 + a_{1,1}1 = 1$$
$$a_{0,1}0 + a_{1,1}1 = 1/2$$

- When $n = 2$

$$a_{0,2}1 + a_{1,2}1 + a_{2,2}1 = 1$$
$$a_{0,2}0 + a_{1,2}(1/2) + a_{2,2}1 = 1/2$$
$$a_{0,2}0 + a_{1,2}(1/4) + a_{2,2}1 = 1/3$$

- When $n = 3$

$$a_{0,3}1 + a_{1,3}1 + a_{2,3}1 + a_{3,3}1 = 1$$
$$a_{0,3}0 + a_{1,3}(1/3) + a_{2,3}(2/3) + a_{3,3}1 = 1/2$$
$$a_{0,3}0 + a_{1,3}(1/9) + a_{2,3}(4/9) + a_{3,3}1 = 1/3$$
$$a_{0,3}0 + a_{1,3}(1/27) + a_{2,3}(8/27) + a_{3,3}1 = 1/4.$$

The system of equations for $n = 3$ could be written as the augmented matrix $[A|\vec{\mathbf{b}}]$:

$$\begin{bmatrix} 1 & 1 & 1 & 1 & 1 \\ 0 & 1/3 & 2/3 & 1 & 1/2 \\ 0 & 1/9 & 4/9 & 1 & 1/3 \\ 0 & 1/27 & 8/27 & 1 & 1/4 \end{bmatrix}.$$

b. These wouldn't be too bad to solve by hand (although already the last would be distinctly unpleasant). We wrote a little MATLAB m-file to do it systematically:

```
function [N,b,c] = EqSp(n)

N = zeros(n+1); % make an n+1 × n+1 matrix of zeros
c=linspace(1,n+1,n+1); % make a place holder for the right side
for i=1:n+1
    for j=1:n+1
    N(i,j)= ((j-1)/n)^(i-1); % put the right coefficients in the matrix
end
c(i)=1/c(i); % put the right entries in the right side
end
b=c'; % our c was a row vector, take its transpose
c=N% this solves the system of linear equations
```

If you write and save this file as 'EqSp.m', and then type

$[A,b,c]=EqSp(5)$, for the case when $n = 5$,

you will get

$$A = \begin{bmatrix} 1 & 1 & 1 & 1 & 1 & 1 \\ 0 & 1/5 & 2/5 & 3/5 & 4/5 & 1 \\ 0 & 1/25 & 4/25 & 9/25 & 16/25 & 1 \\ 0 & 1/125 & 8/125 & 27/125 & 64/125 & 1 \\ 0 & 1/625 & 16/625 & 81/625 & 256/625 & 1 \\ 0 & 1/3125 & 32/3125 & 243/3125 & 541/1651 & 1 \end{bmatrix}, \quad b = \begin{bmatrix} 1 \\ 1/2 \\ 1/3 \\ 1/4 \\ 1/5 \\ 1/6 \end{bmatrix}, \quad c = \begin{bmatrix} 19/288 \\ 25/96 \\ 25/144 \\ 25/144 \\ 25/96 \\ 19/288 \end{bmatrix}.$$

This corresponds to the equation $Ac = b$, where the matrix $A$ is the matrix of coefficients for $n = 5$, and the vector $c$ is the desired set of coefficients – the solutions when $n = 5$.

When $n = 1, 2, 3$, the coefficients – i.e., the solutions to the systems of equations in part a – are

For instance, for $n = 2$, we have

$$\begin{bmatrix} 1 & 1 \\ 0 & 1 \end{bmatrix} \begin{bmatrix} 1/2 \\ 1/2 \end{bmatrix} = \begin{bmatrix} 1 \\ 1/2 \end{bmatrix}.$$

$$\begin{bmatrix} 1/2 \\ 1/2 \end{bmatrix}, \quad \begin{bmatrix} 1/6 \\ 2/3 \\ 1/6 \end{bmatrix}, \quad \begin{bmatrix} 1/8 \\ 3/8 \\ 3/8 \\ 1/8 \end{bmatrix}.$$

The approximations to $\int_0^1 \frac{dx}{1+x} = \log 2 = 0.69314718055995\ldots$ obtained with these coefficients are $.75$ for $n = 1$, $\frac{25}{36} = .6944\ldots$ for $n = 2$, and $\frac{111}{160} = .69375$ for $n = 3$.

c. If you compute

$$\sum_{i=0}^5 a_{i,5} \frac{1}{(i/5)+1} \approx \int_0^1 \frac{dx}{1+x} = \log 2 = 0.69314718055995\ldots$$

you will find $0.69316302910053$, which is a pretty good approximation for a Riemann sum with six terms. For instance, the midpoint Riemann sum gives

$$\frac{1}{5} \sum_{i=1}^5 \frac{1}{((2i-1)/10)} \approx 0.69190788571594,$$

which is a much worse approximation. But this scheme runs into trouble. All the coefficients are positive up to $n = 7$, but for $n = 8$ they are

$$\begin{bmatrix} 0.0118 \\ 0.1141 \\ -0.2362 \\ 1.2044 \\ -3.7636 \\ 10.3135 \\ -22.6521 \\ 41.7176 \\ -63.9006 \\ 82.5706 \\ -89.7629 \\ 82.5829 \\ -63.9189 \\ 41.7345 \\ -22.6633 \\ 10.3191 \\ -3.7656 \\ 1.2050 \\ -0.2363 \\ 0.1141 \\ 0.0118 \end{bmatrix}.$$

Coefficients when $n = 20$.

$$\begin{bmatrix} 248/7109 \\ 578/2783 \\ -111/3391 \\ 97/262 \\ -454/2835 \\ 97/262 \\ -111/3391 \\ 578/2783 \\ 248/7109 \end{bmatrix} \approx \begin{bmatrix} 0.0349 \\ 0.2077 \\ -0.0327 \\ 0.3702 \\ -0.1601 \\ 0.3702 \\ -0.0327 \\ 0.2077 \\ 0.0349 \end{bmatrix}$$

and the approximation scheme starts depending on cancellations. This is much worse when $n = 20$, where the coefficients are as shown in the margin.

Despite these bad sign variations, the Riemann sum works pretty well: the approximation to the integral above gives $0.69314718055995$, which is $\ln 2$ to the precision of the machine.

**2.4.13** In the process of row reducing $A = \begin{bmatrix} 1 & a & a & a \\ 1 & 1 & a & a \\ 1 & 1 & 1 & a \\ 1 & 1 & 1 & 1 \end{bmatrix}$, you will come to the matrix

$$\begin{bmatrix} 1 & a & a & a \\ 0 & 1-a & 0 & 0 \\ 0 & 1-a & 1-a & 0 \\ 0 & 1-a & 1-a & 1-a \end{bmatrix}.$$

If $a = 1$, the matrix will not row reduce to the identity, because you can't choose a pivotal 1 in the second column, so one necessary condition for $A$ to be invertible is that $a \neq 1$. Let us suppose that this is the case. We can now row reduce two steps further to find

$$\begin{bmatrix} 1 & 0 & a & a \\ 0 & 1 & 0 & 0 \\ 0 & 0 & 1-a & 0 \\ 0 & 0 & 1-a & 1-a \end{bmatrix}, \quad \text{and then} \quad \begin{bmatrix} 1 & 0 & 0 & a \\ 0 & 1 & 0 & 0 \\ 0 & 0 & 1 & 0 \\ 0 & 0 & 0 & 1-a \end{bmatrix}$$

The next step row reduces the matrix to the identity, so the matrix is invertible if and only if $a \neq 1$.

**2.5.1** a. The vectors $\vec{v}_1$ and $\vec{v}_3$ are in the kernel of $A$, since $A\vec{v}_1 = \mathbf{0}$ and $A\vec{v}_3 = \mathbf{0}$. But $\vec{v}_2$ is not, since $A\vec{v}_2 = \begin{bmatrix} 2 \\ 3 \\ 3 \end{bmatrix}$. The vector $\begin{bmatrix} 2 \\ 3 \\ 3 \end{bmatrix}$ is in the image of $A$.

b. The matrix $T$ represents a transformation from $\mathbb{R}^5$ to $\mathbb{R}^3$; it takes a vector in $\mathbb{R}^5$ and gives a vector in $\mathbb{R}^3$. Therefore, $\vec{w}_4$ has the right height to be in the kernel (although it isn't), and $\vec{w}_1$ and $\vec{w}_3$ have the right height to be in its image.

Since the sum of the second and fifth columns of $T$ is $\vec{0}$, one element of the kernel is $\begin{bmatrix} 0 \\ 1 \\ 0 \\ 0 \\ 1 \end{bmatrix}$.

**2.5.3** nullity $T = \dim \ker T =$ number of nonpivotal columns of $T$;

rank of $T = \dim \operatorname{image} T$

$=$ number of linearly independent columns of $T$

$=$ number of pivotal columns of $T$.

rank $T +$ nullity $T = \dim \operatorname{domain} T$

**2.5.5** By definition 1.1.5 of a subspace, we need to show that the kernel and the image of a linear transformation $T$ are closed under addition and multiplication by scalars. These are straightforward computations, using the linearity of $T$.

*The kernel of T:* If $\vec{v}, \vec{w} \in \ker T$, i.e., if $T(\vec{v}) = \vec{0}$ and $T(\vec{w}) = \vec{0}$, then

$$T(\vec{v} + \vec{w}) = T(\vec{v}) + T(\vec{w}) = \vec{0} + \vec{0} = \vec{0} \quad \text{and} \quad T(a\vec{v}) = aT(\vec{v}) = a\vec{0} = \vec{0},$$

so $\vec{v} + \vec{w} \in \ker T$ and $a\vec{v} \in \ker T$.

*The image of T:* If $\vec{v} = T(\vec{v}_1)$, $\vec{w} = T(\vec{w}_1)$, then

$$\vec{v} + \vec{w} = T(\vec{w}_1) + T(\vec{v}_1) = T(\vec{w}_1 + \vec{v}_1) \quad \text{and} \quad a\vec{v} = aT(\vec{v}_1) = T(a\vec{v}_1),$$

So the image is also closed under addition and multiplication by scalars.

**2.5.7 a.** $n = 3$. The last three columns of the matrix are clearly linearly independent, so the matrix has rank at least 3, and it has rank at most 3 because there can be at most three linearly independent vectors in $\mathbb{R}^3$.

b. Yes. For example, the first three columns are linearly independent, since the matrix composed of just those columns row reduces to the identity.

c. The 3rd, 4th, and 6th columns are linearly dependent.

d. You cannot choose freely the values of $x_1, x_2, x_5$. Since the rank of the matrix is 3, three variables must correspond to pivotal (linearly independent) columns. For the variables $x_1, x_2, x_5$ to be freely chosen, i.e., nonpivotal, $x_3$, $x_4$, $x_6$ would have to correspond to linearly independent columns.

**2.5.9 a.** The matrix of a linear transformation has as its columns the images of the standard basis vectors, in this case identified to the polynomials $p_1(x) = 1$, $p_2(x) = x$, $p_3(x) = x^2$. Since

$$T(p_1)(x) = 0, \quad T(p_2)(x) = x, \quad T(p_3)(x) = 4x^2,$$

the matrix of $T$ is $\begin{bmatrix} 0 & 0 & 0 \\ 0 & 1 & 0 \\ 0 & 0 & 4 \end{bmatrix}$.

b. The matrix $\begin{bmatrix} 0 & 0 & 0 \\ 0 & 1 & 0 \\ 0 & 0 & 4 \end{bmatrix}$ row reduces to $\begin{bmatrix} 0 & 0 & 0 \\ 1 & 0 & 0 \\ 0 & 1 & 0 \end{bmatrix}$. Thus the image has dimension 2 (the number of pivotal columns) and has a basis made up of the polynomials $ax + 4bx^2$, the linear combinations of the second and third columns of the matrix of $T$. The kernel has dimension 1 (the number of nonpivotal columns), and consists precisely of the constant polynomials.

**2.5.11** The sketch is shown at left.

a. If $ab \neq 2$, then $\dim(\ker(A)) = 0$, so in that case the image has dimension 2. If $ab = 2$, the image and the kernel have dimension 1.

b. This is more complicated. By row operations, we can bring the matrix $B$ to

$$\begin{bmatrix} 1 & 2 & a \\ 0 & b & ab - a \\ 0 & 2a - b & a \end{bmatrix}.$$

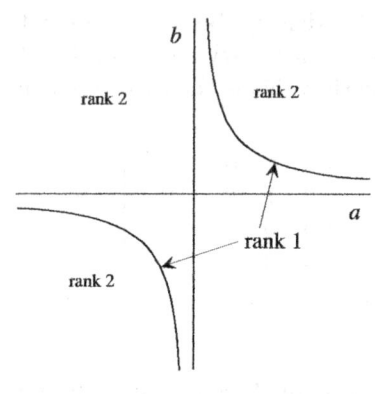
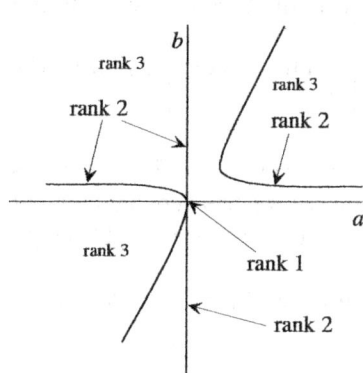

FIGURE FOR SOLUTION 2.5.11
    TOP: On the curves, the kernel of $A$ has dimension 1 and its image has dimension 1. Elsewhere, the rank (dimension of the image) is 2, so by the dimension formula the kernel has dimension 0. The rank is never 0 or 3.

    BOTTOM: On the $b$-axis and on the hyperbola, the image of $B$ has dimension 2, i.e., its kernel has dimension 1. At the origin the rank is 1 and the dimension of the kernel is 2. Elsewhere, the kernel has dimension 0 and the rank is 3.

We now separate the case $b \neq 0$ and $b = 0$.

- If $b \neq 0$, then we can do further row operations to bring the matrix to the form

$$\begin{bmatrix} 1 & 0 & a - 2\frac{ab-a}{b} \\ 0 & 1 & \frac{ab-a}{b} \\ 0 & 0 & -a - (b - 2a)\frac{ab-a}{b} \end{bmatrix}.$$ The entry in the 3rd row, 3rd column is

$$-\frac{a}{b}(b^2 - 2ab + 2a).$$

So if $b \neq 0$, and the point $\begin{pmatrix} a \\ b \end{pmatrix}$ is neither on the line $a = 0$ nor on the hyperbola of equation $b^2 - 2ab + 2a = 0$, the matrix has rank 3, whereas if $b \neq 0$ and the point $\begin{pmatrix} a \\ b \end{pmatrix}$ is on one of these curves, the matrix has rank 2.

- If $b = 0$, the matrix is $\begin{bmatrix} 1 & 2 & a \\ 0 & -2a & -a \\ 0 & 0 & a \end{bmatrix}$, which evidently has rank 3 unless $a = 0$, in which case it has rank 1.

**2.5.13** a. As in the example following proposition 2.5.14, we need to put the right side on a common denominator and consider the resulting system of linear equations. Row reduction then tells us for what values of $a$ the system has no solutions. So:

$$\frac{x-1}{(x+1)(x^2+ax+5)} = \frac{A_0}{x+1} + \frac{B_1 x + B_0}{x^2+ax+5}$$
$$= \frac{A_0 x^2 + aA_0 x + 5A_0 + B_1 x^2 + B_1 x + B_0 x + B_0}{(x+1)(x^2+ax+5)}.$$

This gives

$$x - 1 = A_0 x^2 + aA_0 x + 5A_0 + B_1 x^2 + B_1 x + B_0 x + B_0,$$

i.e.,

$$\begin{aligned} 5A_0 + B_0 &= -1 \\ aA_0 + B_1 + B_0 &= 1 \\ A_0 + B_1 &= 0, \end{aligned} \quad \text{which we can write as} \quad \begin{bmatrix} 5 & 0 & 1 & -1 \\ a & 1 & 1 & 1 \\ 1 & 1 & 0 & 0 \end{bmatrix} \begin{bmatrix} A_0 \\ B_1 \\ B_0 \end{bmatrix}.$$

Row reduction gives $\begin{bmatrix} 1 & 0 & 0 & \frac{2}{a-6} \\ 0 & 1 & 0 & \frac{-2}{a-6} \\ 0 & 0 & 1 & \frac{-4-a}{a-6} \end{bmatrix}$, so the fraction in question cannot

be written as a partial fraction when $a = 6$.

b. This does not contradict proposition 2.5.14 because that proposition requires that $p$ be factored as

$$p(x) = (x - a_1)^{n_1} \cdots (x - a_k)^{n_k}.$$

with the $a_i$ distinct. If you substitute 6 for $a$ in $x^2 + ax + 5$ you get $x^2 + 6x + 5 = (x+1)(x+5)$, so factoring $p/q$ as

$$\frac{p(x)}{q(x)} = \frac{x-1}{(x+1)(x^2+ax+5)} = \frac{A_0}{x+1} + \frac{B_1 x + B_0}{x^2+ax+5} = \frac{A_0}{x+1} + \frac{B_1 x + B_0}{(x+1)(x+5)}$$

does not meet that requirement; both terms contain $(x + 1)$ in the denominator.

You could avoid this by using a different factorization:

$$\frac{p(x)}{q(x)} = \frac{x - 1}{(x + 1)(x^2 + 6x + 5)} = \frac{x - 1}{(x + 1)^2(x + 5)} = \frac{A_1 x + A_0}{(x + 1)^2} + \frac{B_0}{x + 5}$$

**2.5.15**  Note first the following results: if $T_1, T_2 : \mathbb{R}^n \to \mathbb{R}^n$ are linear transformations, then

    1. the image of $T_1$ contains the image of $T_1 \circ T_2$, and

    2. the kernel of $T_1 \circ T_2$ contains the kernel of $T_2$.

The first is true because, by the definition of image, for any vector $\vec{v}$ in $\mathrm{img}\, T_1 \circ T_2$, there exists a vector $\vec{w}$ such that $(T_1 \circ T_2)\vec{w} = \vec{v}$. Since $T_1(T_2(\vec{w})) = \vec{v}$, the vector $\vec{v}$ is also in the image of $T_1$.

The second is true because for any $\vec{v} \in \ker T_2$, we have $T_2(\vec{v}) = \vec{0}$. Since $T_1(\vec{0}) = \vec{0}$, we see that

$$(T_1 \circ T_2)(\vec{v}) = T_1\big(T_2(\vec{v})\big) = T_1(\vec{0}) = \vec{0},$$

so $\vec{v}$ is also in the kernel of $T_1 \circ T_2$.

Recall that the *nullity* of a linear transformation is the dimension of its kernel.

If $AB$ is invertible, then the image of $A$ contains the image of $AB$ by statement 1. So $A$ has rank $n$, hence nullity 0 by the dimension formula, so $A$ is invertible. Since $B = A^{-1}(AB)$, we have $B^{-1} = (AB)^{-1}A$.

By proposition 1.2.15, since the matrices $A^{-1}$ and $AB$ are invertible, so is the product

$$B = (A^{-1}A)B = A^{-1}(AB),$$

and $B^{-1} = (AB)^{-1}A$. Note that this uses associativity of matrix mutiplication (corollary 1.3.12)

For $B$, one could argue that $\ker B \subset \ker AB = \{\mathbf{0}\}$, so $B$ has nullity 0, and thus rank $n$, so $B$ is invertible.

**\*2.5.17**  a.    Since $p(0) = a + 0b + 0^2 c = 1$, we have $a = 1$. Since $p(1) = a + b + c = 4$, we have $b + c = 3$. Since $p(3) = a + 3b + 9c = -2$, we have $3b + 9c = -3$. It follows that $c = -2$ and $b = 5$.

b.    Let $M_{\mathbf{x}}$ be the linear transformation from the space of $P_n$ of polynomials of degree at most $n$ to $\mathbb{R}^{n+1}$ given by

$$p \mapsto \begin{bmatrix} p(x_0) \\ \vdots \\ p(x_n) \end{bmatrix}, \quad \text{where } \mathbf{x} = \begin{bmatrix} x_0 \\ \vdots \\ x_n \end{bmatrix}.$$

Solution 2.5.17, part b : The polynomials themselves are almost certainly nonlinear, but $M_{\mathbf{x}}$ is linear since it handles polynomials that have already been evaluated at the given points.

Assume that the polynomial $q$ is in the kernel of $M_{\mathbf{x}}$. Then $q$ vanishes at $n + 1$ distinct points, so that either $q$ is the zero polynomial or it has degree at least $n + 1$. Since $q$ cannot have degree greater than $n$, it must be the zero polynomial. So $\ker (M_{\mathbf{x}}) = \{\mathbf{0}\}$, hence $M_{\mathbf{x}}$ is injective, so by corollary 2.5.10 it is surjective. It follows that a solution of $M_{\mathbf{x}}(p) = \begin{bmatrix} a_0 \\ \vdots \\ a_n \end{bmatrix}$ exists and is unique.

c. Take the linear transformation $M'_{\mathbf{x}}$ from $P_k$ to $R^{2n+2}$ defined by

$$p \mapsto \begin{bmatrix} p(x_0) \\ \vdots \\ p(x_n) \\ p'(x_0) \\ \vdots \\ p'(x_n) \end{bmatrix} . \text{ If } \ker (M'_{\mathbf{x}}) = 0, \text{ then a solution of } M'_{\mathbf{x}}(p) = \begin{bmatrix} a_0 \\ \vdots \\ a_n \\ b_0 \\ \vdots \\ b_n \end{bmatrix}$$

exists and so a value for $k$ is $2n+2$ (as shown above). In fact this is the lowest value for $k$ that always has a solution.

**2.5.19** a. $\qquad H_2 = \begin{bmatrix} 1 & 1 \\ 1/2 & 1/4 \end{bmatrix}, \quad H_3 = \begin{bmatrix} 1 & 1 & 1 \\ 1/2 & 1/4 & 1/8 \\ 1/3 & 1/9 & 1/27 \end{bmatrix}.$

b. $\qquad H_n = \begin{bmatrix} 1 & 1 & 1 & \cdots & 1 \\ 1/2 & 1/4 & 1/8 & \cdots & 1/2^n \\ 1/3 & 1/9 & 1/27 & \cdots & 1/3^n \\ \vdots & \vdots & \vdots & \ddots & \vdots \\ 1/n & 1/n^2 & 1/n^3 & \cdots & 1/n^n \end{bmatrix}.$

c. If $H_n$ is not invertible, then there exist numbers $a_1, \ldots, a_n$ not all zero such that

$$f_{\vec{\mathbf{a}}}(1) = \cdots = f_{\vec{\mathbf{a}}}(n) = 0.$$

But we can write

$$f_{\vec{\mathbf{a}}}(x) = \frac{a_1 x^{n-1} + a_2 x^{n-2} + \cdots + a_n}{x^n},$$

and the only way this function can vanish at the integers $1, \ldots, n$ is if the numerator vanishes at all these points. But it is a polynomial of degree $n-1$, and cannot vanish at $n$ different points without vanishing identically.

**2.5.21** a. If $P_{[\mathbf{v}]}$ is one to one, then $P_{[\mathbf{v}]}$ has kernel $\{\mathbf{0}\}$. It then follows that $\sum(a_i\vec{\mathbf{v}}_i) = \mathbf{0}$ has as its only solution $a_i = 0, \forall i$, so $\vec{\mathbf{v}}_1, \ldots, \vec{\mathbf{v}}_n$ are linearly independent.

Conversely, if the vectors $\vec{\mathbf{v}}_1, \ldots, \vec{\mathbf{v}}_n$ are linearly independent, then the equation $\sum(a_i\vec{\mathbf{v}}_i) = \mathbf{0}$ has as its only solution $a_i = 0, \forall i$. This means that $P_{[\mathbf{v}]}$ has kernel $\{\mathbf{0}\}$ and so is one to one.

b. If $P_{[\mathbf{v}]}$ is onto, then $\forall \vec{\mathbf{w}} \in \mathbb{R}^m, \exists \vec{\mathbf{a}} \in \mathbb{R}^n$ such that

$$P_{[\mathbf{v}]}(\vec{\mathbf{a}}) = \sum(a_i\vec{\mathbf{v}}_i) = \vec{\mathbf{w}},$$

so the vectors $\vec{\mathbf{v}}_1, \ldots \vec{\mathbf{v}}_n$ span $\mathbb{R}^m$.

Conversely, if $\vec{\mathbf{v}}_1, \ldots, \vec{\mathbf{v}}_n$ span $\mathbb{R}^m$ then

$$\forall \vec{\mathbf{w}} \in \mathbb{R}^m, \exists a_1, \ldots a_n \quad \text{such that} \quad \sum(a_i\vec{\mathbf{v}}_i) = \vec{\mathbf{w}},$$

so $\forall \vec{\mathbf{w}} \in \mathbb{R}^m, \exists \vec{\mathbf{a}} \in \mathbb{R}^n$ such that $P_{[\mathbf{v}]}(\vec{\mathbf{a}}) = \vec{\mathbf{w}}$. Therefore, $P_{[\mathbf{v}]}$ is onto.

c.   The vectors $\vec{v}_1, \ldots, \vec{v}_n$ form a basis of $\mathbb{R}^m$ if and only if they are linearly independent and they span $\mathbb{R}^m$. By parts a and b, this is equivalent to $P_{[\mathbf{v}]}$ being one to one (part a) and onto (part b).

**2.6.1**  a.  It corresponds to the basis $\underline{\mathbf{v}}_1 = \begin{bmatrix} 1 & 0 \\ 0 & 0 \end{bmatrix}$, $\underline{\mathbf{v}}_2 = \begin{bmatrix} 0 & 1 \\ 0 & 0 \end{bmatrix}$, $\underline{\mathbf{v}}_3 = \begin{bmatrix} 0 & 0 \\ 1 & 0 \end{bmatrix}$, $\underline{\mathbf{v}}_4 = \begin{bmatrix} 0 & 0 \\ 0 & 1 \end{bmatrix}$. We have $\begin{bmatrix} 2 & 1 \\ 5 & 4 \end{bmatrix} = 2\underline{\mathbf{v}}_1 + \underline{\mathbf{v}}_2 + 5\underline{\mathbf{v}}_3 + 4\underline{\mathbf{v}}_4$.

b.   It corresponds to the basis $\underline{\mathbf{v}}_1 = \begin{bmatrix} 1 & 0 \\ 0 & 0 \end{bmatrix}$, $\underline{\mathbf{v}}_2 = \begin{bmatrix} 0 & 0 \\ 1 & 0 \end{bmatrix}$, $\underline{\mathbf{v}}_3 = \begin{bmatrix} 0 & 1 \\ 0 & 0 \end{bmatrix}$, $\underline{\mathbf{v}}_4 = \begin{bmatrix} 0 & 0 \\ 0 & 1 \end{bmatrix}$. We have $\begin{bmatrix} 2 & 1 \\ 5 & 4 \end{bmatrix} = 2\underline{\mathbf{v}}_1 + 5\underline{\mathbf{v}}_2 + \underline{\mathbf{v}}_3 + 4\underline{\mathbf{v}}_4$.

**2.6.3**

$$\Phi_{\{\underline{\mathbf{v}}\}}\left( \begin{bmatrix} a \\ b \\ c \\ d \end{bmatrix} \right) = a\begin{bmatrix} 1 & 0 \\ 0 & 1 \end{bmatrix} + b\begin{bmatrix} 1 & 0 \\ 0 & -1 \end{bmatrix} + c\begin{bmatrix} 0 & 1 \\ 1 & 0 \end{bmatrix} + d\begin{bmatrix} 0 & -1 \\ 1 & 0 \end{bmatrix} = \begin{bmatrix} a+b & c-d \\ c+d & a-b \end{bmatrix}$$

**2.6.5**  a. The $i$th column of $[R_A]$ is $[R_A]\vec{e}_i$:

$$[R_A]\vec{e}_1 = \begin{bmatrix} a \\ b \\ 0 \\ 0 \end{bmatrix}, \quad \text{which corresponds to } \begin{bmatrix} a & b \\ 0 & 0 \end{bmatrix} = \begin{bmatrix} 1 & 0 \\ 0 & 0 \end{bmatrix}\begin{bmatrix} a & b \\ c & d \end{bmatrix}.$$

$$[R_A]\vec{e}_2 = \begin{bmatrix} c \\ d \\ 0 \\ 0 \end{bmatrix}, \quad \text{which corresponds to } \begin{bmatrix} c & d \\ 0 & 0 \end{bmatrix} = \begin{bmatrix} 0 & 1 \\ 0 & 0 \end{bmatrix}\begin{bmatrix} a & b \\ c & d \end{bmatrix}.$$

$$[R_A]\vec{e}_3 = \begin{bmatrix} 0 \\ 0 \\ a \\ b \end{bmatrix}, \quad \text{which corresponds to } \begin{bmatrix} 0 & 0 \\ a & b \end{bmatrix} = \begin{bmatrix} 0 & 0 \\ 1 & 0 \end{bmatrix}\begin{bmatrix} a & b \\ c & d \end{bmatrix}.$$

$$[R_A]\vec{e}_4 = \begin{bmatrix} 0 \\ 0 \\ c \\ d \end{bmatrix}, \quad \text{which corresponds to } \begin{bmatrix} 0 & 0 \\ c & d \end{bmatrix} = \begin{bmatrix} 0 & 0 \\ 0 & 1 \end{bmatrix}\begin{bmatrix} a & b \\ c & d \end{bmatrix}.$$

Similarly, the first column of $[L_A]$ corresponds to $\begin{bmatrix} a & b \\ c & d \end{bmatrix}\begin{bmatrix} 1 & 0 \\ 0 & 0 \end{bmatrix}$; the second column to $\begin{bmatrix} a & b \\ c & d \end{bmatrix}\begin{bmatrix} 0 & 1 \\ 0 & 0 \end{bmatrix}$, the third to $\begin{bmatrix} a & b \\ c & d \end{bmatrix}\begin{bmatrix} 0 & 0 \\ 1 & 0 \end{bmatrix}$, and the fourth to $\begin{bmatrix} a & b \\ c & d \end{bmatrix}\begin{bmatrix} 0 & 0 \\ 0 & 1 \end{bmatrix}$.

b. From part a. we have

$$|R_A| = |L_A| = \sqrt{2a^2 + 2b^2 + 2c^2 + 2d^2} = \sqrt{2}|A|.$$

**2.6.7** a. This is not a subspace, since 0 is not in it.

b. This is a subspace: If $f, g$ satisfy the differential equation, then so does $af + bg$:

$$(af + bg)(x) = af(x) + bg(x) = axf'(x) + bxg'(x) = x(af + bg)'(x).$$

c. This is not a vector space: the function $f(x) = x^2/4$ is in it, but $x^2 = 4(x^2/4)$ is not, so it isn't closed under multiplication by scalars.

**2.6.9** a. Take any basis $\vec{\mathbf{w}}_1, \ldots, \vec{\mathbf{w}}_n$ of $V$, and discard from the ordered set of vectors

$$\vec{\mathbf{v}}_1, \ldots \vec{\mathbf{v}}_k, \vec{\mathbf{w}}_1, \ldots, \vec{\mathbf{w}}_n$$

any vectors $\vec{\mathbf{w}}_i$ that are linear combinations of earlier vectors. At all stages, the set of vectors obtained will span $V$, since they do when you start and discarding a vector that is a linear combination of others doesn't change the span. When you are through, the vectors obtained will be linearly independent, so they satisfy condition 3 of definition 2.4.12.

b. The approach is identical: eliminate from $\vec{\mathbf{v}}_1, \ldots \vec{\mathbf{v}}_k$ any vectors that depend linearly on earlier vectors; this never changes the span, and you end up with linearly independent vectors that span $V$.

**2.6.11** We have

$$AB = \begin{bmatrix} 1 + ab & a \\ b & 1 \end{bmatrix}, \quad BA = \begin{bmatrix} 1 & a \\ b & 1 + ab \end{bmatrix}.$$

Remember that the rank of a linear transformation is the dimension of its image; see definition 2.5.9.

Thus we are asking about the rank of the matrix

$$\begin{bmatrix} 1 & 1 & 1 + ab & 1 \\ a & 0 & a & a \\ 0 & b & b & b \\ 1 & 1 & 1 & 1 + ab \end{bmatrix}.$$

We need to row reduce this matrix, but before starting let us see what happens if $a = 0$, or $b = 0$, or both. If $a = 0$, the matrix is $\begin{bmatrix} 1 & 1 & 1 & 1 \\ 0 & 0 & 0 & 0 \\ 0 & b & b & b \\ 1 & 1 & 1 & 1 \end{bmatrix}$,

which evidently has rank 2 if $b \neq 0$, and rank 1 if $b = 0$. Similarly, if $b = 0$ and $a \neq 0$, the matrix has rank 2. Now let us suppose that $ab \neq 0$. Then row reduction gives

$$\begin{bmatrix} 1 & 1 & 1 + ab & 1 \\ a & 0 & a & a \\ 0 & b & b & b \\ 1 & 1 & 1 & 1 + ab \end{bmatrix} \rightarrow \begin{bmatrix} 1 & 1 & 1 + ab & 1 \\ 0 & -a & -a^2b & 0 \\ 0 & b & b & b \\ 0 & 0 & -ab & ab \end{bmatrix}$$

$$\begin{bmatrix} 1 & 0 & ab & 0 \\ 0 & a & 1 & 1 \\ 0 & 0 & a - a^2b & a \\ 0 & 0 & 1 & -1 \end{bmatrix} \rightarrow \begin{bmatrix} 1 & 0 & 0 & ab \\ 0 & 1 & 0 & 2 \\ 0 & 0 & 1 & -1 \\ 0 & 0 & 0 & a(2 - ab) \end{bmatrix}.$$

Thus we see that the four matrices are linearly independent in Mat $(2,2)$ if $ab \neq 0$ and $ab \neq 2$, and

- if $a = b = 0$, they span a space of dimension 1
- if $a = 0$ and $b \neq 0$, or $a \neq 0$ and $b = 0$, they span a space of dimension 2
- if $ab = 2$, they span a space of dimension 3
- if $ab \neq 0$ and $ab \neq 2$, they span a space of dimension 4

**2.7.1** The successive vectors $\vec{e}_1$, $A\vec{e}_1$, $A^2\vec{e}_1, \dots$ are

$$\begin{bmatrix} 1 \\ 0 \end{bmatrix}, \begin{bmatrix} 1 \\ 1 \end{bmatrix}, \begin{bmatrix} 1 \\ 2 \end{bmatrix}, \dots.$$

The vectors $\begin{bmatrix} 1 \\ 0 \end{bmatrix}, \begin{bmatrix} 1 \\ 1 \end{bmatrix}$ are linearly independent, and

$$\begin{bmatrix} 1 \\ 2 \end{bmatrix} = -\begin{bmatrix} 1 \\ 0 \end{bmatrix} + 2\begin{bmatrix} 1 \\ 1 \end{bmatrix}, \quad \text{or better,} \quad A^2\vec{e}_1 - 2A\vec{e}_1 + \vec{e}_1 = \vec{0}.$$

Thus the polynomial $p_1$ is $p_1(t) = t^2 - 2t + 1 = (t-1)^2$.

**2.7.3 a.** We will give two solutions.

The notation $T(p) = p + xp''$ is standard, but it may be confusing, because when applied (for example) to the polynomial $x$, we have $x$ playing a double role – it is both the polynomial and the name of the variable. It would be heavier but perhaps clearer to write

$$(T(p))(x) = p(x) + xp''(x).$$

Then, applied to the basis vectors $1, x, x^2, x^3$, we have

$(T(1))(x) = 1 + x \cdot 0 = 1,$

$(T(x))(x) = x + x \cdot 0 = x,$

$(T(x^2))(x) = x^2 + x \cdot 2$

$(T(x^3))(x) = x^3 + x \cdot 6x.$

*First solution:* In the basis $1, x, x^2, x^3$, the matrix of $T$ is

$$A = \begin{bmatrix} 1 & 0 & 0 & 0 \\ 0 & 1 & 2 & 0 \\ 0 & 0 & 1 & 6 \\ 0 & 0 & 0 & 1 \end{bmatrix},$$

since

$$T(1) = 1, \ T(x) = x, \ T(x^2) = x^2 + 2x, \ T(x^3) = x^3 + 6x^2.$$

The change of basis matrix is

$$S = \begin{bmatrix} 1 & 1 & 1 & 1 \\ 0 & 1 & 1 & 1 \\ 0 & 0 & 1 & 1 \\ 0 & 0 & 0 & 1 \end{bmatrix}, \quad \text{with inverse} \quad S^{-1} = \begin{bmatrix} 1 & -1 & 0 & 0 \\ 0 & 1 & -1 & 0 \\ 0 & 0 & 1 & -1 \\ 0 & 0 & 0 & 1 \end{bmatrix},$$

giving for our desired matrix

$$S^{-1}AS = \begin{bmatrix} 1 & -1 & 0 & 0 \\ 0 & 1 & -1 & 0 \\ 0 & 0 & 1 & -1 \\ 0 & 0 & 0 & 1 \end{bmatrix} \begin{bmatrix} 1 & 0 & 0 & 0 \\ 0 & 1 & 2 & 0 \\ 0 & 0 & 1 & 6 \\ 0 & 0 & 0 & 1 \end{bmatrix} \begin{bmatrix} 1 & 1 & 1 & 1 \\ 0 & 1 & 1 & 1 \\ 0 & 0 & 1 & 1 \\ 0 & 0 & 0 & 1 \end{bmatrix} = \begin{bmatrix} 1 & 0 & -2 & -2 \\ 0 & 1 & 2 & -4 \\ 0 & 0 & 1 & 6 \\ 0 & 0 & 0 & 1 \end{bmatrix}.$$

*Second solution:* The other approach is to compute directly:

$$T(1) = 1$$

$$T(1 + x) = 1 + x$$

$$T(1 + x + x^2) = 1 + x + x^2 + 2x = (1 + x + x^2) + 2(x + 1) - 2$$

$$T(1 + x + x^2 + x^3) = 1 + x + x^2 + 2x + x^3 + 6x$$

$$= (1 + x + x^2 + x^3) + 6(x^2 + x + 1) - 4(x + 1) - 2.$$

This expresses the images of the basis vectors under $T$ as linear combinations of the basis vectors, so the coefficients are the columns of the desired matrix, giving again

$$\begin{bmatrix} 1 & 0 & -2 & -2 \\ 0 & 1 & 2 & -4 \\ 0 & 0 & 1 & 6 \\ 0 & 0 & 0 & 1 \end{bmatrix}.$$

b. We give a solution analogous to the first above. In the "standard basis" $1, x, \ldots, x^k$ of $P_k$, the matrix of $T$ is the diagonal matrix

$$A = \begin{bmatrix} 1 & 0 & \ldots & 0 \\ 0 & 2 & \ldots & 0 \\ \vdots & \vdots & \ddots & \vdots \\ 0 & 0 & \ldots & k+1 \end{bmatrix}.$$

The change of basis is analogous to the above, and

$$S^{-1}AS = \begin{bmatrix} 1 & -1 & \ldots & 0 \\ 0 & 1 & \ldots & 0 \\ \vdots & \vdots & \ddots & \vdots \\ 0 & 0 & \ldots & 1 \end{bmatrix} \begin{bmatrix} 1 & 0 & \ldots & 0 \\ 0 & 2 & \ldots & 0 \\ \vdots & \vdots & \ddots & \vdots \\ 0 & 0 & \ldots & k+1 \end{bmatrix} \begin{bmatrix} 1 & 1 & \ldots & 1 \\ 0 & 1 & \ldots & 1 \\ \vdots & \vdots & \ddots & \vdots \\ 0 & 0 & \ldots & 1 \end{bmatrix} = \begin{bmatrix} 1 & -1 & -1 & \ldots & -1 \\ 0 & 2 & -1 & \ldots & -1 \\ 0 & 0 & 3 & \ldots & -1 \\ \vdots & \vdots & \vdots & \ddots & \vdots \\ 0 & 0 & 0 & \ldots & k+1 \end{bmatrix}.$$

**2.7.5** The recursive formula for the $b_n$ can be written

$$\begin{bmatrix} b_n \\ b_{n+1} \end{bmatrix} = \begin{bmatrix} 0 & 1 \\ 1 & 2 \end{bmatrix} \begin{bmatrix} b_{n-1} \\ b_n \end{bmatrix} = \begin{bmatrix} 0 & 1 \\ 1 & 2 \end{bmatrix} \begin{bmatrix} 1 \\ 1 \end{bmatrix}.$$

To compute the power of the matrix, we use eigenvalues and eigenvectors. The eigenvalues are $1 \pm \sqrt{2}$, and a basis of eigenvectors is

$$\begin{bmatrix} 1 \\ 1+\sqrt{2} \end{bmatrix}, \quad \begin{bmatrix} 1 \\ 1+\sqrt{2} \end{bmatrix}.$$

This leads to the change of basis matrix

$$S = \begin{bmatrix} 1 & 1 \\ 1+\sqrt{2} & 1-\sqrt{2} \end{bmatrix}, \quad \text{with inverse} \quad S^{-1} = \frac{\sqrt{2}}{4} \begin{bmatrix} \sqrt{2}-1 & 1 \\ \sqrt{2}+1 & -1 \end{bmatrix}.$$

Indeed, you can check by multiplication that

$$\frac{\sqrt{2}}{4} \begin{bmatrix} \sqrt{2}-1 & 1 \\ \sqrt{2}+1 & -1 \end{bmatrix} \begin{bmatrix} 0 & 1 \\ 1 & 2 \end{bmatrix} \begin{bmatrix} 1 & 1 \\ 1+\sqrt{2} & 1-\sqrt{2} \end{bmatrix} = \begin{bmatrix} 1+\sqrt{2} & 0 \\ 0 & 1-\sqrt{2} \end{bmatrix}.$$

Thus

$$S^{-1} \begin{bmatrix} 0 & 1 \\ 1 & 2 \end{bmatrix}^n S = \begin{bmatrix} 1+\sqrt{2} & 0 \\ 0 & 1-\sqrt{2} \end{bmatrix}^n = \begin{bmatrix} (1+\sqrt{2})^n & 0 \\ 0 & (1-\sqrt{2})^n \end{bmatrix},$$

and finally

$$\begin{bmatrix} 0 & 1 \\ 1 & 2 \end{bmatrix}^n = S \begin{bmatrix} (1+\sqrt{2})^n & 0 \\ 0 & (1-\sqrt{2})^n \end{bmatrix} S^{-1}$$

$$= \frac{\sqrt{2}}{4} \begin{bmatrix} (1+\sqrt{2})^n - (1-\sqrt{2})^n & (1+\sqrt{2})^{n+1} - (1-\sqrt{2})^{n+1} \\ (1+\sqrt{2})^{n+1} - (1-\sqrt{2})^{n+1} & (1+\sqrt{2})^{n+2} - (1-\sqrt{2})^{n+2} \end{bmatrix}.$$

Multiplying this by $\begin{bmatrix} b_0 \\ b_1 \end{bmatrix} = \begin{bmatrix} 1 \\ 1 \end{bmatrix}$ we find

$$b_n = \frac{\sqrt{2}}{4}\left((2+\sqrt{2})(1+\sqrt{2})^n - (2-\sqrt{2})(1-\sqrt{2})^n\right).$$

b. Since $|1 - \sqrt{2}| < 1$, it will contribute practically nothing to $b_n$, and the relevant number is

$$\frac{\sqrt{2}}{4}(2+\sqrt{2})(1+\sqrt{2})^{1000} = \frac{1}{2}(1+\sqrt{2})^{1001}.$$

You will find that your calculator will refuse to evaluate this, but using logarithms base 10 for a change, you find

$$\log_{10} b_n \sim 382.857,$$

so $b_n$ has 383 digits, starting with $719449\ldots$.

c. This time the relevant matrix description of the recursion is

$$\begin{bmatrix} c_n \\ c_{n+1} \\ c_{n+2} \end{bmatrix} = \begin{bmatrix} 0 & 1 & 0 \\ 0 & 0 & 1 \\ 1 & 1 & 1 \end{bmatrix}\begin{bmatrix} c_{n-1} \\ c_n \\ c_{n+1} \end{bmatrix} = \begin{bmatrix} 0 & 1 & 0 \\ 0 & 0 & 1 \\ 1 & 1 & 1 \end{bmatrix}^n\begin{bmatrix} 1 \\ 1 \\ 1 \end{bmatrix}.$$

Set

$$A = \begin{bmatrix} 0 & 1 & 0 \\ 0 & 0 & 1 \\ 1 & 1 & 1 \end{bmatrix}.$$

We have

$$A^2 = \begin{bmatrix} 0 & 0 & 1 \\ 1 & 1 & 1 \\ 1 & 2 & 2 \end{bmatrix}$$

$$A^3 = \begin{bmatrix} 1 & 1 & 1 \\ 1 & 2 & 2 \\ 2 & 3 & 4 \end{bmatrix}$$

One can easily check that

$$A^3 - A^2 - A = I,$$

so the roots of the polynomial $\lambda^3 - \lambda^2 - \lambda - 1$ are eigenvalues of the matrix.[1] These eigenvalues are approximately

$$\lambda_1 \sim 1.83926\ldots, \quad \text{and} \quad \lambda_{2,3} \sim -.419643 \pm .6062907.$$

A basis of eigenvectors is

$$S = \begin{bmatrix} 1 & 1 & 1 \\ \lambda_1 & \lambda_2 & \lambda_3 \\ \lambda_1^2 & \lambda_2^2 & \lambda_3^2 \end{bmatrix}$$

---

[1]How did we find the coefficients $x = y = z = -1$ for the equation

$$A^3 + xA^2 + yA + zI = 0?$$

This equation corresponds to

$$\begin{bmatrix} 1+z & 1+y & 1+x \\ 1+x & 2+x+z & 2+x+y \\ 2+x+y & 3+2x+y & 4+2x+y+z \end{bmatrix} = \begin{bmatrix} 0 & 0 & 0 \\ 0 & 0 & 0 \\ 0 & 0 & 0 \end{bmatrix}.$$

A glance just at the top line immediately gives the answer. Normally one would not expect a system of nine equations in three unknowns to have a solution! The Cayley-Hamilton theorem (theorem 4.8.24) says that every $n \times n$ matrix satisfies a polynomial equation of degree at most $n$, whereas you would expect that it would only satisfy an equation of degree $n^2$.

and we can compute the high powers as above by

$$A^n = S \begin{bmatrix} \lambda_1^n & 0 & 0 \\ 0 & \lambda_2^n & 0 \\ 0 & 0 & \lambda_3^n \end{bmatrix} S^{-1}.$$

Computing $S^{-1}$ is what computers are for, but when the dust has settled, we find

$$S^{-1} \begin{bmatrix} 1 \\ 1 \\ 1 \end{bmatrix} \sim \begin{bmatrix} .4356 \\ .28219 - .359i \\ .28219 + .359i \end{bmatrix},$$

and, as above,

$$c_n \sim .4356 \cdot + \lambda_1^n + (.28219 - .359i) \cdot \lambda_2^n + (.28219 + .359i) \cdot \lambda_3^n.$$

Finally, to find the number of digits of $c_{1000}$, only the term involving $\lambda_1$ is relevant, and we resort to logarithms base 10:

$$\log .4356 + 1000 \log 1.83926 \sim 264.282.$$

Thus $c_{1000}$ has 265 digits.

**2.8.1** The partial derivatives are

$$D_1 F_1 = -\sin(x - y), \ D_2 F_1 = \sin(x - y) - 1,$$
$$D_1 F_2 = \cos(x + y) - 1, \ D_2 F_2 = \cos(x + y),$$

so

$$D_{1,1} F_1 = D_{2,2} F_1 = -\cos(x - y); \qquad D_{1,2} F_1 = D_{2,1} F_1 = \cos(x - y)$$
$$D_{1,1} F_2 = D_{2,2} F_2 = D_{1,2} F_2 = D_{2,1} F_2 = -\sin(x + y).$$

Therefore, the absolute value of each second partial is bounded by 1, and there are eight second partials, so

$$\big| [\mathbf{Df(u)}] - [\mathbf{Df(v)}] \big| \le \sqrt{8 \cdot 1^2} |\mathbf{u} - \mathbf{v}|;$$

using this method we get $M = 2\sqrt{2}$, the same result as equation 2.8.57.

**2.8.3** a. Use the triangle inequality:

$$\begin{aligned} |x| = |x - y + y| \le |x - y| + |y| \\ |y| = |y - x + x| \le |y - x| + |x| \end{aligned} \quad \text{imply} \quad \begin{aligned} |x| - |y| \le |x - y| \\ |y| - |x| \le |x - y| \end{aligned} \quad \text{imply} \quad \big| |x| - |y| \big| \le |x - y|.$$

b. Choose $C > 0$, and find $x$ such that $1/2\sqrt{x} > C$, which will happen for $x > 0$ sufficiently small. From the definition of the derivative,

$$\frac{1}{2\sqrt{x}} = \lim_{y \to x} \frac{\sqrt{y} - \sqrt{x}}{y - x}.$$

Thus taking $y > x$ sufficiently close to $x$, we will find

$$|\sqrt{y} - \sqrt{x}| \ge (C - 1)|y - x|.$$

Since this is possible for every $C > 0$, we see that there cannot exist a number $M$ such that $|\sqrt{y} - \sqrt{x}| \le M|y - x|$ for all $x, y > 0$.

**2.8.5**  a.  Newton's method gives $x_{n+1} = \dfrac{2x_n s + 9}{3x^2}$, which leads to

$$x_0 = 2, \ x_1 = \frac{25}{12} = 2.08\overline{3}, \ x_2 = \frac{46802}{22500} = 2.0800\overline{8},$$

$$x_3 = \frac{307548373003216}{147853836270000} \sim 2.08008382306424.$$

b.  In this case,

$$h_0 = -\frac{2^3 - 9}{2 \cdot 2^2} = \frac{1}{12}, \quad U_1 = \left[2, 2 + \frac{1}{6}\right].$$

We have $f(x_0) = 2^3 - 9 = -1$, $f'(x_0) = 3 \cdot 2^2 = 12$, and finally

$$M = \sup_x \in U_1 |f''(x)| = 6\left(2 + \frac{1}{6}\right) = 13.$$

c.  Just compute:

$$\frac{|f(x_0)|M}{|f'(x_0)|^2} = \frac{1 \cdot 13}{12^2} = 0.0902\overline{7} < \frac{1}{2},$$

so we know that Newton's method will converge – not that there was much doubt after part a.

**2.8.7**  a.  The MATLAB command

$$\text{newton}([\cos(x1)+x2-1.1; \ x1+\cos(x1+x2)-.9], \ [0; 0], \ 5)$$

produces the output

$$\begin{pmatrix} -0.10000000000000 \\ 0.10000000000000 \end{pmatrix}, \ \begin{pmatrix} -0.10000000000000 \\ 0.10499583472197 \end{pmatrix}, \ \begin{pmatrix} -0.09998746447058 \\ 0.10499458325724 \end{pmatrix},$$

$$\begin{pmatrix} -0.09998746440624 \\ 0.10499458332900 \end{pmatrix}, \ \begin{pmatrix} -0.09998746440624 \\ 0.10499458332900 \end{pmatrix}.$$

We see that the first nine significant digits are unchanged from the third to the fourth step, and that all 15 digits are preserved from the fourth to the fifth step.

b.  Set

$$\mathbf{f}\begin{pmatrix} x \\ y \end{pmatrix} = \begin{bmatrix} \cos x + y - 1.1 \\ x + \cos(x + y) - .9 \end{bmatrix}.$$

Then

$$\left[D\mathbf{f}\begin{pmatrix} x \\ y \end{pmatrix}\right] = \begin{bmatrix} -\sin x & 1 \\ 1 - \sin(x + y) & -\sin(x + y) \end{bmatrix},$$

so

$$\left[D\mathbf{f}\begin{pmatrix} 0 \\ 0 \end{pmatrix}\right] = \begin{bmatrix} 0 & 1 \\ 1 & 0 \end{bmatrix}, \quad \mathbf{h}_0 = \mathbf{a}_1 = \begin{bmatrix} .1 \\ -.1 \end{bmatrix}.$$

Now we estimate the Lipschitz ratio:

$$\left\|\begin{bmatrix} -\sin u_1 & 1 \\ 1-\sin(u_1+v_1) & -\sin(u_1+v_1) \end{bmatrix} - \begin{bmatrix} -\sin u_2 & 1 \\ 1-\sin(u_2+v_2) & -\sin(u_2+v_2) \end{bmatrix}\right\|$$

$$\leq \left\|\begin{bmatrix} |u_1-u_2| & 0 \\ |(u_1-u_2)+(v_1-v_2)| & |(u_1-u_2)+(v_1-v_2)| \end{bmatrix}\right\|$$

$$= \left((u_1-u_2)^2 + 2((u_1-u_2)+(v_1-v_2))^2\right)^{1/2} \leq \sqrt{5}\left((u_1-u_2)^2+(v_1-v_2)^2\right)^{1/2},$$

so we may take $M = \sqrt{5}$.

The quantity

$$M|\mathbf{f}(\mathbf{a}_0)| \left|[\mathbf{D}\vec{\mathbf{f}}(\mathbf{a}_0)]^{-1}\right|^2 = \sqrt{5}\frac{\sqrt{2}}{10}(\sqrt{2})^2 = \frac{2\sqrt{10}}{10} \sim .632\ldots$$

is not smaller than $1/2$, so we cannot simply invoke Kantorovich.

Using the norm rather than the length is justified by theorem 2.9.8. The norm is defined in definition 2.9.6.

There are at least two ways around this. One is to use the norm of the derivative rather than the length: the norm of $\begin{bmatrix} 0 & 1 \\ 1 & 0 \end{bmatrix}$ is 1, not $\sqrt{2}$, so this improves the estimate by a factor of 2, which is enough.

Another is to look at the next step of Newton's method, which gives $|\mathbf{f}(\mathbf{a}_1)| \approx .005$. We can use the same Lipschitz ratio. The derivative $L = [\mathbf{Df}(\mathbf{a}_1)]$ is, by the same Lipschitz estimate, of the form

$$\begin{bmatrix} a & 1+b \\ 1+c & d \end{bmatrix}$$

with $|a|, |b|, |c|, |d|$ all less than $.1$. So $|\det L| > .75$, and we have $|L^{-1}| < 2.2/.75 < 3$. This gives

$$M|\mathbf{f}(\mathbf{a}_1)|\left|L^{-1}\right|^2 < \sqrt{5}\cdot(.005)\cdot 9,$$

which is much less than $1/2$.

**2.8.9** We will start Newton's method at the origin, and try to solve the system of equation

$$F\begin{pmatrix} x \\ y \\ z \end{pmatrix} = \begin{pmatrix} x+y^2-a \\ y+z^2-b \\ z+x^2-c \end{pmatrix} = \begin{pmatrix} 0 \\ 0 \\ 0 \end{pmatrix}.$$

Solution 2.8.9: This is an exceptionally easy case in which to apply Kantorovich's theorem.

What would we get if we used the norm (discussed in section 2.9) instead of the length? Since the norm of the identity is 1, we would get $\sqrt{a^2+b^2+c^2} \leq \frac{1}{4}$. We don't know how sharp this is, but $1/2$ is certainly too big.

First, a bit of computation will show you that $M = 2$ is a global Lipschitz ratio for $[\mathbf{D}F(\mathbf{x})]$. Next, the derivative at the origin is the identity, so its inverse is also the identity, with length $\sqrt{3}$. So we find that the condition for Kantorovich's theorem to work is

$$\sqrt{a^2+b^2+c^2}\cdot(\sqrt{3})^2\cdot 2 \leq \frac{1}{2}, \quad \text{i.e.,} \quad \sqrt{a^2+b^2+c^2} \leq \frac{1}{12}.$$

**2.8.11** a. Set $p(x) = x^5 - x - 6$. One step of Newton's method to solve $p(x) = 0$ is

$$N(a) = a - \frac{a^5-a-6}{5a^4-1} = \frac{4a^5+6}{5a^4-1}.$$

In particular, $x_1 = N(2) = \frac{134}{79}$, with a first Newton step $h_0 = -24/79$.

b. To apply Kantorovitch's theorem, we need to compute, we need to show that $Mp(2)/(p'(2)^2) < 1/2$, where

$$M = \sup_{x \in [110/79, 2]} |p''(x)|.$$

The second derivative of $p$ is $20x^3$; the supremum $M$ is $20 \cdot 2^3 = 160$. The other two numbers of interest are $p(2) = 24$ and $p'(2) = 79$. Unfortunately,

$$\frac{24 \cdot 160}{79^2} \approx .6152860119 > \frac{1}{2},$$

so Kantorovitch's theorem does not apply. But it does apply at $x_1$. Exact computation with rational numbers gets out of hand, but decimal approximations are no problem. We find $p(x_1) \approx 1.09777$ and $p'(x_1) \approx 27.0579$. We can use the value for $M1 = p''(x_1)$, and indeed

$$\frac{Mp(x_1)}{(p''(x_1)^2} \approx .157 < 1/2.$$

**2.8.13** a. Direct calculation gives

$$\|[\mathbf{D}F(\mathbf{u})] - [\mathbf{D}F(\mathbf{v})]\| = \left\| \begin{bmatrix} 2u_1 - 2 & 2u_2 \\ u_1 - 1 & u_2 - 1 \end{bmatrix} - \begin{bmatrix} 2v_1 - 2 & 2v_2 \\ v_1 - 1 & v_2 - 1 \end{bmatrix} \right\|$$

$$= \left\| \begin{bmatrix} 2(u_1 - v_1) & 2(u_2 - v_2) \\ u_1 - v_1 & u_2 - v_2 \end{bmatrix} \right\| = \sqrt{5}|\mathbf{u} - \mathbf{v}|.$$

So $\sqrt{5}$ is a global Lipschitz ratio. The second derivative approach gives $\sqrt{10}$.

b. We have

$$F\begin{pmatrix} 5 \\ 1 \end{pmatrix} = \begin{bmatrix} 1 \\ -1 \end{bmatrix}, \quad \left[\mathbf{D}F\begin{pmatrix} 5 \\ 1 \end{pmatrix}\right] = \begin{bmatrix} 8 & 2 \\ 0 & 4 \end{bmatrix}, \quad \left[\mathbf{D}F\begin{pmatrix} 5 \\ 1 \end{pmatrix}\right]^{-1} = \frac{1}{32}\begin{bmatrix} 4 & -2 \\ 0 & 8 \end{bmatrix},$$

so

$$\mathbf{h}_0 = -\frac{1}{32}\begin{bmatrix} 4 & -2 \\ 0 & 8 \end{bmatrix}\begin{bmatrix} 1 \\ -1 \end{bmatrix} = \begin{bmatrix} -3/16 \\ 1/4 \end{bmatrix}, \quad \mathbf{x}_1 = \begin{pmatrix} 5 \\ 1 \end{pmatrix} + \begin{bmatrix} -3/16 \\ 1/4 \end{bmatrix} = \begin{pmatrix} 77/16 \\ 5/4 \end{pmatrix},$$

and $|\mathbf{h}_0| = 5/16$. For Kantorovich's theorem to guarantee convergence, we must have

$$\sqrt{2}\frac{84}{32^2}\sqrt{5} \le \frac{1}{2},$$

which is true (and is still true if we use $\sqrt{10}$ instead of $\sqrt{5}$).

c. By part b, there is a unique root in the disc of radius 5/16 around $\begin{pmatrix} 77/16 \\ 5/4 \end{pmatrix} \approx \begin{pmatrix} 4.8 \\ 1.25 \end{pmatrix}$. See the picture in the margin.

**2.9.1** For the polynomial $f(x) = (x-1)^2$, it is always true that

$$\left|f'(a) - f'(b)\right| = \left|2(a-1) - 2(b-1)\right| = 2|a-b|,$$

so $M = 2$ is a Lipschitz ratio, and certainly the best ratio, realized at every pair of points.

FIGURE FOR SOLUTION 2.8.13. The curve of equation

$$x^2 + y^2 - 2x - 15 = 0$$

is the circle of radius 4 centered at $\begin{pmatrix} 1 \\ 0 \end{pmatrix}$. The curve of equation

$$xy - x - y = 0$$

is a hyperbola with asymptotes the lines $x = 1$ and $y = 1$. Exercise 2.8.13 uses Newton's method to find an intersection of these curves. There are four such intersections, but starting at $\begin{pmatrix} 5 \\ 1 \end{pmatrix}$ leads to a sequence that rapidly converges to the intersection in the small shaded circle.

Since

$$\frac{f(0)M}{f'(0)^2} = \frac{1 \cdot 2}{(-2)^2} = \frac{1}{2},$$

inequality 2.8.51 of Kantorovich's theorem is satisfied as an equality.

One step of Newton's method is

$$a_{n+1} = a_n + \frac{(1-a_n)^2}{2(1-a_n)} = \frac{a_n+1}{2},$$

so if we assume by induction that $a_n = 1 - 1/2^n$, which is true for $n = 0$, we find

$$a_{n+1} = \frac{1 - 1/2^n + 1}{2} = 1 - 1/2^{n+1},$$

and $h_n = 1/2^{n+1}$, which is exactly the worst case in Kantorovich's theorem.

Solution 2.9.3: In the second inequality, we choose one unit $\vec{v}$ to make $|A\vec{v}|$ as large as possible, and another to make $|B\vec{v}|$ as large as possible. The sum we get this way is certainly at least as large as the sum we would get if we chose the same $\vec{v}$ for both terms.

**2.9.3** Just apply the definition of the norm, and the triangle inequality in the codomain:

$$\|A + B\| = \sup_{|\vec{v}|=1} |(A+B)\vec{v}| = \sup_{|\vec{v}|=1} |A\vec{v} + B\vec{v}|$$

$$\leq \sup_{|\vec{v}|=1} |A\vec{v}| + |B\vec{v}| \leq \sup_{|\vec{v}|=1} |A\vec{v}| + \sup_{|\vec{v}|=1} |B\vec{v}| = \|A\| + \|B\|.$$

**2.9.5** a. In part b, we will solve this equation by Newton's method. For now, you might guess that there exists a solution of the form $\begin{bmatrix} x & x \\ x & x \end{bmatrix}$. Substitute that into the equation, to find

$$\begin{bmatrix} 2x^2 & 2x^2 \\ 2x^2 & 2x^2 \end{bmatrix} + \begin{bmatrix} x & x \\ x & x \end{bmatrix} = \begin{bmatrix} 1 & 1 \\ 1 & 1 \end{bmatrix}, \quad \text{i.e.,} \quad x^2 + x - 1 = 0.$$

This has two solutions, $-1$ and $1/2$, which lead to two solutions of the original problem:

$$\begin{bmatrix} -1 & -1 \\ -1 & -1 \end{bmatrix} \quad \text{and} \quad \begin{bmatrix} 1/2 & 1/2 \\ 1/2 & 1/2 \end{bmatrix}.$$

b. We will use the *length* of matrices, but the *norm* for elements of $\mathcal{L}(\text{Mat}\,(2,2), \text{Mat}\,(2,2))$ – i.e., for linear maps from $\text{Mat}\,(2,2)$ to $\text{Mat}\,(2,2)$.

The equation we want to solve is $X^2 + X - \begin{bmatrix} 1 & 1 \\ 1 & 1 \end{bmatrix} = \begin{bmatrix} 0 & 0 \\ 0 & 0 \end{bmatrix}$, so the relevant mapping is the (nonlinear) mapping $F : \text{Mat}\,(2,2) \to \text{Mat}\,(2,2)$ given by

$$F(X) = X^2 + X - \begin{bmatrix} 1 & 1 \\ 1 & 1 \end{bmatrix}.$$

Clearly $|F(I)| = \left\| \begin{bmatrix} 1 & -1 \\ -1 & 1 \end{bmatrix} \right\| = 2$. Since

$$[DF(I)](H) = HI + IH + H = 3H,$$

we have $[DF(I)]^{-1}H = H/3$, i.e., $[DF(I)]^{-1}$ is $1/3$ times the identity function $\mathrm{id} : \mathrm{Mat}\,(2,2) \to \mathrm{Mat}\,(2,2)$. Thus

$$\left\| [DF(I)]^{-1} \right\| = \frac{1}{3},$$

which squared gives $1/9$.

Moreover, $\left\| [DF(A)] - [DF(B)] \right\| \le 2|A - B|$:

$$
\begin{aligned}
\left\| [DF(A)] - [DF(B)] \right\| &= \sup_{|H|=1} \left| (AH + HA + H) - (BH + HB + H) \right| \\
&= \sup_{|H|=1} \left| H(A - B) + (A - B)H \right| \\
&\le \sup_{|H|=1} \left( \left| H(A - B) \right| + \left| (A - B)H \right| \right) \\
&\le \sup_{|H|=1} \left( |H||A - B| + |A - B||H| \right) = 2|A - B|.
\end{aligned}
$$

So the conditions of Kantorovich's theorem are satisfied: $2 \cdot 1/9 \cdot 2 < 1/2$.

**2.9.7** First, let us compute the norm of $f(A_0) = \begin{bmatrix} 1 & -1 \\ 1 & -1 \end{bmatrix}$:

$$
\begin{aligned}
\| f(A_0) \| &= \sup_{|\vec{x}|=1} \left| \begin{bmatrix} 1 & -1 \\ 1 & -1 \end{bmatrix} \begin{bmatrix} x_1 \\ x_2 \end{bmatrix} \right| = \sup_{|\vec{x}|=1} \left| \begin{bmatrix} x_1 - x_2 \\ x_1 - x_2 \end{bmatrix} \right| \\
&= \sup_{|\vec{x}|=1} \sqrt{2(x_1 - x_2)^2} = \sup_{|\vec{x}|=1} \sqrt{2(x_1^2 + x_2^2) - 4 x_1 x_2}.
\end{aligned}
$$

If we substitute $\cos\theta$ for $x_1$ and $\sin\theta$ for $x_2$ (since $x_1^2 + x_2^2 = 1$) this gives

$$\sup \sqrt{2(\cos^2\theta + \sin^2\theta) - 4\cos\theta\sin\theta} = \sup \sqrt{2 - 2\sin 2\theta} = \sqrt{4} = 2.$$

The following (virtually identical to equation 2.9.24) shows that the derivative of $f$ is Lipschitz with Lipschitz ratio 2:

$$
\begin{aligned}
\left\| [\mathbf{D}f(A_1)] - [\mathbf{D}f(A_2)] \right\| &= \sup_{|B|=1} \left| \left( [\mathbf{D}f(A_1)] - [\mathbf{D}f(A_2)] \right) B \right| \\
&= \sup_{|B|=1} \left| A_1 B + B A_1 + \begin{bmatrix} 0 & 1 \\ 1 & 0 \end{bmatrix} - A_2 B - B A_2 - \begin{bmatrix} 0 & 1 \\ 1 & 0 \end{bmatrix} \right| \\
&= \sup_{|B|=1} \left| (A_1 - A_2)B + B(A_1 - A_2) \right| \\
&\le \sup_{|B|=1} |A_1 - A_2||B| + |B||A_1 - A_2| \le \sup_{|B|=1} 2|B||A_1 - A_2| \\
&= 2|A_1 - A_2|.
\end{aligned}
$$

Solution 2.9.5, part b: If we wanted to consider $[DF(I)]$ and $[DF(I)]^{-1}$ as matrices, we would have to consider an element of $\mathrm{Mat}\,(2,2)$ as an element of $\mathbb{R}^4$ and the matrices in question would be $4 \times 4$, with 16 entries, making computing the length cumbersome.

The first inequality is the triangle inequality; the second inequality is Schwarz. Remember that we are taking the sup of $H$ such that $|H| = 1$.

Finally we need to compute the norm of $[\mathbf{D}f(A_0)]^{-1} = \dfrac{1}{63}\begin{bmatrix} 8 & -1 \\ -1 & 8 \end{bmatrix}$:

$$\left\| \frac{1}{63}\begin{bmatrix} 8 & -1 \\ -1 & 8 \end{bmatrix} \right\| = \sup_{|\vec{x}|=1} \left\| \begin{bmatrix} \dfrac{8x_1}{63} - \dfrac{x_2}{63} \\ -\dfrac{x_1}{63} + \dfrac{8x_2}{63} \end{bmatrix} \right\|$$

$$= \sup_{|\vec{x}|=1} \sqrt{\left(\frac{8x_1 - x_2}{63}\right)^2 + \left(\frac{8x_2 - x_1}{63}\right)^2}$$

$$= \sup_{|\vec{x}|=1} \frac{1}{63}\sqrt{64x_1^2 - 16x_1x_2 + x_2^2 + 64x_2^2 - 16x_2x_1 + x_1^2}$$

$$= \sup_{|\vec{x}|=1} \frac{1}{63}\sqrt{(x_1^2 + x_2^2) + 64(x_1^2 + x_2^2) - 32x_1x_2}.$$

Switching again to sines and cosines, we get

$$\sup \frac{1}{63}\sqrt{65(\cos^2\theta + \sin^2\theta) - 16\sin 2\theta} = \frac{1}{63}\sqrt{65 + 16} = \frac{9}{63} = \frac{1}{7}.$$

Squaring this norm gives $\|[\mathbf{D}f(A_0)]^{-1}\|^2 \approx 0.0204$.

Thus instead of

$$\underbrace{2}_{|f(A_0)|} \cdot \underbrace{2.8}_{M} \cdot \underbrace{(.256)^2}_{|[\mathbf{D}f(A_0)]^{-1}|^2} = .367 < .5,$$

obtained using the length (see equation 2.8.68 in the text), we have

$$\underbrace{2}_{|f(A_0)|} \cdot \underbrace{2}_{M} \cdot \underbrace{0.0204}_{\|[\mathbf{D}f(A_0)]^{-1}\|^2} = 0.08166.$$

Solution 2.10.1: Remember, the inverse function theorem says that *if the derivative of a mapping is invertible, the mapping is locally invertible.*

**2.10.1** Parts a, c, and d: The theorem does not guarantee that these functions are locally invertible with differentiable inverse, since the derivative is not invertible: for the first two, because the derivative is not square; for the third, because all entries of the derivative are 0. (In fact, exercise 2.10.4 shows that the functions are definitely not locally invertible with differentiable inverse.)

b. Invertible: the derivative at $\begin{pmatrix} 1 \\ 1 \end{pmatrix}$ is $\begin{bmatrix} 2 & 1 \\ -2 & 0 \end{bmatrix}$, which is invertible; hence the mapping is invertible.

e. The same mapping as in part d is invertible at $\begin{pmatrix} 1 \\ 1 \\ 1 \end{pmatrix}$; the derivative at that point is $\begin{bmatrix} 1 & 1 & 1 \\ 2 & 0 & 0 \\ 0 & 0 & 2 \end{bmatrix}$, which is invertible.

**2.10.3** a. Because $\sin 1/x$ isn't defined at $x = 0$, to see that this function is differentiable at 0, you *must* apply the definition of the derivative:

$$f'(0) = \lim_{h \to 0} \frac{h/2 + h^2 \sin(1/h)}{h} = \frac{1}{2} + \lim_{h \to 0} h \sin\frac{1}{h} = \frac{1}{2}.$$

b. Away from 0, you can apply the ordinary rules of differentiation, to find

$$f'(x) = \frac{1}{2} + 2x \sin \frac{1}{x} + x^2 \cos \frac{1}{x}\left(-\frac{1}{x^2}\right) = \frac{1}{2} + 2x \sin \frac{1}{x} - \cos \frac{1}{x}.$$

As $x$ approaches 0, the term $2x \sin \frac{1}{x}$ approaches 0, but the term $\cos \frac{1}{x}$ oscillates infinitely many times between 1 and $-1$. Even if you add $1/2$, there still are $x$ arbitrarily close to 0 where $f'(x) < 0$. Since $f$ is decreasing at these points, the function $f$ is not monotone in any neighborhood of 0, and thus it cannot have an inverse on any neighborhood of 0.

c. It doesn't contradict theorem 2.10.2 because $f$ is not increasing or decreasing. More important, it doesn't contradict theorem 2.10.7 (the inverse function theorem), because that theorem requires that the derivative be continuous, and although $f$ is differentiable, the derivative is not continuous.

**2.10.5** a. Two different implicit functions exist, both for $x \leq 1/4$. We find them by computing

$$y^2 + y + 3x + 1 = (y^2 + y + 1/4) + 3x + 3/4 = 0, \quad \text{so that}$$
$$(y + 1/2)^2 = -3(x + 1/4),$$

giving the implicit functions $y = \pm\sqrt{-3x - 3/4} - 1/2$. Since $-3x - 3/4$ can't be negative, we have $x \leq -1/4$.

b. To check that this result agrees with the implicit function theorem, we compute the derivative of $f\begin{pmatrix} x \\ y \end{pmatrix} = y^2 + y + 3x + 1$, getting

$$\left[\mathbf{D}f\begin{pmatrix} x \\ y \end{pmatrix}\right] = [3 \quad 2y + 1].$$

Defining $y$ implicitly as a function of $x$ means considering $y$ the pivotal variable, so the matrix corresponding to equation 2.10.25 is $[2y + 1]$, which is invertible if $y \neq -1/2$. This value for $y$ gives $x \neq -1/4$. Given any point $\begin{pmatrix} x_0 \\ y_0 \end{pmatrix}$ satisfying $y_0^2 + y_0 + 3x_0 + 1 = 0$ *except* the point $\begin{pmatrix} -1/4 \\ -1/2 \end{pmatrix}$, there is a neighborhood of $f\begin{pmatrix} x_0 \\ y_0 \end{pmatrix}$ in which $f\begin{pmatrix} x \\ y \end{pmatrix} = 0$ expresses $y$ implicitly as a function of $x$.

Part b: Does this seem contradictory? After all, we know that given $x = -1/4$ we can use

$$4y = \pm\sqrt{-3x - 3/4} - 1/2$$

to compute $y = -1/2$, and here is the implicit function theorem telling us there's no implicit function there! But the implicit function theorem doesn't talk about implicit functions at points, it talks about neighborhoods. At $x = -1/4 + \epsilon$, with $\epsilon > 0$, there is no implicit function.

c. Since $y$ is the pivotal unknown, we write

$$L = \begin{bmatrix} 2y + 1 & 3 \\ 0 & 1 \end{bmatrix}, \quad L^{-1} = \frac{1}{2y + 1}\begin{bmatrix} 1 & -3 \\ 0 & 2y + 1 \end{bmatrix}.$$

So at the point with coordinates $x = -\frac{1}{2}, y = \frac{\sqrt{3}-1}{2}$, we have

$$L^{-1} = \frac{1}{\sqrt{3}}\begin{bmatrix} 1 & -3 \\ 0 & \sqrt{3} \end{bmatrix}, \qquad |L^{-1}| = \frac{1}{\sqrt{3}}\sqrt{13}, \quad \text{giving } |L^{-1}|^2 = \frac{13}{3}.$$

The derivative of $f$ is Lipschitz with Lipschitz ratio 2, so the largest radius $R$ of a ball on which the implicit function $g$ is guaranteed to exist by the implicit function theorem satisfies

$$2 = \frac{1}{2R|L^{-1}|^2}, \quad \text{so } R = \frac{3}{52}. \tag{1}$$

In equation (1), we are using equation 2.10.27.

d. If we choose the point with coordinates $x = -\frac{13}{4}$, $y = \frac{5}{2}$, we do a bit better; in this case

$$L^{-1} = \frac{1}{6}\begin{bmatrix} 1 & -3 \\ 0 & 6 \end{bmatrix}, \quad \text{giving } |L^{-1}|^2 = \frac{23}{18} \text{ and } R = \frac{18}{92}.$$

**2.10.7** Yes: there is a function $g(t) = \begin{pmatrix} 0 \\ t \end{pmatrix}$ such that $\mathbf{F}\begin{pmatrix} g(t) \\ t \end{pmatrix} = \begin{pmatrix} 0 \\ 0 \end{pmatrix}$ in a neighborhood of $\begin{pmatrix} 0 \\ 0 \\ 0 \end{pmatrix}$. But the implicit function theorem does not say that an implicit function exists: the derivative of $\mathbf{F}\begin{pmatrix} x \\ y \\ t \end{pmatrix} = \begin{pmatrix} x^2 \\ y - t \end{pmatrix}$ is

$$\left[\mathbf{DF}\begin{pmatrix} x \\ y \\ t \end{pmatrix}\right] = \begin{bmatrix} 2x & 0 & 0 \\ 0 & 1 & -1 \end{bmatrix}, \text{ which at } \begin{pmatrix} 0 \\ 0 \\ 0 \end{pmatrix} \text{ is } \begin{bmatrix} 0 & 0 & 0 \\ 0 & 1 & -1 \end{bmatrix}; \text{ this matrix}$$

is not onto $\mathbb{R}^2$.

**2.10.9** The function $F\begin{pmatrix} x \\ y \end{pmatrix} = \begin{pmatrix} x + y + \sin(xy) \\ \sin(x^2 + y) \end{pmatrix}$ maps the origin to the origin, and $[\mathbf{D}F(0)] = I$. Certainly $F$ is differentiable with Lipschitz derivative. So it is locally invertible near the origin, and the point $\begin{pmatrix} a \\ 2a \end{pmatrix}$ will be in the domain of the inverse for $|a|$ sufficiently small.

**2.10.11** a. If you set up the two multiplications $AH_{i,j}$ and $H_{i,j}A$, you will see that

$$T(H_{i,j}) = (a_i + a_j)H_{i,j}.$$

b. The $H_{i,j}$ form a basis for Mat $(n, n)$. Their images are all multiples of themselves, so they are linearly independent unless any multiple is 0, i.e., unless $a_i = -a_j$ for some $i$ and $j$.

It follows from the chain rule that the derivative of the inverse is the inverse of the derivative.

c. False. If there were such a differentiable map, its derivative would be the inverse of $T : \text{Mat}\,(n, n) \to \text{Mat}\,(n, n)$, where $T(H) = BH + HB$, and

$$B = \begin{bmatrix} -1 & 0 & 0 \\ 0 & 1 & 0 \\ 0 & 0 & 1 \end{bmatrix}.$$

But $T$ is not invertible, since $a_1 = -1$ and $a_2 = 1$, so $a_1 = -a_2$.

**2.10.13** False. For example, the function $f\begin{pmatrix} x \\ y \\ x \end{pmatrix} = y + z$ satisfies the hypotheses but does not express $x$ in any way. For the implicit function theorem to apply, we would need $D_1 f \neq 0$.

**2.10.15**  a. The derivative of $F$ is given by $\left[\mathbf{D}F\begin{pmatrix} x \\ y \end{pmatrix}\right] = \begin{bmatrix} e^x & e^y \\ e^x & -e^{-y} \end{bmatrix}$.

The columns of this matrix are linearly independent (for instance because the two entries of the first column have the same sign and the two entries of the second column have opposite signs, or because the determinant is $-e^x(e^y + e^{-y})$, which never vanishes). So the inverse function theorem guarantees that $F$ is locally invertible.

b. Differentiate $F^{-1} \circ F = I$ using the chain rule:

$$[\mathbf{D}(F^{-1} \circ F)(\mathbf{a})] = [\mathbf{D}F^{-1}(F(\mathbf{a}))][\mathbf{D}F(\mathbf{a})] = [\mathbf{D}F^{-1}(\mathbf{b})][\mathbf{D}F(\mathbf{a})] = I,$$

so    $[\mathbf{D}F^{-1}(\mathbf{b})] = [\mathbf{D}F(\mathbf{a})]^{-1} = \dfrac{-1}{e^{a_1}(e^{a_2} + e^{-a_2})} \begin{bmatrix} -e^{-a_2} & -e^{a_2} \\ -e^{a_1} & e^{a_1} \end{bmatrix}.$

**2.1** a. Row reduction gives

$$\begin{bmatrix} 1 & 1 & -1 & a \\ 1 & 0 & 2 & b \\ 1 & a & 1 & b \end{bmatrix} \rightarrow \begin{bmatrix} 1 & 1 & -1 & a \\ 0 & -1 & 3 & b-a \\ 0 & a-1 & 2 & b-a \end{bmatrix} \rightarrow \begin{bmatrix} 1 & 0 & 2 & b \\ 0 & 1 & -3 & a-b \\ 0 & 0 & 3a-1 & a(b-a) \end{bmatrix}.$$

We will consider the cases $a = 1/3$ and $a \neq 1/3$. If $a = 1/3$, we get

$$\begin{bmatrix} 1 & 0 & 2 & b \\ 0 & 1 & -3 & \frac{1}{3}-b \\ 0 & 0 & 0 & \frac{1}{3}(b-\frac{1}{3}) \end{bmatrix},$$

If $a = 1/3$ and $b = 1/3$, there is no pivotal 1 in the 4th column or the 3rd column. If $a = 1/3$ and $b \neq 1/3$, then by further row reducing we can get a pivotal 1 in the 4th column.

so if $a = 1/3$ and $b = 1/3$, there are infinitely many solutions. If $a = 1/3$ and $b \neq 1/3$, then there are no solutions.

If $a \neq 1/3$, we can further row reduce our original matrix to get a pivotal 1 in the third column. In that case, the system of equations has a unique solution.

b. We have already done all the work: the matrix of coefficients (i.e., the matrix consisting of the first three columns) is invertible if and only if $a \neq 1/3$.

**2.3** a. The first statement is true; the others are false.

b. The false statements are true if $n = m$.

**2.5** a. Saying that a matrix is invertible is equivalent to saying that it can be row reduced to the identity. The matrix A is invertible, so $[A, \ C]$ can be reduced to $[I_n, \widetilde{C}]$; likewise, $[0, \ B]$ can be reduced to $[0, \ I_m]$, so

Solution 2.5, part b: If you did not assume that the inverse is upper triangular, you could argue that it is of the form $\begin{bmatrix} X & Y \\ Z & W \end{bmatrix}$, where $X$, like $A$, is $n \times n$, and $W$, like $B$, is $m \times m$. Then the multiplication

$$\begin{bmatrix} A & C \\ 0 & B \end{bmatrix}\begin{bmatrix} X & Y \\ Z & W \end{bmatrix} = \begin{bmatrix} I & 0 \\ 0 & I \end{bmatrix}$$

says that $BZ = 0$; since $B$ is invertible, this tells us that $Z = 0$. It also says that $BW = I$, i.e., that $W = B^{-1}$. So now we have

$$\begin{bmatrix} A & C \\ 0 & B \end{bmatrix}\begin{bmatrix} X & Y \\ 0 & B^{-1} \end{bmatrix}$$
$$= \begin{bmatrix} AX & AY+CB^{-1} \\ 0 & I \end{bmatrix},$$

where

$$AX = I \quad \text{and} \quad AY+CB^{-1} = 0,$$

giving $Y = -A^{-1}CB^{-1}.$

$$M = \begin{bmatrix} A & C \\ 0 & B \end{bmatrix} \text{ can be transformed into } \begin{bmatrix} I_n & \widetilde{C} \\ 0 & I_m \end{bmatrix}$$

by row operations. We can now add multiples of the last $m$ rows to the first $m$ to cancel the entries of $\widetilde{C}$, which row reduces the whole matrix to the identity.

b. If you guess that there is an inverse of the form $\begin{bmatrix} A^{-1} & Y \\ 0 & B^{-1} \end{bmatrix}$ and multiply out, you find

$$\begin{bmatrix} A^{-1} & Y \\ 0 & B^{-1} \end{bmatrix}\begin{bmatrix} A & C \\ 0 & B \end{bmatrix} \begin{bmatrix} I & A^{-1}C+YB \\ 0 & I \end{bmatrix}.$$

So if $A^{-1}C + YB = 0$, you will have found an inverse; this occurs when $Y = -A^{-1}CB^{-1}.$

69

**2.7** a. A couple of row operations will bring

$$\begin{bmatrix} 1 & -1 & -1 \\ 0 & a & 1 \\ 2 & a+2 & a+2 \end{bmatrix} \quad \text{to} \quad \begin{bmatrix} 1 & -1 & -1 \\ 0 & 1 & (a+3)/4 \\ 0 & 0 & (1-a)(a+4)/4 \end{bmatrix}.$$

Thus the matrix is invertible if and only if $(1-a)(a+4) \neq 0$, i.e., if $a \neq 1$ and $a \neq -4$.

b. This is an unpleasant row reduction of

$$\begin{bmatrix} 1 & -1 & -1 & 1 & 0 & 0 \\ 0 & a & 1 & 0 & 1 & 0 \\ 2 & a+2 & a+2 & 0 & 0 & 1 \end{bmatrix}.$$

The final result is

$$\frac{1}{(a-1)(a+4)} \begin{bmatrix} (a+2)(a+1) & 0 & a-1 \\ 2 & a+4 & -1 \\ -2a & -(a+4) & a \end{bmatrix}.$$

**2.9** a. The vectors $\begin{bmatrix} \cos\theta \\ \sin\theta \end{bmatrix}$ and $\begin{bmatrix} -\sin\theta \\ \cos\theta \end{bmatrix}$ are orthogonal since

$$\begin{bmatrix} \cos\theta \\ \sin\theta \end{bmatrix} \cdot \begin{bmatrix} -\sin\theta \\ \cos\theta \end{bmatrix} = 0$$

(corollary 1.4.8). By proposition 2.4.17, they are linearly independent, and two linearly independent vectors in $\mathbb{R}^2$ form a basis (corollary 2.4.18). The length of each basis vector is 1:

$$\sqrt{\cos^2\theta + \sin^2\theta} = 1 \quad \text{and} \quad \sqrt{\cos^2\theta + (-\sin\theta)^2} = 1,$$

so the basis is orthonormal, not just orthogonal.

b. The argument in part a applies.

Recall that an $n \times n$ matrix whose columns form an orthonormal basis of $\mathbb{R}^n$ is called an orthogonal matrix.

c. Let $A = [\vec{v}_1, \vec{v}_2]$ be any orthogonal matrix. Then $\vec{v}_1$ is a unit vector, so it can be written $\begin{bmatrix} \cos\theta \\ \sin\theta \end{bmatrix}$ for some angle $\theta$. The vector $\vec{v}_2$ must be orthogonal to $\vec{v}_1$ and of unit length. There are only two choices: $\begin{bmatrix} -\sin\theta \\ \cos\theta \end{bmatrix}$ and $\begin{bmatrix} \sin\theta \\ -\cos\theta \end{bmatrix}$. The first gives positive determinant and corresponds to rotation. Note that proposition 1.4.14 says that to have positive determinant,

We saw in example 1.3.9 that

$$\begin{bmatrix} \cos\theta & -\sin\theta \\ \sin\theta & \cos\theta \end{bmatrix}$$

corresponds to rotation by $\theta$ counterclockwise around the origin.

$\begin{bmatrix} \cos\theta \\ \sin\theta \end{bmatrix}$ must be clockwise from $\begin{bmatrix} -\sin\theta \\ \cos\theta \end{bmatrix}$, which it always will be.

The second choice gives a negative determinant and corresponds to reflection with respect to a line through the origin that forms angle $\theta/2$ with the $x$-axis (see example 1.3.5).

**2.11**   a.   Write $\begin{bmatrix} a & b \\ c & d \end{bmatrix}$ as $\begin{bmatrix} a \\ b \\ c \\ d \end{bmatrix}$; the question is whether the columns of

Here we write

$$I = \begin{bmatrix} 1 & 0 \\ 0 & 1 \end{bmatrix} \text{ as } \begin{bmatrix} 1 \\ 0 \\ 0 \\ 1 \end{bmatrix}$$

$$A = \begin{bmatrix} 1 & 2 \\ 2 & 1 \end{bmatrix} \text{ as } \begin{bmatrix} 1 \\ 2 \\ 2 \\ 1 \end{bmatrix},$$

$$A^2 = \begin{bmatrix} 5 & 4 \\ 4 & 5 \end{bmatrix} \text{ as } \begin{bmatrix} 5 \\ 4 \\ 4 \\ 5 \end{bmatrix},$$

and so on.

$$\begin{bmatrix} 1 & 1 & 5 & 13 \\ 0 & 2 & 4 & 14 \\ 0 & 2 & 4 & 14 \\ 1 & 1 & 5 & 13 \end{bmatrix}$$ are linearly independent. Since this matrix row reduces

to $\begin{bmatrix} 1 & 0 & 3 & 6 \\ 0 & 1 & 2 & 7 \\ 0 & 0 & 0 & 0 \\ 0 & 0 & 0 & 0 \end{bmatrix}$, it follows that $I$ and $A$ are linearly independent, and

$$A^2 = 2A + 3I, \quad A^3 = 7A + 6I.$$

The subspace spanned by $I$, $A$, $A^2$, $A^3$ has dimension 2.

  b. Suppose $B_1$ and $B_2$ satisfy $AB_1 = B_1 A$ and $AB_2 = B_2 A$. Then

$$(xB_1 + yB_2)A = xB_1 A + yB_2 A = xAB_1 + yAB_2 = A(xB_1 + yB_2).$$

So $W$ is a subspace of $\mathrm{Mat}\,(2,2)$. If you set $B = \begin{bmatrix} a & b \\ c & d \end{bmatrix}$ and write out $AB = BA$ as equations in $a, b, c, d$, you find

$$a + 2c = a + 2b \quad, \quad b + 2d = 2a + b \quad, \quad 2a + c = c + 2d \quad, \quad 2b + d = 2c + d.$$

These equations pretty obviously describe a 2-dimensional subspace of $\mathrm{Mat}\,(2,2)$, but to make it clear, write the equations

$$b - c = 0 \quad, \quad a - d = 0 \quad, \quad a - d = 0 \quad, \quad b - c = 0.$$

In other words, $W$ is the kernel of the matrix

$$\begin{bmatrix} 0 & 1 & -1 & 0 \\ 1 & 0 & 0 & -1 \\ 1 & 0 & 0 & -1 \\ 0 & 1 & -1 & 0 \end{bmatrix}. \quad \text{This matrix row reduces to} \quad \begin{bmatrix} 1 & 0 & 0 & -1 \\ 0 & 1 & -1 & 0 \\ 0 & 0 & 0 & 0 \\ 0 & 0 & 0 & 0 \end{bmatrix},$$

so $W$ has dimension 2, the number of nonpivotal columns.

  c. If $B = aI + bA$, then

$$BA = (aI + bA)A = aA + bA^2 = A(aI + bA) = AB,$$

so clearly $V \subset W$. But they have the same dimension, so they are equal.

Solution 2.13: The pivotal columns of $T$ form a basis for $\mathrm{Im}\,T$ (see theorem 2.5.4). By definition 2.2.6, a column of $T$ is pivotal if the corresponding column of the row-reduced matrix $\widetilde{T}$ contains a pivotal 1.

**2.13**   a.   The matrix

$$\begin{bmatrix} 1 & 1 & 3 & 6 & 2 \\ 2 & -1 & 0 & 4 & 1 \\ 4 & 1 & 6 & 16 & 5 \end{bmatrix} \quad \text{row reduces to} \quad \begin{bmatrix} 1 & 0 & 1 & 10/3 & 1 \\ 0 & 1 & 2 & 8/3 & 1 \\ 0 & 0 & 0 & 0 & 0 \end{bmatrix}.$$

The first two columns are a basis for the image, and the vectors

$$\begin{bmatrix} -1 \\ -2 \\ 1 \\ 0 \\ 0 \end{bmatrix}, \begin{bmatrix} -10/3 \\ -8/3 \\ 0 \\ 1 \\ 0 \end{bmatrix}, \begin{bmatrix} -1 \\ -1 \\ 0 \\ 0 \\ 1 \end{bmatrix}$$

form a basis for the kernel (see theorem 2.5.6). Indeed we find

$$\dim(\ker\ ) + \dim(\mathrm{img}) = 3 + 2 = 5.$$

b. The matrix

$$\begin{bmatrix} 2 & 1 & 3 & 6 & 2 \\ 2 & -1 & 0 & 4 & 1 \end{bmatrix} \quad \text{row reduces to} \quad \begin{bmatrix} 1 & 0 & 3/4 & 5/2 & 3/4 \\ 0 & 1 & 3/2 & 1 & 1/2 \end{bmatrix}.$$

If we had followed to the letter the procedure outlined in example 2.5.7, and had not "cleared denominators," we would have 1 instead of $-4$ in the first and third vector, and 1 instead of $-2$ in the second vector. The other nonzero entries would then be fractions.

So the image is all of $\mathbb{R}^2$, and a basis for the kernel (we cleared denominators) is

$$\begin{bmatrix} 3 \\ 6 \\ -4 \\ 0 \\ 0 \end{bmatrix}, \begin{bmatrix} 5 \\ 2 \\ 0 \\ -2 \\ 0 \end{bmatrix}, \begin{bmatrix} 3 \\ 2 \\ 0 \\ 0 \\ -4 \end{bmatrix}, \quad \text{which has } \dim = 3.$$

Again, we find $\dim(\ker\ ) + \dim(\mathrm{img}) = 3 + 2 = 5$.

**2.15** Let $P_{2k-1}$ be the space of polynomials of degree at most $2k - 1$; it can be identified with $\mathbb{R}^{2k}$, by identifying $c_0 + c_1 x + \cdots + c_{2k-1} x^{2k-1}$

with $\begin{bmatrix} c_0 \\ \vdots \\ c_{2k-1} \end{bmatrix}$. We will apply the dimension formula to the linear mapping

$$T(p) = \begin{bmatrix} p(1) \\ \vdots \\ p(k) \\ p'(1) \\ \vdots \\ p'(k) \end{bmatrix}$$

Map $T$ for solution 2.15

$T : P_{2k-1} \to \mathbb{R}^{2k}$ defined in the margin. The kernel of $T$ is 0: indeed, a polynomial in the kernel can be factored as

$$p(x) = q(x)(x-1)^2 \ldots (x-k)^2,$$

and such a polynomial has degree at least $2k$ unless $q = 0$. Since $p$ has degree at most $2k - 1$, it follows that $q$ must be 0, hence $p$ must be 0, so the kernel of $T$ is $\{0\}$. So from the dimension formula, the image of $T$ has dimension $2k$, hence is all of $\mathbb{R}^{2k}$. So there is a unique solution to the equations $p(1) = a_1, \ldots, p(k) = a_k$ and $p'(1) = b_1, \ldots, p'(k) = b_k$.

**2.17** Just compute away:

$$\begin{aligned} \big(T(af+bg)\big)(x) &= (x^2+1)(f+g)''(x) - x(f+g)'(x) + 2(f+g)(x) \\ &= (x^2+1)f''(x) + (x^2+1)g''(x) - xf'(x) - xg'(x) + 2f(x) + 2g(x) \\ &= (x^2+1)f''(x) - xf'(x) + 2f(x) + (x^2+1)g''(x) - xg'(x) + 2g(x) \\ &= \big(aT(f) + bT(g)\big)(x). \end{aligned}$$

**2.19** Suppose that $\mathbf{v}_1, \ldots, \mathbf{v}_n$ form a basis of $V$, and $\mathbf{w}_1, \ldots, \mathbf{w}_m$ are linearly independent. Then (by proposition 2.6.17), $\Phi_{\{\mathbf{v}\}}$ is one to one and onto, hence invertible, and $\Phi_{\{\mathbf{w}\}}$ is one to one, so $\Phi_{\{\mathbf{v}\}}^{-1} \circ \Phi_{\{\mathbf{w}\}} : \mathbb{R}^m \to \mathbb{R}^n$ is a one-to-one linear transformation, so the columns of its matrix are linearly independent. But these are $m$ vectors in $\mathbb{R}^n$, so $m \leq n$.

Similarly, if $\mathbf{v}_1, \ldots, \mathbf{v}_n$ form a basis of $V$, and $\mathbf{w}_1, \ldots, \mathbf{w}_m$ span $V$, then $\Phi_{\{\mathbf{v}\}}^{-1} \circ \Phi_{\{\mathbf{w}\}} : \mathbb{R}^m \to \mathbb{R}^n$ is onto, so the columns of its matrix are $m$ vectors that span $\mathbb{R}^n$. Then $m \geq n$.

**2.21** a. We need to show that if $\vec{v}_1$ and $\vec{v}_2$ are in $V \cap W$, then so is $a\vec{v}_1 + b\vec{v}_2$ for any $a, b \in \mathbb{R}$. But $a\vec{v}_1 + b\vec{v}_2$ is in $V$, since $V$ is a subspace, and in $W$ also for the same reason. So it is in $V \cap W$.

b. Suppose $\vec{v}$ belongs to $V$ but not to $W$, and that $\vec{w}$ belongs to $W$ but not to $V$. Then $\vec{v} + \vec{w}$ belongs to neither $V$ nor $W$. Indeed, if it belongs to $V$, then so does $\vec{v} + \vec{w} - \vec{v} = \vec{w}$, contrary to our hypothesis.

**2.23** It takes a bit of patience entering this problem into MATLAB, much helped by using the computer to compute $A^3$ symbolically. The answer ends up being

$$\begin{bmatrix} 2.08008382305190 & -0.00698499417571 & 0.08008382305190 \\ 0 & 1.91293118277239 & 0 \\ 0 & 0.17413763445522 & 2 \end{bmatrix}.$$

Note that one step of Newton's method gives

$$\begin{bmatrix} 2 + 1/12 & 0 & 1/12 \\ 0 & 2 + 1/12 & 0 \\ 0 & 1/6 & 2 \end{bmatrix} = \begin{bmatrix} 2.08\overline{3} & 0 & 0.08\overline{3} \\ 0 & 1.91\overline{6} & 0 \\ 0 & 0.1\overline{6} & 2 \end{bmatrix}$$

which is very close.

**2.25** a. Since $\left| \begin{bmatrix} 2x_1 & -1 \\ -1 & 2y_1 \end{bmatrix} - \begin{bmatrix} 2x_2 & -1 \\ -1 & 2y_2 \end{bmatrix} \right| = 2|\mathbf{x} - \mathbf{y}|$, we see that 2 is a Lipschitz constant for the derivative of $F$.

b. We have

$$\vec{h}_0 = -\frac{1}{23} \begin{bmatrix} 6 & 1 \\ 1 & 4 \end{bmatrix} \begin{bmatrix} 0 \\ 1 \end{bmatrix} = \begin{bmatrix} -1/23 \\ -4/23 \end{bmatrix}, \quad \text{so} \quad \mathbf{x}_1 = \begin{bmatrix} 2 - 1/23 \\ 3 - 4/23 \end{bmatrix}.$$

The length of the inverse of the derivative is $\sqrt{54}/23$, so

$$|f(\mathbf{x}_0)| \; \left|[\mathbf{D}f(\mathbf{x}_0)]^{-1}\right|^2 \; M = 1 \cdot \frac{54}{23^2} \cdot 2 = \frac{108}{529} < \frac{1}{2},$$

so Newton's method converges, to a point of the disc of radius

$$|\vec{h}_0| = \frac{\sqrt{17}}{23} \approx .2 \quad \text{around } \mathbf{x}_1.$$

c. The disc centered at $\begin{pmatrix} 45/23 \\ 65/23 \end{pmatrix}$ and with radius $\frac{1}{23}\sqrt{17}$ is guaranteed to contain a unique root; see the figure in the margin.

**2.27** First, note that

$$\|A\| = \sup_{|\vec{v}|=1} |A\vec{v}| = \sup_{|\vec{w}|=1,|\vec{v}|=1} |\vec{w}^\top| \|A\vec{v}| \geq \sup_{|\vec{w}|=1,|\vec{v}|=1} |\vec{w}^\top A\vec{v}|; \quad (1)$$

similarly,

$$\|A^\top\| = \sup_{|\vec{w}|=1} |A^\top\vec{w}| = \sup_{|\vec{v}|=1,|\vec{w}|=1} |\vec{v}^\top| \|A^\top\vec{w}| \geq \sup_{|\vec{w}|=1,|\vec{v}|=1} |\vec{v}^\top A^\top\vec{w}|. \quad (2)$$

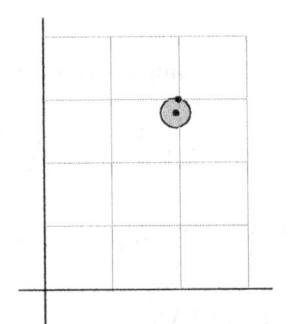

FIGURE FOR SOLUTION 2.25.

The shaded region, described in part c, is guaranteed to contain a unique root.

Note also that

$$|\vec{\mathbf{w}}^\top A \vec{\mathbf{v}}| = |((A\vec{\mathbf{v}})^\top \vec{\mathbf{w}})^\top| = |(A\vec{\mathbf{v}})^\top \vec{\mathbf{w}}| = |\vec{\mathbf{v}}^\top A^\top \vec{\mathbf{w}}|.$$

Since $(A\vec{\mathbf{v}})^\top \vec{\mathbf{w}}$ is a number, it equals its transpose.

Therefore, if we can show that the inequalities (1) and (2) are equalities for appropriate $\vec{\mathbf{v}}, \vec{\mathbf{w}}$, we will have shown that $\|A\| = \|A^\top\|$.

By definition,

$$\|A\| = \sup_{|\vec{\mathbf{v}}|=1} |A\vec{\mathbf{v}}|,$$

and since the set $\{\, \vec{\mathbf{v}} \in \mathbb{R}^n \mid |\vec{\mathbf{v}}| = 1 \,\}$ is compact, the maximum is realized: there exists a unit vector $\vec{\mathbf{v}}_0$ such that $|A\vec{\mathbf{v}}_0| = \|A\|$. Let $\vec{\mathbf{w}}_0 = \dfrac{A\vec{\mathbf{v}}_0}{|A\vec{\mathbf{v}}_0|}$. Then

$$\|A\| = |A\vec{\mathbf{v}}_0| = \underbrace{\vec{\mathbf{w}}_0^\top \vec{\mathbf{w}}_0}_{1} |A\vec{\mathbf{v}}_0| = \vec{\mathbf{w}}_0^\top \frac{A\vec{\mathbf{v}}_0}{|A\vec{\mathbf{v}}_0|} |A\vec{\mathbf{v}}_0| = \vec{\mathbf{w}}_0^\top A\vec{\mathbf{v}}_0 = |\vec{\mathbf{w}}_0^\top A\vec{\mathbf{v}}_0|.$$

Therefore,

$$\|A\| = \sup_{|\vec{\mathbf{w}}|=1, |\vec{\mathbf{v}}|=1} |\vec{\mathbf{w}}^\top A \vec{\mathbf{v}}|$$

and by a similar argument,

$$\|A^\top\| = \sup_{|\vec{\mathbf{w}}|=1, |\vec{\mathbf{v}}|=1} |\vec{\mathbf{v}}^\top A^\top \vec{\mathbf{w}}|.$$

It follows that $\|A\| = \|A^\top\|$.

**2.29**  a. False. Suppose $\sin(xyz) - z = 0$ expresses $x$ implicitly as $g\left(\dfrac{y}{z}\right)$

Since $F \circ G$ goes from $\mathbb{R}^3$ to $\mathbb{R}$, its derivative $[\mathbf{D}F \circ G(\mathbf{a})]$ is a $3 \times 1$ (row) matrix.

near $\mathbf{a} = \begin{pmatrix} \pi/2 \\ 1 \\ 1 \end{pmatrix}$. Set $F\begin{pmatrix} x \\ y \\ z \end{pmatrix} = \sin(xyz) - z$ and $G\begin{pmatrix} x \\ y \\ z \end{pmatrix} = \begin{pmatrix} g\left(\dfrac{y}{z}\right) \\ y \\ z \end{pmatrix}$.

Then $F \circ G = 0$, so, by the chain rule, $[\mathbf{D}F(G(\mathbf{a}))][\mathbf{D}G(\mathbf{a})] = [0 \quad 0 \quad 0]$. Writing this out in terms of matrices gives

$$\Big[\underbrace{1 \cdot 1 \cdot \cos \tfrac{\pi}{2}}_{D_1 F = yz\cos xyz} \quad \underbrace{\tfrac{\pi}{2} \cdot 1 \cdot \cos \tfrac{\pi}{2}}_{D_2 F = xz\cos xyz} \quad \underbrace{\tfrac{\pi}{2} \cos \tfrac{\pi}{2} - 1}_{D_3 F = xy\cos xyz - 1}\Big] \begin{bmatrix} 0 & D_1 g\left(\tfrac{1}{1}\right) & D_2 g\left(\tfrac{1}{1}\right) \\ 0 & 1 & 0 \\ 0 & 0 & 1 \end{bmatrix} = [0 \quad 0 \quad 0].$$

This is a contradiction, since the third entry of the product is $-1$.

We could not simply invoke the implicit function theorem in part a. Since $D_1 F(\mathbf{a}) = 0$, the theorem does not guarantee existence of a differentiable implicit function expressing $x$ and in terms of $y$ and $z$, but it does not guarantee nonexistence either.

b.   True.   Since the partial derivative with respect to $z$ at $\mathbf{a}$ is $D_3 F(\mathbf{a}) = -1$, which is invertible, the implicit function theorem guarantees that $F(\mathbf{x}) = 0$ expresses $z$ implicitly as a function $g$ of $x$ and $y$ near $\mathbf{a}$, and that it is differentiable, with derivative given by equation 2.10.30.

**2.31**   False.  This is a problem most easily solved using the chain rule, not the inverse function theorem. Call $S$ the map $S(A) = A^2$. If such a mapping $g$ exists, then $g$ is an inverse function of $S$, so $S \circ g = I$, and since

$g$ is differentiable, we can apply the chain rule:

$$\overbrace{\left[\mathbf{D}S\left(g\left(\begin{bmatrix} -3 & 0 \\ 0 & -3 \end{bmatrix}\right)\right)\right]}^{\left[\mathbf{D}S\left(\begin{bmatrix} 1 & 2 \\ -2 & -1 \end{bmatrix}\right)\right]}\left[\mathbf{D}g\left(\begin{bmatrix} -3 & 0 \\ 0 & -3 \end{bmatrix}\right)\right] = \left[\mathbf{D}(S \circ g)\left(\begin{bmatrix} -3 & 0 \\ 0 & -3 \end{bmatrix}\right)\right]$$

$$= \left[\mathbf{D}I\left(\begin{bmatrix} -3 & 0 \\ 0 & -3 \end{bmatrix}\right)\right] = I.$$

The derivative of the identity is the identity everywhere; this uses theorem 1.8.1, part 2.

Thus if the desired function $g$ exists, then $\left[\mathbf{D}S\left(\begin{bmatrix} 1 & 2 \\ -2 & -1 \end{bmatrix}\right)\right]$ is invertible; if the derivative is not invertible, either $g$ does not exist, or it exists but is not differentiable. From example 1.7.18, we know that $[\mathbf{D}S(A)]H = AH + HA$, so

$$\left[\mathbf{D}S\left(\begin{bmatrix} 1 & 2 \\ -2 & -1 \end{bmatrix}\right)\right]B = \begin{bmatrix} 1 & 2 \\ -2 & -1 \end{bmatrix}B + B\begin{bmatrix} 1 & 2 \\ -2 & -1 \end{bmatrix}.$$

Written out in coordinates, setting $B = \begin{bmatrix} a & b \\ c & d \end{bmatrix}$, this gives

If you want to find out whether a map $\mathbf{f}$ is locally invertible, the first thing to do is to see whether the derivative $[\mathbf{D}f]$ (here, $[\mathbf{D}S]$) is invertible. If it is not, then the chain rule tells you that there are two possibilities. Either there is no implicit function (the usual case), or there is one and it is not differentiable.

$$\begin{bmatrix} 1 & 2 \\ -2 & -1 \end{bmatrix}\begin{bmatrix} a & b \\ c & d \end{bmatrix} + \begin{bmatrix} a & b \\ c & d \end{bmatrix}\begin{bmatrix} 1 & 2 \\ -2 & -1 \end{bmatrix} = 2\begin{bmatrix} a - b + c & a + d \\ -a - d & -b + c - d \end{bmatrix}.$$

The transformation

If $[\mathbf{D}f]$ is invertible, the next step is to show that $\mathbf{f}$ is continuously differentiable, so that the inverse function theorem applies. In this case, it is an easy consequence of theorem 1.8.1 (rules for computing derivatives) that the squaring function $S$ is continuously differentiable.

$$\begin{bmatrix} a & b \\ c & d \end{bmatrix} \mapsto \begin{bmatrix} a - b + c & a + d \\ -a - d & -b + c - d \end{bmatrix} \tag{1}$$

is not invertible, since the entries are not linearly independent.[2] Therefore $g$ does not exist.

Alternatively, you could show that the derivative $\left[\mathbf{D}S\left(\begin{bmatrix} 1 & 2 \\ -2 & -1 \end{bmatrix}\right)\right]$ is not invertible by noticing that

$$\begin{bmatrix} 1 & 2 \\ -2 & -1 \end{bmatrix}\begin{bmatrix} 1 & 1 \\ 0 & -1 \end{bmatrix} + \begin{bmatrix} 1 & 1 \\ 0 & -1 \end{bmatrix}\begin{bmatrix} 1 & 2 \\ -2 & -1 \end{bmatrix} = 0,$$

so that

$$\begin{bmatrix} 1 & 1 \\ 0 & -1 \end{bmatrix} \in \ker\left[\mathbf{D}f\left(\begin{bmatrix} 1 & 2 \\ -2 & -1 \end{bmatrix}\right)\right].$$

---

[2]Note that the matrix of the linear transformation (1) is *not* the matrix $\begin{bmatrix} a - b + c & a + d \\ -a - d & -b + c - d \end{bmatrix}$; it is $\begin{bmatrix} 1 & -1 & 1 & 0 \\ 1 & 0 & 0 & 1 \\ -1 & 0 & 0 & -1 \\ 0 & -1 & 1 & -1 \end{bmatrix}$, since

$$\begin{bmatrix} 1 & -1 & 1 & 0 \\ 1 & 0 & 0 & 1 \\ -1 & 0 & 0 & -1 \\ 0 & -1 & 1 & -1 \end{bmatrix}\begin{bmatrix} a \\ b \\ c \\ d \end{bmatrix} = \begin{bmatrix} a - b + c \\ a + d \\ -a - d \\ -b + c - d \end{bmatrix}.$$

This matrix is clearly noninvertible, since the third column is a multiple of the second column.

**2.33** If the matrix $Df(x_n)$ is not invertible, one can pick any invertible matrix $A$, and use it instead of $[\mathbf{D}f(x_n)]^{-1}$ at the troublesome iterate $x_n$:

$$x_{n+1} = x_n - Af(x_n).$$

This procedure cannot produce a false root $x^*$, since then we would have

$$x^* = x^* - Af(x^*),$$

which means that $Af(x^*) = 0$. The requirement that $A$ be invertible then implies that $f(x^*) = 0$, so $x^*$ is indeed a true root. Of course, we can not longer hope to have superconvergence, but the fact that our algorithm doesn't break down is probably worth the sacrifice.

**2.35** a. Suppose first that $p_1$ and $p_2$ have no common factors, and that

$$p_1(x)q_1(x) + p_2(x)q_2(x) = 0;$$

we need to show that then $q_1$ and $q_2$ are the zero polynomial.

Write

$$p_1(x) = (x - a_1) \ldots (x - a_{k_1}) \quad \text{and} \quad p_2(x) = (x - b_1) \ldots (x - b_{k_2}).$$

By the fundamental theorem of algebra (theorem 1.6.13), this factoring is possible, but the roots may be complex.

Since

$$p_1(x)q_1(x) + p_2(x)q_2(x) = 0 \text{ and } p_1(b_i) \neq 0 \quad \text{for all } i = 1, \ldots, k_2,$$

we see that $q_1(b_1) = \cdots = q_1(b_{k_2}) = 0$, and that $q_2(a_1) = \cdots = q_2(a_{k_1}) = 0$.

If all the $a_i$ are distinct, this shows that $q_2$ is a polynomial of degree $< k_1$ that vanishes at $k_1$ points; so it must be the zero polynomial. The same argument applies to $q_1$ if the $b_i$ are distinct.

Actually, this still holds if the roots have multiplicities. Suppose that $q_2$ is not the zero polynomial. In the equation

$$(x - a_1) \ldots (x - a_{k_1})q_1(x) + (x - b_1) \ldots (x - b_{k_2})q_2(x) = 0,$$

cancel all the powers of $(x - a_i)$ that appear in both $p_1(x)$ and $q_2(x)$. There will be some such term $(x - a_j)$ left from $p_1(x)$ after all cancellations, since $p_1$ has higher degree than $q_2$, so if after cancellation we evaluate what is left at $a_j$ the first summand will give 0 and the second will not. This is a contradiction and shows that $q_2 = 0$. The same argument shows that $q_1 = 0$.

This proves the "if" part of a. For the "only if" part, suppose that

$$p_1(x) = (x - c)\tilde{p}_1(x) \quad \text{and} \quad p_2(x) = (x - c)\tilde{p}_2(x),$$

i.e., that $p_1$ and $p_2$ have the common factor $(x - c)$. Then

$$p_1(x)\tilde{p}_2(x) - p_2(x)\tilde{p}_1(x) = (x - c)(\tilde{p}_1(x)\tilde{p}_2(x) - \tilde{p}_2(x)\tilde{p}_1(x)) = 0,$$

so $q_1 = \tilde{p}_2$ and $q_2 = -\tilde{p}_1$ provide a nonzero element of the kernel of $T$.

b. If $p_1$ and $p_2$ are relatively prime, then $T$ is an injective (one to one) linear transformation between spaces of the same dimension, hence by the dimension formula (theorem 2.5.8) it is onto. In particular, the polynomial 1 is in the image. But if $p_1$ and $p_2$ have the common factor $(x - c)$, i.e.,

$$p_1(x) = (x - c)\tilde{p}_1(x) \quad \text{and} \quad p_2(x) = (x - c)\tilde{p}_2(x),$$

then any polynomial of the form $p_1(x)q_1(x) + p_2(x)q_2(x)$ will vanish at $c$, and hence will not be a nonzero constant.

**3.1.1** The derivative of $F\begin{pmatrix} x \\ y \end{pmatrix} = x^2 + y^2 + z^2 - 1$ is $[2x \quad 2y \quad 2z]$, which is $[0 \quad 0 \quad 0]$ only at the origin, which does not satisfy $F\begin{pmatrix} x \\ y \\ z \end{pmatrix} = 0$. So the unit sphere given by $x^2 + y^2 + z^2 = 1$ is a smooth surface.

**3.1.3** If the straight line is not vertical, it is the graph of the function $f : \mathbb{R} \to \mathbb{R}$ given by $f(x) = a + bx$, i.e., $y = a + bx$. Since $b$ is a constant, every value of $x$ gives one and only one value of $y$. If the straight line is vertical, it is the graph of the function $g : \mathbb{R} \to \mathbb{R}$ given by $g(y) = a$, i.e., $x = a$; every value of $y$ gives one and only one value of $x$.

**3.1.5** a. The derivative of $x^2 + y^3$ is $[2x, \; 3y^2]$. Since this is a transformation $\mathbb{R}^2 \to \mathbb{R}$, the only way it can fail to be onto is for it to vanish, which happens only if $x = y = 0$. This point is on $X_0$, so all $X_c$ are smooth curves for $c \neq 0$. But $X_0$ is not a smooth curve near the origin. The equation $x^2 + y^3 = 0$ can be "solved" for $x$: $x = \pm\sqrt{-y^3}$, and this is not a function near $y = 0$, for instance it isn't defined for $y > 0$. It can also be solved for $y$: $y = x^{2/3}$. This is a function, but it is not differentiable at 0. Thus $X_0$ is not a smooth curve.

b. These curves are shown in the figure at left.

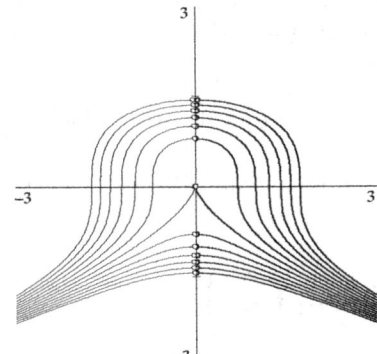

FIGURE FOR SOLUTION 3.1.5.

The curves $X_c$ for

$c = -3, -2.5, \ldots, 3.$

Note that $X_0$ has a cusp at the origin, which is why it is not a smooth curve.

**3.1.7** a. Both $X_a$ and $Y_b$ are graphs of smooth functions representing $z$ as functions of $\begin{pmatrix} x \\ y \end{pmatrix}$, so they are smooth surfaces.

b. The intersection $X_a \cap Y_b$ is a smooth curve if the derivative of

$$F\begin{pmatrix} x \\ y \\ z \end{pmatrix} = \begin{pmatrix} x^2 + y^3 + z - a \\ x + y + z - b \end{pmatrix}$$

is onto (i.e., has rank 2) at every point of $X_a \cap Y_b$. The derivative is

$$[\mathbf{DF}] = \begin{bmatrix} 2x & 3y^2 & 1 \\ 1 & 1 & 1 \end{bmatrix},$$

and the only way it can have rank $< 2$ is if all three columns are $\begin{bmatrix} 1 \\ 1 \end{bmatrix}$, i.e., if $2x = 3y^2 = 1$. So all the intersections $X_a \cap Y_b$ are smooth curves except the ones that go through the two vertical lines $x = 1/2$, $y = \pm 1/\sqrt{3}$. The values of $a$ and $b$ for the surfaces $X_a$ and $Y_b$ that contain such a point

$$\begin{pmatrix} 1/2 \\ \pm 1\sqrt{3} \\ z \end{pmatrix} \quad \text{are} \quad \begin{aligned} a &= z + \frac{1}{4} \pm \left(\frac{1}{\sqrt{3}}\right)^3 \\ b &= z + \frac{1}{2} \pm \left(\frac{1}{\sqrt{3}}\right) \end{aligned},$$

which occurs precisely when (if and only if) $a - b = -1/4 \pm 2/(3\sqrt{3})$.

"Precisely when" means "if and only if".

At the points $\begin{pmatrix} 1/2 \\ +1\sqrt{3} \\ z \end{pmatrix}$, $\begin{pmatrix} 1/2 \\ -1\sqrt{3} \\ z \end{pmatrix}$, the plane $Y_b$ is the tangent plane to the surface $X_a$.

**3.1.9**  a. We have

$$x^4 + x^2 + y^4 - y^2 = \left(x^2 + \frac{1}{2}\right)^2 + \left(y^2 - \frac{1}{2}\right)^2 - \frac{1}{2}.$$

So we can take

$$p(x) = x^2 + \frac{1}{2} \quad \text{and} \quad q(y) = y^2 - \frac{1}{2}.$$

b. The graphs of $p$, $p^2$, $q$, and $q^2$ are shown at top and middle of the figure in the margin. The graph of the function $F$ is shown at the bottom. You can see that the slices at $x$ constant look like the graph of $q^2$, raised or lowered, and similarly, slices at $y$ constant look like the graph of $p^2$, raised or lowered, all with a single minimum. (Looking ahead to section 3.6, we see that the critical point $\begin{pmatrix} 0 \\ 0 \end{pmatrix}$ is a saddle point, where $p$ goes up and $q$ goes down. The other critical points are $x = 0$, $y = \pm 1/\sqrt{2}$, which are minima.)

If you slice this graph parallel to the $(x, y)$-plane, you will get the curves of figure 3.1.10 in the text.

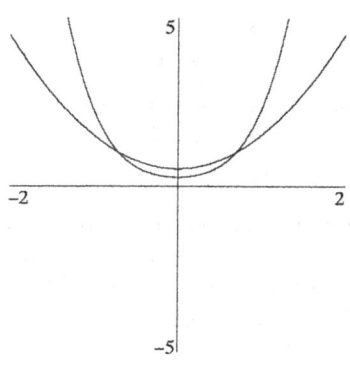

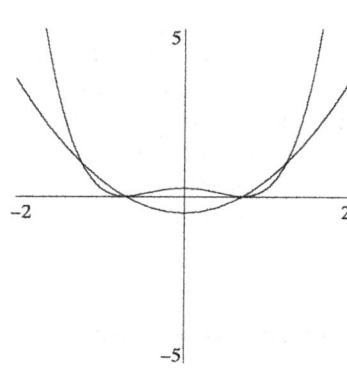

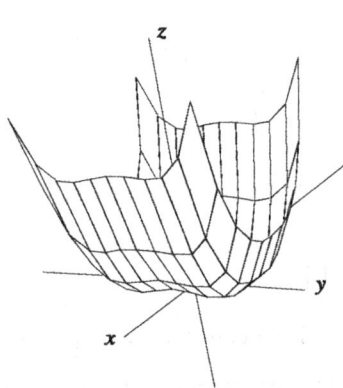

FIGURE FOR SOLUTION 3.1.9, part b. TOP: The graphs of $p$ and $p^2$. MIDDLE: The graphs of $q$ and $q^2$. BOTTOM: Graph of $F$.

**3.1.11**  a. The union is parametrized by $\begin{pmatrix} s \\ t \end{pmatrix} \mapsto \begin{pmatrix} st \\ st^2 \\ st^3 \end{pmatrix}$.

b. This means eliminating $s$ and $t$ from the equations $x = st$, $y = st^2$, and $z = st^3$.

In this case, it is easy to see that $t = y/x$, $s = x^2/y$, so that an equation for $X$ is

$$x = \frac{y^2}{x} \quad \text{or} \quad xz = y^2.$$

The set defined by this equation contains two lines that are not in $X$: the $x$-axis and the $z$-axis.

c. The derivative of the function $f \begin{pmatrix} x \\ y \\ z \end{pmatrix} = xz - y^2$ is

$$[\mathbf{D}f(v)] = [z, \; -2y, \; x],$$

which evidently never vanishes except at the origin.

d. Just substitute into the equation $xz = y^2$:

$$r(1 + \sin\theta)r(1 - \sin\theta) = r^2(\cos\theta)^2.$$

The figure below shows the two parametrizations. The parametrization of part a, shown to the right, doesn't look like much. The left side shows the parametrization of part d; it certainly looks like a cone of revolution. Indeed, if you switch to the coordinates $x = u + v$, $z = u - v$, the equation becomes $u^2 - v^2 = y^2$. In the form $u^2 = y^2 + v^2$, you should recognize it as a cone of revolution around the $u$-axis.

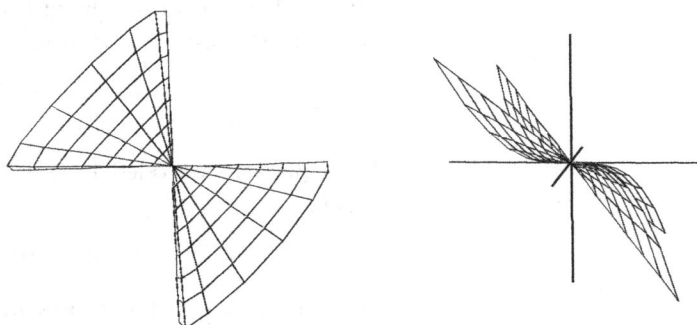

FIGURE FOR SOLUTION 3.1.11. LEFT: The parametrization of part d. RIGHT: That of part a.

e. The set of noninvertible symmetric matrices $\begin{bmatrix} a & b \\ b & d \end{bmatrix}$ has equation $ad - b^2 = 0$, so it is the same cone.

**3.1.13** a. The equation is $(\mathbf{x} - \mathbf{a}) \cdot \vec{\mathbf{v}} = 0$.

b. The equation of $P_t$ is

$$\left[ \begin{pmatrix} x \\ y \\ z \end{pmatrix} - \begin{pmatrix} t \\ t^2 \\ t^3 \end{pmatrix} \right] \cdot \begin{bmatrix} 1 \\ 2t \\ 3t^2 \end{bmatrix} = 0.$$

c. Writing the above equations in full, we find that the equations of $P_{t_1} \cap P_{t_2}$ are

$$x + 2t_1 y + 3t_1^2 z = t_1 + 2t_1^3 + 3t_1^5$$
$$x + 2t_2 y + 3t_2^2 z = t_2 + 2t_2^3 + 3t_2^5.$$

Two planes $x + a_1 y + b_1 z = c_1$ and $x + a_2 y + b_2 z = c_2$ fail to intersect if they coincide or are parallel. They coincide if the equations are the same; they are parallel if $x + a_1 y + b_1 z$ is a multiple of $x + a_2 y + b_2 z$. In either case, the matrix of coefficients

$$\begin{bmatrix} 1 & a_1 & b_1 \\ 1 & a_2 & b_2 \end{bmatrix}$$

has row rank 1, so by proposition 2.5.11 it has rank 1.

To be sure that the planes intersect in a line, we must check that the matrix of coefficients

$$\begin{bmatrix} 1 & 2t_1 & 3t_1^2 \\ 1 & 2t_2 & 3t_2^2 \end{bmatrix}$$

has exactly two pivotal columns (i.e., rank 2). But if $t_1 \neq t_2$, the first two columns are linearly independent, so that is true.

d. The intersection $P_1 \cap P_{1+h}$ has equations

$$x + 2y + 3z = 6$$
$$x + 2(1 + h)y + 3(1 + h)^2 z = (1 + h) + 2(1 + h)^3 + 3(1 + h)^5$$
$$= 6 + 22h + 36h^2 + 32h^3 + 15h^4 + 3h^5.$$

To solve this system, we row reduce the matrix

$$\begin{bmatrix} 1 & 2 & 3 & 6 \\ 1 & 2+2h & 3+6h+3h^2 & 6+22h+36h^2+32h^3+15h^4+3h^5 \end{bmatrix}.$$

Note that when row reducing this matrix, we can get rid of one power of $h$ (subtract the second row from the first and divide by $h$). This leads to

$$\begin{bmatrix} 1 & 0 & -3-3h & -16-36h-32h^2-15h^3-3h^4 \\ 0 & 1 & 3+\frac{3}{2}h & 11+18h+16h^2+\frac{15}{2}h^3+\frac{3}{2}h^4 \end{bmatrix}.$$

The equations encoded by this matrix have a perfectly good limit when $h \to 0$:

$$x = 3z - 16 \quad \text{and} \quad y = -3z + 11,$$

which are a parametric representation of the required limiting line.

**3.1.15**  Clearly if $l_1 = l_2 + l_3 + l_4$, the only positions are those where the vertices are aligned. In that case, the position of $\mathbf{x}_1$ and the polar angle of the linkage determine the position in $X_2$; thus $X_2$ is a 3-dimensional manifold, which looks like $\mathbb{R}^2 \times S^1$. Similarly, $X_3$ is a 5-dimensional manifold, which looks like $\mathbb{R}^3 \times S^2$.

**3.1.17**  Two circles with radii $r_1$ and $r_2$, whose centers are a distance $d$ apart, intersect at two points exactly if

$$|r_1 - r_2| < d, \quad |r_1 + r_2| > d.$$

If one of these inequalities is satisfied and the other is an equality, the circles are tangent and intersect in one point.

The point $\mathbf{x}_2$ is on the circle of radius $l_1$ around $\mathbf{x}_1$ and on the circle of radius $l_2$ around $\mathbf{x}_3$. These two circles therefore intersect at two points precisely if

$$|l_1 - l_2| < |\mathbf{x}_1 - \mathbf{x}_3|, \quad |l_1 + l_2| > |\mathbf{x}_1 - \mathbf{x}_3|.$$

Similarly, the point $\mathbf{x}_4$ is on the circle of radius $l_3$ around $\mathbf{x}_3$ and the circle of radius $l_4$ around $\mathbf{x}_1$. These two circles intersect in two points precisely if

$$|l_3 - l_4| < |\mathbf{x}_1 - \mathbf{x}_3|, \quad |l_3 + l_4| > |\mathbf{x}_1 - \mathbf{x}_3|.$$

Under these conditions, there are two choices for $\mathbf{x}_2$, and two choices for $\mathbf{x}_4$, leading to four positions in all.

**3.1.19**  a. The space of $2 \times 2$ matrices of rank 1 is precisely the set of matrices $A = \begin{bmatrix} a & b \\ c & d \end{bmatrix}$ such that $\det A = ad - bc = 0$, but $A \neq \begin{bmatrix} 0 & 0 \\ 0 & 0 \end{bmatrix}$. Compute the derivative:

$$\left[ \mathbf{D} \det \left( \begin{bmatrix} a & b \\ c & d \end{bmatrix} \right) \right] = \begin{bmatrix} d & -c & -b & a \end{bmatrix},$$

so that the derivative is nonzero on $M_1(2,2)$.

b. This is similar but longer. The space $M_2(3, 3)$ is the space of $3 \times 3$ matrices whose determinant vanishes, but which have two linearly independent columns. If two vectors

$$\vec{\mathbf{a}} = \begin{bmatrix} a_1 \\ a_2 \\ a_3 \end{bmatrix} \quad \text{and} \quad \vec{\mathbf{b}} = \begin{bmatrix} b_1 \\ b_2 \\ b_3 \end{bmatrix},$$

are linearly independent, then $\vec{\mathbf{a}} \times \vec{\mathbf{b}} \neq \vec{\mathbf{0}}$ (see exercise 1.4.13). Now we have

$$\det \begin{bmatrix} a_1 & b_1 & c_1 \\ a_2 & b_2 & c_2 \\ a_3 & b_3 & c_3 \end{bmatrix} = a_1 b_2 c_3 - a_1 c_2 b_3 - b_1 a_2 c_3 + b_1 c_2 a_3 + c_1 a_2 b_3 - c_1 b_2 a_3,$$

so taking the variables in the order $a_1, a_2, \ldots, c_3$, the derivative of the determinant is

$$[b_2 c_3 - c_3 b_2, \ -b_1 c_3 + c_1 b_3, \ b_1 c_2 - c_1 b_2, \ \ldots, \ a_1 b_2 - b_1 a_2],$$

precisely the determinants of all the $2 \times 2$ submatrices you can make from the $3 \times 3$ matrix, or alternatively, precisely the coordinates of all the cross products of the columns. We just saw that at least one coordinate of such a cross product must be nonzero at a point in $M_2(3, 3)$.

**3.1.21** a. Yes, the theorem tells us that they are smooth surfaces. We define $f \begin{pmatrix} x \\ y \\ z \end{pmatrix} = e^x + 2e^y + 3e^z - a$, so that $X_a$ is defined by $f \begin{pmatrix} x \\ y \\ z \end{pmatrix} = 0$. This has solutions only for $a > 0$; for $a \leq 0$, the set $X_a$ is the empty set and hence is a smooth surface by our definition. If $a > 0$, we have to check that the derivative is onto at all points:

$$\left[ \mathbf{D}f \begin{pmatrix} x \\ y \\ z \end{pmatrix} \right] = [e^x \ 2e^y \ 3e^z] \neq [0 \ 0 \ 0] \quad \text{for all} \quad \begin{pmatrix} x \\ y \\ z \end{pmatrix} \in \mathbb{R}^3.$$

Thus $X_a$ is a smooth surface.

b. No, theorem 3.1.10 does not guarantee that they are smooth curves. We define $F : \mathbb{R}^3 \to \mathbb{R}^2$ as

Solution 3.1.21, part b: Recall that we cannot use theorem 3.1.10 to determine that a locus is *not* a smooth manifold; a locus defined by $\mathbf{F}(\mathbf{z}) = 0$ may be a smooth manifold even though $[\mathbf{D}F(\mathbf{z})]$ is not onto. But in this case, at these values of $a$ and $b$ the locus $X_{a,b}$ is a single point, which is not a smooth curve.

$$F \begin{pmatrix} x \\ y \\ z \end{pmatrix} = \begin{pmatrix} e^x + 2e^y + 3e^z - a \\ x + y + z - b \end{pmatrix},$$

so that the subsets in question are given by $F(\mathbf{x}) = 0$. Then

$$[\mathbf{D}F(\mathbf{x})] = \begin{bmatrix} e^x & 2e^y & 3e^z \\ 1 & 1 & 1 \end{bmatrix},$$

which has a single pivotal column if and only if $e^x = 2e^y = 3e^z$, i.e.,

$x = y + \ln(2) = z + \ln(3)$. A triple $\begin{pmatrix} x \\ x - \ln(2) \\ x - \ln(3) \end{pmatrix}$ satisfies both equations

of $F(\mathbf{x}) = 0$ when

$$e^x = a/3 \text{ and } 3x - \ln(2) - \ln(3) = b, \quad \text{i.e.,}$$

$$x = \ln(a/3) \quad \text{and} \quad x = \frac{b + \ln(6)}{3}.$$

Setting these two equations equal, we find that the subsets $X_{a,b}$ may not be smooth curves when $a = 3e^{(b+\ln(6))/3}$.

**3.1.23** Manifolds are invariant under rotations (see corollary 3.1.17), so we may assume that our linkage is on the $x$-axis, say at points

$$\mathbf{a}_1 \stackrel{\text{def}}{=} \begin{pmatrix} a_1 \\ 0 \end{pmatrix}, \quad \mathbf{a}_2 \stackrel{\text{def}}{=} \begin{pmatrix} a_2 \\ 0 \end{pmatrix}, \quad \mathbf{a}_3 \stackrel{\text{def}}{=} \begin{pmatrix} a_3 \\ 0 \end{pmatrix}, \quad \mathbf{a}_4 \stackrel{\text{def}}{=} \begin{pmatrix} a_4 \\ 0 \end{pmatrix},$$

with $a_1 < a_2 < a_4 < a_3$ and

$$a_2 - a_1 = l_1, \; a_3 - a_2 = l_2, \; a_4 - a_3 = l_3, \; a_1 - a_4 = l_4.$$

Recall that $X_2 \subset (\mathbb{R}^2)^4$ is the set defined by the equations

$$|\mathbf{x}_1 - \mathbf{x}_2| = l_1, \quad |\mathbf{x}_2 - \mathbf{x}_3| = l_2, \quad |\mathbf{x}_3 - \mathbf{x}_4| = l_3, \quad |\mathbf{x}_4 - \mathbf{x}_1| = l_4.$$

We will denote by $A \in X_2$ the position where $\mathbf{x}_i = \mathbf{a}_i$, for $i = 1 \dots 4$. Let $X_2^\epsilon \subset X_2$ be the subset where $|\mathbf{x}_i - \mathbf{a}_i| < \epsilon$, for $i = 1, 2, 3, 4$. We will call our variables $x_1, \dots, x_4, y_1, \dots, y_4$, where $\mathbf{x}_i = \begin{pmatrix} x_i \\ y_i \end{pmatrix}$.

The question is whether there exists $\epsilon > 0$ such that $X_2^\epsilon$ is the graph of a $C^1$ map representing four of the eight variables in terms of the other four. We will show that this is not the case: indeed, $X_2^\epsilon$ is not the graph of *any* map ($C^1$ or otherwise) representing four of the eight variables in terms of the other four.

There are many (8 choose 4, i.e., 70) possible choices of candidate "domain variables" (the active variables), which we will call *candidate coordinates*; they will be *coordinates* if the other variables are indeed functions of these. Let us make some preliminary simplifications, which will eliminate many of the candidates. Clearly $y_1$, $y_2$, $y_3$, $y_4$ are not a possible set of coordinates, since we can add any fixed constant to $x_1, x_2, x_3, x_4$ and stay in $X_2$. Also, no two linked $x_i$ can belong to any set of possible coordinates, since $|x_{i+1} - x_i| \leq l_i$. Any choice of three $x_i$-variables will contain a pair of linked ones, so any set of coordinates must include either two $x$-variables and two $y$-variables, or one $x$-variable and three $y$-variables.

The two cases are different. If our candidate coordinates consist of two $x$-variables (easily seen to be necessarily $x_2$ and $x_4$), and two $y$-variables, then in every neighborhood of the position $A$, there exist a position $X$ where the two coordinate $y$-variables are 0, and the two non-coordinate $y$-variables are nonzero. Indeed, as shown by the figure in the margin, if you can choose $x_2$ and $x_4$ so that $|x_2 - x_4| > l_2 - l_3$, and then the linkage will not lie on a line. If

$$X = \left( \begin{pmatrix} x_1 \\ y_1 \end{pmatrix}, \begin{pmatrix} x_2 \\ y_2 \end{pmatrix}, \begin{pmatrix} x_3 \\ y_3 \end{pmatrix}, \begin{pmatrix} x_4 \\ y_4 \end{pmatrix} \right) \in X_2^\epsilon$$

then

$$X' = \left( \begin{pmatrix} x_1 \\ -y_1 \end{pmatrix}, \begin{pmatrix} x_2 \\ -y_2 \end{pmatrix}, \begin{pmatrix} x_3 \\ -y_3 \end{pmatrix}, \begin{pmatrix} x_4 \\ -y_4 \end{pmatrix} \right) \in X_2^\epsilon$$

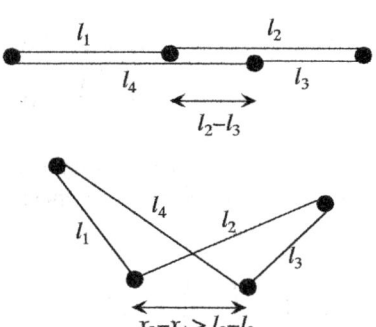

FIGURE FOR SOLUTION 3.1.23.

The case where the candidate coordinates consist of $x_2$ and $x_4$ and two $y$-variables.

We can choose the values of the coordinate functions to be whatever we like; in particular, they can be 0. At the base position $A$, all the $y_i$ vanish.

also and has the same values for the candidate coordinates. Thus these candidate coordinates fail: their values do not determine the other four variables.

In the case where the candidate consists of one $x$-coordinate and three $y$-coordinates, two $y$-coordinates must belong to non-linked points, say $y_1$ and $y_3$ ($y_2$ and $y_4$ would do just as well), and without loss of generality, we may assume that the third coordinate $y$-variable is $y_2$. There is then a position $X$ where $y_1 = y_3 = 0$, $y_2 \neq 0$; it then follows that $y_4 \neq 0$ also. If

$$X = \left( \begin{pmatrix} x_1 \\ 0 \end{pmatrix}, \begin{pmatrix} x_2 \\ y_2 \end{pmatrix}, \begin{pmatrix} x_3 \\ 0 \end{pmatrix}, \begin{pmatrix} x_4 \\ y_4 \end{pmatrix} \right) \in X_2^\epsilon$$

then

$$X' = \left( \begin{pmatrix} x_1 \\ 0 \end{pmatrix}, \begin{pmatrix} x_2 \\ y_2 \end{pmatrix}, \begin{pmatrix} x_3 \\ 0 \end{pmatrix}, \begin{pmatrix} x_4 \\ -y_4 \end{pmatrix} \right) \in X_2^\epsilon$$

also and has the same values for the candidate coordinates. Thus the candidate $y_1$, $y_2$, $y_3$, and $x_j$ are not coordinates for any $j$, as they do not determine $y_4$.

The upshot of all this is that there is no successful candidate for coordinate variables, so $X_2$ is not a manifold near $A$.

**3.1.25** a. Set $F \begin{pmatrix} x \\ y \\ z \end{pmatrix} = z - (x - \sin z)(\sin z - x + y)$. We need to check that the equation $F \left( \mathbf{g} \begin{pmatrix} u \\ v \end{pmatrix} \right) = 0$ is satisfied for all $\begin{pmatrix} u \\ v \end{pmatrix} \in \mathbb{R}^2$. This is straightforward:

$$F \left( \mathbf{g} \begin{pmatrix} u \\ v \end{pmatrix} \right) = uv - (\sin uv + u - \sin uv)(\sin uv - (\sin uv + u) + (u + v))$$

$$= uv - (u)(v) = 0.$$

b. This requires showing that $[\mathbf{D}F(\mathbf{x})] \neq [0, 0, 0]$ for all $\mathbf{x} = \begin{pmatrix} x \\ y \\ z \end{pmatrix} \in S$; it is actually true for all $\mathbf{x} \in \mathbb{R}^3$. Indeed, using the product rule, we compute

$$\left[ \mathbf{D}F \left( \begin{pmatrix} x \\ y \\ z \end{pmatrix} \right) \right] = [\underbrace{-(\sin z - x + y) + (x - \sin z)}_{D_1 F(\mathbf{x})} \quad \underbrace{-(x - \sin z)}_{D_2 F(\mathbf{x})} \quad \underbrace{1 + \cos z(\sin z - x + y) - \cos z(x - \sin z)}_{D_3 F(\mathbf{x})}].$$

When will all three entries be 0? For $D_2 F(\mathbf{x})$ to be 0, we need $x - \sin z = 0$, but then (looking at the first entry) $y = 0$, and then (looking at the third entry) $1 = 0$, which is a contradiction. So the derivative of $F$ never vanishes on $\mathbb{R}^3$.

Part c illustrates that showing that a map is onto is equivalent to showing that some equation has a solution.

c. Given $\mathbf{x} = \begin{pmatrix} x \\ y \\ z \end{pmatrix} \in S$, we need to show that $\mathbf{g} \begin{pmatrix} u \\ v \end{pmatrix} = \begin{pmatrix} x \\ y \\ z \end{pmatrix}$ for $u, v$ has a solution. Clearly we can just solve the equations

$$x = \sin uv + u$$

$$y = u + v$$

$$z = uv.$$

The first and third equations give $u = x - \sin z$, and the second then gives $v = y - x + \sin z$. We still need to check that this actually is a solution. Recall that the point $\mathbf{x}$ is a point of $S$, in particular

$$F(\mathbf{x}) = z - (x - \sin z)(\sin z - x + y) = z - uv = 0,$$

so the third equation is satisfied, and then from the definition of $u$ the first is satisfied, and from the definition of $v$ the second is satisfied.

d. To see that $\mathbf{g}$ is injective, suppose that $\mathbf{g}\begin{pmatrix} u_1 \\ v_1 \end{pmatrix} = \mathbf{g}\begin{pmatrix} u_2 \\ v_2 \end{pmatrix}$, i.e.,

$$\sin u_1 v_1 + u_1 = \sin u_2 v_2 + u_2$$
$$u_1 + v_1 = u_2 + v_2$$
$$u_1 v_1 = u_2 v_2.$$

The first and third equations imply $u_1 = u_2$, and then the second equation implies $v_1 = v_2$. Thus $\mathbf{g}$ is injective.

We have

$$\left[\mathbf{Dg}\begin{pmatrix} u \\ v \end{pmatrix}\right] = \begin{bmatrix} v \cos uv + 1 & u \cos uv \\ 1 & 1 \\ v & u \end{bmatrix}.$$

For the projection of the column vectors onto the last two coordinates to be linearly dependent, we must have $u = v$, and then for the projection onto the first two coordinates to be linearly dependent we must have

$$u \cos u^2 + 1 = u \cos u^2,$$

a contradiction. Thus the derivative of $\mathbf{g}$ is everywhere injective.

**3.2.1** a. At $\begin{pmatrix} 1 \\ 0 \end{pmatrix}$ the derivative is $[3\,,\,0]$, so the equation for the tangent line is $3(x - 1) = 0$, i.e., $x = 1$.

b. At the same point, the tangent space is the kernel of $[3\,,\,0]$, which is the line of equation $\dot{x} = 0$, i.e., the $y$-axis.

**3.2.3** A point $\begin{pmatrix} x \\ y \end{pmatrix}$ is in the tangent line to $X_c$ at $\begin{pmatrix} u \\ v \end{pmatrix}$ if and only if $\begin{pmatrix} x - u \\ y - v \end{pmatrix} \in \ker [2u,\ 3v^2]$, i.e., when $2u(x - u) + 3v^2(y - v) = 0$.

The tangent space $T_{\begin{pmatrix} u \\ v \end{pmatrix}} X_c$ is the kernel of $[2u,\ 3v^2]$, i.e., the 1-dimensional subspace of $\mathbb{R}^2$ of equation $2u\dot{x} + 3v^2\dot{y} = 0$.

**3.2.5** a. At the point $\begin{pmatrix} u \\ v \\ Au^2 + Bv^2 \end{pmatrix}$, the tangent plane to the surface of equation $Ax^2 + By^2 - z = 0$ has equation

$$[2Au \quad 2Bv \quad -1] \begin{bmatrix} x - u \\ y - v \\ z - (Au^2 + Bv^2) \end{bmatrix} = 0.$$

Applied to our three points, this means that $P_1, P_2, P_3$ have equations

$$[0 \quad 0 \quad -1] \begin{bmatrix} x \\ y \\ z \end{bmatrix} = 0, \quad [2Aa \quad 0 \quad -1] \begin{bmatrix} x - a \\ y \\ z - Aa^2 \end{bmatrix} = 0, \quad [0 \quad 2Bb \quad -1] \begin{bmatrix} x \\ y - b \\ z - Bb^2 \end{bmatrix} = 0,$$

which expand to

$$z = 0$$
$$2aAx - z = aA^2$$
$$2bBy - z = bB^2.$$

This is easy to solve; we find $\mathbf{q} = \begin{pmatrix} A/2 \\ B/2 \\ 0 \end{pmatrix}$.

b. The volume of the tetrahedron is

$$\frac{1}{3} \left| \det \begin{bmatrix} a & 0 & A/2 \\ 0 & b & B/2 \\ a^2 A & b^2 B & 0 \end{bmatrix} \right| = \frac{ab}{6}(A^2 a + B^2 b).$$

**3.2.7** For any $\begin{pmatrix} x \\ y \end{pmatrix}$ near $\begin{pmatrix} a \\ b \end{pmatrix}$, the point $f\begin{pmatrix} x \\ y \end{pmatrix}$ is the unique $z$ such that $\begin{pmatrix} x \\ y \\ z \end{pmatrix}$ is on the surface $S$. Similarly, $\begin{pmatrix} x \\ g(x,z) \\ z \end{pmatrix}$, and $\begin{pmatrix} h(y,z) \\ y \\ z \end{pmatrix}$ are on $S$.

Note that $f\begin{pmatrix} h(b,z) \\ b \end{pmatrix} = z$ and $g\begin{pmatrix} h(y,c) \\ c \end{pmatrix} = y$. After differentiating these equations with respect to the appropriate variables ($z$ for the first, $y$ for the second) and using the chain rule, we discover that $(D_1 f)(D_2 h) = 1$ and $(D_1 g)(D_1 h) = 1$, where all derivatives are evaluated at $\mathbf{a} = \begin{pmatrix} a \\ b \\ c \end{pmatrix}$. These equations imply that

$$D_1 f \neq 0, \quad D_1 g \neq 0, \quad D_1 h = \frac{1}{D_1 g}, \quad D_2 h = \frac{1}{D_1 f}.$$

We can also combine the functions like this:

$$f\begin{pmatrix} h(y,c) \\ y \end{pmatrix} = c \quad \text{and} \quad g\begin{pmatrix} h(b,z) \\ z \end{pmatrix} = b.$$

After differentiating the first equation with respect to $y$ and the second with respect to $z$ (and using the chain rule) we get $(D_1 f)(D_1 h) + (D_2 f) = 0$ and $(D_1 g)(D_2 h) + (D_2 g) = 0$. Since $D_1 f$ and $D_1 g$ do not equal zero,

$$D_1 h = -\frac{D_2 f}{D_1 f} \quad \text{and} \quad D_2 h = -\frac{D_2 g}{D_1 g}.$$

Since we know that $D_1 f$ and $D_1 g$ do not equal 0, the equations for the tangent planes at $\mathbf{a}$ may be written as follows by linearizing the equations

$x = h(y, z)$, $y = g(x, z)$, and $z = f(x, y)$ and rearranging terms:

$$\dot{x} = -\frac{D_2 f}{D_1 f}\dot{y} + \frac{1}{D_1 f}\dot{z}$$

$$\dot{x} = \frac{1}{D_1 g}\dot{y} - \frac{D_2 g}{D_1 g}\dot{z}$$

$$\dot{x} = (D_1 h)\dot{y} + (D_2 h)\dot{z}.$$

From the equations we derived above, we see that the first and third tangent planes are the same, as are the second and third. So all tangent planes are the same.

**3.2.9** The tangent plane to $CX$ through a point $\begin{pmatrix} x_0 \\ y_0 \\ z_0 \end{pmatrix}$ has equation

$$D_1 f\begin{pmatrix} x_0/z_0 \\ y_0/z_0 \end{pmatrix}\frac{x - x_0}{z_0} + D_2 f\begin{pmatrix} x_0/z_0 \\ y_0/z_0 \end{pmatrix}\frac{y - y_0}{z_0} = \frac{z - z_0}{z_0^2}\left(x_0 D_1 f\begin{pmatrix} x_0/z_0 \\ y_0/z_0 \end{pmatrix} + y_0 D_2 f\begin{pmatrix} x_0/z_0 \\ y_0/z_0 \end{pmatrix}\right).$$

Note that this is the plane through the origin that contains the tangent line to the curve $X$ at the point $\begin{pmatrix} x_0/z_0 \\ y_0/z_0 \\ 1 \end{pmatrix}$.

**3.2.11** a. If $A = [\vec{v}_1, \ldots, \vec{v}_n]$, then the $i, j$th entry of $A^\top A$ is $\vec{v}_i^\top \vec{v}_j = \vec{v}_i \cdot \vec{v}_j$. If $\vec{v}_1, \ldots, \vec{v}_n$ form an orthonormal basis of $\mathbb{R}^n$, this dot product is 1 when $i = j$ and 0 otherwise. That means that $A^\top A = I$.

Conversely, if $A = [\vec{v}_1, \ldots, \vec{v}_n]$ and $A^\top A = I$, then $\vec{v}_i \cdot \vec{v}_j$ is 1 when $i = j$ and 0 otherwise, i.e., $\vec{v}_1, \ldots, \vec{v}_n$ form an orthonormal set. Since there are $n$ of them, they are a basis.

b. By part a, if $A \in O(n)$, then $A^\top A = I$, so $A^\top = A^{-1}$. Moreover, if $A, B \in O(n)$, then $(AB)^\top(AB) = B^\top A^\top AB = B^\top B = I$; this shows that a product of orthogonal matrices is orthogonal.

$$(A^{-1})^\top A^{-1} = (A^\top)^\top A^{-1} = AA^{-1} = I,$$

so $A^{-1}$ is also orthogonal.

c. Just compute: $(A^\top A - I)^\top = (A^\top A)^\top - I = A^\top (A^\top)^\top - I = A^\top A - I$.

d. We have $[\mathbf{D}F(A)]H = A^\top H + H^\top A$; the problem is to show that given any $M \in S(n, n)$, there exists $H \in \mathrm{Mat}\,(n, n)$ such that $A^\top H + H^\top A = M$. Try $H = \frac{1}{2}AM$:

**Solution 3.2.11, part d:** There is an error in the first printing of the text: $A$ is orthogonal, not just invertible.

To compute the derivative $[\mathbf{D}F(A)]$, compute

$$F(A + H) - F(A)$$

and discard all terms in the increment $H$ that are quadratic or higher; see the remark after example 1.7.18.

$$A^\top\left(\frac{1}{2}AM\right) + \left(\frac{1}{2}AM\right)^\top A = \frac{1}{2}A^\top AM + \frac{1}{2}M^\top A^\top A = \frac{1}{2}M + \frac{1}{2}M^\top = M,$$

since $M = M^\top$.

e. That $O(n)$ is a manifold follows immediately from theorem 3.1.10, and the characterization of the tangent space follows from theorem 3.2.4: it is $\ker[\mathbf{D}F(I)]$, which is the set of $H \in \mathrm{Mat}\,(n, n)$ such that $H + H^\top = 0$, which is precisely the set of antisymmetric matrices.

**3.3.1**

$$D_1(D_2 f) = D_1(x^2 + 2xy + z^2) = 2x + 2y$$
$$D_2(D_3 f) = D_2(2zy) = 2z$$
$$D_3(D_1 f) = D_3(2xy + y^2) = 0$$
$$D_1(D_2(D_3 f)) = D_1(D_2((2yz)) = D_1(2z) = 0.$$

**3.3.3** We have

$$P_{f,\mathbf{a}}^2(\mathbf{a} + \vec{\mathbf{h}}) = \sum_{m=0}^{2} \sum_{I \in \mathcal{I}_3^m} \frac{1}{I!} D_I f(\mathbf{a}) \vec{\mathbf{h}}^I = \underbrace{\frac{1}{0!0!} D_{(0,0,0)} f(\mathbf{a}) h_1^0 h_2^0 h_3^0}_{f(\mathbf{a})} +$$

$$+ \frac{1}{1!0!0!} D_{(1,0,0)} f(\mathbf{a}) h_1^1 h_2^0 h_3^0 + \frac{1}{0!1!0!} D_{(0,1,0)} f(\mathbf{a}) h_1^0 h_2^1 h_3^0 + \frac{1}{0!0!1!} D_{(0,0,1)} f(\mathbf{a}) h_1^0 h_2^0 h_3^1$$

$$+ \frac{1}{2!0!0!} D_{(2,0,0)} f(\mathbf{a}) h_1^2 h_2^0 h_3^0 + \frac{1}{1!1!0!} D_{(1,1,0)} f(\mathbf{a}) h_1 h_2 h_3^0 + \frac{1}{1!0!1!} D_{(1,0,1)} f(\mathbf{a}) h_1 h_2^0 h_3$$

$$+ \frac{1}{0!1!1!} D_{(0,1,1)} f(\mathbf{a}) h_1^0 h_2 h_3 + \frac{1}{0!2!0!} D_{(0,2,0)} f(\mathbf{a}) h_1^0 h_2^2 h_3^0 + \frac{1}{0!0!2!} D_{(0,0,2)} f(\mathbf{a}) h_1^0 h_2^0 h_3^2;$$

i.e.,

$$P_{f,\mathbf{a}}^2(\mathbf{a} + \vec{\mathbf{h}}) = f(\mathbf{a}) + D_{(1,0,0)} f(\mathbf{a}) h_1 + D_{(0,1,0)} f(\mathbf{a}) h_2 + D_{(0,0,1)} f(\mathbf{a}) h_3$$
$$+ \frac{1}{2} D_{(2,0,0)} f(\mathbf{a}) h_1^2 + D_{(1,1,0)} f(\mathbf{a}) h_1 h_2 + D_{(1,0,1)} f(\mathbf{a}) h_1 h_3$$
$$+ D_{(0,1,1)} f(\mathbf{a}) h_2 h_3 + \frac{1}{2} D_{(0,2,0)} f(\mathbf{a}) h_2^2 + \frac{1}{2} D_{(0,0,2)} f(\mathbf{a}) h_3^2.$$

Solution 3.3.5: Equation (1) is reminiscent of *Pascal's triangle recursion for binomials*,

$$\binom{m}{r} = \binom{m-1}{r-1} + \binom{m-1}{r}.$$

Indeed, setting $m = n + k - 1$, $r = n - 1$ gives

$$\binom{n+k-1}{n-1} = \binom{n+k-2}{n-1} + \binom{n+k-2}{n-2}.$$

Thus the formula

$$\text{cardinality } \mathcal{I}_n^k = \binom{n-1+k}{n-1}$$

satisfies equation (1). Moreover, cardinality $\mathcal{I}_n^k$ satisfies the boundary conditions of Pascal's triangle, namely,

$$\text{cardinality } \mathcal{I}_1^k = 1 = \binom{k}{0}$$

$$\text{cardinality } \mathcal{I}_n^0 = 1 = \binom{n-1}{n-1}.$$

These together with Pascal's triangle recursion for binomials completely specify the binomial coefficients.

**3.3.5**

$$\text{cardinality } \mathcal{I}_1^k = \text{cardinality } \{(k)\} = 1;$$
$$\text{cardinality } \mathcal{I}_2^k = \text{cardinality } \{(0,k),(1,k-1),\ldots,(k,0)\}, = k+1;$$
$$\text{cardinality } \mathcal{I}_3^k = \text{cardinality } \{(0,\mathcal{I}_2^k),(1,\mathcal{I}_2^{k-1}),\ldots,(k,\mathcal{I}_2^0 k)\}$$

$$= (k+1) + k + \cdots + 2 + 1 = \frac{(k+1)(k+2)}{2}$$

In general,

$$\text{cardinality } \mathcal{I}_n^k = \text{cardinality } \mathcal{I}_{n-1}^0 + \text{cardinality } \mathcal{I}_{n-1}^1 + \cdots$$
$$\cdots + \text{cardinality } \mathcal{I}_{n-1}^{k-1} + \text{cardinality } \mathcal{I}_{n-1}^k,$$

i.e.,

$$\text{cardinality } \mathcal{I}_n^k = \text{cardinality } \mathcal{I}_n^{k-1} + \text{cardinality } \mathcal{I}_{n-1}^k. \tag{1}$$

**3.3.7** If $I \in \mathcal{I}_n^m$, then $\vec{\mathbf{h}}^I = h_1^{i_1} h_2^{i_2} \ldots h_n^{i_n}$, where $i_1 + i_2 + \cdots i_n = m$, the total degree. Therefore

$$(x\vec{\mathbf{h}})^I = x^{i_1} h_1^{i_1} x^{i_2} h_2^{i_2} \ldots x^{i_n} h_n^{i_n}, \quad \text{i.e.,} \quad x^{i_1 + \cdots + i_n} \vec{\mathbf{h}}^I = x^m \vec{\mathbf{h}}^I.$$

**3.3.9  a.** Clearly this function has partial derivatives of all orders everywhere, except perhaps at the origin. It is also clear that both first partials exist at the origin, and are 0 there, since $f$ vanishes identically on the axes.

Elsewhere, the partials are given by

$$D_1 f \begin{pmatrix} x \\ y \end{pmatrix} = \frac{x^4 y + 3x^2 y^3 - 2xy^4}{(x^2 + y^2)^2} \quad \text{and} \quad D_2 f \begin{pmatrix} x \\ y \end{pmatrix} = \frac{x^5 - x^3 y^2 - 2x^4 y}{(x^2 + y^2)^2}.$$

These partials are continuous on all of $\mathbb{R}^2$, since the limits of the expressions above are 0 as $x, y \to 0$. In particular, $f$ is of class $C^1$.

**b. and c.** Since $D_1 f \begin{pmatrix} x \\ 0 \end{pmatrix} = 0$ for all $x$, and $D_2 f \begin{pmatrix} 0 \\ y \end{pmatrix} = 0$ for all $y$,

$$D_1(D_1 f) \begin{pmatrix} 0 \\ 0 \end{pmatrix} = 0 \quad \text{and} \quad D_2(D_2 f) \begin{pmatrix} 0 \\ 0 \end{pmatrix} = 0.$$

Since $D_1 f \begin{pmatrix} 0 \\ y \end{pmatrix} = 0$ for all $y$ and $D_2 f \begin{pmatrix} x \\ 0 \end{pmatrix} = x$ for all $x$,

$$D_2 \left( D_1 f \begin{pmatrix} 0 \\ y \end{pmatrix} \right) = 0 \quad \text{and} \quad D_1 \left( D_2 f \begin{pmatrix} x \\ 0 \end{pmatrix} \right) = 1,$$

so the crossed partials are not equal.

**d.** This does not contradict theorem 3.3.9 because the crossed partials are not continuous.

Solution 3.3.11: $f^{(i)}$ denotes the $i$th derivative of $f$.

**3.3.11**  The $j$th derivative of the top of the fraction is

$$f^{(j)}(a + h) - \sum_{i=j}^{k} \frac{f^{(i)}(a)}{(i-j)!} h^{i-j},$$

and the $j$th derivative of the bottom is

$$\frac{k!}{(k-j)!} h^{k-j}.$$

If we evaluate these two expressions at $h = 0$, they yield 0 for $0 \le j < k$. Thus the hypotheses of l'Hôpital's rule are satisfied until we take the $k$th derivative, at which point the top equation is 0 and the bottom is $k!$ when evaluated at $h = 0$. Thus the limit in question is 0.

For uniqueness, suppose $p_1$ and $p_2$ are two polynomials of degree $\le k$ such that

$$\lim_{h \to 0} \frac{f(a+h) - p_1(a+h)}{h^k} = 0, \quad \lim_{h \to 0} \frac{f(a+h) - p_2(a+h)}{h^k} = 0.$$

By subtraction, we see that

$$\lim_{h \to 0} \frac{p_1(a+h) - p_2(a+h)}{h^k} = 0.$$

If $p_1 \ne p_2$, then $p_1(a+h) - p_2(a+h) = b_0 + \cdots + b_k h^k$ has a first nonvanishing term, i.e., there exists $l \le k$ such that $b_0 = \cdots = b_{l-1} = 0$ and $b_l \ne 0$. Then the limit

$$\lim_{h \to 0} \frac{p_1(a+h) - p_2(a+h)}{h^k} = \lim_{h \to 0} \frac{b_l h^l + \cdots + b_k h^k}{h^k}$$

$$= \lim_{h \to 0} \frac{1}{h^{k-l}} (b_l + \cdots + b_k h^{k-l})$$

is zero only if $\lim_{h \to 0} \frac{1}{h^{k-l}} = 0$, and that is not the case when $l \leq k$.

**3.3.13** First, we compute the partial derivatives and second partials:

$$D_{(1,0)}f = \frac{1+y}{2\sqrt{x+y+xy}} \qquad D_{(0,1)}f = \frac{1+x}{2\sqrt{x+y+xy}}$$

$$D_{(2,0)}f = \frac{\sqrt{x+y+xy} \cdot 0 - (1+y)\frac{1+y}{2\sqrt{x+y+xy}}}{2(x+y+xy)} = \frac{-(1+y)^2}{4(x+y+xy)^{3/2}}$$

$$D_{(1,1)}f = \frac{\sqrt{x+y+xy} \cdot 1 - \frac{(1+y)(1+x)}{2\sqrt{x+y+xy}}}{2(x+y+xy)} = \frac{2(x+y+xy) - (1+y)(1+x)}{4(x+y+xy)^{3/2}}$$

$$D_{(0,2)}f = \frac{\sqrt{x+y+xy} \cdot 0 - (1+x)\frac{1+x}{2\sqrt{x+y+xy}}}{2(x+y+xy)} = -\frac{(1+x)^2}{4(x+y+xy)^{3/2}}.$$

At the point $\begin{pmatrix} -2 \\ -3 \end{pmatrix}$ these are

$$D_{(1,0)}f\begin{pmatrix} -2 \\ -3 \end{pmatrix} = \frac{1-3}{2\sqrt{-2-3+6}} = -1, \qquad D_{(0,1)}f\begin{pmatrix} -2 \\ -3 \end{pmatrix} = \frac{1-2}{2 \cdot 1} = -\frac{1}{2},$$

$$D_{(2,0)}f\begin{pmatrix} -2 \\ -3 \end{pmatrix} = \frac{-(1-3)^2}{4 \cdot 1} = -1, \qquad D_{(1,1)}f\begin{pmatrix} -2 \\ -3 \end{pmatrix} = \frac{2 \cdot 1 - (-2)(-1)}{4} = 0,$$

$$D_{(0,2)}f\begin{pmatrix} -2 \\ -3 \end{pmatrix} = -\frac{(-1)^2}{4 \cdot 1} = -\frac{1}{4}.$$

Write $x = -2 + u$, $y = -3 + v$; i.e., the increment $\vec{h}$ is $\begin{pmatrix} u \\ v \end{pmatrix}$. The Taylor polynomial of degree 2 of $f$ at $\begin{pmatrix} -2 \\ -3 \end{pmatrix}$ is then $1 - u - \frac{1}{2}v + \frac{1}{2}(-u^2 - \frac{1}{4}v^2)$:

$$P^2_{f,\begin{pmatrix} -2 \\ -3 \end{pmatrix}}\begin{pmatrix} -2+u \\ -3+v \end{pmatrix} = \underbrace{1}_{f\begin{pmatrix} -2 \\ -3 \end{pmatrix}} + \underbrace{-u}_{D_{1,0}f\begin{pmatrix} -2 \\ -3 \end{pmatrix}u^1v^0} + \underbrace{-\frac{1}{2}v}_{D_{(0,1)}f\begin{pmatrix} -2 \\ -3 \end{pmatrix}u^0v^1} + \underbrace{-\frac{1}{2}u^2}_{\frac{1}{2}D_{(2,0)}f\begin{pmatrix} -2 \\ -3 \end{pmatrix}u^2v^0} + \underbrace{\frac{1}{2}\left(-\frac{1}{4}v^2\right)}_{\frac{1}{2}D_{(0,2)}f\begin{pmatrix} -2 \\ -3 \end{pmatrix}u^0v^2}.$$

**3.4.1** Using equation 3.4.9, with $m = 1/2$ and $\sin(x+y)$ playing the role of $x$ in that equation, we have

$$\left(1 + \sin(x+y)\right)^{1/2} = 1 + \frac{1}{2}\sin(x+y) + \frac{-\frac{1}{4}}{2!}\left(\sin(x+y)\right)^2 + \cdots.$$

Equation 3.4.6 tells us that $\sin(x+y) = x + y - \frac{(x+y)^3}{3!}$ plus higher degree terms; discarding terms greater than 2 leaves just $x + y$. Therefore:

$$\sqrt{1 + \sin(x+y)} = 1 + \frac{1}{2}(x+y) - \frac{1}{8}(x+y)^2 + \cdots.$$

We could also use the chain rule in the other direction, first developing the sine, then the square root:

$$\sqrt{1 + \sin(x+y)} = \sqrt{1 + (x+y) + \cdots} = 1 + \frac{1}{2}(x+y) - \frac{1}{8}(x+y)^2 + \cdots.$$

**3.4.3** Set $\mathbf{a} = \begin{pmatrix} -2 \\ -3 \end{pmatrix}$ and $\mathbf{u} = \begin{pmatrix} u \\ v \end{pmatrix}$. Then

$$f(\mathbf{a} + \mathbf{u}) = \left( -2 + u - 3 + v + (-2+u)(-3+v) \right)^{1/2} = (1 - 2u - v + uv)^{1/2}$$

Going from the first to the second line uses the binomial formula, equation 3.4.9.

$$= 1 + \frac{1}{2}(-2u - v + uv) - \frac{1}{8}(-2u - v + uv)^2 + \cdots$$

$$= 1 - u - \frac{1}{2}v + \frac{1}{2}uv - \frac{1}{8}(4u^2 + v^2 + 4uv + \cdots) + \cdots$$

Therefore,

$$P^2_{f,a}(\mathbf{a}+\mathbf{u}) = 1 - u - \frac{1}{2}v + \frac{1}{2}uv - \frac{1}{2}u^2 - \frac{1}{8}v^2 - \frac{1}{2}uv = 1 - u - \frac{1}{2}v - \frac{1}{2}u^2 - \frac{1}{8}v^2.$$

**3.4.5** Write $f(x) = A + Bx + Cx^2 + R(x)$ with $R(x) \in o(h^2)$. Then

$$h\Big(af(0) + bf(h) + cf(2h)\Big) = h\Big(aA + b(A + Bh + Ch^2) + c(A + 2Bh + 4Ch^2)\Big) + h\Big(bR(h) + cR(2h)\Big)$$

$$= hA(a + b + c) + h^2 B(b + 2c) + h^3 C(b + 4c) + h\Big(bR(h) + cR(2h)\Big);$$

note that $h\Big(bR(h) + cR(2h)\Big) \in o(h^3)$.

On the other hand,

$$\int_0^{2h} f(t)\,dt = 2Ah + 4B\frac{h^2}{2} + 8C\frac{h^3}{3} + \int_0^{2h} R(t)\,dt.$$

Note that $\int_0^{2h} R(t)\,dt \in o(h^3)$. Thus we find

$$
\begin{aligned}
a + b + c &= 2 \\
b + 2c &= 2 \\
b + 4c &= \frac{8}{3}
\end{aligned}
\qquad \text{with solution} \qquad
\begin{aligned}
a &= \frac{1}{3} \\
b &= \frac{4}{3} \\
c &= \frac{1}{3}
\end{aligned}.
$$

Our error terms above show that with these numbers, the equation

$$h\Big(af(0) + bf(h) + cf(2h)\Big) - \int_0^h f(t)\,dt \in o(h^3)$$

is satisfied by all functions $f$ of class $C^3$.

**3.4.7** Let $f\begin{pmatrix} x \\ y \\ z \end{pmatrix} = \sin(xyz) - z = 0$ and $\mathbf{a} = \begin{pmatrix} \pi/2 \\ 1 \\ 1 \end{pmatrix}$. Then

$$D_3 f\begin{pmatrix} x \\ y \\ z \end{pmatrix} = xy\cos(xyz) - 1,$$

$f(\mathbf{a}) = 0$, and $D_3 f(\mathbf{a}) = -1 \neq 0$, so $[\mathbf{D}f(\mathbf{a})]$ is onto, and by the implicit function theorem (short version), the equation $f = 0$ expresses one variable as a function of the other two. Since $D_3 f(\mathbf{a}) = -1$ is invertible, the long version of the implicit function theorem says that we can choose $z$ as the "pivotal" or "passive" variable that is a function of the others: $g\begin{pmatrix} x \\ y \end{pmatrix} = z$ in a neighborhood of $\mathbf{p} = \begin{pmatrix} \pi/2 \\ 1 \end{pmatrix}$, where $g(\mathbf{p}) = 1$.

By definition,

$$P^2_{g,\mathbf{p}} \begin{pmatrix} \pi/2 + \dot{x} \\ 1 + \dot{y} \end{pmatrix} = \underbrace{1}_{g(\mathbf{p})} + D_1 g(\mathbf{p})\dot{x} + D_2 g(\mathbf{p})\dot{y} + \frac{1}{2}D_1^2 g(\mathbf{p})\dot{x}^2 + D_1 D_2 g(\mathbf{p})\dot{x}\dot{y} + \frac{1}{2}D_2^2 g(\mathbf{p})\dot{y}^2,$$

Of course you could denote the various components of the increment by $h_1, h_2, h_3$, or by $u, v, w$. But we recommend against writing $x, y, z$ for those increments, which leads to confusion.

where $\dot{x}$ denotes the increment in the $x$ direction and $\dot{y}$ the increment in the $y$ direction. To simplify notation, denote the various coefficients by $\alpha_1, \alpha_2$, etc.:

$$P^2_{g,\mathbf{p}} \begin{pmatrix} \pi/2 + \dot{x} \\ 1 + \dot{y} \end{pmatrix} = 1 + \alpha_1 \dot{x} + \alpha_2 \dot{y} + \alpha_3 \dot{x}^2 + \alpha_4 \dot{x}\dot{y} + \alpha_5 \dot{y}^2, \qquad (1)$$

and note further that equation 2.10.30 tells us that $D_1 g = D_2 g = 0$:

If you didn't use this argument, you would discover that

$$\alpha_1 = \alpha_2 = 0$$

anyway, on coming to equation (3).

$$[\mathbf{D}g(\mathbf{p})] = -[D_3 f(\mathbf{a})]^{-1}[D_1 f(\mathbf{a}), D_2 f(\mathbf{a})] = -[-1]^{-1}[0\ 0] = [0\ 0].$$

Thus equation (1) becomes

$$1 + \dot{z} = P^2_{g,\mathbf{p}} \begin{pmatrix} \pi/2 + \dot{x} \\ 1 + \dot{y} \end{pmatrix} = \underbrace{1}_{g(\mathbf{p})} + \underbrace{\alpha_3 \dot{x}^2 + \alpha_4 \dot{x}\dot{y} + \alpha_5 \dot{y}^2}_{\dot{z}}. \qquad (2)$$

Compute

$$f\begin{pmatrix} \pi/2 + \dot{x} \\ 1 + \dot{y} \\ 1 + \dot{z} \end{pmatrix} = -\underbrace{(1 + \dot{z})}_{z\text{-coordinate}} + \sin\underbrace{\left(\left(\frac{\pi}{2} + \dot{x}\right)(1 + \dot{y})(1 + \dot{z})\right)}_{\text{product of } x,y,z \text{ coordinates}}$$

$$= -1 - \dot{z} + \sin\left(\frac{\pi}{2} + \left(\dot{x} + \frac{\pi}{2}\dot{y} + \frac{\pi}{2}\dot{z} + \dot{x}\dot{y} + \dot{x}\dot{z} + \frac{\pi}{2}\dot{y}\dot{z} + \dot{x}\dot{y}\dot{z}\right)\right)$$

$$= -1 - \dot{z} + \cos\left(\dot{x} + \frac{\pi}{2}\dot{y} + \frac{\pi}{2}\dot{z} + \dot{x}\dot{y} + \dot{x}\dot{z} + \frac{\pi}{2}\dot{y}\dot{z} + \dot{x}\dot{y}\dot{z}\right)$$  (3)

$$= -1 - \dot{z} + 1 - \underbrace{\frac{1}{2}\left(\dot{x} + \frac{\pi}{2}\dot{y} + \frac{\pi}{2}\dot{z} + \dot{x}\dot{y} + \dot{x}\dot{z} + \frac{\pi}{2}\dot{y}\dot{z} + \dot{x}\dot{y}\dot{z}\right)^2 + \cdots}_{\text{Taylor polynomial of cosine}},$$

which gives

Note the handy fact that you don't need to compute $\frac{\pi^2}{8}\dot{z}^2$, $\frac{\pi}{2}\dot{x}\dot{z}$, or $\frac{\pi^2}{4}\dot{y}\dot{z}$, since all terms are higher than quadratic.

$$P^2_{f,\mathbf{a}}\begin{pmatrix} \pi/2 + \dot{x} \\ 1 + \dot{y} \\ 1 + \dot{z} \end{pmatrix} = -\dot{z} - \frac{1}{2}\dot{x}^2 - \frac{\pi^2}{8}\dot{y}^2 - \frac{\pi^2}{8}\dot{z}^2 - \frac{\pi}{2}\dot{x}\dot{y} - \frac{\pi}{2}\dot{x}\dot{z} - \frac{\pi^2}{4}\dot{y}\dot{z}. \quad (4)$$

Substituting $\dot{z} = \alpha_3 \dot{x}^2 + \alpha_4 \dot{x}\dot{y} + \alpha_5 \dot{y}^2$ from equation (2) into (4) gives

$$P^2_{f,\mathbf{a}}\begin{pmatrix} \pi/2 + \dot{x} \\ 1 + \dot{y} \\ P^2_{g,\mathbf{p}} \end{pmatrix} = -(\alpha_3 \dot{x}^2 + \alpha_4 \dot{x}\dot{y} + \alpha_5 \dot{y}^2) - \frac{1}{2}\dot{x}^2 - \frac{\pi^2}{8}\dot{y}^2 - \frac{\pi}{2}\dot{x}\dot{y}.$$

Since $f \begin{pmatrix} x \\ y \\ z \end{pmatrix} = 0$ for any values of $x, y, z$, then $P^2_{f,\mathbf{a}} \begin{pmatrix} \pi/2 + \dot{x} \\ 1 + \dot{y} \\ P^2_{g,\mathbf{p}} \end{pmatrix} = 0$;

setting all coefficients equal to zero gives

$$\alpha_1 = \alpha_2 = 0, \quad \alpha_3 = -\frac{1}{2}, \quad \alpha_4 = -\frac{\pi}{2} \quad \alpha_5 = -\frac{\pi^2}{8},$$

**Solution 3.4.7:** We said above that we recommend using some other notation than $x, y, z$ to denote increments to $x, y, z$.

so

$$P^2_{g,\mathbf{p}}(\mathbf{p} + \dot{\mathbf{x}}) = 1 - \frac{1}{2}\dot{x}^2 - \frac{\pi}{2}\dot{x}\dot{y} - \frac{\pi^2}{8}\dot{y}^2,$$

which we could also write as

$$P^2_{g,\mathbf{p}}(\mathbf{x}) = 1 - \frac{1}{2}\left(x - \frac{\pi}{2}\right)^2 - \frac{\pi}{2}\left(x - \frac{\pi}{2}\right)(y - 1) - \frac{\pi^2}{8}(y - 1)^2.$$

The real danger is in confusing $z$ and the increment to $z$. If you do that you may be tempted to substitute $z = 1 + \dot{z}$ for $\dot{z}$ in equation (4), counting the 1 twice. In example 3.4.9 we do insert the entire Taylor polynomial (degree 2) of $g$ into $x^2 + y^3 + xyz^3 - 3 = 0$, but note that we did not count the 1 twice; $x^2$ became $(1 + u)^2$ and $y^3$ became $(1 + v)^3$, but $z^3$ became $(1 + a_{1,0}u + a_{0,1}v + \frac{a_{2,0}}{2}u^2 + \cdots)^3$, not $(1 + 1 + a_{1,0}u + \cdots)^3$. As long as you can keep variables and increments straight, it does not matter which you do.

**3.4.9** Compute

$$\mathrm{erf}(x) = \frac{2}{\sqrt{\pi}}\int_0^x \overbrace{\sum_0^\infty \frac{(-t^2)^k}{k!}}^{\text{Taylor series of } e^{-t^2}} dt = \frac{2}{\sqrt{\pi}}\sum_0^\infty \frac{(-1)^k}{k!}\int_0^x t^{2k}\,dt \tag{5}$$
$$= \frac{2}{\sqrt{\pi}}\sum_{k=0}^\infty \frac{(-1)^k(x)^{2k+1}}{(2k+1)k!}.$$

Set $P_n(x) = \dfrac{2}{\sqrt{\pi}}\displaystyle\sum_{k=0}^{n-1}\frac{(-1)^k x^{2k+1}}{(2k+1)k!}$ and

**Solution 3.4.9:** The function "erf" computes the area under the bell or "Gaussian" curve and is thus important in probability. The exchange of sum and integral equation (5) is justified because the series $\frac{(-t^2)^k}{k!}$ is absolutely uniformly convergent on $[0, x]$.

$$E_n = \mathrm{erf}(1/2) - P_n(1/2) = \overbrace{\frac{2}{\sqrt{\pi}}\sum_{k=n}^\infty \frac{(-1)^k(1/2)^{2k+1}}{(2k+1)k!}}^{\text{alternating series because of } (-1)^k}.$$

Since $E_n$ is an alternating series with decreasing terms tending to 0, it converges, and for each $n$,

$$|E_n| \le \big|\text{first term of the series } E_n\big| = \frac{2}{\sqrt{\pi}}\frac{(1/2)^{2n+1}}{(2n+1)n!}.$$

Trying different values of $n$, we find that $E_2$ may be too big but $E_3$ is definitely small enough:

$$|E_2| \le \frac{1}{160\sqrt{\pi}} > 10^{-3} \quad \text{and} \quad |E_3| = \frac{1}{2688\sqrt{\pi}} < 10^{-3}.$$

So take $n = 3$; then

$$P_3(x) = \frac{2}{\sqrt{\pi}}\left(x - \frac{1}{3}x^3 + \frac{1}{10}x^5\right).$$

**3.4.11 a.** If we systematically ignore all terms of degree $> 3$, we can write

$$\left(1 + \frac{x+y}{1+xz}\right)^{1/2} \approx (1 + (x+y)(1-xz))^{1/2} = \left(1 + x + y - x^2 z - xyz\right)^{1/2}$$

$$\approx 1 + \frac{1}{2}\left(x + y - x^2 z - xyz\right) - \frac{1}{8}\left(x + y - x^2 z - xyz\right)^2$$

$$+ \frac{1}{6}\cdot\frac{1}{2}\left(-\frac{1}{2}\right)\left(-\frac{3}{2}\right)(x+y)^3$$

$$\approx 1 + \frac{1}{2}(x+y) - \frac{1}{8}(x+y)^2 - \frac{1}{2}x^2 z - \frac{1}{2}xyz$$

$$+ \frac{1}{16}(x^3 + y^3 + 3x^2 y + 3xy^2).$$

**b.** The number $D_{(1,1,1)}f\begin{pmatrix}0\\0\\0\end{pmatrix}$ is exactly the coefficient of $xyz$, so it is $-1/2$.

**3.5.1** We find signature $(2,1)$, since

$$-4z^2 + 2yz - 4xz + 2xy + x^2 = -4\left(z - \frac{y}{4} + \frac{x}{2}\right)^2 + \frac{1}{4}(y+2x)^2 + x^2.$$

**3.5.3 a.** $x^2 + xy - 3y^2 + \frac{y^2}{4} - \frac{y^2}{4} = \left(x + \frac{y}{2}\right)^2 - \left(\frac{y\sqrt{13}}{2}\right)^2$; signature $(1,1)$.

**b.** $x^2 + 2xy - y^2 + y^2 - y^2 = (x+y)^2 - (\sqrt{2}y)^2$; signature $(1,1)$.

**c.**

$$x^2 + xy + zy = x^2 + xy + \frac{y^2}{4} - \frac{y^2}{4} + zy$$

$$= \left(x + \frac{y}{2}\right)^2 - \left(\frac{y^2}{4} - yz + z^2\right) + z^2$$

$$= \left(x + \frac{y}{2}\right)^2 - \left(\frac{y}{2} - z\right)^2 + z^2; \text{ signature } (2,1).$$

**3.5.5 a.** The signature is $(1,1)$, since

$$x^2 + xy = \left(x^2 + xy + \frac{y^2}{4}\right) - \frac{y^2}{4} = \left(x + \frac{y}{2}\right)^2 - \left(\frac{y}{2}\right)^2.$$

**b.** Let us introduce $u = x + y$, so that $x = u - y$. In these variables, our quadratic form is

$$(u-y)y + yz = -\left(y^2 - uy - yz + \frac{uz}{2} + \frac{u^2}{4} + \frac{z^2}{4}\right) + \left(\frac{uz}{2} + \frac{u^2}{4} + \frac{z^2}{4}\right)$$

$$= -\left(y - \frac{u}{2} - \frac{z}{2}\right)^2 + \left(\frac{u}{2} + \frac{z}{2}\right)^2.$$

Again the quadratic form has signature $(1,1)$, but this time it is degenerate.

Solution 3.4.11: First we use equation 3.4.9 with $m = -1$ to write

$$\frac{1}{1+zy} \approx 1 - xz.$$

Then we use equation 3.4.9 with $m = 1/2$ to compute the Taylor polynomial of

$$(1 + (1 - xz)(x+y))^{1/2}.$$

**3.5.7** First, assume $Q$ is a positive definite quadratic form on $\mathbb{R}^n$ with signature $(k, l)$:

$$Q(\mathbf{x}) = \big(\alpha_1(\mathbf{x})\big)^2 + \cdots + \big(\alpha_k(\mathbf{x})\big)^2 - \big(\alpha_{k+1}(\mathbf{x})\big)^2 - \cdots - \big(\alpha_{k+l}(\mathbf{x})\big)^2.$$

Each linear function

$$\alpha_i : \mathbb{R}^n \to \mathbb{R}$$

is a $1 \times n$ row matrix, the $i$th row of the matrix $T$.

We want to show that $k = n$ and $l = 0$. First let us see that $l = 0$. The $\alpha_i$ are linearly independent, so the matrix $T = \begin{bmatrix} \alpha_1 \\ \vdots \\ \alpha_{k+l} \end{bmatrix}$ has rank $k + l$, and is onto $\mathbb{R}^{k+l}$.

Thus there exists $\vec{\mathbf{x}}_0$ such that $T\vec{\mathbf{x}}_0 = \vec{\mathbf{e}}_{k+1}$. Then $Q(\vec{\mathbf{x}}_0) = -1$, so $Q$ is not positive definite. Therefore if $Q$ is positive definite, we must have $l = 0$.

Now assume $Q$ is positive definite with signature $(k, 0)$ and $k < n$. Then the $k \times n$ matrix $T = \begin{bmatrix} \alpha_1 \\ \vdots \\ \alpha_k \end{bmatrix}$ satisfies $\ker T \neq \{\mathbf{0}\}$. Let $\vec{\mathbf{y}} \neq \mathbf{0}$ be in $\ker T$.

Then $Q(\vec{\mathbf{y}}) = 0$, so $Q$ is not positive definite.

In the other direction, if $Q$ has signature $(n, 0)$, then the linear transformation $T : \mathbb{R}^n \to \mathbb{R}^n$ given by the $n \times n$ matrix $T(\vec{\mathbf{x}}) = \begin{bmatrix} \alpha_1(\vec{\mathbf{x}}) \\ \vdots \\ \alpha_n(\vec{\mathbf{x}}) \end{bmatrix}$ is onto $\mathbb{R}^n$, so by the dimension formula, $\ker T = 0$. So for all $\vec{\mathbf{x}} \neq \mathbf{0}$ in $\mathbb{R}^n$, at least one of the $\alpha_i(\vec{\mathbf{x}}) \neq 0$, so squaring the functions ensures that

$$Q(\vec{\mathbf{x}}) = \big(\alpha_1(\vec{\mathbf{x}})\big)^2 + \cdots + \big(\alpha_n(\vec{\mathbf{x}})\big)^2 > 0.$$

**3.5.9** As in example 3.5.7, we use the substitution $u = a - d$ in the following computation:

This result has important ramifications for physics: according to the theory of relativity, spacetime naturally carries a quadratic form of signature $(1, 3)$. Thus this space of Hermitian matrices is a natural model for spacetime.

$$\det H = ad - b^2 - c^2$$

$$= (u + d)d - b^2 - c^2 = d^2 + ud + \left(\frac{u}{2}\right)^2 - \left(\frac{u}{2}\right)^2 - b^2 - c^2$$

$$= \left(d + \frac{u}{2}\right)^2 - \left(\frac{u}{2}\right)^2 - b^2 - c^2$$

$$= \left(\frac{a + d}{2}\right)^2 - \left(\frac{a - d}{2}\right)^2 - b^2 - c^2.$$

The signature of $\det H$ is thus $(1, 3)$.

**3.5.11**  a. Set $A = \begin{bmatrix} a & b \\ c & d \end{bmatrix}$. Then

$$\operatorname{tr}\big(A^2\big) = a^2 + d^2 + 2bc \quad \text{and} \quad \operatorname{tr}\big(A^\top A\big) = a^2 + b^2 + c^2 + d^2.$$

b.  The signature of $\operatorname{tr}(A^\top A)$ is $(4, 0)$. The trace of $A^2$ can be written $a^2 + d^2 + \frac{1}{2}(b + c)^2 - \frac{1}{2}(b - c)^2$, so $\operatorname{tr}(A^2)$ has signature $(3, 1)$.

**3.5.13** a. By the associative and distributive properties of matrix multiplication,

$$(a_1\mathbf{v}_1 + a_2\mathbf{v}_2)^\top A\mathbf{w} = a_1\mathbf{v}_1^\top A\mathbf{w} + a_2\mathbf{v}_2^\top A\mathbf{w}.$$

Since $A$ is symmetric, and the transpose of a number equals the number,

$$\underbrace{\mathbf{v}^\top A\mathbf{w}}_{\text{a number}} = (\mathbf{v}^\top A\mathbf{w})^\top = \mathbf{w}^\top (\mathbf{v}^\top A)^\top = \mathbf{w}^\top A^\top \mathbf{v} = \mathbf{w}^\top A\mathbf{v}.$$

b. Given a symmetric bilinear function $B$, define the matrix $A$ by

$$a_{i,j} = B(\mathbf{e}_i, \mathbf{e}_j).$$

This matrix is symmetric since $B$ is symmetric, and

$$B_A(\mathbf{e}_i, \mathbf{e}_j) = \mathbf{e}_i^\top A\mathbf{e}_j = a_{i,j} = B(\mathbf{e}_i, \mathbf{e}_j),$$

so the desired relation $B = B_A$ holds if applied to the standard basis vectors. But then it holds for the linear combinations of standard basis vectors, i.e., for all pairs of vectors.

c. Again, there isn't very much to show:

$$\int_0^1 (a_1 p_1(t) + a_2 p_2(t)) q(t)\, dt = a_1 \int_0^1 p_1(t) q(t)\, dt + a_2 \int_0^1 p_2(t) q(t)\, dt,$$

and

$$\int_0^1 p(t) q(t)\, dt = \int_0^1 q(t) p(t)\, dt.$$

d. If $T : V \to W$ is any linear transformation, and $B$ is a symmetric bilinear function on $W$, it is easy to check that the function $B(T(\mathbf{v}_1), T(\mathbf{v}_2))$ is always a symmetric bilinear function on $V$. To find the matrix, compute

$$B\Big(\Phi_p(\mathbf{e}_i), \Phi_p(\mathbf{e}_j)\Big) = \int_0^1 t^{i+j-2} dt = \frac{1}{i+j-1}.$$

Solution 3.5.13, part d: This matrix is called the

$$(k+1) \times (k+1)$$

*Hilbert matrix*; it is a remarkable fact that its inverse has only integer entries. You might compute this in a couple of cases, but proving it is quite tricky.

Thus the matrix is

$$\begin{bmatrix} 1 & \frac{1}{2} & \cdots & \frac{1}{k+1} \\ \frac{1}{2} & \frac{1}{3} & \cdots & \frac{1}{k+2} \\ \vdots & \vdots & \ddots & \vdots \\ \frac{1}{k+1} & \frac{1}{k+2} & \cdots & \frac{1}{2k+1} \end{bmatrix}.$$

**3.5.15** a. Both $(p(t))^2$ and $(p'(t))^2$ are polynomials, whose coefficients are homogeneous quadratic polynomials in the coefficients of $p$. After integrating, the coefficient of $t^k$ in $(p(t))^2 - (p'(t))^2$ simply gets multiplied by $1/(k+1)$ in $Q$.

Explicitly: setting $p = \sum_{i=0}^{k} a_i t^i$ we get

$$Q(p) = \sum_{i=0}^{k}\sum_{j=0}^{k} a_i a_j \int_0^1 \left(t^{i+j} - ij\, t^{i+j-2}\right)\, dt$$

$$= \sum_{i=0}^{k}\sum_{j=0}^{k} a_i a_j \left(\frac{1}{i+j+1} - \frac{ij}{i+j-1}\right).$$

b. We find

$$\int_0^1 \left((at^2 + bt + c)^2 - (2at+b)^2\right)dt = -\frac{17}{15}a^2 - \frac{2}{3}b^2 + c^2 - \frac{3}{2}ab + \frac{2}{3}ac + bc.$$

It seems easiest to remove $c$, to find

$$Q(p) = \left(c + \frac{b}{2} + \frac{a}{3}\right)^2 - \frac{17}{15}a^2 - \frac{2}{3}b^2 - \frac{3}{2}ab - \frac{1}{9}a^2 - \frac{1}{4}b^2 - \frac{1}{3}ab.$$

The terms other than the first square are

$$-\frac{56}{45}a^2 - \frac{11}{12}b^2 - \frac{11}{6}ab.$$

It seems easier to remove $b$ next, writing

$$-\frac{11}{12}(a+b)^2 + \frac{11}{12}a^2 - \frac{56}{45}a^2.$$

We are left with $\dfrac{-59}{180}a^2$. So the quadratic form is

$$\left(c + \frac{a}{3} + \frac{b}{2}\right)^2 - \frac{11}{12}(a+b)^2 - \frac{59}{180}a^2, \quad \text{with signature } (1,2).$$

**3.5.17**  a. The quadratic form corresponding to this matrix is

$$[x\ y]\begin{bmatrix} 2 & 1 \\ 1 & 3 \end{bmatrix}\begin{bmatrix} x \\ y \end{bmatrix} = 2x^2 + 2xy + 3y^2.$$

Completing squares gives $2x^2 + 2xy + 3y^2 = 2(x + y/2)^2 + 5y^2/2$, so the quadratic form has signature $(2,0)$, and the curve of equation

$$2x^2 + 2xy + 3y^2 = 1$$

is an ellipse, drawn in the top figure in the margin.

b. The quadratic form corresponding to the matrix

$$\begin{bmatrix} 2 & 1 & 0 \\ 1 & 2 & 1 \\ 0 & 1 & 2 \end{bmatrix} \quad \text{is} \quad x^2 + y^2 + z^2 - 2xy - 2yz.$$

This can be written

$$x^2 + y^2 + z^2 - 2xy - 2yz = (x + y + z)^2 + \frac{1}{2}(y - z)^2 - \frac{1}{2}(y + z)^2,$$

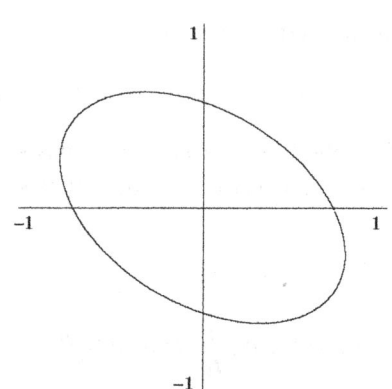

FIGURE FOR SOLUTION 3.5.17, part a. The ellipse of equation

$$2x^2 + 2xy + 3y^2 = 1.$$

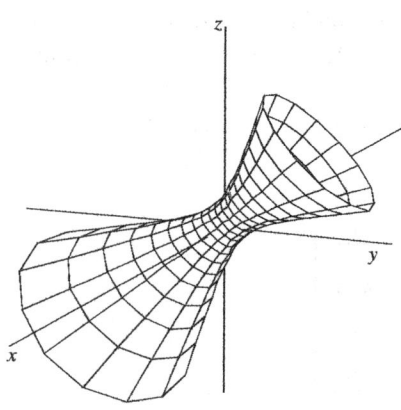

FIGURE FOR SOLUTION 3.5.17 The hyperboloid of part b.

which has signature $(2,1)$; the surface of equation

$$x^2 + y^2 + z^2 - 2xy - 2yz = 1$$

is a hyperboloid of one sheet.

We will specify how the surface is parametrized. A parametrization of the surface of equation $u^2 + v^2 - w^2 = 1$ is given by setting

$$\begin{pmatrix} x \\ y \\ z \end{pmatrix} = \begin{pmatrix} \cosh s \cos t \\ \cosh s \sin t \\ \sinh s \end{pmatrix}, \quad s \in \mathbb{R}, \quad 0 \le v \le 2\pi.$$

Thus if we solve the equations

$$x + y + z = \cosh s \cos t$$

$$\frac{y - z}{\sqrt{2}} = \cosh s \sin t$$

$$\frac{y + z}{\sqrt{2}} = \sinh s$$

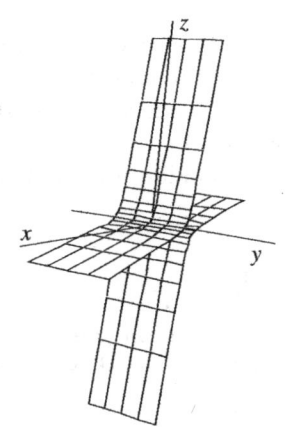

FIGURE FOR SOLUTION 3.5.17, part c. The hyperbolic cylinder.

for $x, y$, and $z$ in terms of $s$ and $t$, we will have parametrized the surface. This is straightforward, giving

$$x = \cosh s \cos t - \sqrt{2}\sinh s$$

$$y = \frac{\sqrt{2}}{2}(\sinh s + \cosh s \sin t)$$

$$z = \frac{\sqrt{2}}{2}(\sinh s - \cosh s \sin t).$$

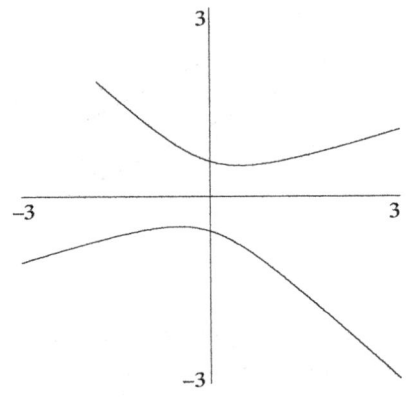

FIGURE FOR SOLUTION 3.5.17, part c. The hyperbola in the $(x, z)$-plane over which we are constructing the cylinder.

c. The quadratic form corresponding to the matrix $\begin{bmatrix} 2 & 0 & 3 \\ 0 & 0 & 0 \\ 3 & 0 & -1 \end{bmatrix}$ is $2x^2 + 6xz - z^2 = (x + 3z)^2 - (\sqrt{8}z)^2$. It is degenerate, of signature $(1,1)$, and the surface of equation

$$2x^2 + 6xz - z^2 = 1$$

is a hyperbolic cylinder.

As above this can be parametrized by

$$x = \cosh s - \frac{3\sinh s}{\sqrt{8}}$$

$$y = t$$

$$z = \frac{\sinh s}{\sqrt{8}}.$$

The resulting surface is plotted at left.

d. The quadratic form corresponding to the matrix $\begin{bmatrix} 2 & 4 & -3 \\ 4 & 1 & 3 \\ -3 & 3 & -1 \end{bmatrix}$ is $2x^2 + y^2 - z^2 + 8xy - 6xz + 6yz$. Writing this as a sum or difference of squares is a bit more challenging; we find

$$\left(y + 4x + \frac{3}{2}y\right)^2 - 14\left(x - \frac{9}{14}z\right)^2 + \frac{71}{28}z^2,$$

so the quadratic form is of signature $(2,1)$. Again the surface of equation $2x^2 + y^2 - z^2 + 8xy - 6xz + 6yz = 1$ is a hyperboloid of one sheet, in the bottom left figure. (Parametrizing this hyperboloid was done as in part b, but the computations are quite a bit more unpleasant.)

e. This is easy again; the quadratic form corresponding to the matrix $\begin{bmatrix} 1 & 2 \\ 2 & 4 \end{bmatrix}$ is

$$x^2 + 4xy + 4y^2 = (x + 2y)^2.$$

It is degenerate, of signature $(1,0)$, and the curve of equation $(x+2y)^2 = 1$ is the union of the two lines $x + 2y = 1$ and $x + 2y = -1$, shown in the bottom right figure.

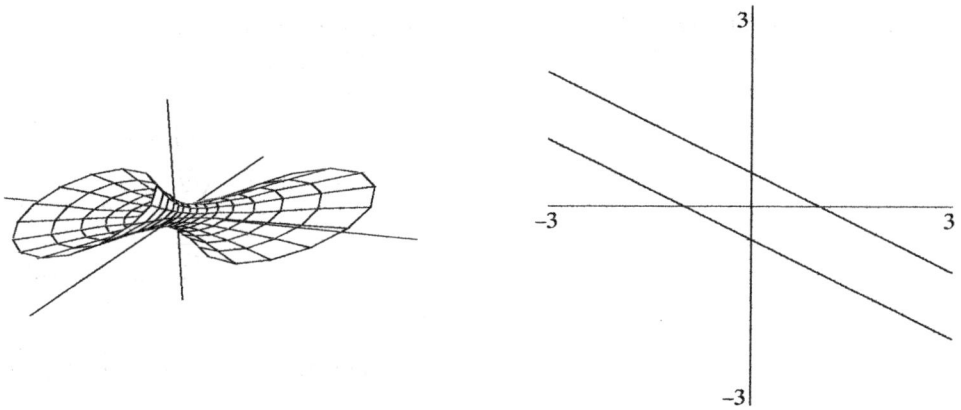

FIGURE FOR SOLUTION 3.5.17. LEFT: The hyperboloid of one sheet of part d. RIGHT: The two lines $x + 2y = 1$ and $x + 2y = -1$ of part e.

**3.6.1**   a. Since all the partials of $f$ are continuous, $f$ is differentiable on all of $\mathbb{R}^3$. In fact,

$$\left[\mathbf{D}f\begin{pmatrix} x \\ y \\ z \end{pmatrix}\right] = \begin{bmatrix} 2x + y & x + \sin y & 2z \end{bmatrix}, \quad \text{which vanishes at the origin,}$$

and is thus a critical point by definition 3.6.4.

b. We have

$$P^2_{f,0}(\mathbf{h}) = -1 + h_1^2 + h_1 h_2 + \frac{h_2^2}{2} + h_3^2$$

and

$$h_1^2 + h_1 h_2 + \frac{h_2^2}{2} + h_3^2 = \left(h_1 + \frac{h_2}{2}\right)^2 + \frac{h_2^2}{4} + h_3^2,$$

so the critical point has signature $(3,0)$ and is a local minimum by theorem 3.6.7.

**3.6.3** By proposition 3.5.11 there exists a subspace $W$ of $\mathbb{R}^n$ such that $Q$ is negative definite on $W$. If a quadratic form on $W$ is negative definite, then by proposition 3.5.14, there exists a constant $C > 0$ such that

$$Q(\vec{\mathbf{x}}) \leq -C|\mathbf{x}|^2 \quad \text{for all } \vec{\mathbf{x}} \in \mathbb{R}^n.$$

Thus if $\vec{\mathbf{h}} \in W$, and $t > 0$, there exists $C > 0$ such that

For $|t| \neq 0$ sufficiently small,

$$\left| \frac{r(t\vec{\mathbf{h}})}{t^2} \right| < C|\vec{\mathbf{h}}|^2.$$

so we have

$$\frac{f(\mathbf{a} + t\vec{\mathbf{h}}) - f(\mathbf{a})}{t^2} < 0.$$

$$\frac{f(\mathbf{a} + t\vec{\mathbf{h}}) - f(\mathbf{a})}{t^2} = \frac{t^2 Q(\vec{\mathbf{h}}) + r(t\vec{\mathbf{h}})}{t^2} \leq -C|\vec{\mathbf{h}}|^2 + \frac{r(t\vec{\mathbf{h}})}{t^2},$$

where $t^2 Q(\vec{\mathbf{h}})$ is the quadratic term of the Taylor polynomial, and $r(\vec{\mathbf{h}})$ is the remainder. Since

$$\lim_{t \to 0} \frac{r(t\vec{\mathbf{h}})}{t^2} = 0,$$

it follows that $f(\mathbf{a} + t\vec{\mathbf{h}}) < f(\mathbf{a})$ for $t > 0$ sufficiently small.

**3.6.5**  a. The critical points are the points where $D_1 f$, $D_2 f$, and $D_3 f$ vanish, i.e., the points satisfying the three equations

$$y - z + yz = 0$$
$$x + z + xz = 0$$
$$y - x + xy = 0.$$

If we use the first two to express $x$ and $y$ in terms of $z$ and substitute into the third equation, we get

$$\frac{z}{z+1}\left(2 - \frac{z}{z+1}\right) = 0.$$

A polynomial function like $f$ is its own Taylor polynomial at the origin (but not elsewhere).

This leads to the two roots $z = 0$ and $z = -2$, and finally to the critical points $\begin{pmatrix} 0 \\ 0 \\ 0 \end{pmatrix}$ and $\begin{pmatrix} -2 \\ 2 \\ -2 \end{pmatrix}$.

b. The quadratic terms at the first point are evidently $uv + vw - uw$. Set $u = s - v$, and get rid of $u$, to find

$$sv - v^2 + vw - sw + vw = -\left(v - w - \frac{s}{2}\right)^2 + w^2 + \frac{1}{4}s^2.$$

The origin is a saddle of type $(2,1)$.

Here we turn the polynomial into a new polynomial, whose variables are the increments from the point $\begin{pmatrix} -2 \\ 2 \\ -2 \end{pmatrix}$.

At the other critical point, set

$$x = -2 + u, \quad y = 2 + v, \quad z = -2 + w$$

and substitute into the function. We find

$$-4 - uv - vw + uw + uvw.$$

The quadratic terms are $-uv - vw + uw$, exactly the opposites of the quadratic terms at the first point, so again we have a saddle, this time of type $(1,2)$.

**3.6.7 a.** We compute

$$\left[\mathbf{D}f\begin{pmatrix} x \\ y \end{pmatrix}\right] = \left[2x(1 + x^2 + y^2)e^{x^2 - y^2} \ , \ 2y\big(1 - (x^2 + y^2)\big)e^{x^2 - y^2}\right].$$

Thus $D_1 f = 0$ when $x = 0$, and $D_2 f = 0$ when $y = 0$ or $x^2 + y^2 = 1$. Thus the critical points occur when $x = 0$ and either $y = 0$ or $x^2 + y^2 = 1$, that is, at the points $\begin{pmatrix} 0 \\ 0 \end{pmatrix}, \begin{pmatrix} 0 \\ 1 \end{pmatrix}$, and $\begin{pmatrix} 0 \\ -1 \end{pmatrix}$.

**b.** The second-degree terms of the Taylor approximation of $f$ about the origin are $x^2$ and $y^2$. (Just multiply $x^2 + y^2$ and $e^u = 1 + u + u^2/2 + \cdots$, where $u = x^2 - y^2$, and take the second-degree terms.) Since $x^2 + y^2$ is a positive definite quadratic form, $f$ has a local minimum at the origin. In order to characterize the other two critical points, we first compute the three second partial derivatives:

$$D_1^2 f \begin{pmatrix} x \\ y \end{pmatrix} = (2 + 10x^2 + 2y^2 + 4x^2y^2 + 4x^4)e^{(x^2 - y^2)}$$

$$D_1 D_2 f \begin{pmatrix} x \\ y \end{pmatrix} = -4xy(x^2 + y^2)e^{(x^2 - y^2)}$$

$$D_2^2 f \begin{pmatrix} x \\ y \end{pmatrix} = (2 - 2x^2 - 10y^2 + 4x^2y^2 + 4y^2)e^{(x^2 - y^2)}.$$

At both $\begin{pmatrix} 0 \\ 1 \end{pmatrix}$ and $\begin{pmatrix} 0 \\ -1 \end{pmatrix}$, we have $D_1^2 = 4/e$, $D_1 D_2 f = 0$, and $D_2^2 f = -4/e$. Thus the second-degree terms of the Taylor polynomial are

$$\frac{D_1^2 f h_1^2}{2!} = \frac{2h_1^2}{e} \quad \text{and} \quad \frac{D_2^2 f h_2^2}{2!} = \frac{-2h_2^2}{e},$$

which give the quadratic form

$$\frac{2h_1^2}{e} - \frac{2h_2^2}{e}$$

at both points. This form has signature $(1,1)$, so $f$ has a saddle at both of these critical points.

**3.7.1** Let us call the three sides of the box $x$, $2x$, and $y$. Then the problem is to maximize the volume $2x^2y$, subject to the constraint that the surface area is $2(2x^2 + xy + 2xy) = 10$. This leads to the Lagrange multiplier problem

$$\begin{bmatrix} 4xy & 2x^2 \end{bmatrix} = \lambda \begin{bmatrix} 4x + 3y & 3x \end{bmatrix}.$$

In the resulting two equations

$$4xy = \lambda(4x + 3y) \tag{1}$$

$$2x^2 = \lambda(3x) \tag{2}$$

we see that equation (2) has $x = 0$ as one solution, which is certainly not the maximum. The other solution is $2x = 3\lambda$. Substitute this value of $\lambda$

Solution 3.6.7: Remember that
$$(fg)' = gf' + fg'$$
$$(e^f)' = e^f f'.$$

In part b, we use equation 3.4.5 for the Taylor polynomial of $e^x$.

into equation (1), to find $4x = 3y$, and then the corresponding value of $y$ in terms of $x$ into the constraint equation, to find $x = \sqrt{\frac{5}{6}}$. This leads to the maximum

$$V = \frac{20}{9}\sqrt{\frac{5}{6}}.$$

**3.7.3** Suppose we use Lagrange multipliers, computing the derivative of $\varphi = x^3 + y^3 + z^3$ and the derivatives of the constraint functions $F_1 = x + y + z = 2$ and $F_2 = x + y - z = 3$. Then at a critical point,

$$D\varphi = [3x^2, 3y^2, 3z^2] = \lambda_1[1,1,1] + \lambda_2[1,1,-1];$$

i.e.,

$$x = \pm y = \pm\sqrt{\frac{\lambda_1 + \lambda_2}{3}} \quad \text{and} \quad z = \pm\sqrt{\frac{\lambda_1 - \lambda_2}{3}}.$$

Adding the constraints gives $2(x + y) = 5$, hence we must use the positive square root for both $x$ and $y$. Similarly, subtracting the constraints gives $-2z = 1$, so for $z$ we must use the negative square root:

$$x = y = \sqrt{\frac{\lambda_1 + \lambda_2}{3}} \quad \text{and} \quad z = -\sqrt{\frac{\lambda_1 - \lambda_2}{3}}.$$

Substituting these values in the constraint functions tells us that

$$\lambda_1 = \frac{87}{32} \quad \text{and} \quad \lambda_2 = \frac{63}{32},$$

which we can use to determine that there is a critical point at $\begin{pmatrix} 5/4 \\ 5/4 \\ -1/2 \end{pmatrix}$.

But is it a minimum of the constrained function? Our treatment of Lagrange multipliers doesn't allow us to say.

Instead we can use a parametrization that incorporates the constraints. If we write the constraint functions in matrix form and row reduce, we get

$$\begin{bmatrix} 1 & 1 & 1 & 2 \\ 1 & 1 & -1 & 3 \end{bmatrix}, \quad \text{which row reduces to} \quad \begin{bmatrix} 1 & 1 & 0 & 5/2 \\ 0 & 0 & 1 & -1/2 \end{bmatrix};$$

i.e.,

$$x + y = 5/2 \quad \text{and} \quad z = -1/2.$$

Substituting $x = 5/2 - y$ and $z = -1/2$ in the function $x^3 + y^3 + z^3$ gives

$$(5/2 - y)^3 + y^3 - 1/8 = (5/3)^3 - 3(5/2)^2 y + 3(5/2)y^2 - 1/8,$$

which is the equation for a parabola; the fact that the coefficient for $y^2$ is positive tells us that the parabola is shaped like a cup, and therefore has a minimum.

**3.7.5** The volume of the parallelepiped in the first octant (where $x \geq 0$, $y \geq 0$, $z \geq 0$) is gotten by maximizing $xyz$ subject to the constraint

$$x^2 + 4y^2 + 9z^2 = 9.$$

---

Solution 3.7.3: This is a case where using Lagrange multipliers does not give all the information the problem asks for.

Our treatment of Lagrange multipliers doesn't allow us to say whether the critical point is a minimum of the constrained function because the domain of that function is not compact. In contrast, in example 3.7.10, the domain is compact.

In example 3.7.11, we discuss a local maximum on a compact part of the domain. There might be a similar way to analyze this problem, looking for an appropriate compact subset of the constraint space.

Solution 3.7.5: We will compute the volume in one octant, then multiply that volume by 8.

The Lagrange multiplier theorem says that at such a maximum, there exists a number $\lambda$ such that

$$[yz, \ xz, \ xy] = \lambda[2x, \ 8y, \ 18z],$$

i.e., we have the three equations $yz = 2\lambda x$, $xz = 8\lambda y$, $xy = 18\lambda z$. Multiply the first by $x$, the second by $y$, and the third by $z$, to find

$$xyz = 2\lambda x^2 = 8\lambda y^2 = 18\lambda z^2.$$

This leads to $x^2 = 4y^2 = 9z^2$, hence $3x^2 = 9$, or

$$x = \sqrt{3}, \ y = \frac{\sqrt{3}}{2}, \ z = \frac{\sqrt{3}}{3},$$

and finally the maximum of $xyz$ is $\sqrt{3}/2$. There are eight octants, so the total volume is $4\sqrt{2}$.

**3.7.7**   a.  Using the hint, write $z = \frac{1}{c}(1 - ax - by)$. The function now becomes

$$xyz = \frac{1}{c}xy(1 - ax - by) = F\begin{pmatrix} x \\ y \end{pmatrix},$$

with partial derivatives

$$D_1 F\begin{pmatrix} x \\ y \end{pmatrix} = \frac{1}{c}(y - 2axy - by^2) = \frac{y}{c}(1 - 2ax - by),$$

$$D_2 F\begin{pmatrix} x \\ y \end{pmatrix} = \frac{1}{c}(x - ax^2 - 2bxy) = \frac{x}{c}(1 - ax - 2by).$$

There are clearly four critical points:

$$\begin{pmatrix} 0 \\ 0 \\ 1/c \end{pmatrix}, \quad \begin{pmatrix} 0 \\ 1/b \\ 0 \end{pmatrix}, \quad \begin{pmatrix} 1/a \\ 0 \\ 0 \end{pmatrix}, \quad \begin{pmatrix} 1/(3a) \\ 1/(3b) \\ 1/(3c) \end{pmatrix}.$$

b.  At these four points the Hessian matrices (matrices whose entries are second derivatives) are

$$\frac{1}{c}\begin{bmatrix} 0 & 1 \\ 1 & 0 \end{bmatrix}, \ \frac{1}{c}\begin{bmatrix} -2a/b & -1 \\ -1 & 0 \end{bmatrix}, \ \frac{1}{c}\begin{bmatrix} 0 & -1 \\ -1 & -2b/a \end{bmatrix}, \ \frac{1}{c}\begin{bmatrix} -(2a)/(3b) & -1/3 \\ -1/3 & -(2b)/(3a) \end{bmatrix}.$$

By proposition 3.5.16, each of these symmetric matrices uniquely determines a quadratic form; by theorem 3.6.7, these quadratic forms can be used to analyze the critical points.

The quadratic form corresponding to the first matrix is

$$\frac{2}{c}xy = \frac{1}{2c}\left((x+y)^2 - (x-y)^2\right),$$

with signature $(1,1)$, so the first point is a saddle point. Since nothing made the $z$-coordinate special, the second and third points are also saddles. (This could also be computed directly.)

The fourth corresponds to the quadratic form

$$-\frac{2}{3c}\left(\frac{a}{b}x^2 + \frac{b}{a}y^2 + xy\right) = -\frac{2}{3c}\left(\left(\sqrt{\frac{a}{b}}x + \frac{1}{2}\sqrt{\frac{b}{a}}y\right)^2 + \frac{3b}{4a}y^2\right).$$

Thus the quadratic form is negative definite, and the function is a maximum. (You could also do this conceptually: The region

$$ax + by + cz = 1, \ x, y, z \geq 0$$

is compact, and the function is nonnegative there. Thus it must have a positive maximum; since the function is equal to 0 at the other critical points, this maximum must be the fourth critical point.)

**3.7.9** The definition of the derivative tells us that

$$[\mathbf{D}Q_A(\vec{\mathbf{a}})]\vec{\mathbf{h}} = \lim_{k \to 0} \frac{1}{k} \left( Q_A(\vec{\mathbf{a}} + k\vec{\mathbf{h}}) - Q_A(\vec{\mathbf{a}}) \right),$$

and the definition of $Q_A$ tells us that

$$\begin{aligned}
\lim_{k \to 0} \frac{1}{k} \left( Q_A(\vec{\mathbf{a}} + k\vec{\mathbf{h}}) - Q_A(\vec{\mathbf{a}}) \right) &= \lim_{k \to 0} \frac{1}{k} \left( (\vec{\mathbf{a}} + k\vec{\mathbf{h}}) \cdot (A(\vec{\mathbf{a}} + k\vec{\mathbf{h}})) - \vec{\mathbf{a}} \cdot (A\vec{\mathbf{a}}) \right) \\
&= \lim_{k \to 0} \left( \vec{\mathbf{a}} \cdot (A\vec{\mathbf{h}}) + \vec{\mathbf{h}} \cdot (A\vec{\mathbf{a}}) + k\vec{\mathbf{h}} \cdot (A\vec{\mathbf{h}}) \right) \qquad (1) \\
&= \vec{\mathbf{a}} \cdot (A\vec{\mathbf{h}}) + \vec{\mathbf{h}} \cdot (A\vec{\mathbf{a}}) \\
&= \vec{\mathbf{a}}^\top A\vec{\mathbf{h}} + \vec{\mathbf{h}}^\top A\vec{\mathbf{a}}.
\end{aligned}$$

Using the properties of transposes on products and noting that $A$ is symmetric, we see that

$$(\vec{\mathbf{h}}^\top A\vec{\mathbf{a}})^\top = \vec{\mathbf{a}}^\top A^\top \vec{\mathbf{h}} = \vec{\mathbf{a}}^\top A\vec{\mathbf{h}}.$$

Since $\vec{\mathbf{h}}^\top A\vec{\mathbf{a}}$ is a number, it is symmetric, so we also have

$$\vec{\mathbf{h}}^\top A\vec{\mathbf{a}} = \vec{\mathbf{a}}^\top A\vec{\mathbf{h}}.$$

Substituting this into the last line of equation (1) yields

$$\lim_{k \to 0} \frac{1}{k} \left( Q_A(\vec{\mathbf{a}} + k\vec{\mathbf{h}}) - Q_A(\vec{\mathbf{a}}) \right) = 2\vec{\mathbf{a}}^\top A\vec{\mathbf{h}}.$$

Solution 3.7.11: Exercise 1.2.16 asked you to show that $A^\top A$ is symmetric. If you did not do that exercise, just look at equation 5.1.9.

**3.7.11** a. Since the matrix $M = A^\top A$ is symmetric, we have

$$Q_M(\vec{\mathbf{x}}) = Q_{A^\top A}(\vec{\mathbf{x}}) = \vec{\mathbf{x}} \cdot (A^\top A\vec{\mathbf{x}}) = \vec{\mathbf{x}}^\top A^\top A\vec{\mathbf{x}} = (A\vec{\mathbf{x}})^\top A\vec{\mathbf{x}} = (A\vec{\mathbf{x}}) \cdot (A\vec{\mathbf{x}}),$$

so if the quadratic form is expressed as a sum of squares of linearly independent functions, they all are positive. By definition 3.5.12, the rank of a quadratic form is the number of linearly independent squares that appear when the quadratic form is represented as a sum of linearly independent squares. So write

$$Q_M(\mathbf{x}) = (\alpha_1(\mathbf{x}))^2 + \cdots + (\alpha_m(\mathbf{x}))^2,$$

Proposition 2.5.11: The number of linearly independent columns of a matrix $A$ equals the number of linearly independent rows.

where the condition that the $\alpha_i$ are linearly independent row matrices means that the $m \times n$ matrix

$$B = \begin{bmatrix} \alpha_1 \\ \vdots \\ \alpha_m \end{bmatrix}$$

has rank $m$. But $\ker B = \ker A$; indeed,

$$\mathbf{x} \in \ker B \iff \underbrace{(\alpha_1(\mathbf{x}))^2 + \cdots + (\alpha_m(\mathbf{x}))^2}_{Q_M(\mathbf{x})} = 0 \iff \underbrace{(A\mathbf{x}) \cdot A\mathbf{x}}_{Q_M(\mathbf{x})} = 0 \iff A\mathbf{x} = \mathbf{0}.$$

By the dimension formula,

$$m = n - \dim \ker B = n - \dim \ker A = \operatorname{rank} A.$$

b. If $A$ is invertible, its rank is $n$, so by part a, the signature of $Q_M$ is $(n, 0)$, so $Q_M$ is positive definite (see exercise 3.5.7).

c. To show that $\vec{\mathbf{w}}_1, \ldots, \vec{\mathbf{w}}_n$ is an orthonormal basis of $\mathbb{R}^n$ we must show that $\vec{\mathbf{w}}_i \cdot \vec{\mathbf{w}}_j$ is 0 if $i \neq j$ and 1 if $i = j$:

$$\vec{\mathbf{w}}_i \cdot \vec{\mathbf{w}}_j = \frac{A\vec{\mathbf{v}}_i}{\sqrt{\lambda_i}} \cdot \frac{A\vec{\mathbf{v}}_j}{\sqrt{\lambda_j}} = \frac{1}{\sqrt{\lambda_i \lambda_j}} \left( \vec{\mathbf{v}}_i^\top A^\top A \vec{\mathbf{v}}_j \right) = \frac{1}{\sqrt{\lambda_i \lambda_j}} (\vec{\mathbf{v}}_i^\top \lambda_j \vec{\mathbf{v}}_j)$$

$$= \frac{\lambda_j}{\sqrt{\lambda_i \lambda_j}} \vec{\mathbf{v}}_i \cdot \vec{\mathbf{v}}_j.$$

Since $\vec{\mathbf{v}}_1, \ldots, \vec{\mathbf{v}}_n$ is an orthonormal basis, this gives 0 when $i \neq j$ and $\dfrac{\lambda_j}{\sqrt{\lambda_j^2}} = 1$ when $i = j$.

d. This follows immediately from the fact that the $i$th column of $[\vec{\mathbf{w}}_1, \ldots, \vec{\mathbf{w}}_n]$ is $A\dfrac{\vec{\mathbf{v}}_i}{\sqrt{\lambda_i}}$.

e. Multiply both sides of the equation

$$[\vec{\mathbf{w}}_1, \ldots, \vec{\mathbf{w}}_n] = A[\vec{\mathbf{v}}_1, \ldots, \vec{\mathbf{v}}_n] \begin{bmatrix} \dfrac{1}{\sqrt{\lambda_1}} & & 0 \\ & \ddots & \\ 0 & & \dfrac{1}{\sqrt{\lambda_n}} \end{bmatrix}$$

A matrix $Q$ is orthogonal if and only if $Q^\top Q = I$, i.e., $Q^{-1} = Q^\top$ (exercise 2.4.7). If $Q$ is orthogonal, then $Q^{-1}$ is orthogonal:

$$(Q^{-1})^\top = (Q^\top)^\top = Q = (Q^{-1})^{-1}$$

(you were asked to show this in exercise 3.2.11).

So since $[\vec{\mathbf{v}}_1, \ldots, \vec{\mathbf{v}}_n]$ is orthogonal, $[\vec{\mathbf{v}}_1, \ldots, \vec{\mathbf{v}}_n]^{-1}$ is orthogonal. We showed that $[\vec{\mathbf{w}}_1, \ldots, \vec{\mathbf{w}}_n]$ is orthogonal in part c.

first by $\begin{bmatrix} \dfrac{1}{\sqrt{\lambda_1}} & & 0 \\ & \ddots & \\ 0 & & \dfrac{1}{\sqrt{\lambda_n}} \end{bmatrix}^{-1} = \begin{bmatrix} \sqrt{\lambda_1} & & 0 \\ & \ddots & \\ 0 & & \sqrt{\lambda_n} \end{bmatrix}$, then by $[\vec{\mathbf{v}}_1, \ldots, \vec{\mathbf{v}}_n]^{-1}$

to get

$$A = \underbrace{[\vec{\mathbf{w}}_1, \ldots, \vec{\mathbf{w}}_n]}_{\text{orthogonal}} \begin{bmatrix} \sqrt{\lambda_1} & & 0 \\ & \ddots & \\ 0 & & \sqrt{\lambda_n} \end{bmatrix} \underbrace{[\vec{\mathbf{v}}_1, \ldots, \vec{\mathbf{v}}_n]^{-1}}_{\text{orthogonal}}.$$

**3.7.13** The curve $C$ is a hyperbola. Indeed, take a point of the plane curve $C'$ of equation $2xy + 2x + 2y + 1 = 0$, and set $z = 1 + x + y$. Then $z^2 = x^2 + y^2 + 1 + 2xy + 2x + 2y = x^2 + y^2$, so this point is both on the plane and on the cone. The equation for $C'$ can be written $(x+1)(y+1) = 1/2$, so $C'$ is the hyperbola of equation $xy = 1/2$, translated by $\begin{bmatrix} -1 \\ -1 \end{bmatrix}$.

In particular, $C$ stretches out to infinity, and there is no point furthest from the origin. But the hyperbola is a closed subset of $\mathbb{R}^3$, and each branch contains a point closest to the origin, i.e., a local minimum.

We are interested in the extrema of the function $x^2 + y^2 + z^2$ (measuring squared distance from the origin), constrained to the two equations defining $C$. At a critical point of the constrained function, we must have

$$[2x,\ 2y, 2z] = \lambda_1[2x,\ 2y,\ -2z] + \lambda_2[1,\ 1,\ -1].$$

The first two of these equations, $2x = \lambda_1 2x + \lambda_2$ and $2y = \lambda_1 2y + \lambda_2$, imply that either $\lambda_1 = 1$ and $\lambda_2 = 0$ or that $x = y$.

The hypothesis $\lambda_2 = 0$ gives $z = 0$, hence the first constraint (equation of the cone) becomes $x^2 + y^2 = 0$, i.e., $x = y = 0$. This is incompatible with the second constraint.

So we must have $x = y$. Then the two equations $2x^2 - z^2 = 0$ and $2x - z = -1$ give $2x^2 + 4x + 1 = 0$. Solving this quadratic equation, we find the two roots

$$x_1 = -\frac{2 + \sqrt{2}}{2} \quad \text{and} \quad x_2 = \frac{-2 + \sqrt{2}}{2}.$$

These give the points

$$\begin{pmatrix} (-2 - \sqrt{2})/2 \\ (-2 - \sqrt{2})/2 \\ -1 - \sqrt{2} \end{pmatrix} \quad \text{and} \quad \begin{pmatrix} (-2 + \sqrt{2})/2 \\ (-2 + \sqrt{2})/2 \\ -1 + \sqrt{2} \end{pmatrix}.$$

The distances squared from the origin to these points are $6 + 4\sqrt{2}$ and $6 - 4\sqrt{2}$ respectively. Since both of the local minima of the distance function are critical points, these two points are both local minima.

**3.7.15 a.** The partial derivatives of $f$ are

$$D_1 f \begin{pmatrix} x \\ y \\ z \end{pmatrix} = y - 1, \qquad D_2 f \begin{pmatrix} x \\ y \\ z \end{pmatrix} = x + 1, \qquad D_3 f \begin{pmatrix} x \\ y \\ z \end{pmatrix} = 2z.$$

The only point at which all three vanish is $\mathbf{x}_0$, so this is the only critical point of $f$. Since $f$ is a quadratic polynomial, its Taylor polynomial of degree two at $\mathbf{x}_0$ is itself:

$$P^2_{f,\mathbf{x}_0} \begin{pmatrix} -1 + h_x \\ 1 + h_y \\ 0 + h_z \end{pmatrix} = f \begin{pmatrix} -1 + h_x \\ 1 + h_y \\ 0 + h_z \end{pmatrix}$$

$$= (-1 + h_x)(1 + h_y) - (-1 + h_x) + (1 + h_y) + (h_z)^2$$

$$= -1 + h_x - h_y + h_x h_y + 1 - h_x + 1 + h_y + h_z^2$$

$$= 1 + h_x h_y + h_z^2.$$

The quadratic terms yield the quadratic form

$$\frac{1}{4}\left((h_x + h_y)^2 - (h_x - h_y)^2\right) + h_z^2,$$

which has signature $(2, 1)$. Therefore $\mathbf{x}_0$ is a saddle of $f$ (definition 3.6.8).

b. The function $F$ is our constraint function. Its derivative is

$$\left[\mathbf{D}F\begin{pmatrix} x \\ y \\ z \end{pmatrix}\right] = \begin{bmatrix} 2(x + 1) & 2(y - 1) & 2z \end{bmatrix}.$$

By the Lagrange multiplier theorem, a constrained critical point on $S_{\sqrt{2}}(\mathbf{x}_0)$ occurs at a point where

$$\left[\mathbf{D}f\begin{pmatrix} x \\ y \\ z \end{pmatrix}\right] = \lambda \left[\mathbf{D}F\begin{pmatrix} x \\ y \\ z \end{pmatrix}\right], \qquad \text{i.e.,} \qquad \begin{cases} y - 1 = 2\lambda(x + 1) \\ x + 1 = 2\lambda(y - 1) \\ 2z = 2\lambda z \end{cases}.$$

The final equation says either $\lambda = 1$ or $z = 0$.

If $z \neq 0$, solving the other equations yields $x = -1$, $y = 1$. Then the constraint function $F$ shows that $z = \pm\sqrt{2}$. Thus $\begin{pmatrix} -1 \\ 1 \\ \sqrt{2} \end{pmatrix}, \begin{pmatrix} -1 \\ 1 \\ -\sqrt{2} \end{pmatrix}$ are constrained critical points of $f$.

If $z = 0$, then the first two equations imply that $\lambda = \pm 1/2$ or $\lambda = 0$. But if $\lambda = 0$, then $y = 1$ and $x = -1$, yielding the point $\mathbf{x}_0$, which is not on $S_{\sqrt{2}}(\mathbf{x}_0)$. Therefore $\lambda = \pm 1/2$. In the case $\lambda = 1/2$, we get $y = x + 2$, while if $\lambda = -1/2$, we have $y = -x$. These, plus the equation of the sphere, lead to $x = 0$ or $x = -2$. Thus

$$\begin{pmatrix} 0 \\ 2 \\ 0 \end{pmatrix}, \quad \begin{pmatrix} -2 \\ 0 \\ 0 \end{pmatrix}, \quad \begin{pmatrix} 0 \\ 0 \\ 0 \end{pmatrix}, \quad \begin{pmatrix} -2 \\ 2 \\ 0 \end{pmatrix}$$

are also constrained critical points of $f$.

c. We showed in part a that the only critical point in the interior of the ball is at $\mathbf{x}_0$, and that this point is a saddle; therefore it cannot be an extremum. We compute $f$ at the constrained critical points (on the surface of the ball) found in part b:

$$f\begin{pmatrix} -1 \\ 1 \\ \sqrt{2} \end{pmatrix} = f\begin{pmatrix} -1 \\ 1 \\ -\sqrt{2} \end{pmatrix} = 3, \quad f\begin{pmatrix} 0 \\ 2 \\ 0 \end{pmatrix} = 2, \quad f\begin{pmatrix} -2 \\ 0 \\ 0 \end{pmatrix} = 2, \quad f\begin{pmatrix} 0 \\ 0 \\ 0 \end{pmatrix} = 0, \quad f\begin{pmatrix} -2 \\ 2 \\ 0 \end{pmatrix} = 0$$

Therefore the maximum value of $f$ on $\overline{B_{\sqrt{2}}(\mathbf{x}_0)}$ is 3 and the minimum value is 0. Note that at the critical point $\mathbf{x}_0$ we have $f(\mathbf{x}_0) = 1$, which does indeed lie between the maximum and minimum values.

**3.8.1**  It's probably easiest here to write $y$ as an explicit function of $x$:

$$y = f(x) = b\left(1 - \frac{x^2}{a^2}\right)^{1/2}.$$

Then

$$f'(x) = b \cdot \frac{1}{2}\left(1 - \frac{x^2}{a^2}\right)^{-1/2}\left(\frac{-2x}{a^2}\right) = \frac{-bx}{a^2}\frac{a}{(a^2 - x^2)^{1/2}} = \frac{-bx}{a\sqrt{a^2 - x^2}}$$

and

$$f''(x) = -\frac{b}{a}\frac{\sqrt{a^2 - x^2} - x\frac{-2x}{2\sqrt{a^2 - x^2}}}{a^2 - x^2} = -\frac{b}{a}\frac{\sqrt{a^2 - x^2} + \frac{x^2}{\sqrt{a^2 - x^2}}}{a^2 - x^2}$$

$$= \frac{-ab}{(a^2 - x^2)^{3/2}}.$$

Inserting these values into the equation

$$\kappa = \frac{|f''(a)|}{\left(1 + (f'(a))^2\right)^{3/2}}$$

given in proposition 3.8.2 gives

$$\kappa = \frac{ab}{(a^2 - x^2)^{3/2}} \cdot \frac{1}{\left(1 + \frac{b^2 x^2}{a^2(a^2 - x^2)}\right)^{3/2}} = \frac{ab}{(a^2 - x^2)^{3/2}} \cdot \frac{a^3(a^2 - x^2)^{3/2}}{(a^4 - a^2 x^2 + b^2 x^2)^{3/2}}$$

$$= \frac{ba^4}{(a^4 - x^2(a^2 - b^2))^{3/2}}. \tag{1}$$

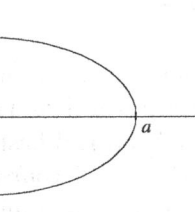

FIGURE FOR SOLUTION 3.8.1
The ellipse given by

$$\frac{x^2}{a^2} + \frac{y^2}{b^2} = 1$$

Since we know that $\frac{x^2}{a^2} + \frac{y^2}{b^2} = 1$ gives the ellipse shown at left, we may wish to see whether this value for the curvature seems reasonable for certain values of $a$, $b$, and $x$. For example, if $a = b = 1$, the ellipse is the unit circle, which should have curvature 1. This is indeed what equation (1) gives. At $x = 0$ the curvature is $\frac{b}{a^2}$, while at $x = a$ it is $\frac{a}{b^2}$, a reciprocal relationship that seems reasonable. If $a = b = r$, so that the ellipse is a circle with radius $r$, the equation gives $\kappa = 1/r$: the larger the circle, the smaller the curvature.

**3.8.3** First, note that any point can be rotated to the point $\mathbf{a} = \begin{pmatrix} 0 \\ 0 \\ 1 \end{pmatrix}$, so it is enough to compute the curvatures there.

Let us set $z = f\begin{pmatrix} x \\ y \end{pmatrix} = \sqrt{1 - x^2 - y^2}$. Then near $\mathbf{a}$ the "best coordinates" are $X = x$, $Y = y$, $Z = z - 1$, so that $Z = z - 1 = \sqrt{1 - X^2 - Y^2} - 1$. To determine the Taylor polynomial of $f$ at the origin, we use equation 3.4.9:

$$(1 + x)^m = 1 + mx + \frac{m(m - 1)}{2!}x^2 + \frac{m(m - 1)(m - 2)}{3!}x^3 + \cdots \tag{1}$$

with $m = 1/2$, to get

$$Z \approx 1 + \frac{1}{2}(-X^2 - Y^2) + \frac{1}{2}\left(1 - \frac{1}{2}\right)\frac{1}{2!}(-X^2 - Y^2)^2 + \cdots - 1. \tag{2}$$

Solution 3.8.3: We discuss the effect of the choice of point on the sign of the mean curvature at the end of the solution.

Note that the quadratic terms of equation (2) do not correspond to the quadratic terms of equation (1).

The quadratic terms of the Taylor polynomial are $-\dfrac{X^2}{2} - \dfrac{Y^2}{2}$. Since in an adapted coordinate system the quadratic terms of the Taylor polynomial are

$$\frac{1}{2}(A_{2,0}X^2 + 2A_{1,1}XY + A_{0,2}Y^2),$$

we have $A_{2,0} = A_{0,2} = -1$ and $A_{1,1} = 0$. Definition 3.8.7 of mean curvature then gives

$$\text{mean curvature} = \frac{1}{2}(A_{2,0} + A_{0,2}) = -1$$

and definition 3.8.8 of Gaussian curvature gives

$$\text{Gaussian curvature} = A_{2,0}A_{0,2} - A_{1,1}^2 = 1.$$

What if you had chosen to work near $\begin{pmatrix} 0 \\ 0 \\ -1 \end{pmatrix}$? In a neighborhood of this point, $z = f\begin{pmatrix} x \\ y \end{pmatrix} = -\sqrt{1 - x^2 - y^2}$, so setting $X = x$, $Y = y$, $Z = z + 1$, we would have $Z = z + 1 = -\sqrt{1 - X^2 - Y^2} + 1$, so that

$$Z \approx -1 - \frac{1}{2}(-X^2 - Y^2) - \frac{1}{2}\left(1 - \frac{1}{2}\right)\frac{1}{2!}(-X^2 - Y^2)^2 - \cdots + 1.$$

This gives a mean curvature of $+1$.

In this simple case, the "best coordinates" are automatic. Even if you did not use the $X, Y, Z$ notation when you computed the Taylor polynomial you would have found no first-degree terms. Thus if you tried to apply proposition 3.8.9 you would have run into

$$c = \sqrt{a_1^2 + a_2^2} = 0,$$

and all the expressions that involve dividing by $c$ or by $c^2$ would not make sense. That proposition only applies to cases where at least one of $a_1$ and $a_2$ is nonzero.

Recall that in the discussion of "best coordinates for surfaces," we said that an adapted coordinate system for a surface $S$ at $\mathbf{a}$ "is a system where $X$ and $Y$ are coordinates with respect to an orthonormal basis of the tangent plane, and the $Z$-axis is the normal direction." Of course there are two normal directions: pointing up on the $Z$-axis or pointing down. If we assume that the normal is pointing up, then when we chose $\mathbf{a} = \begin{pmatrix} 0 \\ 0 \\ 1 \end{pmatrix}$, we chose a point at which the sphere is curving away from the direction of the normal, giving a negative mean curvature. When we redid the computation at $\mathbf{a} = \begin{pmatrix} 0 \\ 0 \\ -1 \end{pmatrix}$, then the sphere was curving towards the direction of the normal, giving a positive mean curvature.

**3.8.5**  By equation 3.8.6, the curvature of the curve at $\begin{pmatrix} x \\ f(x) \end{pmatrix}$ is $\kappa = \dfrac{|f''(x)|}{(1 + (f'(x))^2)^{3/2}}$. The surface satisfies

$$a_1 = f'(x), \ a_2 = 0, \ a_{2,0} = f''(x), \ a_{1,1} = a_{0,2} = 0,$$

hence $c^2 = (f'(x))^2$, so equation 3.8.38 says that the absolute value of the mean curvature is

$$|H| = \frac{|f''(x)|}{2(1 + (f'(x))^2)^{3/2}},$$

which is indeed half the curvature as given by proposition 3.8.2.

Solution 3.8.7 is intended to illustrate figure 3.8.7. A goat tethered at the North Pole with a chain 2533 kilometers long would have less grass to eat (assuming grass would grow) than a goat tethered at the North Pole on a "flat" earth.

**3.8.7**  a. Remember (or look up) that the radius of the earth is 6366 km.[3] Remember also that the inclination of the axis of the earth to the ecliptic is 23.45°. (Again, this isn't exact, and the inclination varies periodically with a period of about 30 000 years and amplitude $\approx .5°$.)

Thus the real radius of the arctic circle (the radius as measured in the plane containing the arctic circle) is $6366 \sin(23.45°) = 2533$ km, and its circumference is $2\pi 2533 \approx 15930$ km. If the earth were flat, the circumference of a circle with the radius of the arctic circle as measured from the pole would be $2\pi 2607.5 \approx 16 383$ km.

b. By the same computation as above, the equation for the angle (in radians) between the pole and a point of the circumference, with vertex at the center of the earth, is

$$2\pi 6366\theta - 2\pi 6366 \sin\theta = 1, \quad \text{i.e.,} \quad \theta - \sin\theta = \frac{1}{2\pi \cdot 6366}.$$

A computer will tell you that the solution to this equation is approximately $\theta \approx .05341$ radians. If you don't have a computer handy, a very good approximation is obtained by using two terms of the Taylor expansion for sine, which gives $(6/(2\pi \cdot 6366))^{1/3} \approx .05313$. The corresponding circle has radius on earth about 340 km.

**3.8.9**  a. The hypocycloid looks like the solid curve in the margin (top).

b. If you ride a 2-dimensional bicycle on the inside of the dotted circle, and the radius of your wheels is $1/4$ of the radius of the circle, a dot on the rim will describe a hypocycloid, as shown in the bottom figure at left.

c. Using definition 3.8.5, the length is given by

$$4a \int_0^{\pi/2} \sqrt{(3\sin^2 t \cos t)^2 + (3\sin t \cos^2 t)^2}\, dt$$

$$= 12a \int_0^{\pi/2} \sqrt{\sin^2 t \cos^2 t(\sin^2 t + \cos^2 t)}\, dt$$

$$= 6a \int_0^{\pi/2} 2\sin t \cos t\, dt = 6a.$$

The hypocycloid has four arcs, so the length of one arc is $3a/2$.

Note that the circle has circumference $2\pi a$, so that the lengths of the circle and of the hypocycloid are close.

**3.8.11**  The map

$$\varphi : s \mapsto \left[\vec{\mathbf{t}}(s), \vec{\mathbf{n}}(s), \vec{\mathbf{b}}(s)\right], \quad s \in I$$

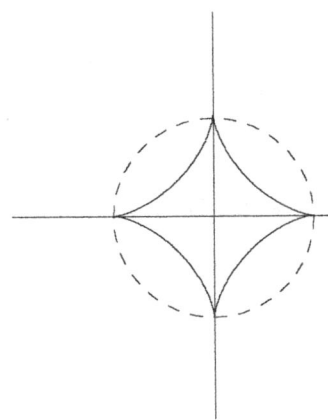

FIGURE FOR SOLUTION 3.8.9
TOP: The hypocycloid of part a is the curve inscribed in the dotted circle. BOTTOM: A dot on the rim of this 2-dimensional bicycle describes a hypocycloid.

---

[3]Actually, the earth is flattened at the poles, and the radius ranges from 6356 kilometers at the poles to 6378 km at the equator, but the easy number to remember is that the circumference of the earth is 40 000 km, so the radius is approximately $40\,000/(2\pi) \approx 6366$ km.

($\varphi$ for "phrenet") is a mapping $I \to O(3)$. Exercise 3.2.11 says that the tangent space to $O(3)$ at the identity is the space of antisymmetric matrices. If we adapt our Frenet map $\varphi$ to consider the map

$$\psi : s \mapsto \left[\vec{\mathbf{t}}(s_0), \vec{\mathbf{n}}(s_0), \vec{\mathbf{b}}(s_0)\right]^{-1} \left[\vec{\mathbf{t}}(s), \vec{\mathbf{n}}(s), \vec{\mathbf{b}}(s)\right], \quad s \in I,$$

part b of exercise 3.2.11 tells us that this represents a parametrized curve in $O(3)$, such that $\psi(s_0) = I$. Thus its derivative $\psi'(s_0)$ is antisymmetric; temporarily let us call it

$$A = \left[\vec{\mathbf{t}}(s_0), \vec{\mathbf{n}}(s_0), \vec{\mathbf{b}}(s_0)\right]^{-1} \left[\vec{\mathbf{t}}'(s), \vec{\mathbf{n}}'(s), \vec{\mathbf{b}}'(s)\right].$$

This can be rewritten

$$\left[\vec{\mathbf{t}}'(s), \vec{\mathbf{n}}'(s), \vec{\mathbf{b}}'(s)\right] = \left[\vec{\mathbf{t}}(s_0), \vec{\mathbf{n}}(s_0), \vec{\mathbf{b}}(s_0)\right] A,$$

which, if $A$ is the antisymmetric matrix

$$A = \begin{bmatrix} 0 & -\kappa & 0 \\ \kappa & 0 & -\tau \\ 0 & \tau & 0 \end{bmatrix},$$

corresponds to

$$\vec{\mathbf{t}}'(s_0) = \qquad\qquad \kappa(s_0)\vec{\mathbf{n}}(s_0)$$
$$\vec{\mathbf{n}}'(s_0) = -\kappa(s_0)\vec{\mathbf{t}}(s_0) \qquad\qquad + \tau(s_0)\vec{\mathbf{b}}(s_0)$$
$$\vec{\mathbf{b}}'(s_0) = \qquad\qquad -\tau(s_0)\vec{\mathbf{n}}(s_0).$$

# SOLUTIONS FOR REVIEW EXERCISES, CHAPTER 3

**3.1** a. The derivative of the mapping $x^3 + xy^2 + yz^2 + z^3 - 4$ is

$$[\, 3x^2 + y^2 \quad 2xy + z^2 \quad 2yz + 3z^2 \,],$$

which only vanishes if $x = y = z = 0$. (Look at the first entry, then the second.) The origin isn't on $X$, so $X$ is a smooth surface.

b. At the point $\begin{pmatrix} 1 \\ 1 \\ 1 \end{pmatrix}$, the derivative of $F$ is $[\,4 \quad 3 \quad 5\,]$. The tangent plane to the surface is the plane of equation $4x + 3y + 5z = 12$.

The tangent space to the surface is the kernel of the derivative, i.e., it is the plane of equation $4\dot{x} + 3\dot{y} + 5\dot{z} = 0$.

**3.3** a. The space $X$ is given by the equations

$$(x_1 - 1)^2 + y_1^2 + z_1^2 - 1 = 0$$
$$(x_2 + 1)^2 + y_2^2 + z_2^2 - 1 = 0$$
$$(x_1 - x_2)^2 + (y_1 - y_2)^2 + (z_1 - z_2)^2 - 4 = 0.$$

b. The derivative of the mapping $\mathbb{R}^6 \to \mathbb{R}^3$ defining $X$ is

$$\begin{bmatrix} 2(x_1 - 1) & 2y_1 & 2z_1 & 0 & 0 & 0 \\ 0 & 0 & 0 & 2(x_2 + 1) & 2y_2 & 2z_2 \\ 2(x_1 - x_2) & 2(y_1 - y_2) & 2(z_1 - z_2) & 2(x_2 - x_1) & 2(y_2 - y_1) & 2(z_2 - z_1) \end{bmatrix}.$$

When

$$\begin{pmatrix} x_1 \\ y_1 \\ z_1 \end{pmatrix} = \begin{pmatrix} 1 \\ 1 \\ 0 \end{pmatrix}, \quad \begin{pmatrix} x_2 \\ y_2 \\ z_2 \end{pmatrix} = \begin{pmatrix} -1 \\ 1 \\ 0 \end{pmatrix}, \tag{1}$$

this matrix is

$$\begin{bmatrix} 0 & 2 & 0 & 0 & 0 & 0 \\ 0 & 0 & 0 & 0 & 2 & 0 \\ 4 & 0 & 0 & -4 & 0 & 0 \end{bmatrix}, \tag{2}$$

which has rank 3 since the first, second, and fifth columns are linearly independent. So $X$ is a manifold of dimension 3 near the points in equation (1).

c. The tangent space is the kernel of the matrix (2) computed in part b, i.e., the set of

$$\begin{pmatrix} u_1 \\ v_1 \\ w_1 \end{pmatrix}, \quad \begin{pmatrix} u_2 \\ v_2 \\ w_2 \end{pmatrix}$$

such that $2v_1 = 0$, $2v_2 = 0$, $4(u_1 - u_2) = 0$, which is a 3-dimensional subspace of $\mathbb{R}^6$.

111

**3.5** The best way to deal with this is to use the Taylor series for sine and cosine given in proposition 3.4.2. Set

$$x = \frac{\pi}{6} + u, \quad y = \frac{\pi}{4} + v, \quad z = \frac{\pi}{3} + w,$$

and compute

$$\sin\left(\frac{3\pi}{4} + u + v + w\right) = \sin\left(\frac{3\pi}{4}\right)\cos(u+v+w) + \cos\left(\frac{3\pi}{4}\right)\sin(u+v+w)$$

$$= \frac{\sqrt{2}}{2}\left(1 - \frac{1}{2}(u+v+w)^2\right) - \frac{\sqrt{2}}{2}\left(u+v+w - \frac{1}{6}(u+v+w)^3\right) + \cdots$$

$$= \frac{\sqrt{2}}{2}\left(1 - (u+v+w) - \frac{1}{2}(u+v+w)^2 + \frac{1}{6}(u+v+w)^3\right) + \cdots.$$

**Moral:** If you can, compute Taylor polynomials from known expansions rather than by computing partial derivatives.

Thus the required Taylor polynomial is

$$\frac{\sqrt{2}}{2}\left(1 - (u+v+w) - \frac{1}{2}(u+v+w)^2 + \frac{1}{6}(u+v+w)^3\right).$$

**Solution 3.7:** The wrong way to go about this problem is to say that $\sin(x^2 + y) = x^2 + y - \frac{1}{6}y^3$ plus higher-degree terms, and then apply equation 3.4.7, saying that

$$\cos\left(1 + \sin(x^2 + y)\right)$$

$$\approx \cos(1 + x^2 + y - \frac{1}{6}y^3)$$

$$= 1 - \frac{1}{2}\left(1 + x^2 + y - \frac{1}{6}y^3\right)^2$$

$$+ \cdots.$$

This is WRONG (not just a harder way of going about things), because equation 3.4.7 is true only near the origin. At the origin,

$$1 + \sin(x^2 + y) = 1,$$

so we cannot treat it as the $x$ of equation 3.4.7, which must be 0 at the origin.

**3.7** The margin note discusses the incorrect way to go about this. What we can do is use the formula

$$\cos(a + b) = \cos a \cos b - \sin a \sin b,$$

which gives

$$\cos\left(1 + \sin(x^2 + y)\right) = \cos 1\left(\cos(\sin(x^2 + y))\right) - \sin 1\left(\sin(\sin(x^2 + y))\right).$$

At the origin, $\sin(x^2 + y) = 0$, so now we can apply equations 3.4.6 and 3.4.7 (discarding terms higher than degree 3), to get

$$\cos\left(1 + \sin(x^2 + y)\right) = \cos 1\left(1 - \frac{(x^2 + y)^2}{2}\right)$$

$$- \sin 1\left((x^2 + y) - \frac{1}{6}(x^2 + y)^3 - \frac{1}{6}(x^2 + y)^3\right)$$

$$= \cos 1 - (\sin 1)y - (\sin 1)x^2 - \frac{\cos 1}{2}y^2 - (\cos 1)x^2 y - \frac{\sin 1}{3}y^3.$$

**3.9** Both first partials exist and are continuous for all homogeneous polynomials of degree 4, and

$$D_2\left(D_1 f\begin{pmatrix}0\\y\end{pmatrix}\right) = d \quad \text{and} \quad D_1\left(D_2 f\begin{pmatrix}x\\0\end{pmatrix}\right) = b.$$

It follows that

$$D_2\left(D_1 f\begin{pmatrix}0\\0\end{pmatrix}\right) - D_1\left(D_2 f\begin{pmatrix}0\\0\end{pmatrix}\right) = d - b,$$

and that the condition for the crossed partials to be equal is that $d = b$.

We are evaluating

$$D_z(y\cos z - x\sin z)$$

at $\begin{pmatrix} r \\ 0 \\ 0 \end{pmatrix}$, not multiplying.

**3.11**  a. The implicit function theorem says that this will happen if

$$D_z(y\cos z - x\sin z)\begin{pmatrix} r \\ 0 \\ 0 \end{pmatrix} = -y\sin z - x\cos z \begin{pmatrix} r \\ 0 \\ 0 \end{pmatrix}$$

$$= -0\cdot\sin 0 - r\cos 0 = -r \neq 0.$$

b. The hint gives it away: since the $x$-axis is contained in the surface, we have $g_r\begin{pmatrix} x \\ 0 \end{pmatrix} = 0$ for all $x$, so

$$D_x^1 g_r\begin{pmatrix} r \\ 0 \end{pmatrix} = D_x^2 g_r\begin{pmatrix} r \\ 0 \end{pmatrix} = 0.$$

For a picture of this surface, see figure 3.8.9.

**3.13**  a. We have

$$\det\begin{bmatrix} 1 & x & y \\ 1 & y & z \\ 1 & z & x \end{bmatrix} = xy + xz + yz - x^2 - y^2 - z^2,$$

which certainly is a quadratic form. But $\det\begin{bmatrix} 0 & x & y \\ x & 0 & z \\ y & z & 0 \end{bmatrix} = 2xyz$ certainly is not.

b. We have $xy + xz + yz - x^2 - y^2 - z^2 = -\left(x - \frac{y}{2} - \frac{z}{2}\right)^2 - \frac{3}{4}(y-z)^2$, so the signature is $(0,2)$, and the quadratic form is degenerate.

**3.15**  The quadratic form represented is $ax^2 + 2bxy + dy^2$.

(if) If $a+d > 0$, at least one of $a$ and $d$ is positive, and then if $ad - b^2 > 0$, we must have both $a > 0$, $d > 0$. So we can write

$$ax^2 + 2bxy + dy^2 = \left(ax^2 + 2bxy + \frac{b^2}{a}y^2\right) + \frac{ad-b^2}{a}y^2$$

$$= a\left(x + \frac{b}{a}y\right)^2 + \frac{ad-b^2}{a}y^2,$$

which is strictly positive if $\begin{bmatrix} x \\ y \end{bmatrix} \neq \begin{bmatrix} 0 \\ 0 \end{bmatrix}$.

(only if) If you apply the quadratic form to $\begin{pmatrix} 1 \\ 0 \end{pmatrix}$ and $\begin{pmatrix} 0 \\ 1 \end{pmatrix}$, you get

$$a = \begin{bmatrix} 1 \\ 0 \end{bmatrix} \cdot G \begin{bmatrix} 1 \\ 0 \end{bmatrix} \quad \text{and} \quad d = \begin{bmatrix} 0 \\ 1 \end{bmatrix} \cdot G \begin{bmatrix} 0 \\ 1 \end{bmatrix},$$

so if the quadratic form is positive definite, you find $a > 0$, $d > 0$, hence $a + d > 0$.

Now apply the quadratic form to the vector $\begin{bmatrix} -b \\ a \end{bmatrix}$, to find

$$\begin{bmatrix} -b \\ a \end{bmatrix} \cdot \left(\begin{bmatrix} a & b \\ b & d \end{bmatrix}\begin{bmatrix} -b \\ a \end{bmatrix}\right) = a(ad - b^2).$$

Since this must also be positive, we find $ad - b^2 > 0$.

**3.17**   a.   The critical points are $\begin{pmatrix} 0 \\ 0 \end{pmatrix}$ and $\begin{pmatrix} 1 \\ 1 \end{pmatrix}$, since at those points the partial derivatives $D_1 f = 6x - 6y$ and $D_2 f = -6x + 6y^2$ vanish.

   b. We have $D_1^2 f = 6$, $D_1 D_2 f = -6$, and $D_2^2 f = 12y$, so the second-degree term of the Taylor polynomial is $3h_1^2 - 6h_1 h_2 + 6yh_2^2$. At the critical point $\begin{pmatrix} 0 \\ 0 \end{pmatrix}$, the term $6yh_2^2$ vanishes, so

$$Q(\vec{\mathbf{h}}) = 3(h_1^2 - 2h_1 h_2) = 3\big((h_1 - h_2)^2 - (h_2)^2\big);$$

the signature of this quadratic form is $(1, 1)$, and the critical point is a saddle.

   At the critical point $\begin{pmatrix} 1 \\ 1 \end{pmatrix}$, we have $6yh_2^2 = 6h_2^2$, so

$$Q(\vec{\mathbf{h}}) = 3\big(h_1^2 - 2h_1 h_2 + 2h_2^2\big) = 3\big((h_1 - h_2)^2 + h_2^2\big);$$

the signature of $Q$ is $(2, 0)$, and the critical point is a minimum.

**3.19**   a.   The function is $1 + 2xyz - x^2 - y^2 - z^2$. The critical points occur where all three partials vanish, i.e., where

$$2yz - 2x = 0$$
$$2xz - 2y = 0$$
$$2xy - 2z = 0.$$

Since $x = yz$, we have

$$x^2 = x(yz),$$

and so on.

It follows from these equations that $x^2 = y^2 = z^2 = xyz$. One solution is $\begin{pmatrix} 0 \\ 0 \\ 0 \end{pmatrix}$. The other solutions are of the form

$$\begin{pmatrix} a \\ a \\ a \end{pmatrix}, \quad \begin{pmatrix} a \\ a \\ -a \end{pmatrix}, \quad \begin{pmatrix} a \\ -a \\ -a \end{pmatrix}, \quad \begin{pmatrix} a \\ -a \\ a \end{pmatrix},$$

for an appropriate value of $a$. These are solutions of $a = a^2$ in the first and third, and $a = -a^2$ in the second and fourth, so the critical points are

$$\begin{pmatrix} 0 \\ 0 \\ 0 \end{pmatrix}, \quad \begin{pmatrix} 1 \\ 1 \\ 1 \end{pmatrix}, \quad \begin{pmatrix} -1 \\ -1 \\ 1 \end{pmatrix}, \quad \begin{pmatrix} 1 \\ -1 \\ -1 \end{pmatrix}, \quad \begin{pmatrix} -1 \\ 1 \\ -1 \end{pmatrix}.$$

   b.   At the origin, the quadratic terms are evidently $-x^2 - y^2 - z^2$, so that point is a maximum.

   At the point $\begin{pmatrix} 1 \\ 1 \\ 1 \end{pmatrix}$, if we set $x = 1 + u$, $y = 1 + v$, $z = 1 + w$, and multiply out, keeping only the quadratic terms, we find

$$2(1+u)(1+v)(1+w) - (1+u)^2 - (1+v)^2 + (1+w)^2 = -(u-v-w)^2 + (v+w)^2 - (v-w)^2.$$

This point is a $(1, 2)$-saddle.

At the point $\begin{pmatrix} 1 \\ -1 \\ -1 \end{pmatrix}$, if we set $x = 1+u$, $y = -1+v$, $z = -1+w$, and multiply out, keeping only the quadratic terms, we find

$$2(1+u)(-1+v)(-1+w)-(1+u)^2-(-1+v)^2+(-1+w)^2 = -(u+v+w)^2+(v+w)^2-(v-w)^2.$$

This point is also a $(1,2)$-saddle. Since the expression for the function is symmetric with respect to the variables, evidently the other two points behave the same way.

**3.21**  a. Consider the figure at left. The cosine law says

$$e^2 = a^2 + d^2 - 2ad\cos\varphi = b^2 + c^2 - 2bc\cos\psi.$$

b. The area of the quadrilateral is given by $\frac{1}{2}(ad\sin\varphi + bc\sin\psi)$.

c. Our quadrilateral satisfies the constraint

$$a^2 + d^2 - 2ad\cos\varphi - b^2 - c^2 + 2bc\cos\psi = 0.$$

So the Lagrange multiplier theorem asserts that at the maximum of the area function, there is a number $\lambda$ such that

$$[ad\cos\varphi, \ bc\cos\psi] = \lambda[2ad\sin\varphi, \ -2bc\sin\psi].$$

(To find the critical point of the area, we can ignore the $1/2$ in the formula for the area.) This immediately gives $\cot\varphi + \cot\psi = 0$, i.e., opposite angles are supplementary (i.e., sum to $180°$). It follows from high school geometry that the quadrilateral can be inscribed in a circle; see the figure below.

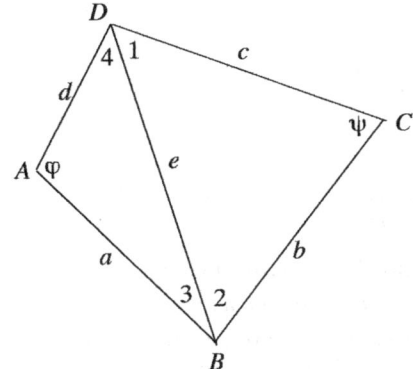

FIGURE FOR SOLUTION 3.21.

We compute the area of the quadrilateral $ABCD$ by computing the area of the two triangles, using the formula area $= \frac{1}{2}ab\sin\theta$, where $\theta$ is the angle between sides $a$ and $b$.

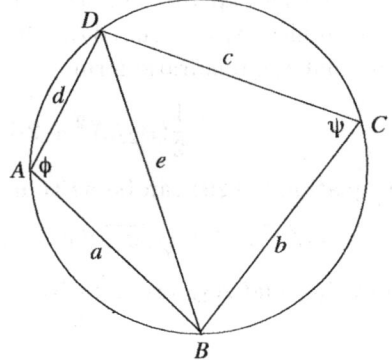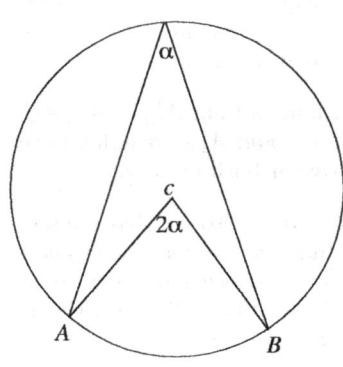

FIGURE FOR PART C, SOLUTION 3.21. LEFT: Our quadrilateral, inscribed in a circle. RIGHT: An angle inscribed in a circle is half the corresponding angle at the center of the circle. This is key to proving that a quadrilateral can be inscribed in a circle if and only if opposite angles are supplementary, a result you may remember from high school.

**3.23** Computing the derivatives of $f\begin{pmatrix} x \\ y \end{pmatrix} = \sqrt{x^2 + y^2}$ gives the following quantities (see equation 3.8.33), which enter in proposition 3.8.10:

$$D_1 f\begin{pmatrix} a \\ b \end{pmatrix} = a_1 = \frac{a}{\sqrt{a^2 + b^2}}, \quad D_2 f\begin{pmatrix} a \\ b \end{pmatrix} = a_2 = \frac{b}{\sqrt{a^2 + b^2}}, \quad c = 1,$$

$$a_{2,0} = \frac{b^2}{(a^2 + b^2)^{3/2}}, \quad a_{1,1} = -\frac{ab}{(a^2 + b^2)^{3/2}}, \quad a_{0,2} = \frac{a^2}{(a^2 + b^2)^{3/2}}.$$

Inserting these values into equation 3.8.37, we see that the Gaussian curvature is

$$K = \frac{a^2 b^2 - a^2 b^2}{2^2 (a^2 + b^2)^3} = 0.$$

By equation 3.8.38, the mean curvature is

$$H = \frac{1}{2\sqrt{2}\sqrt{a^2 + b^2}}.$$

Thus the mean curvature decreases as $|a|$ and $|b|$ increase. This is what one would expect: our surface is a cone, so it becomes flatter as we move away from the vertex of the cone. If we draw a circle on the cone near its vertex, and determine the minimal surface bounded by that circle, then repeat the procedure further and further away from the vertex, the difference between the area of the disc on the cone and the area of the minimal surface will decrease with distance from the vertex.

One way to explain why the Gaussian curvature is 0 is to look at equation 3.8.32. Our surface is a cone, which can be made out of a flat piece of paper, with no distortion. Thus the area of a disc around a point on the surface (a point not the vertex of the cone) is the same as the area of a flat disc with the same radius. Therefore, $K$ must be 0.

For another explanation, note that by definition 3.8.8, Gaussian curvature 0 implies $A_{1,1}^2 = A_{0,2} A_{2,0}$. If both $A_{0,2}$ and $A_{2,0}$ are positive, this means that the quadratic form

$$\frac{1}{2}\left(A_{2,0} X^2 + 2A_{1,1} XY + A_{0,2} Y^2\right) \tag{1}$$

(see equation 3.8.29) can be written

$$\left(\sqrt{A_{2,0}}\, X + \sqrt{A_{0,2}}\, Y\right)^2 = A_{2,0} X^2 + 2A_{1,1} XY + A_{0,2} Y^2.$$

If both $A_{0,2}$ and $A_{2,0}$ are negative, it can be written

$$-\left(\sqrt{-A_{2,0}}\, X - \sqrt{-A_{0,2}}\, Y\right)^2$$

Thus the quadratic form (second fundamental form) is degenerate.

Why should the second fundamental form of a cone be degenerate? In adapted coordinates, a surface is locally the graph of a function whose domain is the tangent space and whose values are in the normal line; see equation 3.8.29. Thus the quadratic terms making up the second fundamental form are also a map from the tangent plane to the normal line. In

---

A goat grazing on the flank of a conical volcano, constrained by a chain of a given length, too short for him to reach the top, has access to the same amount of grass as a goat grazing on a flat surface, and constrained by the same length chain; see figure 3.8.7.

The quadratic form (1) is called the *second fundamental form* of a surface at a point.

It follows from $A_{1,1}^2 = A_{0,2} A_{2,0}$ that $A_{0,2}$ and $A_{2,0}$ are either both positive or both negative.

We could also see that the second fundamental form is degenerate by considering the first equation in solution 3.25, substituting $A_{2,0}$ for $a$, $A_{1,1}$ for $2b$, and $A_{0,2}$ for $c$: the only way the quadratic form

$$a\left(x + \frac{b}{a} y\right)^2 + \frac{ac - b^2}{a} y^2$$

in the third line can be degenerate is if $ac = b^2$, i.e., $A_{1,1}^2 = A_{0,2} A_{2,0}$

the case of a cone, the tangent plane to a cone contains a line of the cone, so the quadratic terms vanish on that line.

**3.25** If $a \neq 0$ we can write

$$ax^2 + 2bxy + cy^2 = a\left(x^2 + \frac{2b}{a}xy\right) + cy^2$$

$$= a\left(x^2 + \frac{2b}{a}xy + \frac{b^2}{a^2}y^2\right) + \frac{ac - b^2}{a}y^2$$

$$= a\left(x + \frac{b}{a}y\right)^2 + \frac{ac - b^2}{a}y^2.$$

This represents the quadratic form as a sum of squares of linearly independent linear functions, with coefficients

$$a \quad \text{and} \quad \frac{ac - b^2}{a}.$$

This shows that if $a \neq 0$, then

$$a > 0 \quad \text{and} \quad ac - b^2 > 0 \iff \text{signature } (2, 0)$$

$$a < 0 \quad \text{and} \quad ac - b^2 > 0 \iff \text{signature } (0, 2)$$

$$a \neq 0 \quad \text{and} \quad ac - b^2 < 0 \iff \text{signature } (1, 1).$$

Conversely, if $ac - b^2 > 0$, then $a \neq 0$, so the only thing left to show is that if $ac - b^2 < 0$ and $a = 0$, then the quadratic form has signature $(1, 1)$. Exactly as above, if $c \neq 0$ we can write

$$2xy + cy^2 = c\left(\frac{b^2}{c^2}x^2 + 2\frac{b}{c}xy + y^2\right) - \frac{b^2}{c}x^2 = c\left(y + \frac{b}{c}x\right)^2 - \frac{b^2}{c}x^2,$$

and the coefficients do have opposite signs.

Finally, if $a = c = 0$, and $ac - b^2 < 0$, we have

$$2bxy = \frac{b}{2}\Big((x + y)^2 - (x - y)^2\Big),$$

so the quadratic form has signature $(1, 1)$.

**4.1.1**  a. The area of a dyadic cube $C \in \mathcal{D}_3(\mathbb{R}^2)$ is $(\frac{1}{2^3})^2 = \frac{1}{64}$; the area of a cube $C \in \mathcal{D}_4(\mathbb{R}^2)$ is $(\frac{1}{2^4})^2 = (\frac{1}{16})^2$; the area of a cube $C \in \mathcal{D}_5(\mathbb{R}^2)$ is $(\frac{1}{2^5})^2$.

b. The volume of a dyadic cube $C \in \mathcal{D}_3(\mathbb{R}^3)$ is $(\frac{1}{2^3})^3$; the volume of a cube $C \in \mathcal{D}_4(\mathbb{R}^3)$ is $(\frac{1}{2^4})^3$; the volume of $C \in \mathcal{D}_5(\mathbb{R}^3)$ is $(\frac{1}{2^5})^3$.

**4.1.3**  a. 2-dimensional volume (i.e., area) $(\frac{1}{2^3})^2 = 1/64$.

b. 3-dimensional volume $(\frac{1}{2^2})^3 = 1/64$

c. 4-dimensional volume $(\frac{1}{2^3})^4 = (\frac{1}{8})^4$

d. 3-dimensional volume $(\frac{1}{2^3})^3 = (\frac{1}{8})^3$

For each cube, the first index gives unnecessary information. This index is $\mathbf{k}$, which tells where the cube is. The number $n$ of entries of $\mathbf{k}$ gives necessary information: if it is 2, the cube is in $\mathbb{R}^2$, if it is 3, the cube is in $\mathbb{R}^3$, and so on. But we do not need to know what the entries are to compute the $n$-dimensional volume.

**4.1.5**  a. $\qquad \displaystyle\sum_{i=0}^{n} i = \frac{n(n+1)}{2}.$

b.

$$\int_{\mathbb{R}} x\mathbf{1}_{[0,1)}(x)|dx| = \int_{\mathbb{R}} x\mathbf{1}_{[0,1]}(x)|dx| = \int_{\mathbb{R}} x\mathbf{1}_{(0,1]}(x)|dx| = \int_{\mathbb{R}} x\mathbf{1}_{(0,1)}(x)|dx| = \frac{1}{2}.$$

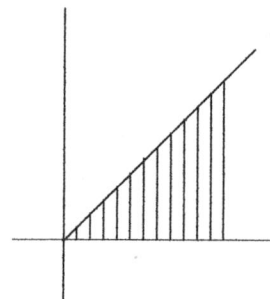

We will write out the first one in detail. Since our theory of integration is based on dyadic decompositions, we divide the interval from 0 to 1 into $2^N$ pieces, as suggested by the figure at left.

Let us compare upper and lower sums when the interval is $[0,1)$. For the lower sum (which corresponds to the left Riemann sum, since the function is increasing),the width of each piece is $\frac{1}{2^N}$, and the height of the first is 0, that of the second is $\frac{1}{2^N}$, and so on, ending with $\frac{2^N-1}{2^N}$. This gives the lower sum

$$L_N\left(x\mathbf{1}_{[0,1)}(x)\right) = \frac{1}{2^N}\sum_{i=0}^{2^N-1}\frac{i}{2^N} = \frac{1}{2^{2N}}\sum_{i=0}^{2^N-1} i.$$

For the upper sum (corresponding to the right Riemann sum), the height of the first piece is $\frac{1}{2^N}$, that of the second is $\frac{2}{2^N}$, and so on, ending with $\frac{2^N}{2^N} = 1$:

$$U_N\left(x\mathbf{1}_{[0,1)}(x)\right) = \frac{1}{2^N}\sum_{i=0}^{2^N}\frac{i}{2^N} = \frac{1}{2^N}\sum_{i=0}^{2^N-1}\frac{i+1}{2^N}.$$

So

$$U_N - L_N = \frac{1}{2^N},$$

which goes to 0 as $N \to \infty$, so the function is integrable. Using the result from part a, we see that it has integral $1/2$:

$$\lim_{N\to\infty} L_N\left(x\mathbf{1}_{[0,1)}(x)\right) = \lim_{N\to\infty}\left(\frac{1}{2^{2N}} \cdot \frac{(2^N-1)(2^N)}{2}\right) = \lim_{N\to\infty}\left(\frac{1}{2} - \frac{1}{2^{N+1}}\right) = \frac{1}{2}.$$

For the interval $[0,1]$, note that

$$U_N\left(x\mathbf{1}_{[0,1]}(x)\right) = U_N\left(x\mathbf{1}_{[0,1)}(x)\right) + \frac{1}{2^N};$$

$$L_N\left(x\mathbf{1}_{[0,1]}(x)\right) = L_N\left(x\mathbf{1}_{[0,1)}(x)\right).$$

Therefore,

$$\int_{\mathbb{R}} x\mathbf{1}_{[0,1)}(x)\,|dx| = \int_{\mathbb{R}} x\mathbf{1}_{[0,1]}(x)\,|dx| = \frac{1}{2}.$$

The upper and lower sums for the interval $(0,1]$ are the same as those for the interval $[0,1]$, and those for the interval $(0,1)$ are the same as those for $[0,1)$, so those integrals are equal also.

c. This is fundamentally easy; the difficulty is all in seeing exactly how many dyadic intervals are contained in $[0,a]$, and fiddling with the little bit at the right that is left over. We will require the notation $\lfloor b \rfloor$, called the *floor* of $b$, and which stands for the largest integer $\leq b$.

Consider first the case of the closed interval, i.e., integrate $x\mathbf{1}_{[0,a]}(x)$. Then we have

$$U_N\left(x\mathbf{1}_{[0,a]}(x)\right) = \sum_{i=0}^{\lfloor 2^N a \rfloor - 1} \frac{i+1}{2^{2N}} + \frac{a}{2^N} \quad \text{and} \quad L_N\left(x\mathbf{1}_{[0,a]}(x)\right) = \sum_{i=0}^{\lfloor 2^N a \rfloor - 1} \frac{i}{2^{2N}}.$$

The last term in the upper sum is the contribution of the right-most interval

$$\left[\frac{\lfloor 2^N a \rfloor}{2^N}, \frac{\lfloor 2^N a \rfloor + 1}{2^N}\right),$$

where the maximum value is $a$ and is achieved at $a$.

So we find

$$U_N\left(x\mathbf{1}_{[0,a]}(x)\right) - L_N\left(x\mathbf{1}_{[0,a]}(x)\right) = \sum_{i=0}^{\lfloor 2^N a \rfloor - 1} \frac{1}{2^{2N}} + \frac{a}{2^N} \leq \frac{2a}{2^N},$$

which clearly tends to 0 as $N \to \infty$, so the function is integrable. To find the integral, we will use the lower sums (which are slightly less messy). We find

$$L_N\left(x\mathbf{1}_{[0,a]}(x)\right) = \sum_{i=0}^{\lfloor 2^N a \rfloor - 1} \frac{i}{2^{2N}} = \frac{1}{2^{2N}} \frac{1}{2}(\lfloor 2^N a \rfloor - 1)\lfloor 2^N a \rfloor$$

$$= \frac{1}{2} \frac{\lfloor 2^N a \rfloor - 1}{2^N} \frac{\lfloor 2^N a \rfloor}{2^N} = \frac{a^2}{2}.$$

For the other integrals, note that whether the interval is open or closed at 0 makes no difference to the upper or lower sums. The interval being open or closed at $a$ makes a difference exactly when $a = k/2^N$, in which case the upper sum contains an extra term when the endpoint is there. But this extra term is $a/2^N$, and does not change the limit as $N \to \infty$.

d. Observe that

$$x\mathbf{1}_{[a,b]}(x) = x\mathbf{1}_{[0,b]}(x) - x\mathbf{1}_{[0,a)}(x),$$
$$x\mathbf{1}_{[a,b)}(x) = x\mathbf{1}_{[0,b)}(x) - x\mathbf{1}_{[0,a)}(x),$$
$$x\mathbf{1}_{(a,b]}(x) = x\mathbf{1}_{[0,b]}(x) - x\mathbf{1}_{[0,a]}(x),$$
$$x\mathbf{1}_{(a,b)}(x) = x\mathbf{1}_{[0,b)}(x) - x\mathbf{1}_{[0,a]}(x).$$

By proposition 4.1.13, all four right sides are integrable, and the integrals are all $(b^2 - a^2)/2$.

**4.1.7** As suggested in the text, write $f_1 = f_1^+ - f_1^-$, $f_2 = f_2^+ - f_2^-$. . Then we have

$$\int_{\mathbb{R}^{n+m}} g \, |d^n\mathbf{x}||d^m\mathbf{y}| = \int_{\mathbb{R}^{n+m}} \left( f_1^+(\mathbf{x}) - f_1^-(\mathbf{x}) \right) \left( f_2^+(\mathbf{y}) - f_2^-(\mathbf{y}) \right) |d^n\mathbf{x}||d^m\mathbf{y}|$$

$$= \int_{\mathbb{R}^{n+m}} \left( f_1^+(\mathbf{x}) f_2^+(\mathbf{y}) - f_1^-(\mathbf{x}) f_2^+(\mathbf{y}) - f_1^+(\mathbf{x}) f_2^-(\mathbf{y}) + f_1^-(\mathbf{x}) f_2^-(\mathbf{y}) \right) |d^n\mathbf{x}||d^m\mathbf{y}|$$

$$= \int_{\mathbb{R}^{n+m}} f_1^+(\mathbf{x}) f_2^+(\mathbf{y}) |d^n\mathbf{x}||d^m\mathbf{y}| - \int_{\mathbb{R}^{n+m}} f_1^-(\mathbf{x}) f_2^+(\mathbf{y}) |d^n\mathbf{x}||d^m\mathbf{y}|$$

$$\quad - \int_{\mathbb{R}^{n+m}} f_1^+(\mathbf{x}) f_2^-(\mathbf{y}) |d^n\mathbf{x}||d^m\mathbf{y}| + \int_{\mathbb{R}^{n+m}} f_1^-(\mathbf{x}) f_2^-(\mathbf{y}) |d^n\mathbf{x}||d^m\mathbf{y}|$$

$$= \left( \int_{\mathbb{R}^n} f_1^+(\mathbf{x}) |d^n\mathbf{x}| \right) \left( \int_{\mathbb{R}^m} f_2^+(\mathbf{y}) |d^m\mathbf{y}| \right) - \left( \int_{\mathbb{R}^n} f_1^-(\mathbf{x}) |d^n\mathbf{x}| \right) \left( \int_{\mathbb{R}^m} f_2^+(\mathbf{y}) |d^m\mathbf{y}| \right)$$

$$\quad - \left( \int_{\mathbb{R}^n} f_1^+(\mathbf{x}) |d^n\mathbf{x}| \right) \left( \int_{\mathbb{R}^m} f_2^-(\mathbf{y}) |d^m\mathbf{y}| \right) + \left( \int_{\mathbb{R}^n} f_1^-(\mathbf{x}) |d^n\mathbf{x}| \right) \left( \int_{\mathbb{R}^m} f_2^-(\mathbf{y}) |d^m\mathbf{y}| \right)$$

$$= \left( \int_{\mathbb{R}^n} (f_1^+(\mathbf{x}) - f_1^-(\mathbf{x})) |d^n\mathbf{x}| \right) \left( \int_{\mathbb{R}^m} (f_2^+(\mathbf{y}) - f_2^-(\mathbf{y})) |d^m\mathbf{y}| \right)$$

$$= \left( \int_{\mathbb{R}^n} f_1 \, |d^n\mathbf{x}| \right) \left( \int_{\mathbb{R}^m} f_2 \, |d^m\mathbf{y}| \right)$$

**4.1.9** Using the inequality $|\sin a - \sin b| \leq |a - b|$ (see the footnote in the text, page 245), we see that within each dyadic square $C \in \mathcal{D}_N(\mathbb{R}^2)$ in the interior of $Q$, we have

$$|\sin(x_1 - y_1) - \sin(x_2 - y_2)| \leq |(x_1 - y_1) - (x_2 - y_2)|$$
$$= |(x_1 - x_2) - (y_2 - y_1)|$$
$$\leq |x_1 - x_2| + |y_2 - y_1| \leq 1/2^N + 1/2^N$$
$$= 2^{1-N}.$$

Furthermore, at most $8(2^N)$ such squares touch the boundary of the square, where the oscillation is at most 1. Thus we have

$$U_N(f) - L_N(f) \leq \frac{1}{2^{N-1}} \frac{2^{2N}}{2^{2N}} + \frac{8 \cdot 2^N}{2^{2N}} = \frac{2+8}{2^N},$$

which evidently goes to 0 as $N \to \infty$.

**4.1.11**  If $X \subset \mathbb{R}^n$ is any set, and $a$ is a number, we will denote

$$aX = \{\, a\mathbf{x} \mid \mathbf{x} \in X \,\}.$$

If $C \in \mathcal{D}_M(\mathbb{R}^n)$, then $2^N C \in \mathcal{D}_{M-N}(\mathbb{R}^n)$. In particular,

$$
\begin{aligned}
U_M(D_{2^N} f) &= \sum_{C \in \mathcal{D}_M(\mathbb{R}^n)} M_C(D_{2^N} f) \operatorname{vol}_n(C) \\
&= \sum_{C \in \mathcal{D}_M(\mathbb{R}^n)} M_{2^{-N}C}(f) \operatorname{vol}_n(C) \\
&= 2^{nN} \sum_{C \in \mathcal{D}_{M+N}(\mathbb{R}^n)} M_C(f) \operatorname{vol}_n(C) = 2^{nN} U_{M+N}(f).
\end{aligned}
$$

An exactly similar computation gives

$$L_M(D_{2^N} f) = 2^{nN} L_{M+N}(f).$$

Putting these inequalities together, we find

$$2^{nN} L_{M+N}(f) = L_M(D_{2^N} f) \leq U_M(D_{2^N} f) = 2^{nN} U_{M+N}(f).$$

The outer terms above can be made arbitrarily close, so the inner terms can also. It follows that $D_{2^N} f$ is integrable, and

$$\int_{\mathbb{R}^n} D_{2^N} f(\mathbf{x}) |d^n \mathbf{x}| = 2^{nN} \int_{\mathbb{R}^n} f(\mathbf{x}) |d^n \mathbf{x}|.$$

**4.1.13**  We have

$$
\begin{aligned}
\frac{2^N |b-a| - 2}{2^N} = \frac{2^N |b-a|}{2^N} - \frac{2}{2^N} &\leq L_N(\mathbf{1}_I) \\
\leq \operatorname{vol}(I) \leq U_N(\mathbf{1}_I) &\leq \frac{2^N |b-a| + 2}{2^N} = \frac{2^N |b-a|}{2^N} + \frac{2}{2^N}.
\end{aligned}
$$

So taking the limit as $N \to \infty$, we find $\operatorname{vol}(I) = |b-a|$.

**4.1.15**  a. $X \cap Y$: If $C \cap (X \cap Y) \neq \emptyset$ then $C \cap X \neq \emptyset$, so for a given $\epsilon$ we can use the same $N$ as we did for $X$ alone in equation 4.1.62.

$X \times Y$: Using proposition 4.1.19 we see that for any $\epsilon_1$, it suffices to choose $N$ to be the larger of

> $N$ in proposition 4.1.22 for $X$ with $\epsilon = \epsilon_1$, and
> $N$ for $Y$ with $\epsilon = 1$.

$X \cup Y$: Using theorem 4.1.20 we see that for any $\epsilon_1$, it is enough to choose $N$ to be the larger of

> $N$ for equation 4.1.17 for $X$ with $\epsilon = \frac{\epsilon_1}{2}$, and
> $N$ for $Y$ with $\epsilon = \frac{\epsilon_1}{2}$.

Observe that the argument for $X \times Y$ proves a stronger statement; it shows that if $X$ has volume 0, and $Y$ has any *finite* volume, then $X \times Y$ has volume 0.)

b. Let $X$ be the set in question. It intersects $2^N + 1$ squares of $\mathcal{D}_N(\mathbb{R}^2)$, so

$$U_N(\mathbf{1}_X) = \frac{2^N + 1}{2^{2N}}, \quad \text{which tends to 0 as } N \to \infty.$$

c. Let $X$ be the set in question. It intersects $2^{2N} + 2^N + 2^N + 1$ cubes of $\mathcal{D}_N(\mathbb{R}^3)$, so

$$U_N(\mathbf{1}_X) = \frac{2^{2N} + 2^{N+1} + 1}{2^{3N}}, \quad \text{which tends to 0 as } N \to \infty.$$

In exercises 4.1.17 and 4.1.19, we will suppose that we have made a decomposition

$$0 = x_0 < x_1 < \cdots < x_n = 1,$$

and will try to see what sort of Riemann sum the proposed "integrand" gives.

**4.1.17**  a.  This is an acceptable but "degenerate" integrand: it will give 0, since

$$\sum_{i=0}^{n-1}(x_{i+1} - x_i)^2 = \underbrace{\sum_{i=0}^{n-1} x_{i+1}(x_{i+1} - x_i)}_{\substack{\text{right Riemann} \\ \text{sum for } \int_0^1 x\,dx}} - \underbrace{\sum_{i=0}^{n-1} x_i(x_{i+1} - x_i)}_{\substack{\text{left Riemann} \\ \text{sum for } \int_0^1 x\,dx}}.$$

b.  This is an acceptable integrand. We have $|\sin(u) - u| < |u|^3/6$ when $|u| < 1$, since the power series for sine is alternating, with decreasing terms tending to 0. So as soon as all $|x_{i+1} - x_i| < 1$, we have

$$\left| \sum_{i=0}^{n-1} \sin(x_{i+1} - x_i) - \sum_{i=0}^{n-1}(x_{i+1} - x_i) \right| \leq \frac{1}{6}\sum_{i=0}^{n-1}(x_{i+1} - x_i)^3.$$

The right side tends to 0 as the decomposition becomes fine (for instance, it is smaller than the sum in part a), so the two terms on the left have the same limit, which is $\int_0^1 1\,dx = 1 - 0$.

c.  This is not an acceptable integrand; it gives an infinite limit. Indeed, if we break up $[0, 1]$ into $n$ intervals of equal length, each will have length $1/n$ and we get

$$\sum_{i=1}^{n} \sqrt{\frac{1}{n}} = \frac{n}{\sqrt{n}} = \sqrt{n},$$

which goes to $\infty$ as $n \to \infty$.

d.  This is an acceptable integrand; it gives the limit is 1. Indeed,

$$\sum_{i=0}^{n-1}\left(x_{i+1}^2 - x_i^2\right) = \sum_{i=0}^{n-1} x_{i+1}(x_{i+1} - x_i) + \sum_{i=0}^{n-1} x_i(x_{i+1} - x_i).$$

As in part a, each of these is a Riemann sum for $\int_0^1 x\,dx$, so the sum tends to

$$\int_0^1 2x\,dx = 1^2 - 0^2.$$

An easier approach is to notice that the sum telescopes:

$$\sum_{i=0}^{n-1} \left(x_{i+1}^2 - x_i^2\right) = x_1^2 - x_0^2 + x_2^2 - x_1^2 + \cdots + x_n^2 - x_{n-1}^2 = x_n^2 - x_0^2 = 1 - 0 = 1.$$

e. This is similar:

$$\sum_{i=0}^{n-1} x_{i+1}^3 - x_i^3 = \sum_{i=0}^{n-1} (x_{i+1}^2 + x_{i+1}x_i + x_i^2)(x_{i+1} - x_i).$$

All three terms of the sum are Riemann sums for $\int_0^1 x^2 \, dx$, so the limit of their sum is

$$\int_0^1 3x^2 \, dx = 1^3 - 0^3.$$

Again, this is easier if you notice that the sum telescopes:

$$\sum_{i=0}^{n-1} x_{i+1}^3 - x_i^3 = x_1^3 - x_0^3 + x_2^3 - x_1^3 + \cdots + x_n^3 - x_{n-1}^3 = x_n^3 - x_0^3 = 1 - 0 = 1.$$

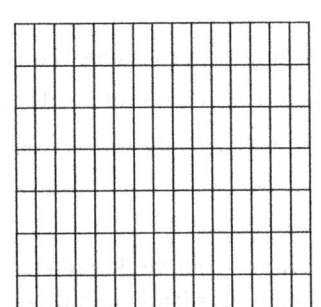

FIGURE FOR SOLUTION 4.1.19.

This is the case $N = 15$ and $M = 7$.

**4.1.19** Suppose we break $[0, 1] \times [0, 1]$ into $NM$ congruent rectangles, as shown in the margin. Then the relevant Riemann sum is

$$\underbrace{NM}_{\text{number of pieces}} \left(\frac{1}{N^2} \sqrt{\frac{1}{M}}\right) = \frac{\sqrt{M}}{N}.$$

The limit depends on how fast $N$ goes to infinity relative to $M$. If $N = \sqrt{M}$, or if $N$ is a multiple of $\sqrt{M}$ ($N = k\sqrt{M}$), then the Riemann sum tends to a finite limit (1 or $1/k$ respectively). If $N$ is much smaller than $\sqrt{M}$, the limit is infinite, and if $N$ is much bigger than $M$, the limit is 0. Therefore, $|b - a|^2 \sqrt{|c - d|}$ is not a reasonable integrand.

**4.2.1** The outcome

"total 2"  has probability 1/100
"total 3"  has probability 2/100
"total 4"  has probability 3/100
"total 5"  has probability 2/20+2/100;
"total 6"  has probability 2/20+ 3/100
"total 7"  has probability 2/20+4/100
"total 8"  has probability 1/4+4/100
"total 9"  has probability 2/20+2/100
"total 10" has probability 2/20+1/100
"total 11" has probability 2/100
"total 12" has probability 1/100

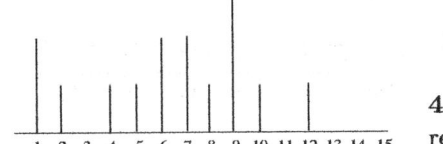

**4.2.3** The quick answer is that these results are not very likely. If we chart results for each integer, the resulting bar graph does not look like a bell curve, as shown in the margin.

To give a more detailed answer, we must compute the standard deviation for the experiment. Figuring out all the possible ways of getting all the various totals would be time-consuming, so instead we use the fact that if an experiment with standard deviation $\sigma$ is repeated $n$ times, the central limit theorem asserts that the average is approximately distributed according to the normal distribution with mean $E$ and standard deviation $\sigma/\sqrt{n}$.

The expectation of getting heads when tossing a coin once is $1/2$. To compute the variance we consider the two cases. If we get heads, then $(f - E(f))^2 = (1 - 1/2)^2 = 1/4$. If we get tails, then

$$(f - E(f))^2 = (0 - 1/2)^2 = 1/4, \quad \text{which gives}$$

$$\text{Var}\,(f) = E(f - E(f))^2 = E(1/4 + 1/4) = 1/2(1/2) = 1/4$$

and $\sigma(f) = 1/2$.

So the standard deviation of the average results when tossing the coin 14 times is $\dfrac{1}{2\sqrt{14}}$. But we are interested in the actual number of heads, not the average. Denote this random variable by $T$; we multiply by 14 to get

$$\sigma(T) = \frac{\sqrt{14}}{2} \approx 1.9.$$

But is $n = 14$ large? If we were asked in a court of law, we would not be willing to affirm that a person reporting those results was lying or cheating. If 300 people each tossed a coin 14 times, one of them might very likely come up with these results.

The expectation of $T$ is 7.5, so (see figure 4.2.9), for $n$ large we should expect 68% of our results to be between 5.6 and 9.4 heads (i.e., within one standard deviation). The actual figure is $8/15 \approx 53.3\%$. For $n$ large we would expect 95% to be within 3.7 and 11.3; the actual figure is $11/15 \approx 73.3\%$.

**4.2.5**

$$E(f) = \int_{-\infty}^{\infty} f(x) \frac{1}{2a} \mathbf{1}_{[-a,a]}(x)\, dx = \int_{-a}^{a} f(x) \frac{1}{2a}\, dx = \int_{-a}^{a} \frac{x}{2a}\, dx$$

$$= \left[\frac{x^2}{4a}\right]_{-a}^{a} = \frac{a^2}{4a} - \frac{a^2}{4a} = 0$$

$$\text{Var}\,(f) = \int_{-a}^{a} \big(f(x) - E(f)\big)^2 \mu(x)\, dx = \int_{-a}^{a} (x - 0)^2 \mu(x)\, dx = \int_{-a}^{a} \frac{x^2}{2a}\, dx$$

$$= \left[\frac{x^3}{6a}\right]_{-a}^{a} = \frac{a^2}{3}.$$

$$\sigma(f) = \sqrt{\frac{a^2}{3}} = \frac{a}{\sqrt{3}}$$

**4.2.7**  The first coordinate of the center of gravity is

$$\frac{\left(\int_{a_1}^{b_1} x_1\, dx_1\right)\left(\int_{a_2}^{b_2} dx_2\right)\ldots\left(\int_{a_n}^{b_n} dx_n\right)}{\left(\int_{a_1}^{b_1} dx_1\right)\left(\int_{a_2}^{b_2} dx_2\right)\ldots\left(\int_{a_n}^{b_n} dx_n\right)} = \frac{\left(\frac{b_1^2 - a_1^2}{2}\right)(b_2 - a_2)\ldots(b_n - a_n)}{(b_1 - a_1)(b_2 - a_2)\ldots(b_n - a_n)}$$

$$= \frac{b_1 + a_1}{2}.$$

The computation for the other coordinates is identical.

**4.3.1 a.** If $f$ is integrable, then $f^+$ and $f^-$ are integrable (corollary 4.3.4) so $|f| = f^+ + f^-$ is integrable (proposition 4.1.13, part a). Now

$$\left| \int f \right| = \left| \int (f^+ - f^-) \right| \overset{\text{prop. 4.1.13, part 1}}{=} \left| \left( \int f^+ \right) - \left( \int f^- \right) \right|$$
$$\leq \left| \int f^+ \right| + \left| \int f^- \right| = \left( \int f^+ \right) + \left( \int f^- \right) = \int (f^+ + f^-)$$
$$= \int |f|.$$

The first equality in the second line is justified because by proposition 4.1.13, part 3,

$$\int f^+ \geq 0 \quad \text{and} \quad \int f^- \geq 0.$$

**b.** Converse counterexample: $f = 1/2 - \mathbf{1}_{\mathbb{Q}}$ restricted to $[0,1]$ has $|f| = 1/2$ (integrable) but $f$ is not integrable.

Solution 4.3.3: A first error to avoid is confusing the unit circle with the unit disc; the problem concerns a curve, not a surface. Second, if you looked carefully at equation 4.1.12 defining dyadic cubes, you may have realized that at level $N = 0$, six cubes are needed, rather than the obvious four, because of the points $\begin{pmatrix} 0 \\ 1 \end{pmatrix}$ and $\begin{pmatrix} 1 \\ 0 \end{pmatrix}$. This is not important, since we are concerned with the number needed as $N$ gets big.

We should have asked you to use this bound to show that the area of the circle is 0, as one would expect (and as required by proposition 4.3.6, since, as we saw in example 3.1.5, the circle is made up of graphs of functions). The area of a cube in $\mathbb{R}^2$ is $\text{vol}_2 = \frac{1}{2^{2N}}$, so multiplying the upper bound by the volume of each cube gives

$$\frac{1}{2^{2N}} 16 \left( \left[ \frac{2^N}{\sqrt{2}} \right] + 1 \right) \approx \frac{1}{2^N} \frac{16}{\sqrt{2}},$$

which goes to 0 as $N$ goes to infinity.

**4.3.3** Since the circle is symmetric, consider only the upper right quadrant, and keep in mind that we are looking just for an upper bound, not for a sharp upper bound. As suggested in the hint, divide that quadrant into two by drawing the diagonal through the origin. This line intersects the circle at the point $\begin{pmatrix} 1/\sqrt{2} \\ 1/\sqrt{2} \end{pmatrix}$. In the part of the circle from $\begin{pmatrix} 0 \\ 1 \end{pmatrix}$ to the diagonal, the slope of the circle is always less than 1, so that if we look at columns of cubes, the circle never intersects more than two cubes in any one column. Starting at the origin, and going to the point $\begin{pmatrix} 1/\sqrt{2} \\ 0 \end{pmatrix}$, there are $\frac{2^N}{\sqrt{2}}$ columns. Since we can't deal in fractions of cubes, take the fractional part of that number (denoted with square brackets, $[\frac{2^N}{\sqrt{2}}]$), and add 1 for good measure. Multiplying by two gives

$$2 \left( \left[ \frac{2^N}{\sqrt{2}} \right] + 1 \right)$$

for that eighth of the circle. This is an over-estimate, since in many columns the circle intersects only one cube, but that doesn't matter.

In the second segment of the upper right quadrant of the circle, the slope is much steeper; indeed, at $\begin{pmatrix} 1 \\ 0 \end{pmatrix}$, the tangent to the circle is vertical. So if we use the same columns, then as $N \to \infty$ there is no limit to the number of cubes the circle will intersect. But if we rotate the circle by 90°, we have the same situation as before, with slope less than 1. (This method counts cubes at the point $\begin{pmatrix} 1/\sqrt{2} \\ 1/\sqrt{2} \end{pmatrix}$ twice, once when we look at the circle the normal way, and again when we have rotated it.) There are eight such segments in all, so for the whole circle we have an (admittedly generous) upper bound of

$$16 \left( \left[ \frac{2^N}{\sqrt{2}} \right] + 1 \right) \approx 2^N \frac{16}{\sqrt{2}}.$$

**4.3.5** Let us apply theorem 4.3.10, which says that if $f$ is bounded with bounded support, and continuous except on a set of volume 0 (area in this case), then it is integrable. The function $\mathbf{1}_P f$ is bounded (by 1), and its support is in $P$, which is bounded. Moreover, it is continuous except on the boundary of $P$. This boundary is the union of the arc of parabola $y = x^2$, $-1 \le x \le 1$, which is the graph of a continuous function, hence has area 0, and the segment $y = 1$, $-1 \le x \le 1$, which is also the graph of a continuous function, hence also has area 0.

Another approach is to claim that for a dyadic square $C \in \mathcal{D}_N(\mathbb{R}^2)$ that doesn't intersect the boundary of $P$, the oscillation of $\sin y$ is at most $\sqrt{2}/2^N$. Now we estimate how many dyadic squares can intersect the boundary of $P$. Since the curve $y = x^2$ has slope at most 2, it will intersect a vertical column of dyadic squares in at most three squares, and the horizontal line $y = 1$ will intersect exactly one square of the column. Moreover, the right edge will intersect $2^N$ squares of $\mathcal{D}_N(\mathbb{R}^2)$, so altogether, at most $5 \cdot 2^N$ dyadic squares will intersect the boundary, with total area $5 \cdot 2^{-N}$. Thus for every $\epsilon > 0$, there exists $N$ such that the function has oscillation $< \epsilon$ except on a set of dyadic cubes of total area $< \epsilon$.

**4.4.1** Clearly, if a set $X$ has measure 0 using open boxes, then it also has measure 0 using closed boxes; the closures of the open boxes used to cover $X$ also cover $X$. The volume of an open box equals the volume of its closure.

The other direction is a little more difficult. For any closed box $B$, let $B'$ be the open box with the same center but twice the sidelength. Note that $\mathrm{vol}_n(B') = 2^n \, \mathrm{vol}_n B$. If $X$ is a set such that there exists a sequence of closed boxes with $X \subset \cup B_i$ and $\sum \mathrm{vol}_n B_i \le \epsilon$, then $X \subset \cup B'_i$ and $\sum \mathrm{vol}_n B'_i \le 2^n \epsilon$. By taking $\epsilon$ sufficiently small, we can make $2^n \epsilon$ arbitrarily small.

**4.4.3**  a.  First, recall that the rationals are countable, i.e., that you can make a list $a_1, a_2, \ldots$ that includes all the rationals (see exercise 0.6.1).

More generally, let $X \subset \mathbb{R}$ be any countable set, with elements listed as $x_1, x_2, \ldots$. Choose $\epsilon > 0$, and $n$ such that $2^{1-n} < \epsilon$. Then the union of intervals $[x_i - 2^{-(n+i)}, x_i + 2^{-(n+i)}]$ certainly contains $X$, and

Equation (1) uses

$$\sum_{i=1}^{\infty} \frac{1}{2^{n+i}} = \frac{1}{2^n},$$

in the form

$$\sum_{i=1}^{\infty} \frac{1}{2^{n+i-1}} = \frac{1}{2^{n-1}}.$$

$$\sum_{i=1}^{\infty} \mathrm{vol}_1\left([x_i - 2^{-(n+i)}, x_i + 2^{-(n+i)}]\right) = \sum_{i=1}^{\infty} \frac{2}{2^{n+i}}$$
$$= \sum_{i=1}^{\infty} \frac{1}{2^{n+i-1}} = 2^{1-n} < \epsilon. \tag{1}$$

b.  Suppose $X = X_1 \cup X_2 \cup \ldots$, and that each $X_i$ has measure 0. Choose $\epsilon > 0$, and for each $i$, find a sequence of cubes $C_{i,j}$ such that

$$X_i = \cup_{j=1}^{\infty} C_{i,j} \quad \text{and} \quad \sum_{j=1}^{\infty} \mathrm{vol}_n(C_{i,j}) \le \frac{\epsilon}{2^i}.$$

Now take all these cubes, say in the order

$$C_1' = C_{1,1}, \ C_2' = C_{2,1}, \ C_3' = C_{1,2}, \ C_4' = C_{3,1}, \ C_5' = C_{2,2}, \ C_6' = C_{1,3},$$
$$C_7' = C_{4,1}, \ C_8' = C_{3,2}, C_9' = C_{2,3}, \ldots$$

so the list contains them all. Clearly,

$$X \subset \cup_{k=1}^\infty C_k' \quad \text{and} \quad \sum_{n=1}^\infty \mathrm{vol}_n(C_k') \le \sum_{i=1}^\infty \sum_{j=1}^\infty \mathrm{vol}_n(C_{i,j}) \le \sum_{i=1}^\infty \frac{\epsilon}{2^i} = \epsilon.$$

**4.4.5** We will prove it false by showing a counterexample. Let $f$ be the function

$$f(x) = \begin{cases} \frac{1}{q} & \text{if } x = \frac{p}{q} \text{ is rational, written in lowest terms, and } x \in [0,1] \\ 0 & \text{if } x \text{ is irrational, or } |x| > 1. \end{cases}$$

and let $g(x) = 0$ for all $x \in \mathbb{R}$. Then $f$ and $g$ are integrable, both with integral 0, and $g \le f$, but

$$\{ x \in \mathbb{R} \mid g(x) < f(x) \} = \mathbb{Q} \cap [0,1],$$

and $\mathbb{Q} \cap [0,1]$ does not have volume.

**4.5.1** By proposition 4.3.7, any bounded part of a line in the plane has 2-dimensional volume (i.e., area) 0, so the values of the function on a line do not affect the value of the integral.

**4.5.3** a. The integral becomes

$$\int_{-1}^0 \left( \int_{-y}^{2+y} f\left(\frac{x}{y}\right) dx \right) dy + \int_0^1 \left( \int_{\sqrt{y}}^{2-\sqrt{y}} f\left(\frac{x}{y}\right) dx \right) dy.$$

b. Below the $x$-axis, the domain of integration would be unchanged, as shown in figure 4.5.4 in the textbook. Above the $x$-axis, it would be the region bounded on the left by the "left branch" of the parabola $y = x^2$, bounded on the right by the "right branch" of the parabola $y = (x-2)^2$, and bounded above by the line $y = 1$. This domain of integration is shown in the figure in the margin.

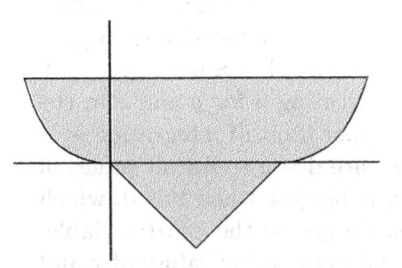

FIGURE FOR SOLUTION 4.5.3.

**4.5.5** a. We use induction over $k$. Suppose we know that $\beta_{2k} = \frac{\pi^k}{k!}$. The relation $\beta_i = c_i \beta_{i-1}$ allows us to express $\beta_{2k+2}$ in terms of $\beta_{2k}$ and $c_i$'s:

$$\beta_{2k+2} = c_{2k+2}\beta_{2k+1} = c_{2k+2}c_{2k+1}\beta_{2k}.$$

The expression for $c_i$ is

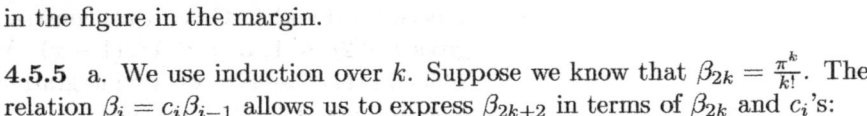

$$c_i = c_{2k} = c_0 \underbrace{\prod_{j=1}^k \frac{2j-1}{2j}}_{\text{for } i \text{ even}} \quad \text{and} \quad c_i = c_{2k+1} = c_1 \underbrace{\prod_{j=1}^k k\frac{2j}{2j+1}}_{\text{for } i \text{ odd}}.$$

Multiplying these together gives

$$c_i c_{i-1} = \frac{(i-1)!}{i!} c_0 c_1 = \frac{2\pi}{i}.$$

Substituting this into the equation for $\beta_{2k+2}$ gives

$$\beta_{2k+2} = c_{2k+2}c_{2k+1}\beta_{2k} = \frac{2\pi}{2k+2}\frac{\pi^k}{k!} = \frac{\pi^{k+1}}{(k+1)!},$$

as desired.

b. Similarly, suppose we know that

$$\beta_{2k-1} = \frac{\pi^{k-1}(k-1)!2^{2k-1}}{(2k-1)!}.$$

Then, as before, we have

$$\beta_{2k+1} = c_{2k+1}c_{2k}\beta_{2k-1} = \frac{2\pi}{2k+1}\frac{\pi^{k-1}(k-1)!2^{2k-1}}{(2k-1)!} = \frac{\pi^k(k-1)!2^{2k}}{(2k+1)(2k-1)!}.$$

Multiplying top and bottom by $2k$ gives

$$\beta_{2k+1} = \frac{\pi^k k! 2^{2k+1}}{(2k+1)!}, \quad \text{as desired.}$$

Solution 4.5.7: You were asked to set up the integral in three ways but there are actually six, for the six permutations of $x, y, z$. Note that since you are not asked to actually compute the integral, whether the function being integrated is $xyz$ or something else is irrelevant.

**4.5.7** We will do $z, y, x$ first:

$$\int_0^1 \left( \int_0^{\frac{1}{2}(1-x)} \left( \int_0^{\frac{1}{3}(1-x-2y)} xyz\,dz \right) dy \right) dx.$$

The integral is *not*

$$\int_0^{1-2y-3z} \left( \int_0^{\frac{1}{2}(1-x-3z)} \left( \int_0^{\frac{1}{3}(1-x-2y)} xyz\,dz \right) dy \right) dx.$$

(or any version with the same limits in a different order). Why not? The above, incorrect integral does not take into account all the available information. When we integrate first with respect to $x$, to determine the upper limit of integration we ask what is the most that $x$ can be. Since $x, y, z \geq 0$, it follows that $x$ is biggest when $y, z = 0$; substituting 0 for $y$ and $z$ in the equation $x + 2y + 3z \leq 1$ gives $x \leq 1$, so the upper limit of integration is 1. When we next integrate with respect to $y$, we are doing so for all values of $x$ between 0 and 1. Given that constraint, $y$ is biggest when $z$ is 0, which gives $x + 2y \leq 1$, or $y \leq 1/2(1-x)$. When we get to the third variable, which in this case is $z$, we are integrating for all permissible values of $x$ and $y$, so the upper limit of integration is $1/3(1-x-2y)$.

The five other ways of setting up the integral are:

1. $\displaystyle\int_0^{1/2} \left( \int_0^{1-2y} \left( \int_0^{\frac{1}{3}(1-x-2y)} xyz\,dz \right) dx \right) dy$     2. $\displaystyle\int_0^{1/3} \left( \int_0^{\frac{1}{2}(1-3z)} \left( \int_0^{1-2y-3z} xyz\,dx \right) dy \right) dz$

3. $\displaystyle\int_0^{1/2} \left( \int_0^{\frac{1}{3}(1-2y)} \left( \int_0^{1-2y-3z} xyz\,dx \right) dz \right) dy$     4. $\displaystyle\int_0^{1/3} \left( \int_0^{1-3z} \left( \int_0^{\frac{1}{2}(1-x-3z)} xyz\,dy \right) dx \right) dz$

5. $\displaystyle\int_0^1 \left( \int_0^{\frac{1}{3}(1-x)} \left( \int_0^{\frac{1}{2}(1-x-3z)} xyz\,dy \right) dz \right) dx.$

**4.5.9** As indicated in the text, the integral for the half above the diagonal is

$$\overbrace{\text{for } x_2 \text{ to the left of the vertical line with } x\text{-coordinate } x_1}$$

$$\int_0^1 \int_{x_1}^1 \int_0^{\frac{x_1}{y_1}} \left( \int_0^{\frac{y_1 x_2}{x_1}} \underbrace{(x_2 y_1 - x_1 y_2)}_{-\det} \, dy_2 + \int_{\frac{y_1 x_2}{x_1}}^1 \underbrace{(x_1 y_2 - x_2 y_1)}_{+\det} \, dy_2 \right) dx_2 \, dy_1 \, dx_1$$

$$+ \int_0^1 \int_{x_1}^1 \int_{\frac{x_1}{y_1}}^1 \underbrace{\left( \int_0^1 (x_2 y_1 - x_1 y_2) \, dy_2 \right)}_{\substack{\text{for } x_2 \text{ to the right of the vertical line} \\ \text{with } x\text{-coordinate } x_1; \, -\det}} dx_2 \, dy_1 \, dx_1. \tag{1}$$

The limit of integration $\frac{y_1 x_2}{x_1}$ is the value of $y_2$ on the line separating the shaded and unshaded regions, the line on which the first dart lands; since that line is given by $y = \frac{y_1}{x_1} x$, we have $y_2 = \frac{y_1}{x_1} x_2$.

We will compute the integral for the half of the square above the diagonal, and then add it to the integral for the bottom half, which by equation 4.5.31 is $\frac{13}{108}$. (Can you think of a way to take advantage of Fubini to make this computation easier?)

Rather surprisingly, when we compute the total integral this way, we encounter some logarithms. The first inner integral in the first line is

$$\int_0^{\frac{y_1 x_2}{x_1}} (x_2 y_1 - x_1 y_2) \, dy_2 = \left[ x_2 y_1 y_2 - x_1 \frac{y_2^2}{2} \right]_0^{\frac{y_1 x_2}{x_1}} = x_2 y_1 \frac{y_1 x_2}{x_1} - x_1 \frac{y_1^2 x_2^2}{2x_1^2}$$

$$= \frac{x_2^2 y_1^2}{2x_1}.$$

The second inner integral in the first line is

$$\int_{\frac{y_1 x_2}{x_1}}^1 (x_1 y_2 - x_2 y_1) \, dy_2 = \left[ x_1 \frac{y_2^2}{2} - x_2 y_1 y_2 \right]_{\frac{y_1 x_2}{x_1}}^1$$

$$= \frac{x_1}{2} - x_2 y_1 - x_1 \frac{x_2^2 y_1^2}{2x_1^2} + x_2 y_1 \frac{x_2 y_1}{x_1} = \frac{x_1}{2} - x_2 y_1 + \frac{y_1^2 x_2^2}{2x_1}.$$

So the iterated integral on the first line of equation (1) is

$$\int_0^1 \int_{x_1}^1 \int_0^{\frac{x_1}{y_1}} \left( \frac{x_1}{2} - x_2 y_1 + \frac{y_1^2 x_2^2}{x_1} \right) dx_2 \, dy_1 \, dx_1 = \int_0^1 \int_{x_1}^1 \left[ \frac{x_1 x_2}{2} - \frac{x_2^2 y_1}{2} + \frac{x_2^3 y_1^2}{3x_1} \right]_0^{\frac{x_1}{y_1}} dy_1 \, dx_1$$

$$= \int_0^1 \int_{x_1}^1 \left( \frac{x_1^2}{2y_1} - \frac{x_1^2 y_1}{2y_1^2} + \frac{x_1^3 y_1^2}{3y_1^3 x_1} \right) dy_1 \, dx_1 = \int_0^1 \left( \int_{x_1}^1 \frac{x_1^2}{3y_1} \, dy_1 \right) dx_1 = \int_0^1 \frac{x_1^2}{3} \Big[ \ln y_1 \Big]_{x_1}^1 dx_1 \tag{2}$$

$$= \int_0^1 \underbrace{\frac{x_1^2}{3}(-\ln x_1)}_{f' \, g} \, dx_1 = \underbrace{\left[ -\frac{x_1^3}{9} \ln x_1 \right]_0^1}_{f \, g} + \int_0^1 \underbrace{\left( \frac{x_1^3}{9} \cdot \frac{1}{x_1} \right)}_{f \, g'} dx_1 = \int_0^1 \frac{x_1^2}{9} \, dx_1 = \left[ \frac{x_1^3}{27} \right]_0^1 = \frac{1}{27}$$

Since $\ln 0 = -\infty$, you might worry that in the third line of equation (2), the first term would lead to $(0^3/9) \cdot (-\infty)$, which is undefined. But as $x_1 \to 0$, $x_1^3$ goes to 0 much faster than $\ln x_1$ goes to $-\infty$.[4]

The iterated integral on the second line of equation (1) is

$$\int_0^1 \int_{x_1}^1 \int_{\frac{x_1}{y_1}}^1 \left( \int_0^1 (x_2 y_1 - x_1 y_2) \, dy_2 \right) dx_2 \, dy_1 \, dx_1$$

$$= \int_0^1 \int_{x_1}^1 \int_{\frac{x_1}{y_1}}^1 \left[ x_2 y_1 y_2 - \frac{x_1 y_2^2}{2} \right]_0^1 dx_2 \, dy_1 \, dx_1$$

$$= \int_0^1 \int_{x_1}^1 \int_{\frac{x_1}{y_1}}^1 \left( x_2 y_1 - \frac{x_1}{2} \right) dx_2 \, dy_1 \, dx_1 = \int_0^1 \int_{x_1}^1 \left[ \frac{x_2^2 y_1}{2} - \frac{x_1 x_2}{2} \right]_{\frac{x_1}{y_1}}^1 dy_1 \, dx_1$$

$$= \int_0^1 \int_{x_1}^1 \left( \frac{y_1}{2} - \frac{x_1}{2} - \frac{x_1^2 y_1}{y_1^2 2} + \frac{x_1^2}{2 y_1} \right) dy_1 \, dx_1 = \int_0^1 \left[ \frac{y_1^2}{4} - \frac{x_1 y_1}{2} \right]_{x_1}^1 dx_1$$

$$= \int_0^1 \left( \frac{1}{4} - \frac{x_1}{2} - \frac{x_1^2}{4} + \frac{x_1^2}{2} \right) dx_1 = \left[ \frac{1}{4} x_1 - \frac{x_1^2}{4} + \frac{x_1^3}{12} \right]_0^1 = \frac{1}{4} - \frac{1}{4} + \frac{1}{12} = \frac{1}{12}.$$

So the integral for the half above the diagonal is $\dfrac{1}{27} + \dfrac{1}{12} = \dfrac{13}{108}$, which gives a total integral of $13/54$.

This computation should convince you that taking advantage of symmetries is worthwhile. But there is a better way to compute the integral for the upper half, as we realized after working through the above computation. The computation is simpler if we integrate with respect to $x_2$ first, rather than $y_2$:

Since the line separating the shaded and unshaded regions is given by $y = \dfrac{y_1}{x_1} x$, we have

$$y_2 = \frac{y_1}{x_1} x_2, \text{ i.e., } x_2 = \frac{x_1 y_2}{y_1}.$$

$$\int_0^1 \int_{x_1}^0 \int_0^1 \left( \underbrace{\int_0^{\frac{x_1 y_2}{y_1}} (x_1 y_2 - x_2 y_1) \, dx_2}_{+\det} + \underbrace{\int_{\frac{x_1 y_2}{y_1}}^1 (x_2 y_1 - x_1 y_2) \, dx_2}_{-\det} \right) dy_2 \, dy_2 \, dx_1.$$

**4.5.11**  a.  We have

$$\int_0^\pi \left( \int_y^\pi \frac{\sin x}{x} \, dx \right) dy = \int_T \frac{\sin x}{x} \, |dx \, dy|,$$

where $T$ is the triangle $0 \le y \le x \le \pi$.

b. Written the other way, this becomes

$$\int_0^\pi \left( \int_0^x \frac{\sin x}{x} dy \right) dx = \int_0^\pi x \frac{\sin x}{x} \, dx = \cos 0 - \cos \pi = 2.$$

**4.5.13**  a.  If $D_2(D_1(f))$ and $D_1(D_2(f))$ both exist and are continuous on $U$, and if

$$D_2(D_1(f))(\mathbf{a}) - D_1(D_2(f))(\mathbf{a}) > 0,$$

---

[4]Powers dominate logarithms, and exponentials dominate powers. This follows almost immediately from l'Hôpital's rule.

then there is a square

$$S = \left\{ \begin{pmatrix} x \\ y \end{pmatrix} \in \mathbb{R}^2 \mid a_1 \leq x \leq a_1 + \delta, a_2 \leq y \leq a_2 + \delta \right\}$$

on which $D_2(D_1(f)) - D_1(D_2(f)) > 0$. In particular,

$$\int_S D_2(D_1(f)) - D_1(D_2(f)) \,|dx\,dy| = \int_S D_2(D_1(f)) \,|dx\,dy| - \int_S D_1(D_2(f)) \,|dx\,dy| > 0.$$

b. Using Fubini's theorem, these two integrals can be computed:

$$\int_S D_2(D_1(f)) \,|dx\,dy| = \int_{a_1}^{a_1+\delta} \left( \int_{a_2}^{a_2+\delta} D_2(D_1(f)) \,dy \right) dx$$

$$= \int_{a_1}^{a_1+\delta} \left( D_1(f) \begin{pmatrix} x \\ a_2 + \delta \end{pmatrix} - D_1(f) \begin{pmatrix} x \\ a_2 \end{pmatrix} \right) dx$$

$$= f \begin{pmatrix} a_1 + \delta \\ a_2 + \delta \end{pmatrix} - f \begin{pmatrix} a_1 \\ a_2 + \delta \end{pmatrix} - f \begin{pmatrix} a_1 + \delta \\ a_2 \end{pmatrix} + f \begin{pmatrix} a_1 \\ a_2 \end{pmatrix}.$$

$$\int_S D_1(D_2(f)) \,|dx\,dy| = \int_{a_2}^{a_2+\delta} \left( \int_{a_1}^{a_1+\delta} D_1(D_2(f)) \,dx \right) dy$$

$$= \int_{a_2}^{a_2+\delta} \left( D_2(f) \begin{pmatrix} a_1 + \delta \\ y \end{pmatrix} - D_1(f) \begin{pmatrix} a_1 \\ y \end{pmatrix} \right) dy$$

$$= f \begin{pmatrix} a_1 + \delta \\ a_2 + \delta \end{pmatrix} - f \begin{pmatrix} a_1 \\ a_2 + \delta \end{pmatrix} - f \begin{pmatrix} a_1 + \delta \\ a_2 \end{pmatrix} + f \begin{pmatrix} a_1 \\ a_2 \end{pmatrix}.$$

So the two integrals are equal.

c. The function

$$f \begin{pmatrix} x \\ y \end{pmatrix} = \begin{cases} xy \dfrac{x^2 - y^2}{x^2 + y^2} & \text{if } \begin{pmatrix} x \\ y \end{pmatrix} \neq \begin{pmatrix} 0 \\ 0 \end{pmatrix} \\ 0 & \text{otherwise} \end{cases}$$

is twice continuously differentiable everywhere except at the origin. Thus you cannot find a square on which one crossed partial is larger than the other.

**Solution 4.5.15:** This region looks roughly like a plastic Easter egg, standing on one end, the bottom half $z \geq x^2 + y^2$, the top half $z \leq 10 - x^2 - y^2$; the two halves join at $z = 5$.

**4.5.15** Since the two halves are symmetric, we will just compute the bottom half and multiply by 2. The volume is given by the following integral, where in the second line we make the change of variables $y = \sqrt{z}\,u$, so that

$dy = \sqrt{z}\, du$:

$$2\int_0^5 \left( \int_{-\sqrt{z}}^{\sqrt{z}} \left( \int_{-\sqrt{z-y^2}}^{\sqrt{z-y^2}} dx \right) dy \right) dz = 2\int_0^5 \left( \int_{-\sqrt{z}}^{\sqrt{z}} \underbrace{\left( [x]_{-\sqrt{z-y^2}}^{\sqrt{z-y^2}} \right)}_{2\sqrt{z-y^2}} dy \right) dz$$

Of course one does not need to use the order $dx\,dy\,dz$.

This exercise would be much easier to do using cylindrical coordinates, discussed in section 4.10.

$$= 4\int_0^5 \left( \int_{-\sqrt{z}}^{\sqrt{z}} \sqrt{z-y^2}\, dy \right) dz = 4\int_0^5 \left( \int_{-1}^{1} \sqrt{z - zu^2}\sqrt{z}\, du \right) dz$$

$$= 4\int_0^5 z\left[ \frac{u}{2}\sqrt{1-u^2} + \frac{1}{2}\arcsin u \right]_{-1}^{1} dz = 4\frac{\pi}{2}\left[ \frac{z^2}{2} \right]_0^5 = 2\pi\frac{25}{2} = 25\pi.$$

**4.5.17** a. Rearranging the order of a set of numbers does not change the value of the $r$th smallest number. Thus $M_r(\mathbf{x})$ satisfies the hypothesis of exercise 4.3.4, and

$$\int_{Q_{0,1}^n} M_r(\mathbf{x})\,|d^n\mathbf{x}| = n!\int_{P_{0,1}^n} M_r(\mathbf{x})\,|d^n\mathbf{x}|.$$

Since $0 \le x_1 \le x_2 \cdots \le x_n \le 1$ in $P_{0,1}^n$, the $r$th smallest coordinate in the second integral is $x_r$, i.e., $M_r(\mathbf{x}) = x_r$. We can integrate with respect to the various $x_i$ in different orders, but the order $x_1, x_2, \ldots x_n$ is convenient. With this order, the integral in question becomes

$$n!\int_0^1 \left( \cdots \left( \int_0^{x_2} x_r\, dx_1 \right) \cdots \right) dx_n.$$

Let us first consider how to do the integral if $r \le n - 2$ (and $n > 2$). After integrating with respect to $x_i$ for $1 \le i < r$, the innermost integral is

$$\int_0^{x_{i+2}} \frac{x_r x_{i+1}^i\, dx_{i+1}}{i!}.$$

(This formula is valid for any $i < r$ and $i \le n - 2$ even if $r > n - 2$.) After we integrate with respect to $x_{r-1}$, the $M_r$ function becomes relevant. The innermost integral in this case will be

$$\int_0^{x_{r+1}} \frac{x_r^r\, dx_r}{(r-1)!} = \int_0^{x_{r+1}} \frac{r x_r^r\, dx_r}{r!}.$$

Thus, after integrating with respect to $x_i$ for $r \le i < n-2$, the innermost integral is

$$\int_0^{x_{i+2}} r\frac{x_{i+1}^{i+1}\, dx_{i+1}}{(i+1)!}.$$

It is now simple to evaluate the whole integral. After using the above formula to get the result of the first $n - 2$ integrations, our integral is

$$n!\int_0^1 \left( \int_0^{x_n} r\frac{x_{n-1}^{n-1}\, dx_{n-1}}{(n-1)!} \right) dx_n = n!\int_0^1 r\frac{x_n^n\, dx_n}{(n)!} = \frac{r(n!)}{(n+1)!} = \frac{r}{n+1}.$$

If $r = n - 1$, then we can use our formula for the innermost integral after $n - 2$ integrations with $i < r$:

$$n! \int_0^1 \left( \int_0^{x_n} \frac{x_{n-1}^{n-2} x_{n-1} \, dx_{n-1}}{(n-2)!} \right) dx_n = n! \int_0^1 \left( \int_0^{x_n} (n-1) \frac{x_{n-1}^{n-1} \, dx_{n-1}}{(n-1)!} \right) dx_n$$

$$= n! \int_0^1 (n-1) \frac{x_n^n \, dx_n}{(n)!} = \frac{(n-1)(n!)}{(n+1)!} = \frac{n-1}{n+1} \left( = \frac{r}{n+1} \right).$$

For $r = n$, we can use the same formula:

$$n! \int_0^1 \left( \int_0^{x_n} \frac{x_{n-1}^{n-2} x_n \, dx_{n-1}}{(n-2)!} \right) dx_n$$

$$= n! \int_0^1 \frac{x_n^{n-1} x_n \, dx_n}{(n-1)!} = n! \int_0^1 n \frac{x_n^n \, dx_n}{(n)!} = \frac{(n)(n!)}{(n+1)!} = \frac{n}{n+1} \left( = \frac{r}{n+1} \right).$$

b. The minimum of a set of numbers is the same, regardless of the order of the numbers, so the min function also satisfies the hypothesis of exercise 4.3.4, and

$$\int_{Q_{0,1}^n} \min \left( 1, \frac{b}{x_1}, \ldots, \frac{b}{x_n} \right) |d^n \mathbf{x}| = n! \int_{P_{0,1}^n} \min \left( 1, \frac{b}{x_1}, \ldots, \frac{b}{x_n} \right) |d^n \mathbf{x}|.$$

In $P_{0,1}^n$, $x_n > x_i$ for $i < n$. Thus $b/x_n < b/x_i$ for $i < n$ (since $b > 0$). With this in mind, it is clear that $\min (1, b/x_1, \ldots, b/x_n) = \min (1, b/x_n)$ in $P_{0,1}^n$. To evaluate the integral it is convenient to use the same order of integration that we chose in part a:

$$n! \int_0^1 \left( \cdots \left( \int_0^{x_2} \min \left( 1, \frac{b}{x_n} \right) dx_1 \right) \cdots \right) dx_n.$$

The expression for the innermost integral after integrating with respect to $x_i$ ($i \leq n - 2$) will be nearly the same as that for part a, since the function we are integrating is only a function of $x_n$ (the expression is the same, except that $\min (1, b/x_n)$ replaces $x_r$):

$$\int_0^{x_{i+2}} \frac{\min \left( 1, \frac{b}{x_n} \right) x_{i+1}^i \, dx_{i+1}}{i!}.$$

It is now simple to evaluate the integral, using the above expression for the innermost integral after $n - 2$ integrations. First perform one additional integration:

$$n! \int_0^1 \left( \int_0^{x_n} \min \left( 1, \frac{b}{x_n} \right) \frac{x_{n-1}^{n-2} \, dx_{n-1}}{(n-2)!} \right) dx_n = n! \int_0^1 \min \left( 1, \frac{b}{x_n} \right) \frac{x_n^{n-1} \, dx_n}{(n-1)!}.$$

To integrate the min function, we have to split the above into two integrals. For $0 \leq x_n \leq b$, we have $1 < b/x_n$ and for $b \leq x_n \leq 1$, we have $b/x_n < 1$.

So

$$n! \int_0^1 \min\left(1, \frac{b}{x_n}\right) \frac{x_n^{n-1}\, dx_n}{(n-1)!} = n! \int_0^b \frac{x_n^{n-1}\, dx_n}{(n-1)!} + n! \int_b^1 \frac{b x_n^{n-2}\, dx_n}{(n-1)!}$$

$$= \frac{(n!)b^n}{n!} + \frac{n!}{(n-1)!}\left(\frac{b}{n-1} - \frac{b^n}{n-1}\right)$$

$$= \frac{b^n(n-1) + nb - nb^n}{n-1} = \frac{nb - b^n}{n-1}.$$

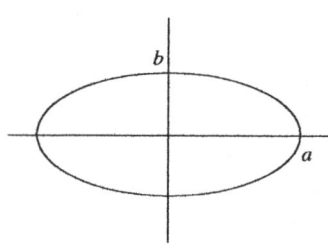

FIGURE FOR SOLUTION 4.5.19.
The ellipse

$$\frac{x^2}{a^2} + \frac{y^2}{b^2} = 1.$$

In the second line we write $(1 - z^3)$ not $(z^3 - 1)$ because $|ab|$ cannot be negative; we can do so, of course, since

$$(1 - z^3)^2 = (z^3 - 1)^2.$$

**4.5.19** Use the fact that the area bounded by the ellipse $\frac{x^2}{a^2} + \frac{y^2}{b^2} = 1$ is $\pi|ab|$. This gives

$$\int_1^1 \left(\text{Area bounded by ellipse given by } \frac{x^2}{(z^3-1)^2} + \frac{y^2}{(z^3+1)^2} = 1\right) dz$$

$$= \int_1^1 \pi(1 - z^3)(1 + z^3)\, dz = \pi \int_{-1}^1 (1 - z^6)\, dz$$

$$= \pi\left[z - \frac{z^7}{7}\right]_{-1}^1 = \frac{12}{7}\pi.$$

**4.6.1**  a.  For the unit square, we find

$$\frac{1}{36}\left( f\begin{pmatrix}0\\0\end{pmatrix} + f\begin{pmatrix}0\\1\end{pmatrix} + f\begin{pmatrix}1\\0\end{pmatrix} + f\begin{pmatrix}1\\1\end{pmatrix} \right.$$

$$\left. + 4\left( f\begin{pmatrix}1/2\\0\end{pmatrix} + f\begin{pmatrix}1/2\\1\end{pmatrix} + f\begin{pmatrix}0\\1/2\end{pmatrix} + f\begin{pmatrix}1\\1/2\end{pmatrix} \right) + 16 f\begin{pmatrix}1/2\\1/2\end{pmatrix} \right).$$

For the unit cube, we find

$$\frac{1}{216}\left( f\begin{pmatrix}0\\0\\0\end{pmatrix} + f\begin{pmatrix}1\\0\\0\end{pmatrix} + f\begin{pmatrix}0\\1\\0\end{pmatrix} + f\begin{pmatrix}1\\1\\0\end{pmatrix} + f\begin{pmatrix}0\\0\\1\end{pmatrix} + f\begin{pmatrix}1\\0\\1\end{pmatrix} + f\begin{pmatrix}0\\1\\1\end{pmatrix}x + f\begin{pmatrix}1\\1\\1\end{pmatrix} \right.$$

$$+ 4\left( f\begin{pmatrix}1/2\\0\\0\end{pmatrix} + f\begin{pmatrix}1/2\\1\\0\end{pmatrix} + f\begin{pmatrix}1/2\\0\\1\end{pmatrix} + f\begin{pmatrix}1/2\\1\\1\end{pmatrix} + f\begin{pmatrix}0\\1/2\\0\end{pmatrix} + f\begin{pmatrix}1\\1/2\\0\end{pmatrix} \right.$$

$$\left. + f\begin{pmatrix}0\\1/2\\1\end{pmatrix} + f\begin{pmatrix}1\\1/2\\1\end{pmatrix} + f\begin{pmatrix}0\\0\\1/2\end{pmatrix} + f\begin{pmatrix}1\\0\\1/2\end{pmatrix} + f\begin{pmatrix}0\\1\\1/2\end{pmatrix} + f\begin{pmatrix}1\\1\\1/2\end{pmatrix} \right)$$

$$+ 16\left( f\begin{pmatrix}0\\1/2\\1/2\end{pmatrix} + f\begin{pmatrix}1\\1/2\\1/2\end{pmatrix} + f\begin{pmatrix}1/2\\0\\1/2\end{pmatrix} + f\begin{pmatrix}1/2\\1\\1/2\end{pmatrix} + f\begin{pmatrix}1/2\\1/2\\0\end{pmatrix} + f\begin{pmatrix}1/2\\1/2\\1\end{pmatrix} \right)$$

$$+ 64 f\begin{pmatrix}1/2\\1/2\\1/2\end{pmatrix} \right).$$

b. The formula for the unit square gives

$$\frac{1}{36}\left( \left(1 + \frac{1}{2} + \frac{1}{2} + \frac{1}{3}\right) + 4\left(\frac{2}{3} + \frac{2}{5} + \frac{2}{3} + \frac{2}{5}\right) + 16\frac{1}{2} \right) = \frac{1}{36}\left(2 + \frac{1}{3} + \frac{128}{15} + \frac{16}{2}\right) \sim .524074.$$

The exact value of the integral is

$$\int_{\text{unit square}} f\left(\begin{matrix} x \\ y \end{matrix}\right) |dx\, dy| = 3\ln 3 - 4\ln 2 \sim .523848.$$

**4.6.3**  If we set

$$X = \frac{b-a}{2}x + \frac{a+b}{2}, \quad dX = \frac{b-a}{2}\, dx$$

and make the corresponding change of variables, we find

$$\int_a^b f(X)\, dX = \int_{-1}^1 f\left(\frac{b-a}{2}x + \frac{a+b}{2}\right)\frac{b-a}{2}\, dx.$$

This integral is approximated by

$$\sum_{i=1}^p w_i f\left(\frac{b-a}{2}x_i + \frac{a+b}{2}\right)\frac{b-a}{2},$$

so to make the "same approximation" for $\int_a^b f(X)\, dX$, we need to use

$$X_i = \frac{b-a}{2}x_i + \frac{a+b}{2} \quad \text{and} \quad W_i = \frac{b-a}{2}w_i.$$

**4.6.5**  a.  By Fubini's theorem,

$$\begin{aligned}
\int_{[a,b]\times[a,b]} f(x)g(y)|dx\, dy| &= \int_{[a,b]}\left(\int_{[a,b]} f(x)g(y)|dx|\right)|dy| \\
&= \left(\int_{[a,b]} g(y)|dy|\right)\left(\int_{[a,b]} f(x)|dx|\right) \\
&= \left(\sum_{i=1}^n c_i f(x_i)\right)\left(\sum_{i=1}^n c_i g(x_i)\right) \\
&= \sum_{i=1}^n\sum_{j=1}^n c_i c_j f(x_i)g(x_j).
\end{aligned}$$

b. Since both quadratic and cubic polynomials are integrated exactly by Simpson's rule, this function is integrated exactly, giving

$$\left[\frac{x^3}{3}\right]_0^1\left[\frac{y^4}{4}\right]_0^1 = \frac{1}{12}.$$

**4.7.1**  a.  Theorem 4.7.4 asserts that you can compute integrals by partitioning the plane into squares with vertices at the points $\left(\begin{matrix} n/N \\ m/N \end{matrix}\right)$ and sidelength $1/N$. The expression

Note that one $N$ goes with the $m$ to make up the $y$; the other two give the area of the square.

$$\lim_{N\to\infty}\frac{1}{N^3}\sum_{0\le n,m<N} me^{-nm/N^2} = \frac{1}{N^2}\sum\sum\frac{m}{N}e^{-nm/N^2}$$

is the left Riemann sum for the integral

$$\int_0^1\int_0^1 ye^{-xy}\, dx\, dy.$$

This comes down to computing the integral of a function on $\mathbb{R}^2$ that is continuous except on the boundary of the unit square; since that boundary has 2-dimensional volume 0, the function is integrable, and the limit exists.

Part b: this is one of those cases where if you apply Fubini one way, the integral can be evaluated in elementary terms, and in the other it can't.

b. The direction in which the integral can be evaluated in elementary terms is

$$\int_0^1 \left( \int_0^1 ye^{-xy}\, dx \right) dy = \int_0^1 \left( \left[ -\frac{y}{y}e^{-xy} \right]_0^1 \right) dy$$

$$= \int_0^1 (1 - e^{-y})\, dy = 1 + e^{-1} - 1 = \frac{1}{e}.$$

**4.8.1**

$$\det A = \det \begin{bmatrix} 0 & 1 & 2 \\ -1 & 2 & 1 \\ 2 & 1 & 0 \end{bmatrix} + 2 \det \begin{bmatrix} 4 & 1 & 2 \\ 5 & 2 & 1 \\ 3 & 1 & 0 \end{bmatrix} + 3 \det \begin{bmatrix} 4 & 0 & 2 \\ 5 & -1 & 1 \\ 3 & 2 & 0 \end{bmatrix}$$

$$= -8 - 6 + 54 = 40.$$

$$\det B = -5; \qquad \det C = 91$$

**4.8.3**  1. Multilinearity: Set $\begin{bmatrix} b \\ d \end{bmatrix} = \alpha \begin{bmatrix} e \\ f \end{bmatrix} + \beta \begin{bmatrix} g \\ h \end{bmatrix}$. Then

$$\det \begin{bmatrix} a & b \\ c & d \end{bmatrix} = \det \begin{bmatrix} a & \alpha e + \beta g \\ c & \alpha f + \beta h \end{bmatrix} = a\alpha f + a\beta h - c\alpha e - c\beta g,$$

which is identical to

$$\alpha \det \begin{bmatrix} a & e \\ c & f \end{bmatrix} + \beta \det \begin{bmatrix} a & g \\ c & h \end{bmatrix} = \alpha a f - \alpha c e + \beta a h - \beta c g.$$

2. Antisymmetry: $\det \begin{bmatrix} a & b \\ c & d \end{bmatrix} = ad - bc$ and $\det \begin{bmatrix} b & a \\ d & c \end{bmatrix} = bc - ad$, so

$$\det \begin{bmatrix} a & b \\ c & d \end{bmatrix} = -\det \begin{bmatrix} b & a \\ d & c \end{bmatrix}.$$

3. Normalization: $\det \begin{bmatrix} 1 & 0 \\ 0 & 1 \end{bmatrix} = 1.$

**4.8.5**    The trace is the sum of the elements on the diagonal.    Set $A = \begin{bmatrix} a & b \\ c & d \end{bmatrix}$ and $B = \begin{bmatrix} e & f \\ g & h \end{bmatrix}$. Then

$$AB = \begin{bmatrix} ae + bg & af + bh \\ ce + dg & cf + dh \end{bmatrix} \quad \text{and} \quad \operatorname{tr}(AB) = ae + bg + cf + dh;$$

$$BA = \begin{bmatrix} ea + fc & eb + fd \\ ga + hc & gb + hd \end{bmatrix} \quad \text{and} \quad \operatorname{tr}(BA) = ea + fc + gb + hd.$$

**4.8.7**  a. Suppose the column of 0's is the $i$th column. The hint suggested multiplying that column by 2 or $-4$. Instead, let's multiply by 0. Then by multilinearity,

$$\det[\vec{a}_1, \ldots, \vec{0}, \ldots, \vec{a}_n] = \det[\vec{a}_1, \ldots, 0 \cdot \vec{0}, \ldots, \vec{a}_n] = 0 \det[\vec{a}_1, \ldots, \vec{0}, \ldots, \vec{a}_n] = 0.$$

b. By antisymmetry, exchanging two columns changes the sign of the determinant, but exchanging two identical columns leaves it unchanged, therefore the determinant must be 0.

**4.8.9** a.  i. $\begin{bmatrix} 0 & 1 & 0 \\ 0 & 0 & 1 \\ 1 & 0 & 0 \end{bmatrix}$;  ii. $\begin{bmatrix} 0 & 0 & 1 & 0 \\ 1 & 0 & 0 & 0 \\ 0 & 0 & 0 & 1 \\ 0 & 1 & 0 & 0 \end{bmatrix}$;  iii. $\begin{bmatrix} 0 & 1 & 0 \\ 0 & 0 & 1 \\ 1 & 0 & 0 \end{bmatrix}$

b. $\begin{bmatrix} 0 & 0 & 1 \\ 0 & 1 & 0 \\ 1 & 0 & 0 \end{bmatrix} \begin{bmatrix} 1 & 0 & 0 \\ 0 & 0 & 1 \\ 0 & 1 & 0 \end{bmatrix} = \begin{bmatrix} 0 & 1 & 0 \\ 0 & 0 & 1 \\ 1 & 0 & 0 \end{bmatrix}$

**4.8.11** Use development of the first column (equation 4.8.5), and work by induction on $n$, the dimension of $A$. It is true when $n = 1$, i.e., when $A = [a]$, since then $\det \begin{bmatrix} a & C \\ 0 & B \end{bmatrix} = a \det B$.

Now suppose it is true for $n - 1$, and that $A$ is an $n \times n$ matrix. Then

$$\det \begin{bmatrix} A & C \\ 0 & B \end{bmatrix} = \sum_{i=1}^{n} (-1)^{i+1} a_{i,1} \det \begin{bmatrix} A_{i,1} & C \\ 0 & B \end{bmatrix} = \sum_{i=1}^{n} (-1)^{i+1} a_{i,1} \det A_{i,1} \det B$$

$$= \det B \sum_{i=1}^{n} (-1)^{i+1} a_{i,1} \det A_{i,1} = \det A \det B.$$

**4.8.13**

$$\operatorname{sgn} \sigma_1 = \det M_{\sigma_1} = \det I = +1$$

$$\operatorname{sgn} \sigma_2 = \det M_{\sigma_2} = \det \begin{bmatrix} 0 & 0 & 1 \\ 1 & 0 & 0 \\ 0 & 1 & 0 \end{bmatrix} = +1$$

$$\operatorname{sgn} \sigma_3 = \det M_{\sigma_3} = \det \begin{bmatrix} 0 & 1 & 0 \\ 0 & 0 & 1 \\ 1 & 0 & 0 \end{bmatrix} = +1$$

$$\operatorname{sgn} \sigma_4 = \det M_{\sigma_4} = \det \begin{bmatrix} 1 & 0 & 0 \\ 0 & 0 & 1 \\ 0 & 1 & 0 \end{bmatrix} = -1$$

$$\operatorname{sgn} \sigma_5 = \det M_{\sigma_5} = \det \begin{bmatrix} 0 & 1 & 0 \\ 1 & 0 & 0 \\ 0 & 0 & 1 \end{bmatrix} = -1$$

$$\operatorname{sgn} \sigma_6 = \det M_{\sigma_6} = \det \begin{bmatrix} 0 & 0 & 1 \\ 0 & 1 & 0 \\ 1 & 0 & 0 \end{bmatrix} = -1.$$

Part a of solution 4.8.15: We do not need to complete the operations to echelon form because any remaining operations add a multiple of one column to another column, which has $\mu = 1$.

**4.8.15** a. We use row operations; column operations would work as well. Each row operation (or column operation) is equivalent to multiplying the determinant by a factor $\mu$; we mark only those $\mu$ that are $\neq 1$:

$$\begin{bmatrix} 2 & 1 & 0 & 1 \\ 1 & 1 & 3 & 2 \\ 2 & 0 & 2 & 1 \\ 1 & 0 & 4 & 2 \end{bmatrix} \underset{\mu_1=1/2}{\rightarrow} \begin{bmatrix} 1 & 1/2 & 0 & 1/2 \\ 1 & 1 & 3 & 2 \\ 2 & 0 & 2 & 1 \\ 1 & 0 & 4 & 2 \end{bmatrix} \rightarrow \begin{bmatrix} 1 & 1/2 & 0 & 1/2 \\ 0 & 1/2 & 3 & 3/2 \\ 0 & -1 & 2 & 0 \\ 0 & -1/2 & 4 & 3/2 \end{bmatrix} \underset{\mu_2=2}{\rightarrow} \begin{bmatrix} 1 & 1/2 & 0 & 1/2 \\ 0 & 1 & 6 & 3 \\ 0 & -1 & 2 & 0 \\ 0 & -1/2 & 4 & 3/2 \end{bmatrix}$$

$$\rightarrow \begin{bmatrix} 1 & 0 & -3 & -1 \\ 0 & 1 & 6 & 3 \\ 0 & 0 & 8 & 3 \\ 0 & 0 & 7 & 3 \end{bmatrix} \underset{\mu_3=1/8}{\rightarrow} \begin{bmatrix} 1 & 0 & -3 & -1 \\ 0 & 1 & 6 & 3 \\ 0 & 0 & 1 & 3/8 \\ 0 & 0 & 7 & 3 \end{bmatrix} \rightarrow \begin{bmatrix} 1 & 0 & 0 & 1/8 \\ 0 & 1 & 0 & 1 \\ 0 & 0 & 1 & 3/8 \\ 0 & 0 & 0 & 3/8 \end{bmatrix} \underset{\mu_4=8/3}{\rightarrow} \begin{bmatrix} 1 & 0 & 0 & 1/8 \\ 0 & 1 & 0 & 1 \\ 0 & 0 & 1 & 3/8 \\ 0 & 0 & 0 & 1 \end{bmatrix}.$$

Thus the determinant is $\dfrac{1}{\mu_1\mu_2\mu_3\mu_4} = 3$.

b. The numbers $1, 2, 3, 4$ have 24 permutations, but only 10 contribute nonzero terms to the sum

$$\det A = \sum_{\sigma \in \mathrm{Perm}(1,\dots,n)} \mathrm{sgn}(\sigma)a_{1,\sigma(1)} \cdots a_{n,\sigma(n)}. \quad \text{of equation 4.8.39:}$$

$$\sigma_1 = (1234) \rightarrow +a_{1,1}a_{2,2}a_{3,3}a_{4,4} = +2 \cdot 1 \cdot 2 \cdot 2 = +8;$$

$$\sigma_2 = (1243) \rightarrow -a_{1,1}a_{2,2}a_{3,4}a_{4,3} = -2 \cdot 1 \cdot 1 \cdot 4 = -8;$$

$$\sigma_3 = (2134) \rightarrow -1 \cdot 1 \cdot 2 \cdot 2 = -4; \quad \sigma_4 = (2143) \rightarrow 1 \cdot 1 \cdot 1 \cdot 4 = +4$$

$$\sigma_5 = (2341) \rightarrow -1 \cdot 3 \cdot 1 \cdot 1 = -3; \quad \sigma_6 = (2314) \rightarrow 1 \cdot 3 \cdot 2 \cdot 2 = +12$$

$$\sigma_7 = (2413) \rightarrow -1 \cdot 2 \cdot 2 \cdot 4 = -16 \quad \sigma_8 = (2431) \rightarrow +1 \cdot 2 \cdot 2 \cdot 1 = +4$$

$$\sigma_9 = (4231) \rightarrow -1 \cdot 1 \cdot 2 \cdot 1 = -2; \quad \sigma_{10} = (4213) \rightarrow +1 \cdot 1 \cdot 2 \cdot 4 = +8.$$

So

$$\det \begin{bmatrix} 2 & 1 & 0 & 1 \\ 1 & 1 & 3 & 2 \\ 2 & 0 & 2 & 1 \\ 1 & 0 & 4 & 2 \end{bmatrix} = +8 - 8 - 4 + 4 - 3 + 12 - 16 + 4 - 2 + 8 = 3.$$

**4.8.17**  a. Let $A$ be an upper triangle matrix with *nonzero* entries on the diagonal. If the entry $a_{1,2}$ (first row, second column) is nonzero, it can be made zero by adding an appropriate multiple of the first column to the second column. This does not change any other entries because every entry of the first column is 0 except the first.

Similarly, if in the third column the two entries above the diagonal are nonzero, they can be made zero: the first by adding an appropriate multiple of the first column, the second by adding an appropriate multiple of the second column (which now has as its only nonzero entry the entry on the diagonal). Working from left to right, one can set any nonzero entries above the diagonal to zero, without affecting any entries on the main diagonal.

The determinant of the resulting diagonal matrix $\widetilde{A}$ is clearly the product of the diagonal entries: if we call those entries $\lambda_1, \dots, \lambda_n$, then

$$\det \widetilde{A} = \det(\lambda_1 \vec{e}_1, \dots, \lambda_n \vec{e}_n) = \lambda_1 \lambda_2 \cdots \lambda_n \det(\vec{e}_1, \dots, \vec{e}_n) = \lambda_1 \lambda_2 \cdots \lambda_n.$$

We do not need to prove the result for lower triangular matrices as well, since it follows from theorem 4.8.7.

We have $\det \widetilde{A} = \det A$ because (by property 1) adding a multiple of one column onto another does not change the determinant (see equation 4.8.11).

b. If some entry on the diagonal of an upper triangle matrix $A$ is 0, say the $i$th entry, then follow the procedure in part a to set to 0 all entries above the main diagonal in the first through $i$th columns. The entries below the main diagonal are 0 because the matrix is upper triangle, so this procedure (which does not change the diagonal entries) results in the $i$th column consisting entirely of 0's. Then $\det A = 0$, which is the product of the diagonal entries.

**4.8.19** Let us first see the following statement, which is interesting in its own right.

> **Proposition** *Any $n \times n$ matrix $A$ of rank $k$ can be written $A = QJ_kP^{-1}$, where $P$ and $Q$ are invertible, and*
> $$J_k = \begin{bmatrix} 0 & 0 \\ 0 & I_k \end{bmatrix},$$
> *where $I_k$ is the $k \times k$ identity matrix and all other entries are 0.*

**Solution 4.8.19:** For instance, if $n = 4$, then

$$J_2 = \begin{bmatrix} 0 & 0 & 0 & 0 \\ 0 & 0 & 0 & 0 \\ 0 & 0 & 1 & 0 \\ 0 & 0 & 0 & 1 \end{bmatrix}$$

$$J_3 = \begin{bmatrix} 0 & 0 & 0 & 0 \\ 0 & 1 & 0 & 0 \\ 0 & 0 & 1 & 0 \\ 0 & 0 & 0 & 1 \end{bmatrix}.$$

Of course if an $n \times n$ matrix $A$ has rank $n$, then $A = QJ_nP^{-1}$ just says that $A = QP^{-1}$, i.e., $A$ is invertible with inverse $PQ^{-1}$; see proposition 1.2.15.

**Proof.** In the domain, choose vectors $\vec{v}_1, \ldots, \vec{v}_{n-k}$ that span the kernel of $A$ (and hence are linearly independent, since the dimension of the kernel is $n - k$, by the dimension formula (theorem 2.5.8). Add further linearly independent vectors to make a basis $\vec{v}_1, \ldots, \vec{v}_{n-k}, \vec{v}_{n-k+1}, \ldots, \vec{v}_n$. In the codomain, define the vectors

$$\vec{w}_{n-k+1} = A\vec{v}_{n-k+1}, \ldots, \vec{w}_n = A\vec{v}_n.$$

The theorem of the incomplete basis (exercise 2.6.9) justifies our adding vectors to make a basis.

These vectors are linearly independent, because

$$\mathbf{0} = \sum_{i=n-k+1}^{n} a_i\vec{w}_i = \sum_{i=n-k+1}^{n} a_i A\vec{v}_i = A\left(\sum_{i=n-k+1}^{n} a_i\vec{v}_i\right)$$

implies that $\sum_{i=n-k+1}^{n} a_i\vec{v}_i$ is in $\ker A$, hence it is in the span of the vectors $\vec{v}_1, \ldots, \vec{v}_{n-k}$. So we can write

$$\sum_{i=n-k+1}^{n} a_i\vec{v}_i = \sum_{i=1}^{n-k} b_i\vec{v}_i,$$

which implies that all $a_i$ (and all $b_i$) vanish.

Hence $\vec{w}_{n-k+1}, \ldots, \vec{w}_n$ can also be completed to form a basis

$$\vec{w}_1, \ldots, \vec{w}_{n-k}, \vec{w}_{n-k+1}, \ldots, \vec{w}_n.$$

Set

$$P = [\vec{v}_1, \ldots, \vec{v}_n] \quad \text{and} \quad Q = [\vec{w}_1, \ldots, \vec{w}_n].$$

The $i$th column of the matrix $Q = [\vec{w}_1, \ldots, \vec{w}_n]$ is $Q\vec{e}_i = \vec{w}_i$, so $Q^{-1}(\vec{w}_i) = Q^{-1}Q\vec{e}_i = \vec{e}_i$.

Both these matrices are invertible, and we have

$$Q^{-1}AP(\vec{e}_i) = Q^{-1}A\vec{v}_i = \begin{cases} 0 & \text{if } i \leq n-k \\ Q^{-1}(\vec{w}_i) = \vec{e}_i & \text{if } i > n-k. \end{cases}$$

So $Q^{-1}AP = J_k$, or equivalently, $A = QJ_kP^{-1}$. $\square$

With this under our belt, the problem is not too hard.

a. Assume $A$ has rank $k = n-1$. We want to show that the linear transformation $[\mathbf{D}\det(A)] : \text{Mat}\,(n, n) \to \mathbb{R}$ is not the zero linear transformation, i.e., that there is a matrix $B$ such that $[\mathbf{D}\det(A)]B \neq 0$.

Write

$$[\mathbf{D}\det(A)]B = [\mathbf{D}\det(QJ_kP^{-1})]B$$

$$= \lim_{h\to 0} \frac{\det(QJ_kP^{-1} + hB) - \det(QJ_kP^{-1})}{h}$$

$$= \lim_{h\to 0} \frac{\det(QJ_kP^{-1} + hQQ^{-1}BPP^{-1}) - \det(QJ_kP^{-1})}{h}$$

$$= \lim_{h\to 0} \frac{\det Q \det P^{-1} \det(J_k + hQ^{-1}BP) - \det(QJ_kP^{-1})}{h}$$

$$= \det Q \det P^{-1} \lim_{h\to 0} \frac{\det(J_k + hQ^{-1}BP)}{h}.$$

Since $Q$ and $P^{-1}$ are invertible, we know that $\det Q \neq 0$ and $\det P^{-1} \neq 0$, so we just need to show that there exists a matrix $B$ satisfying

$$\det(J_k + hQ^{-1}BP) \neq 0.$$

Set $K = Q^{-1}BP$, so that now we need to show that $\det(J_k + hK) \neq 0$, and let $k_{i,j}$ denote the $i, j$th entry of $K$, so that

$$\det(J_k + hQ^{-1}BP) = \det(J_k + hK) = \det \begin{bmatrix} hk_{1,1} & hk_{1,2} & \dots & hk_{1,n} \\ hk_{2,1} & 1 + hk_{2,2} & \dots & hk_{2,n} \\ \vdots & \vdots & \ddots & \vdots \\ hk_{n,1} & hk_{n,2} & \dots & 1 + hk_{n,n} \end{bmatrix},$$

Then (as you can easily confirm by working out an example where $n = 3$), when we compute the determinant, there is a single term – the $hk_{1,1}$ – that is of degree 1 in $h$; all the other terms have at least $h^2$, and thus disappear after dividing by $h$ and taking the limit as $h \to 0$. This yields

$$\lim_{h\to 0} \frac{1}{h} \det(J_k + hQ^{-1}BP) = k_{1,1}.$$

Thus if $B$ is any $n \times n$ matrix such that the $(1,1)$ entry of $K = Q^{-1}BP$ is not 0, then

$$[\mathbf{D}\det(A)]B \neq 0.$$

Clearly a matrix $K'$ exists with $k'_{1,1} \neq 0$; setting $B = QK'P^{-1}$ gives $Q^{-1}BP = Q^{-1}QK'P^{-1}P = K'$.

b. In this case, entries of the first two columns and first two rows of $J_k + hK$ have an $h$. For instance, if $n = 4$ and $k = n - 2 = 2$, we have

$$\det(J_k + hK) = \det \begin{bmatrix} hk_{1,1} & hk_{1,2} & hk_{1,3} & hk_{1,4} \\ hk_{2,1} & hk_{2,2} & hk_{2,3} & hk_{2,4} \\ hk_{3,1} & hk_{3,2} & 1 + hk_{3,3} & hk_{3,4} \\ hk_{4,1} & hk_{4,2} & hk_{4,3} & 1 + hk_{4,4} \end{bmatrix}$$

---

Remember that the determinant of a matrix is a number, so multiplication of determinants is commutative.

We have

$$\det(QJ_kP^{-1})$$
$$= \det Q \det J_k \det P^{-1}$$
$$= 0$$

since $\det J_k = 0$; see (for instance) theorem 4.8.8.

We are assuming $k = n - 1$, so the first column and first row of $J_k$ consist of 0's, so the first column and first row of $J_k + hK$ equal the first column and first row of $hK$.

You can compute

$$\det(J_k + hK) = \det(J_k + hQ^{-1}BP)$$

using development by the first column or theorem 4.8.11, which says that you must choose in each term one element from each row and each column. In the first $n - k$ columns, every entry in the matrix has a factor of $h$, so each term of the determinant has a factor of $h^2$.

When you compute this, the lowest degree term in the expansion will have a factor $h^{n-k}$ (for the case above, $h^2$), giving

$$\lim_{h\to 0}\frac{1}{h}\det(J_k + hQ^{-1}BP) = 0$$

So $[\mathbf{D}\det(A)]B = 0$ for all $B \in \text{Mat}\,(n, n)$.

**4.9.1** It doesn't matter what $A$ is, so long as it has positive and finite volume: since $T$ has a triangular matrix,

$$\frac{\text{vol}_n(T(A))}{\text{vol}_n(A)} = |\det T| = n!,$$

by theorem 4.8.8.

**4.9.3** a. By Fubini's theorem, this volume is

$$\int_0^1 \int_0^{1-z} \int_0^{1-z-y} dx\,dy\,dz = \int_0^1 \int_0^{1-z} 1 - z - y\,dy\,dz$$

$$= \int_0^1 \left[(1-z)y - \frac{y^2}{2}\right]_0^{1-z} dz$$

$$= \int_0^1 \frac{1}{2}(1-z)^2 dz = \left[-\frac{1}{6}(1-z)^3\right]_0^1 = \frac{1}{6}$$

By "maps the tetrahedron $T_1$ onto the tetrahedron $T_2$" we mean that $S$ is an onto map (in fact, it is bijective) that takes a point in $T_1$ and gives a point in $T_2$; for every point $\mathbf{y} \in T_2$ there is a point $\mathbf{x} \in T_1$ such that $S\mathbf{x} = \mathbf{y}$.

b. The matrix $S = \begin{bmatrix} 2 & -1 & -2 \\ 1 & 3 & -5 \\ 1 & 1 & 2 \end{bmatrix}$ maps the tetrahedron $T_1$ onto the tetrahedron $T_2$. Thus

$$\text{vol}\,T_2 = |\det S|\,\text{vol}\,T_1 = \frac{33}{6} = \frac{11}{2}.$$

**4.9.5** The function $A$ is given by

$$A(x) = \det \begin{bmatrix} f(x) & g(x) \\ f'(x) & g'(x) \end{bmatrix} = f(x)g'(x) - f'(x)g(x).$$

Here we use $(ab)' = ab' + ba'$.

So its derivative is given by

$$A'(x) = f'(x)g'(x) + f(x)g''(x) - f'(x)g'(x) - f''(x)g(x)$$

$$= q(x)\big(f(x)g(x) - f(x)g(x)\big) = 0.$$

So the function is constant: $A(x) = A(0)$.

**4.9.7** a. First let us see why there exists such a function and why it is unique. Existence is easy (since we have theorem and definition 4.8.1): the function $|\det T|$ satisfies properties 1–5.

The proof of uniqueness is essentially the same as the proof of uniqueness for the determinant: to compute $\Delta(T)$ from properties 1–5, column reduce $T$. Clearly operations of type 2 and 3 do not change the value of $\Delta$. Operations of type 3 switch two columns, which by property 2 does not change the value of $\Delta$. Operations of type 2 add a multiple of one column onto another, which by property 4 does not change the value either. By property 3, an operation of type 1 multiplies $\Delta(T)$ by $|\mu|$, the factor you multiply a column by. So keep track, for each column operation of type 1, of the numbers $\mu_1, \ldots, \mu_k$ that you multiply columns by.

At the end we see that

$$\Delta(T) = \begin{cases} 0 & \text{if } T \text{ does not column reduce to the identity} \\ \frac{1}{|\mu_1 \cdots \mu_k|} = \frac{1}{|\mu_1 \cdots \mu_k|}\Delta(I) & \text{if } T \text{ column reduces to the identity.} \end{cases}$$

By property 5, $\dfrac{1}{|\mu_1 \cdots \mu_k|}\Delta(I) = \dfrac{1}{|\mu_1 \cdots \mu_k|}$.

b. Now we show that the mapping $T \mapsto \mathrm{vol}_n(T(Q))$ satisfies properties 1–5. Properties 1, 2, and 5 are straightforward, but 3 and 4 are quite delicate.

Property 1: It is not actually necessary to prove that $\mathrm{vol}_n(T(Q)) \geq 0$, but it follows immediately from definition 4.1.16 of $\mathrm{vol}_n$ and from the fact (definition 4.1.1) that the characteristic function is never negative.

Property 2: Since the mapping $T \mapsto T(Q)$ consists of taking each point in $Q$, considering it as a vector, and multiplying it by $T$, changing the order of the vectors does not change the result; it just changes the labeling of the vertices of $T(Q)$. So $\mathrm{vol}_n(T(Q))$ remains unchanged.

Properties 3 and 4 reflect properties that $\mathrm{vol}_n$ obviously ought to have if our definition of $n$-dimensional volume is to be consistent with our intuition about volume (not to mention the various formulas for area and volume you learned in high school). The $n$ vectors $\vec{v}_1, \ldots, \vec{v}_n$ span an $n$-dimensional "parallelogram" (parallelogram in $\mathbb{R}^2$, parallelepiped in $\mathbb{R}^3$, ... ).

Saying that $\mathrm{vol}_n$ should have property 3 says exactly that if you multiply one side of such a parallelogram by a number $a$, keeping the others constant, it should multiply the volume by $|a|$.

Saying that $\mathrm{vol}_n$ should have property 4 says that translating an object should not change its volume: once we know that $\mathrm{vol}_n$ is invariant by translation, property 4 will follow.

It isn't crystal clear that our definition using dyadic decompositions actually has these properties.

The main tool to do this problem is proposition 4.1.21; we will apply it to prove property 4. Consider the following three subsets of $\mathbb{R}^n$

$$A = \{ s_1\vec{v}_1 + t\vec{v}_2 + s_3\vec{v}_3 + \cdots + s_n\vec{v}_n \mid 0 \leq s_i < 1,\ t \in [0, as_1) \}$$
$$B = \{ s_1\vec{v}_1 + t\vec{v}_2 + s_3\vec{v}_3 + \cdots + s_n\vec{v}_n \mid 0 \leq s_i < 1,\ t \in [as_1, 1) \}$$
$$C = \{ s_1\vec{v}_1 + t\vec{v}_2 + s_3\vec{v}_3 + \cdots + s_n\vec{v}_n \mid 0 \leq s_i < 1,\ t \in [1, 1 + as_1) \}.$$

These are shown in the margin for $n = 2$. Then

$$[\vec{v}_1, \ldots, \vec{v}_n](Q) = A \cup B, \qquad [\vec{v}_1 + a\vec{v}_2, \ldots, \vec{v}_n](Q) = B \cup C, \qquad C = A + \vec{v}_2.$$

Property 4 follows.

Part b of exercise 4.9.7 was misstated. The map in question is

$$T \mapsto \mathrm{vol}_n(T(Q));$$

it takes an $n \times n$ matrix $T$ and gives the number $\mathrm{vol}_n T(Q)$, where $Q$ is the $n$-dimensional unit cube. It is not $Q \mapsto \mathrm{vol}_n(T(Q))$.

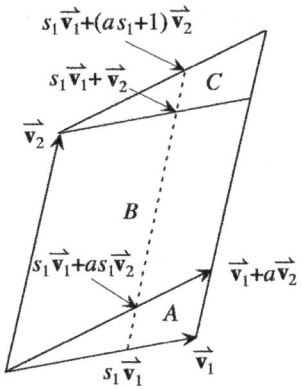

Here $T = [\vec{v}_1, \vec{v}_2]$, and the subset $T(Q)$ is $A \cup B$; the subset $[\vec{v}_1 + a\vec{v}_2, \vec{v}_2](Q)$ is $B \cup C$. We need to show that

$$\mathrm{vol}_2 A \cup B = \mathrm{vol}_2 B \cup C,$$

which it does, since

$$B \cup C = B \cup (A + \vec{v}_2)$$

and (by proposition 4.1.21),

$$\mathrm{vol}_2(A + \vec{v}_2) = \mathrm{vol}_2(A).$$

For property 3, we will start by showing that it is true for $a = 1/2$, $1/3$, $1/4, \ldots$. Choose $m > 0$, and define $A_k$ by

$$A_{k,m} = \left\{ t\vec{v}_1 + s_2\vec{v}_2 + s_3\vec{v}_3 + \cdots + s_n\vec{v}_n \,\middle|\, 0 \le s_i < 1, \frac{k-1}{m} \le t < \frac{k}{m} \right\}.$$

Then for any $m$, the $A_{k,m}$ are disjoint, $A_k = A_1 + \frac{k-1}{m}\vec{v}_1$, so they all have the same volume, and

$$\bigcup_{k=1}^{m} A_k = [\vec{v}_1, \ldots, \vec{v}_n](Q), \quad \text{so} \quad \mathrm{vol}_n(A_k) = \frac{1}{m}[\vec{v}_1, \ldots, \vec{v}_n](Q).$$

Now take $a > 0$ to be any positive real. Choose $m$, and let $p$ be the number such that $p/m \le a < (p+1)/m$. Then

$$\cup_{k=1}^{p} A_k \subset [a\vec{v}_1, \ldots, \vec{v}_n](Q) \subset \cup_{k=1}^{p+1} A_k,$$

giving

$$\frac{p}{m}\,\mathrm{vol}_n[\vec{v}_1, \ldots, \vec{v}_n](Q) \le \mathrm{vol}_n[a\vec{v}_1, \ldots, \vec{v}_n](Q) \le \frac{p+1}{m}\,\mathrm{vol}_n[\vec{v}_1, \ldots, \vec{v}_n](Q).$$

Clearly, the left and right terms can be made arbitrarily close by taking $m$ sufficiently large, so

$$\mathrm{vol}_n[a\vec{v}_1, \ldots, \vec{v}_n](Q) = \lim_{m \to \infty} \frac{p}{m}\,\mathrm{vol}_n[\vec{v}_1, \ldots, \vec{v}_n](Q)$$
$$= a\,\mathrm{vol}_n[\vec{v}_1, \ldots, \vec{v}_n](Q).$$

Finally, if $a < 0$, we have

$$[a\vec{v}_1, \ldots, \vec{v}_n](Q) = \big[\,|a|\vec{v}_1, \ldots, \vec{v}_n\big](Q) - |a|\vec{v}_1,$$

so

$$\mathrm{vol}_n[a\vec{v}_1, \ldots, \vec{v}_n](Q) = \mathrm{vol}_n[\,|a|\vec{v}_1, \ldots, \vec{v}_n\,](Q) = |a|\,\mathrm{vol}_n[\vec{v}_1, \ldots, \vec{v}_n](Q).$$

Property 5: By proposition 4.1.19,

$$I \mapsto \mathrm{vol}_n[\vec{e}_1, \ldots, \vec{e}_n](Q) = \mathrm{vol}_n(Q) = 1.$$

**4.10.1** Since $x^2 + y^2 \le R^2$, $y$ goes from $-R$ to $R$, and $x$ from $-\sqrt{R^2 - y^2}$ to $\sqrt{R^2 - y^2}$. So

$$\int_{D_R} (x^2 + y^2)\, dx\, dy = \int_{-R}^{R} \int_{-\sqrt{R^2-y^2}}^{\sqrt{R^2-y^2}} (x^2 + y^2)\, dx\, dy = \int_{-R}^{R} \left( \left[\frac{x^3}{3} + y^2 x\right]_{-\sqrt{R^2-y^2}}^{\sqrt{R^2-y^2}} \right) dy$$

$$= \int_{-R}^{R} \left( \frac{2(R^2 - y^2)^{3/2}}{3} + 2\sqrt{R^2 - y^2}\, y^2 \right) dy$$

$$= \int_{-R}^{R} \left( \frac{2\left(R^2\left(1 - (\frac{y}{R})^2\right)\right)^{3/2}}{3} + 2\sqrt{R^2(1 - \frac{y^2}{R^2})}\, \frac{R^2 y^2}{R^2} \right) dy$$

$$= \int_{-R}^{R} \left( \frac{2R^3(1 - (\frac{y}{R})^2)^{3/2}}{3} + 2R^3 \sqrt{1 - \left(\frac{y}{R}\right)^2} \left(\frac{y}{R}\right)^2 \right) dy.$$

Now set $\sin\theta = \frac{y}{R}$ (which we can do because $y$ goes from $-R$ to $R$) so that $dy = R\cos\theta\,d\theta$, and continue:

$$\int_{D_R}(x^2+y^2)\,dx\,dy = R^4\int_{-\frac{\pi}{2}}^{\frac{\pi}{2}}\left(\frac{2}{3}\cos^4\theta + 2\cos^2\theta\sin^2\theta\right)d\theta$$

In the second and third lines of this equation, $\cos 2\theta$ and $\cos 4\theta$ integrate to 0 because $2\theta$ goes from $-\pi$ to $\pi$, where cos is as often negative as positive.

$$= R^4\int_{-\frac{\pi}{2}}^{\frac{\pi}{2}}\left(\frac{1}{6}(1+2\,\overbrace{\cos 2\theta}^{\text{integrates to }0}+\cos^2 2\theta) + \frac{1}{2}(1-\cos^2 2\theta)\right)d\theta$$

$$= R^4\int_{-\frac{\pi}{2}}^{\frac{\pi}{2}}\left(\frac{1}{6}+\frac{1}{12}+\overbrace{\frac{\cos 4\theta}{12}}^{\text{integrates to }0}+\frac{1}{2}-\frac{1}{4}-\overbrace{\frac{\cos 4\theta}{4}}^{\text{integrates to }0}\right)d\theta$$

$$= R^4\int_{-\frac{\pi}{2}}^{\frac{\pi}{2}}\frac{1}{2}\,d\theta = \frac{R^4\pi}{2}.$$

**4.10.3** Since $|z^2-\frac{1}{2}| = -\frac{1}{2}$ has no solutions, squaring both sides of $|z^2-\frac{1}{2}| = \frac{1}{2}$ does not add more points to the graph. Squaring both sides gives $|z^2-\frac{1}{2}|^2 = \frac{1}{4}$, which, using de Moivre's formula (equation 0.7.13), can be rewritten as

Remember (see definition 0.7.3) that $|a+ib|^2 = a^2+b^2$; it does not equal $a^2+2iab-b^2$.

$$\frac{1}{4} = \left|r^2\cos^2\theta + ir^2\sin^2\theta - \frac{1}{2}\right|^2 = r^4\cos^2 2\theta - r^2\cos 2\theta + \frac{1}{4} + r^4\sin^2 2\theta,$$

which gives the desired result:

$$r^2(r^2-\cos 2\theta) = 0, \quad\text{i.e.,}\quad r^2 = \cos 2\theta.$$

**4.10.5**  a.  Use the linear change of variables $T\begin{pmatrix}u\\v\end{pmatrix} = \begin{pmatrix}au\\bv\end{pmatrix}$, which maps the unit circle to the ellipse, i.e., the transformation given by $\begin{bmatrix}a&0\\0&b\end{bmatrix}$; see example 4.9.8. The area of the ellipse is the area of the circle multiplied by $|\det T| = |ab|$, so the area is $\pi|ab|$.

Solution 4.10.5, part b: This linear transformation is given by $\begin{bmatrix}a&0&0\\0&b&0\\0&0&c\end{bmatrix}$. By theorem 4.8.8, its determinant is $abc$.

b. This time, use the linear change of variables $T\begin{pmatrix}u\\v\\w\end{pmatrix} = \begin{pmatrix}au\\bv\\cw\end{pmatrix}$.

This maps the unit sphere to the ellipsoid, whose volume is therefore $4\pi/3$ multiplied by $|\det T| = |abc|$.

Solution 4.10.7: For example, if $A$ is the identity, then $X_A$ is the unit ball: $Q_I(\vec{x}) = \vec{x}\cdot\vec{x} = |\vec{x}|^2$. More generally, $X_A$ is an $n$-dimensional ellipsoid; we are computing the $n$-dimensional volume of such ellipsoids.

The volumes of the unit balls are computed in example 4.5.7.

**4.10.7** Let $X_A\subset\mathbb{R}^n$ denote the subset $Q_A(\vec{x})\le 1$. Suppose that there exists a symmetric matrix $C$ such that $C^2 = C^\top C = A$. Note that $|\det C| = \sqrt{\det A}$.

The equation $Q_A(\vec{x})\le 1$ can be rewritten $(C\vec{x})\cdot(C\vec{x})\le 1$:

$$1\ge Q_A(\vec{x}) = \mathbf{x}\cdot A\vec{x} = \vec{x}\cdot C^2\vec{x} = \vec{x}^\top C^\top C\vec{x} = (C\vec{x})^\top C\vec{x} = C\vec{x}\cdot C\vec{x}.$$

In other words, the linear transformation $C$ turns $X_A$ into the unit ball $B_n\subset\mathbb{R}^n$: it takes a point in $\mathbb{R}^n$ satisfying $\mathbf{x}\cdot Ax\le 1$ and returns a point

$C\mathbf{x} \in \mathbb{R}^n$ satisfying $|C\mathbf{x}|^2 \leq 1$. This gives

$$\text{vol}_n(X_A) \underbrace{=}_{\text{thm. 4.9.1}} \frac{\text{vol}_n(B_n)}{|\det C|} = \frac{\text{vol}_n(B_n)}{\sqrt{\det A}}. \tag{1}$$

Checking that $C$ exists requires the spectral theorem (theorem 3.7.14) and also some notions about changes of basis. According to theorem 3.7.14, there exists an orthonormal basis $\vec{v}_1, \ldots \vec{v}_n$ of $\mathbb{R}^n$ such that $A\vec{v}_i = \lambda_i \vec{v}_i$. Since the quadratic form $Q(\vec{v}) = \vec{v} \cdot A\vec{v}$ is positive definite by definition,

$$0 < \vec{v}_i \cdot A\vec{v}_i = \vec{v}_i \cdot \lambda_i \vec{v}_i = \lambda_i |\vec{v}_i|^2,$$

which implies that all the eigenvalues $\lambda_i$ are positive. Let $T = [\vec{v}_1, \ldots, \vec{v}_n]$. We need two properties of $T$:

We saw in exercise 2.4.7 that $T$ is an orthogonal matrix if and only if $T^\top T = I$: if $T = [\vec{v}_1, \ldots, \vec{v}_n]$, then the $i, j$th entry of $T^\top T$ is $\vec{v}_i^\top \vec{v}_j = \vec{v}_i \cdot \vec{v}_j$. Since the $\vec{v}_i$ form an orthonormal basis, this dot product is 1 when $i = j$ and 0 otherwise. Thus $T^\top T = I$.

1. Since $T$ is an orthogonal matrix, $T^\top T = I$.
2. By proposition 2.7.3,

$$T^{-1}AT = \begin{bmatrix} \lambda_1 & 0 & \ldots & 0 \\ 0 & \lambda_2 & \ldots & 0 \\ \vdots & \vdots & \ddots & \vdots \\ 0 & 0 & \ldots & \lambda_n \end{bmatrix}. \tag{2}$$

Let us set

$$C = T \begin{bmatrix} \sqrt{\lambda_1} & 0 & \ldots & 0 \\ 0 & \sqrt{\lambda_2} & \ldots & 0 \\ \vdots & \vdots & \ddots & \vdots \\ 0 & 0 & \ldots & \sqrt{\lambda_n} \end{bmatrix} T^{-1} = T \begin{bmatrix} \sqrt{\lambda_1} & 0 & \ldots & 0 \\ 0 & \sqrt{\lambda_2} & \ldots & 0 \\ \vdots & \vdots & \ddots & \vdots \\ 0 & 0 & \ldots & \sqrt{\lambda_n} \end{bmatrix} T^\top.$$

Then $C$ is symmetric:

$$C^\top = \left( T \begin{bmatrix} \sqrt{\lambda_1} & 0 & \ldots & 0 \\ 0 & \sqrt{\lambda_2} & \ldots & 0 \\ \vdots & \vdots & \ddots & \vdots \\ 0 & 0 & \ldots & \sqrt{\lambda_n} \end{bmatrix} T^\top \right)^\top \underbrace{=}_{\text{theorem 1.2.17}} (T^\top)^\top \begin{bmatrix} \sqrt{\lambda_1} & 0 & \ldots & 0 \\ 0 & \sqrt{\lambda_2} & \ldots & 0 \\ \vdots & \vdots & \ddots & \vdots \\ 0 & 0 & \ldots & \sqrt{\lambda_n} \end{bmatrix}^\top T^\top$$

$$= T \begin{bmatrix} \sqrt{\lambda_1} & 0 & \ldots & 0 \\ 0 & \sqrt{\lambda_2} & \ldots & 0 \\ \vdots & \vdots & \ddots & \vdots \\ 0 & 0 & \ldots & \sqrt{\lambda_n} \end{bmatrix} T^\top = C.$$

Moreover,

$$C^2 = T \begin{bmatrix} \sqrt{\lambda_1} & 0 & \ldots & 0 \\ 0 & \sqrt{\lambda_2} & \ldots & 0 \\ \vdots & \vdots & \ddots & \vdots \\ 0 & 0 & \ldots & \sqrt{\lambda_n} \end{bmatrix} \underbrace{T^\top T}_{I} \begin{bmatrix} \sqrt{\lambda_1} & 0 & \ldots & 0 \\ 0 & \sqrt{\lambda_2} & \ldots & 0 \\ \vdots & \vdots & \ddots & \vdots \\ 0 & 0 & \ldots & \sqrt{\lambda_n} \end{bmatrix} T^\top$$

$$= T \begin{bmatrix} \sqrt{\lambda_1} & 0 & \ldots & 0 \\ 0 & \sqrt{\lambda_2} & \ldots & 0 \\ \vdots & \vdots & \ddots & \vdots \\ 0 & 0 & \ldots & \sqrt{\lambda_n} \end{bmatrix}^2 T^\top = T \begin{bmatrix} \lambda_1 & 0 & \ldots & 0 \\ 0 & \lambda_2 & \ldots & 0 \\ \vdots & \vdots & \ddots & \vdots \\ 0 & 0 & \ldots & \lambda_n \end{bmatrix} T^\top \underbrace{=}_{\text{eq. (2)}} A.$$

Our construction of the square root $C$ of $A$ in solution 4.10.7 is a special case of *functional calculus*. Given any real-valued function of a real variable, we can apply the function to real *symmetric* matrices by the same procedure: diagonalize the matrix using the spectral theorem, apply the function to the eigenvalues, and undiagonalize back. The same idea applied to linear operators in infinite-dimensional vector spaces is a fundamental part of functional analysis.

$$E: \begin{pmatrix} r \\ \theta \\ \varphi \end{pmatrix} \mapsto \begin{pmatrix} x = ar\cos\theta\cos\varphi \\ y = br\sin\theta\cos\varphi \\ z = cr\sin\varphi \end{pmatrix}$$

Elliptical change of coordinates for solution 4.10.9

By equation 4.10.23,

$$\left| \det \left[ \mathbf{D}S \begin{pmatrix} r \\ \theta \\ \varphi \end{pmatrix} \right] \right| = r^2 \cos\varphi.$$

This confirms our initial assumption that there exists a symmetric matrix $C$ such that $C^2 = A$, and equation (1) is correct: the volume of $X_A$ is

$$\mathrm{vol}_n(X_A) = \frac{\mathrm{vol}_n(B_n)}{\sqrt{\det A}}, \quad \text{where } B_n \text{ is the unit ball in } \mathbb{R}^n.$$

**4.10.9** Assume that $a$, $b$, and $c$ are all positive, and use the "elliptical change of coordinates" $E$ shown in the margin. The region $V$ corresponds to the region

$$U = \left\{ \begin{pmatrix} r \\ \theta \\ \varphi \end{pmatrix} \in \mathbb{R}^3 \,\middle|\, 0 < r \le 1,\ 0 \le \theta \le \frac{\pi}{2},\ 0 \le \varphi < \frac{\pi}{2} \right\}$$

in our coordinates. Before we can use these coordinates to integrate the desired function, we must calculate $|\det[\mathbf{D}E]|$. To get $[\mathbf{D}E]$ from $[\mathbf{D}S]$ (calculated in equation 4.10.22), we need only multiply the first, second, and third rows of $[\mathbf{D}S]$ by $a$, $b$, and $c$ respectively, to get

$$\left[ \mathbf{D}E \begin{pmatrix} r \\ \theta \\ \varphi \end{pmatrix} \right] = \begin{bmatrix} a\cos\theta\cos\varphi & -ar\sin\theta\cos\varphi & -ar\cos\theta\sin\varphi \\ b\sin\theta\cos\varphi & br\cos\theta\cos\varphi & -br\sin\theta\sin\varphi \\ c\sin\varphi & 0 & cr\cos\varphi \end{bmatrix}.$$

Theorem 4.8.7, together with equation 4.8.10, tells us that multiplying a row of a matrix by a scalar $a$ scales the determinant of the matrix by $a$, so $\det[\mathbf{D}E] = abc \det[\mathbf{D}S]$, and

$$\left| \det \left[ \mathbf{D}E \begin{pmatrix} r \\ \theta \\ \varphi \end{pmatrix} \right] \right| = abc\, r^2 \cos\varphi.$$

Integration is now a simple matter. First change coordinates under the multiple integral:

$$\int_V xyz |dx\,dy\,dz| = \int_U (ar\cos\theta\cos\varphi)(br\sin\theta\cos\varphi)(cr\sin\varphi)abcr^2\cos\varphi |dr\,d\theta\,d\varphi|.$$

Rearranging terms and changing to an iterated integral yields

$$(abc)^2 \int_0^1 \left( \int_0^{\pi/2} \left( \int_0^{\pi/2} r^5 \sin\theta\cos\theta\cos^3\varphi\sin\varphi\,d\theta \right) d\varphi \right) dr.$$

This is quite simple to evaluate if one notes that

$$\cos\theta\,d\theta = d\sin\theta \quad \text{and} \quad -\sin\varphi\,d\varphi = d\cos\varphi.$$

The result is $(abc)^2/48$.

**4.10.11** Use spherical coordinates, to find

$$\int_0^R \int_0^{2\pi} \int_{-\pi/2}^{\pi/2} r^2 \cos\varphi\,d\varphi\,d\theta\,dr = \int_0^R \int_0^{2\pi} \left[ r^2\sin\varphi \right]_{-\pi/2}^{\pi/2} d\theta\,dr$$

$$= \int_0^R \left[ 2r^2\theta \right]_0^{2\pi} dr = \left[ \frac{4\pi r^3}{3} \right]_0^R = \frac{4R^3\pi}{3}.$$

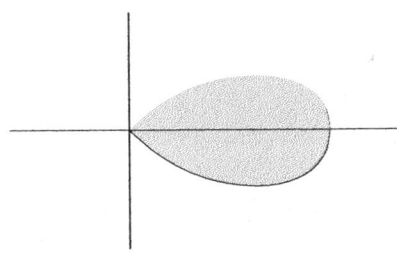

FIGURE FOR SOLUTION 4.10.15
The curve of part a

**4.10.13**  Cylindrical coordinates are appropriate here. We find

$$\int_0^{2\pi}\int_0^1\int_{r^2}^r r\,dz\,dr\,d\theta = 2\pi\int_0^1 r(r-r^2)\,dr = 2\pi\left(\frac{1}{3}-\frac{1}{4}\right) = \frac{\pi}{6}.$$

**4.10.15**　a.　The curve is drawn at left.

b. The $x$-coordinate of the center of gravity is the point

$$\overline{x} = \frac{\int_{-\pi/4}^{\pi/4}\int_0^{\cos 2\theta}(r\cos\theta)r\,dr\,d\theta}{\int_{-\pi/4}^{\pi/4}\int_0^{\cos 2\theta}r\,dr\,d\theta}.$$

We now compute these two integrals.
The numerator gives

$$\frac{1}{3}\int_{-\pi/4}^{\pi/4}\cos^3 2\theta\cos\theta\,d\theta = \frac{1}{12}\int_{-\pi/4}^{\pi/4}(\cos 6\theta\cos\theta + 3\cos 2\theta\cos\theta)\,d\theta$$

$$= \frac{1}{24}\int_{-\pi/4}^{\pi/4}(\cos 7\theta + \cos 5\theta + 3\cos 3\theta + 3\cos\theta)\,d\theta$$

$$= \frac{\sqrt{2}}{24}\left(-\frac{1}{7}-\frac{1}{5}+1+3\right).$$

The denominator gives

$$\int_{-\pi/4}^{\pi/4}\int_0^{\cos 2\theta}r\,dr\,d\theta = \frac{1}{2}\int_{-\pi/4}^{\pi/4}\cos^2 2\theta\,d\theta = \frac{\pi}{8}.$$

After a bit of arithmetic, this comes out to $\overline{x} = (44\sqrt{2})/(35\pi) \sim .600211\ldots$
By symmetry, we have $\overline{y} = 0$.

**4.10.17**　a.　The region $A$ is shown in the figure in the margin. It is bounded by the arcs of the curves $y = e^x + 1$ and $y = e^a(e^x + 1)$ where $0 \le x \le a$, and the arcs of the curves $y = e^{-x} + 1$ and $y = e^a(e^{-x} + 1)$ where $-a \le x \le 0$.

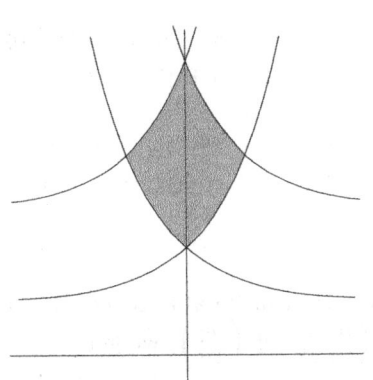

FIGURE FOR SOLUTION 4.10.17
The shaded region is the region $A$, bounded by the arcs of the curves

$$y = e^x + 1 \text{ and } y = e^a(e^x + 1)$$

where $0 \le x \le a$, and the arcs of the curves

$$y = e^{-x} + 1 \text{ and } y = e^a(e^{-x} + 1)$$

where $-a \le x \le 0$.

b. In this case, we can find an explicit inverse of $\Phi$: we need to solve the system of equations

$$u - v = x$$
$$e^u + e^v = y$$

for $u$ and $v$ in terms of $x$ and $y$. From the first equation write $u = x + v$, and substitute in the second equation, to find

$$e^{x+v} + e^v = e^v(e^x + 1) = y.$$

Since $y \ge 0$ in $A$ and $e^x + 1 > 0$ always, this leads to $v = \ln\dfrac{y}{e^x + 1}$.

An exactly analogous argument gives

$$u = \ln\frac{y}{e^{-x} + 1}.$$

c. We have

$$\det\left[\mathbf{D}\Phi\begin{pmatrix} x \\ y \end{pmatrix}\right] = \det\begin{bmatrix} 1 & -1 \\ e^u & e^v \end{bmatrix} = e^u + e^v.$$

The change of variables formula gives

$$\int_A y|dx\,dy| = \int_{Q_a} (e^u + e^v)(e^u + e^v)|\,du\,dv$$

$$= \int_0^a \int_0^a \left(e^{2u} + 2e^u e^v + e^{2v}\right)\,du\,dv$$

$$= a(e^{2a} - 1) + 2(e^a - 1)^2.$$

**4.10.19**  a.  The image of $0 \leq u \leq 1$, $v = 0$, is the arc of parabola parametrized by $u \mapsto \begin{pmatrix} u \\ u^2 \end{pmatrix}$, i.e., the arc of the parabola $y = x^2$ where $0 \leq x \leq 1$.

The image of $0 \leq u \leq 1$, $v = 1$, is the arc of parabola parametrized by $u \mapsto \begin{pmatrix} u - 1 \\ u^2 + 1 \end{pmatrix}$, i.e., the arc of the parabola $y = (x + 1)^2 + 1$ where $-1 \leq x \leq 0$.

The image of $u = 0$, $0 \leq v \leq 1$, is the arc of parabola parametrized by $v \mapsto \begin{pmatrix} -v^2 \\ v \end{pmatrix}$, i.e., the arc of the parabola $x = -y^2$ where $0 \leq y \leq 1$.

The image of $u = 1$, $0 \leq v \leq 1$, is the arc of parabola parametrized by $v \mapsto \begin{pmatrix} 1 - v^2 \\ 1 + v \end{pmatrix}$, i.e., the arc of the parabola $x - 1 = -(y - 1)^2$ where $1 \leq y \leq 2$.

These curves are shown in the margin.

b. We need to show that $\Phi$ is 1–1 so that we will be able to apply the change of variables formula in part c. If $\Phi\begin{pmatrix} u_1 \\ v_1 \end{pmatrix} = \Phi\begin{pmatrix} u_2 \\ v_2 \end{pmatrix}$, we get

$$u_1 - v_1^2 = u_2 - v_2^2$$
$$u_1^2 + v_1 = u_2^2 + v_2.$$

These equations can be rewritten

$$u_1 - u_2 = v_1^2 - v_2^2 = (v_1 - v_2)(v_1 + v_2)$$
$$(u_1 - u_2)(u_1 + u_2) = v_2 - v_1.$$

In the region under consideration, $v_1 + v_2 \geq 0$ and $u_1 + u_2 \geq 0$. It follows from the first equation that if $v_1 \neq v_2$, then $u_1 \neq u_2$, and both $v_1 - v_2$ and $u_1 - u_2$ have the same sign. The same argument for the second equation says that if $u_1 \neq u_2$, then $v_1 \neq v_2$, and $v_1 - v_2$ and $u_1 - u_2$ have opposite sign. So $u_1 = u_2$ and $v_1 = v_2$.

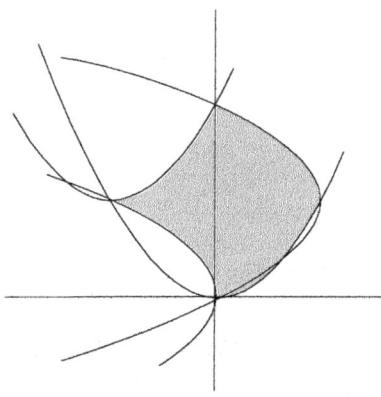

FIGURE FOR SOLUTION 4.10.19.
The shaded area is $\Phi(Q)$. We have two vertical parabolas; the shaded region is inside the lower one and outside the upper one.

There are also two horizontal parabolas; the shaded region is inside the rightmost one and outside the leftmost one.

c. We have $\det\left[\mathbf{D}\Phi\left(\begin{smallmatrix}u\\v\end{smallmatrix}\right)\right] = \det\begin{bmatrix}1 & -2v\\2u & 1\end{bmatrix} = 1+4uv$, which is positive in $Q$, so

$$\int_A x\,|dx\,dy| = \int_Q (u-v^2)(1+4uv)|du\,dv| = \int_0^1\int_0^1 (u+4u^2v - v^2 - 4uv^3)\,du\,dv$$

$$= \int_0^1\left[\frac{u^2}{2}+\frac{4u^3v}{3}-v^2u - 2u^2v^3\right]_{u=0}^{u=1}dv = \int_0^1\left(\frac{1}{2}+\frac{4v}{3}-v^2-2v^3\right)\,dv$$

$$= \left[\frac{v}{2}+\frac{2v^2}{3}-\frac{v^3}{3}-\frac{v^4}{2}\right]_0^1 = \frac{1}{3}.$$

**4.10.21**  a.  Let $\Phi\begin{pmatrix}r\\\theta\\z\end{pmatrix} = \begin{pmatrix}r\cos\theta\\r\sin\theta\\z\end{pmatrix}$, and let the region $B$ in $(r,\theta,z)$-space be a subset on which $\Phi$ is 1–1. Call $A = \Phi(B)$. If $f : A \to \mathbb{R}$ is integrable, then $rf\left(\Phi\begin{pmatrix}r\\\theta\\z\end{pmatrix}\right)$ is integrable over $B$, and

$$\int_B rf\left(\Phi\begin{pmatrix}r\\\theta\\z\end{pmatrix}\right)|dr\,d\theta\,dz| = \int_A f\begin{pmatrix}x\\y\\z\end{pmatrix}|dx\,dy\,dz|.$$

In this case, the appropriate cylindrical coordinates consist of keeping $x$, and writing $\left(\begin{smallmatrix}y\\z\end{smallmatrix}\right)$ in polar coordinates. Then $r$ is the distance to the $x$-axis, so the integral becomes

$$\int_a^b\int_0^{2\pi}\int_0^{f(x)} r\,r^2\,dr\,|d\theta||dx| = \frac{\pi}{2}\int_a^b f(x)^4\,dx.$$

b. Applying the formula above gives that the moment of inertia is

$$\frac{\pi}{2}\int_{-\pi/2}^{\pi/2}\cos^4 x\,dx = \frac{\pi}{2}\int_{-\pi/2}^{\pi/2}\cos^2 x(1-\sin^2 x)\,dx$$

$$= \frac{\pi}{2}\int_{-\pi/2}^{\pi/2}\left(\cos^2 x - \frac{\sin^2 2x}{4}\right)\,dx = \frac{\pi}{2}\left(\frac{\pi}{2}-\frac{\pi}{8}\right)\frac{3\pi^2}{16}.$$

**4.11.1**  Set $f(\mathbf{x}) = |\mathbf{x}|^p\mathbf{1}_D(\mathbf{x})$, where $D$ is the unit disc. Let $A_n \subset \mathbb{R}^2$ be the set $1/n < |\mathbf{x}| \leq 1$ and set $f_n = |\mathbf{x}|^p\mathbf{1}_{A_n}$. Then each $f_n$ is L-integrable (in fact, R-integrable), $0 \leq f_1 \leq f_2 \leq \ldots$, and $\lim_{n\to\infty} f_n \underset{L}{=} f$, so $f$ is integrable precisely if the sequence

$$a_n = \int_{\mathbb{R}^2} f_n(\mathbf{x})|d^2\mathbf{x}| = \int_{A_n}|\mathbf{x}|^p|d^2\mathbf{x}|$$

is bounded.

This sequence is easy to compute explicitly by passing to polar coordinates:

$$a_n = \int_0^{2\pi} \left( \int_{1/n}^1 r^p\, r\, dr \right) d\theta = \begin{cases} \frac{2\pi}{p+2}\left(1 - \frac{1}{n^{p+2}}\right) & \text{if } p \neq -2 \\ 2\pi \ln n & \text{if } p = -2. \end{cases}$$

Thus we see that $f$ is integrable precisely if $p > -2$; in that case, the integral is $2\pi/(p+2)$.

**4.11.3** Using the change of variables for polar coordinates, we have the following, where $A = (1, \infty) \times [0, 2\pi)$:

$$\int_{\mathbb{R}^2 - B_1(0)} \left| \begin{pmatrix} x \\ y \end{pmatrix} \right|^p |dx\, dy| = \int_A \left| \begin{pmatrix} r\cos\theta \\ r\sin\theta \end{pmatrix} \right|^p r\, |dr\, d\theta|$$

$$= \int_0^{2\pi} \int_1^\infty r^{p+1}\, dr\, d\theta = 2\pi \int_1^\infty r^{p+1}\, dr.$$

By theorem 4.11.21, the existence of this last integral is equivalent to the existence of the original integral. If $p \neq -2$, we can write

$$\int_1^\infty r^{p+1}\, dr = \sum_{n=1}^\infty \int_n^{n+1} r^{p+1}\, dr = \sum_{n=1}^\infty \left[ \frac{r^{p+2}}{p+2} \right]_n^{n+1}.$$

Our definition of integrability says that if this series converges, then the integral exists. But the series telescopes:

$$\sum_{n=1}^m \left[ \frac{r^{p+2}}{p+2} \right]_n^{n+1} = \frac{1}{p+2} \left( (m+1)^{p+2} - 1 \right)$$

and converges if $p + 2 < 0$, i.e., if $p < -2$. Moreover, since $|\mathbf{x}|^p > 0$, the integral is then precisely the sum of the series, so if $p < -2$, we have

$$\int_{\mathbb{R}^2 - B_1(0)} \left| \begin{pmatrix} x \\ y \end{pmatrix} \right|^p |dx\, dy| = -\frac{1}{p+2}.$$

We now need to see that if $p \geq -2$, the integral does not exist. By the monotone convergence theorem, the integral will fail to converge if

$$\sup_{m \to \infty} \int_1^m r^{p+1}\, dr = \infty.$$

There are two cases to consider: if $p \neq -2$, we just computed that integral, to find

$$\int_1^m r^{p+1}\, dr = \frac{1}{p+2} \left( (m+1)^{p+2} - 1 \right),$$

which tends to infinity with $m$ if $p + 2 > 0$. If $p = -2$, then

$$\sup_{m \to \infty} \int_1^m \frac{1}{r}\, dr = \sup_{m \to \infty} \ln m = \infty.$$

**4.11.5** Let $A_n = \left\{ \mathbf{x} \in \mathbb{R}^n \mid 2^n \leq |\mathbf{x}| < 2^{n+1} \right\}$, so that

$$\mathbb{R}^n - B_1(0) = A_0 \cup A_1, \cup \dots .$$

Then

$$|x|^p \mathbf{1}_{\mathbb{R}^n - B_1(\mathbf{0})} = \sum_{m=0}^{\infty} |\mathbf{x}|^p \mathbf{1}_{A_m},$$

and so $|\mathbf{x}|^p$ is L-integrable over $\mathbb{R}^n - B_1(\mathbf{0})$ precisely if the series

$$\sum_{m=0}^{\infty} \int_{A_m} |\mathbf{x}|^p |d^n\mathbf{x}|$$

is convergent.

The mapping $\varphi_m(\mathbf{x}) = 2^m\mathbf{x}$ maps $A_1$ to $A_m$, and $\det[\mathbf{D}\varphi_m] = 2^{nm}$. So applying the change of variables formula gives

$$\int_{A_m} |\mathbf{x}|^p |d^n\mathbf{x}| = \int_{A_1} 2^{mp}|\mathbf{x}|^p 2^{nm}|d^n\mathbf{x}| = 2^{m(p+n)} \int_{A_1} |\mathbf{x}|^p |d^n\mathbf{x}|.$$

Thus our series is     *ought this be negative? shouldn't it be positive...*

$$\left( \sum_{m=0}^{\infty} 2^{m(n+p)} \int_{A_1} |\mathbf{x}|^p |d^n\mathbf{x}|, \right]$$

where the integral is some constant. It is convergent exactly when $n+p < 0$, i.e., if $p < -n$.

**4.11.7** The situation is rather different for $m \geq 1$ and $m < 1$. If $m \geq 1$, then $1/(x^2 + y^2)^p$ is integrable over $A_m$ precisely if it is integrable over $\mathbb{R}^n - B_1(\mathbf{0})$, i.e., if $p > 1$. (See exercise 4.11.3; note that the present $p$ is half what is denoted $p$ there.)

This is seen as follows: take the eight subsets $A_{m,1}, \ldots, A_{m,8}$ obtained by reflecting $A_m$ with respect to both axes and both diagonals. Then these eight copies cover the complement of the square $S$ where $|x| \leq 1, |y| \leq 1$, with some points covered twice when $m > 1$. Thus

$$\int_{A_m} \frac{1}{(x^2 + y^2)^p} |dx\, dy| \leq \int_{\mathbb{R}^2 - S} \frac{1}{(x^2 + y^2)^p} |dx\, dy|$$

$$\leq 8 \int_{A_m} \frac{1}{(x^2 + y^2)^p} |dx\, dy|.$$

Clearly, integrability over $\mathbb{R}^2 - B_1(\mathbf{0})$ and over $\mathbb{R}^2 - S$ are equivalent, since $1/(x^2 + y^2)^p$ is always integrable over $S - B_1(\mathbf{0})$.

The case where $m < 1$ is more delicate. In that case, in $A_m$ we have $x^2 \leq x^2 + y^2 \leq 2x^2$, so the integrability of $1/(x^2 + y^2)^p$ is equivalent to integrability of $1/x^{2p}$. But this can be computed explicitly: it gives

$$\int_1^{\infty} \left( \int_0^{x^m} \frac{1}{x^{2p}} dy \right) dx = \int_1^{\infty} x^{m-2p} dx.$$

This last integral is finite precisely if $m - 2p < -1$, i.e., if $p > (m+1)/2$.

**4.11.9** There are no difficulties associated with infinities in this case:

$$\widehat{\mathbf{1}_{[-1,1]}}(\xi) = \int_{\mathbb{R}} \mathbf{1}_{[-1,1]}(x) e^{i\xi x} dx = \int_{-1}^{1} e^{i\xi x} dx$$

$$= \left[\frac{e^{i\xi x}}{i\xi}\right]_{-1}^{1} = \frac{e^{i\xi} - e^{-i\xi}}{i\xi} = 2\frac{\sin\xi}{\xi}.$$

The only fishy part is the case $\xi = 0$; but note that

$$\lim_{\xi \to 0} 2\frac{\sin\xi}{\xi} = 2 = \widehat{\mathbf{1}_{[-1,1]}}(0).$$

**4.11.11** It is clearly enough to show the result for any monomial $x_1^{k_1} \ldots x_n^{k_n}$. Note that $|x_i^{k_i}| \le (1 + |\mathbf{x}|)^{k_i}$, so

*Solution 4.11.11: This is a matter of pinning down the fact that exponentials win over polynomials.*

$$|x_1^{k_1} \ldots x_n^{k_n}| \le (1 + |\mathbf{x}|)^k,$$

where $k = k_1 + \cdots + k_n$. Thus it is enough to show that the integral

$$\int_{\mathbb{R}^n} (1 + |\mathbf{x}|)^k e^{-|\mathbf{x}|^2} |d^n\mathbf{x}|$$

exists for every $k \ge 0$. Developing out the power, it is enough to show that

$$\int_{\mathbb{R}^n} |\mathbf{x}|^k e^{-|\mathbf{x}|^2} |d^n\mathbf{x}| \quad \text{exists for every } k \ge 0.$$

Let us break up $\mathbb{R}^n$ into the unit ball $B$ and the sets

$$A_m = \left\{ \mathbf{x} \in \mathbb{R}^n \mid 2^m < |\mathbf{x}| \le 2^{m+1} \right\}.$$

We can then write

$$|\mathbf{x}|^k = |\mathbf{x}|^k \mathbf{1}_B(\mathbf{x}) + \sum_{m=0}^{\infty} |\mathbf{x}|^k \mathbf{1}_{A_m}(\mathbf{x}) \overset{\text{def}}{=} g(\mathbf{x}) + \sum_{m=0}^{\infty} f_m(\mathbf{x}),$$

and $g$ and all the $f_m$ are R-integrable. Therefore it is enough to prove that the series of numbers

$$\sum_{m=1}^{\infty} \int_{\mathbb{R}^n} f_m(\mathbf{x}) |d^n\mathbf{x}|$$

is convergent. The change of variable $\phi_m : \mathbf{x} \mapsto 2^m \mathbf{x}$ transforms the region $A_1$ into the region $A_m$, and the change of variables formula gives

$$\int_{\mathbb{R}^n} f_m(\mathbf{x}) |d^n\mathbf{x}| = \int_{A_1} 2^{mn} 2^{km} |\mathbf{x}|^k e^{-2^{2m}|\mathbf{x}|^2} |d^n\mathbf{x}| \le 2^{mn} 2^{km} 2^k e^{-2^{2m}} \overset{\text{def}}{=} a_m.$$

The ratio test, applied to this series, gives

$$\frac{a_{m+1}}{a_m} = \frac{2^{(m+1)n} 2^{k(m+1)} 2^k e^{-2^{2(m+1)}}}{2^{mn} 2^{km} 2^k e^{-2^{2m}}} = \frac{2^{n+k}}{e^{2^{2(m+1)} - 2^{2m}}} = \frac{2^{n+k}}{e^{3 \cdot 2^{2m}}},$$

which clearly tends to 0 as $m$ tends to infinity, so the series converges.

Here is another solution, by Vorrapan Chandee, when a freshman at Cornell:

It is clearly enough to show the result for any monomial $x_1^{k_1} \cdots x_n^{k_n}$. By Fubini's theorem, we have

$$\int_{\mathbb{R}^n} x_1^{k_1} \cdots x_n^{k_n} e^{-|\mathbf{x}|^2} |d^n\mathbf{x}| = \int_{\mathbb{R}^n} x_1^{k_1} e^{-x_1^2} \cdots x_n^{k_n} e^{-x_n^2} |d^n\mathbf{x}|$$

$$= \left( \int_{\mathbb{R}} x_1^{k_1} e^{-x_1^2} |dx_1| \right) \cdots \left( \int_{\mathbb{R}} x_n^{k_n} e^{-x_n^2} |dx_n| \right).$$

The original integral exists if each integral in the above product exists.

So it suffices to show that the one-dimensional integral $\int_{\mathbb{R}} x^k e^{-x^2} |dx|$ exists for $k = 0, 1, 2, \ldots$. By a symmetry argument, we can reduce this to showing that $\int_0^{\mathbb{R}} x^k e^{-x^2} dx$ exists. Indeed, $x^k e^{-x^2}$ is eventually dominated by $e^{-x}$, which is easily seen to be integrable over $[0, \infty)$.

**4.11.13** a. For the first sequence, if the $f_k$ converge to anything, they must converge to the function 0. But if you take $\epsilon = 1/2$, then for $x = k + 1/2$ we have $f_k(x) - 0 = 1 > \epsilon$.

For the second example, even if you take $\epsilon = 1$ and $x_k = \frac{1}{2k}$, we have $f_k(x_k) - 0 = k \geq 1$.

For the third, let $f_\infty$ be the function that is 1 on the rationals and 0 on the irrationals. Set $x_k = a_{k+1}$ and $\epsilon = 1/2$. Then

$$f_k(x_k) - f_\infty(x_{k+1} = 1 > \epsilon.$$

b. Write $p = a_0 + a_1 x + \cdots + a_m x^m$. Suppose that $p_k \to p$ uniformly. Choose $\epsilon > 0$; our assumption says that there exists $K$ such that when $k \geq K$, then for any $x$ we have $|p_k(x) - p(x)| < \epsilon$. But $p_k(x) - p(x)$ is a polynomial, and the only bounded polynomials are the constants. Thus $p_k(x) - p(x) = c_k$ for the constant $c_k = a_{0,k} - a_0$ when $k \geq K$. This proves that all the coefficients of the $p_k$ are eventually constant, except perhaps the constant term. Moreover, the inequality

$$|p_k(x) - p(x)| = |a_{0,k} - a_k| < \epsilon$$

clearly implies that $a_{k,0}$ converges to $a_k$.

In the other direction we must show that if all the nonzero coefficients are constant, and the constant coefficient converges, then the sequence $p_k$ converges uniformly to $p$. Suppose the coefficients $a_{i,k}$ are eventually constant for $i > 0$, so we can set

$$a_{i,k} = a_i \quad \text{for } k > K \text{ and } i > 0,$$

and suppose $a_{0,k}$ converges to $a_0$.

Set

$$p = a_0 + a_1 x + \cdots + a_m x^m;$$

clearly, if $k > K$, then

$$p(x) - p_k(x) = a_0 - a_{0,k},$$

which converges to 0; therefore, $p_k$ converges uniformly to $p$.

c. If the two sequences of functions $f_k$ and $g_k$ converge uniformly to $f$ and $g$, then the sequence $f_k + g_k$ converges uniformly to $f + g$. Thus it is enough to show that if the sequence of numbers $a_k$ converges, say to $a_\infty$, then the sequence of functions $a_k x^i \mathbf{1}_A$ converges uniformly on $\mathbb{R}$. Let $M$ be the maximum of $|x|^i$ on $A$. Choose $\epsilon > 0$. There exists $K$ such that if $k > K$, then $|a_k - a_\infty| < \epsilon$, and then

$$|a_k x^i - a_\infty x^i| \leq \epsilon M.$$

The existence of $M$ uses the hypothesis that $A$ is bounded; any polynomial has a maximum on any bounded set.

This can be made arbitrarily small by choosing $\epsilon$ sufficiently small.

**4.11.15**  Since this is a telescoping series, it is enough to show that the last term of any partial sum tends to 0. Since $x(\ln x - 1)$ is an antiderivative of $\ln x$, the last term of

$$\sum_{i=1}^{n-1} \int_{1/2^{i+1}}^{1/2^i} |\ln x|\, dx \quad \text{is} \quad \frac{1}{2^n}\left(\ln\left|\frac{1}{2^n}\right| - 1\right) = -\frac{1}{2^n}\left(n \ln 2 + 1\right).$$

This tends to 0 when $n$ tends to infinity: $2^n$ grows much faster than $n$.

Solution 4.11.17 illustrates the fact that if we allow improper integrals whose existence depends on cancellations, the change of variables formula is not true. The idea of doing integration theory without being able to make changes of variables is ridiculous.

**4.11.17**  After change of variables, this integral becomes $\displaystyle\int_0^\infty \frac{1}{u}\sin\frac{1}{u}\, du.$ As an improper integral, this should mean

$$\lim_{A\to\infty} \int_0^A \frac{1}{u}\sin\frac{1}{u}\, du;$$

see equation 4.11.51. However, the integral inside the limit above does exist; the integrand $\frac{1}{u}\sin\frac{1}{u}$ oscillates between the graph of $1/u$ and the graph of $-1/u$; near $u = 0$ the functions $1/u$ and $-1/u$ are not integrable.

**4.1** It is enough that $U_N(\mathbf{1}_C) = L_N(\mathbf{1}_C)$ because by lemma 4.1.9, for all $M \geq N$,

$$U_N(\chi_C) \geq U_M(\chi_C) \geq L_M(\chi_C) \geq L_N(\chi_C).$$

Thus the upper sums and lower sums are all equal for $M \geq N$, so the function $\chi_C$ is integrable.

But $U_N(\chi_C) = L_N(\chi_C)$ is obviously true.

**4.3** 1. False. For instance, multiplication by 2, i.e., the function $[2]: \mathbb{Z} \to \mathbb{Z}$, is not onto; its image is the even integers.

2. True. If $A$ is onto, the image of $\mathbb{Z}^n$ contains the standard basis vectors, so it spans $\mathbb{R}^n$. In particular, the matrix $A$ viewed as a linear transformation $\mathbb{R}^n \to \mathbb{R}^n$, is onto, hence also injective. But any element of the kernel of $A: \mathbb{Z}^n \to \mathbb{Z}^n$ is also an element of the kernel of $A: \mathbb{R}^n \to \mathbb{R}^n$, so it must be 0.

3. True. The same argument works as for 2: If we view $A$ as representing a linear transformation $\mathbb{R}^n \to \mathbb{R}^n$, then $\det A \neq 0$ implies that $A$ is injective, and so is its restriction to $\mathbb{Z}^n$.

4. False. The same example as part 1 is a counterexample here also: $\det[2] \neq 0$ but $[2]: \mathbb{Z} \to \mathbb{Z}$ is not onto.

**4.5** There are many ways to approach this problem. One is to observe that $\frac{1}{N^2} \sum_{k=1}^{N} \sum_{j=1}^{2N} e^{\frac{k+j}{N}}$ is a Riemann sum (unfortunately not dyadic) for the function

$$f\begin{pmatrix} x \\ y \end{pmatrix} = e^{x+y} \mathbf{1}_{[0,1] \times [0,2]}.$$

By proposition 4.1.15, this function is integrable, with integral

$$\left( \int_0^2 e^y \, dy \right) \left( \int_0^1 e^x \, dx \right) = (e-1)(e^2-1).$$

With the arguments at hand, this is a little shaky; if we only considered the numbers $N = 2^M$ we would be fine, but the other values of $N$ won't quite enter into our dyadic formalism.

Another possibility is to write

$$\left( \frac{1}{N} \sum_{k=1}^{N} e^{k/N} \right) \left( \frac{1}{N} \sum_{j=1}^{2N} e^{j/N} \right)$$

and to observe that both sums are finite geometric series. Using

$$a + ar + \cdots + ar^n = a \frac{1 - r^{n+1}}{1 - r},$$

(see equation 0.5.4), we see that the first sum is

$$\sum_{k=1}^{N} e^{k/N} = e^{1/N} \frac{1 - e^{(N+1)/N}}{1 - e^{1/N}}.$$

We need to evaluate the limit

$$\lim_{N \to \infty} \frac{1}{N} e^{1/N} \frac{1 - e^{(N+1)/N}}{1 - e^{1/N}}.$$

In this expression, $e^{1/N} \to 1$ and $e^{(N+1)/N} \to e$, so the limit that matters is

Equation (1) uses l'Hôpital's rule.

$$\lim_{N \to \infty} \frac{1}{N(1 - e^{1/N})} = \lim_{x \to 0} \frac{x}{1 - e^x} = \lim_{x \to 0} \frac{1}{-e^x} = -1, \qquad (1)$$

giving the limit

$$\lim_{N \to \infty} \frac{1}{N} e^{1/N} \frac{1 - e^{(N+1)/N}}{1 - e^{1/N}} = e - 1.$$

Exactly the same way, we find

$$\lim_{N \to \infty} \sum_{j=1}^{2N} e^{k/N} = e^2 - 1.$$

So we have in all

$$\lim_{N \to \infty} \left( \frac{1}{N} \sum_{k=1}^{N} e^{k/N} \right) \left( \frac{1}{N} \sum_{j=1}^{2N} e^{j/N} \right) = (e - 1)(e^2 - 1).$$

**4.7** Set $\mu(x_i) = \mu_1$ if $x_i \in A$, and $\mu(x_i) = \mu_2$ if $x_i \in B$, and compute as follows:

$$\overline{x_i}(C) = \frac{\int_C x_i \mu(x_i) |d^n \mathbf{x}|}{M(C)} = \frac{1}{M(A) + M(B)} \left( \int_A \mu_1 x_i |d^n \mathbf{x}| + \int_B \mu_2 x_i |d^n \mathbf{x}| \right)$$

$$= \frac{1}{M(A) + M(B)} \left( M(A) \frac{\int_A \mu_1 x_i |d^n \mathbf{x}|}{M(A)} + M(B) \frac{\int_B \mu_2 x_i |d^n \mathbf{x}|}{M(B)} \right)$$

$$= \frac{1}{M(A) + M(B)} \left( M(A) \overline{x_i}(A) + M(B) \overline{x_i}(B) \right).$$

Therefore,

$$\overline{\mathbf{x}}(C) = \frac{M(A) \overline{\mathbf{x}}(A) + M(B) \overline{\mathbf{x}}(B)}{M(A) + M(B)}.$$

**4.9** Choose $\epsilon$, and write $X = \cup_{i=1}^{\infty} B_i$, where the $B_i$ are pavable sets with $\sum_{i=1}^{\infty} \text{vol}_n(B_i) < \epsilon/2$. For each $i$ find a dyadic level $N_i$ such that

$$U_{N_i} \mathbf{1}_{B_i} \le \text{vol}_n(B_i) + \frac{\epsilon}{2^{i+1}}.$$

Then at the level $N_i$, $B_i$ is covered by finitely many dyadic cubes with total volume at most $\text{vol}_n(B_i) + \frac{\epsilon}{2^{i+1}}$. Now list first the dyadic cubes that cover

$B_1$, then the dyadic cubes (at a different level) that cover $B_2$, and so on. The total volume of all these cubes is at most

$$\sum_{i=1}^{\infty} \left( \mathrm{vol}_n(B_i) + \frac{\epsilon}{2^{i+1}} \right) \le \frac{\epsilon}{2} + \frac{\epsilon}{2} = \epsilon.$$

**4.11**  a. We can write

$$\int_{-\sqrt{2}}^{\sqrt{2}} \left( \int_{x^2}^{2} \sin(x+y)\, dy \right) dx \quad \text{or} \quad \int_{0}^{2} \left( \int_{-\sqrt{y}}^{\sqrt{y}} \sin(x+y)\, dx \right) dy.$$

The first leads to

$$\int_{-\sqrt{2}}^{\sqrt{2}} \left( \cos(x+x^2) - \cos(x+2) \right)\, dx;$$

the second leads to

$$\int_{0}^{2} \left( \cos(y-\sqrt{y}) - \cos(y+\sqrt{y}) \right) dy = \int_{0}^{2} 2\sin y \, \sin \sqrt{y}\, dy.$$

We believe that neither can be evaluated in elementary terms. Numerically, using Simpson's rule with 50 subdivisions, the integral comes out to 2.4314344 . . . . MAPLE gives a solution using Fresnel integrals; the numerical value obtained this way is 2.43143285295741 38146 . . . .

b. This is much easier: The integral is

$$\int_{1}^{2} \left( \int_{1}^{2} (x^2+y^2)\, dx \right) dy = \int_{1}^{2} \left( \int_{1}^{2} (x^2+y^2) dy \right) dx = \frac{14}{3}.$$

**4.13**  We have

$$\int_{0}^{1} \int_{0}^{2} f\left( \begin{smallmatrix} x \\ y \end{smallmatrix} \right) |dx\, dy| = \int_{0}^{1} \left[ ax + \frac{bx^2}{2} + cxy \right]_{0}^{2} dy = \int_{0}^{1} (2a + 2b + 2cy) dy$$

$$= \left[ 2ay + 2by + cy^2 \right]_{0}^{1} = 2a + 2b + c.$$

The other integral is longer to compute:

$$\int_{0}^{1} \int_{0}^{2} \left( f\left( \begin{smallmatrix} x \\ y \end{smallmatrix} \right) \right)^2 |dx\, dy| = \int_{0}^{1} \int_{0}^{2} (a^2 + b^2x^2 + c^2y^2 + 2abx + 2acy + 2bcxy)\, dx\, dy$$

$$= \int_{0}^{1} \left( 2a^2 + \frac{8b^2}{3} + 2c^2y^2 + 4ab + 4acy + 4bcy \right) dy$$

$$= 2a^2 + \frac{8b^2}{3} + \frac{2c^2}{3} + 4ab + 2ac + 2bc.$$

The Lagrange multiplier theorem says that at a minimum of the second integral, constrained so that the first integral is 1, there exists a number $\lambda$ such that

$$\left[ 4a + 4b + 2c \quad \frac{16b}{3} + 4a + 2c \quad \frac{4c}{3} + 2a + 2b \right] = \lambda \left[ 2 \quad 2 \quad 1 \right].$$

From this and the constraint we find the system of linear equations

$$4a + 4b + 2c = \frac{16b}{3} + 4a + 2c$$

$$2a + 2b + c = \frac{4c}{3} + 2a + 2b$$

$$2a + 2b + c = 1.$$

This could be solved by row reduction, but it is easier to observe that the first equation says that $b = 0$ and the second equation says that $c = 0$. Thus $a = 1/2$ and the minimum is $\int_0^1 \int_0^2 \left(\frac{1}{2}\right)^2 |dx\,dy| = \frac{1}{2}$.

**4.15** In order for the equality to be true for all polynomials of degree $\leq 3$, it is enough that it should be true for the polynomials $1, x, x^2, x^3$. Thus we want the four equations

1. $\displaystyle\int_{-1}^1 \frac{dx}{\sqrt{1-x^2}} = c(1+1)$    3. $\displaystyle\int_{-1}^1 x^2 \frac{dx}{\sqrt{1-x^2}} = c(u^2 + u^2)$

2. $\displaystyle\int_{-1}^1 x\frac{dx}{\sqrt{1-x^2}} = c(u - u)$    4. $\displaystyle\int_{-1}^1 x^3 \frac{dx}{\sqrt{1-x^2}} = c(u^3 - u^3).$

Equations 2 and 4 are true for all $c$ and all $u$, since both sides are 0. Equation 1 gives

$$2c = \int_{-1}^1 \frac{dx}{\sqrt{1-x^2}} = [\arcsin x]_{-1}^1 = \pi, \quad \text{so } c = \pi/2.$$

Equation 3: Computing the integral (using integration by parts), we get

$$2cu^2 = \int_{-1}^1 -x\frac{-x\,dx}{\sqrt{1-x^2}} = \left[-x\sqrt{1-x^2}\right]_{-1}^1 + \int_{-1}^1 \sqrt{1-x^2}\,dx = \pi.$$

Substituting $c = \pi/2$ in $2cu^2 = \pi$ gives $\pi u^2 = \pi$, i.e., $u = \pm 1$.

**4.17**   $[\mathbf{D}\det(A)]B = \begin{bmatrix} 1 & -2 & -3 & 1 \end{bmatrix}\begin{bmatrix} a \\ b \\ c \\ d \end{bmatrix} = a - 2b - 3c + d.$

Note that $[\mathbf{D}\det(A)]$ is not the derivative of $\det A = -5$. It is the derivative at $A = \begin{bmatrix} 1 & 3 \\ 2 & 1 \end{bmatrix}$ of the function $\det\begin{pmatrix} a \\ b \\ c \\ d \end{pmatrix} = ad - bc$: i.e., it is the row matrix

$$\begin{bmatrix} d & -c & -b & a \end{bmatrix},$$

evaluated at $a = 1$, $b = 3$, $c = 2$, $d = 1$.

$$\det A \operatorname{tr}(A^{-1}B) = -5\operatorname{tr}\left(\begin{bmatrix} -.2 & .6 \\ .4 & -.2 \end{bmatrix}\begin{bmatrix} a & b \\ c & d \end{bmatrix}\right)$$
$$= -5(-.2a + .6c + .4b - .2d) = a - 2b - 3c + d.$$

**4.19** The easiest way to compute this integral (at least if one has already done exercise 4.9.4) is to observe that the volume to be computed is the same as in exercise 4.9.4, after scaling the $x_1$ coordinate by $n$, the $x_2$ coordinate by $n/2$, etc., until the coordinate $x_n$ is scaled by $n/n = 1$. This gives the volume

$$\frac{1}{n!} \cdot \frac{n}{1} \cdot \frac{n}{2} \cdots \frac{n}{n} = \frac{n^n}{(n!)^2}.$$

Here is the solution to exercise 4.9.4:

Let us call $A_n$ the answer to the problem; we will try to set up an inductive formula for $A_n$.

By Fubini's theorem, we find

$$A_1 = \int_0^1 \int_0^{1-x_1} dx_2\, dx_1 = \frac{1}{2}, \quad A_2 = \int_0^1 \int_0^{1-x_1} \int_0^{1-x_1-x_2} dx_3\, dx_2\, dx_1 = \frac{1}{6},$$

and more generally

$$A_n = \int_0^1 \int_0^{1-x_1} \int_0^{1-x_1-x_2} \cdots \int_0^{1-x_1-\cdots-x_{n-1}} dx_n \ldots dx_2 dx_1.$$

We see that if we define

$$B_n(t) = \int_0^t \int_0^{t-x_1} \int_0^{t-x_1-x_2} \cdots \int_0^{t-x_1-\cdots-x_{n-1}} dx_n \ldots dx_2 dx_1,$$

then

$$A_n = B_n(1).$$

Calculate as above $B_1(t) = t$, $B_2(t) = t^2/2$, $B_3(t) = t^3/6$. A natural guess is that $B_n(t) = t^n/n!$. We will show this by induction: a look at the formula will show you that

$$B_{n+1}(t) = \int_0^t B_n(t - x_1)\, dx_1.$$

Thus the following computation does the inductive step:

$$B_{n+1}(t) = \int_0^t B_n(t - x_1)\, dx_1 = -\left[ \frac{(t - x_1)^{n+1}}{(n+1)!} \right]_0^t = \frac{t^{n+1}}{(n+1)!}.$$

Thus

$$A_n = B_n(1) = \frac{1}{n!}.$$

**4.21** This curve is called a *cardioid*, as in "cardiac arrest"). It is shown in the margin. The area, by the change of variables theorem and Fubini's theorem, is

$$\int_0^{2\pi} \left( \int_0^{1+\sin\theta} r\, dr \right) d\theta = \int_0^{2\pi} \left[ \frac{r^2}{2} \right]_0^{1+\sin\theta} d\theta = \int_0^{2\pi} \frac{(1+\sin\theta)^2}{2} d\theta$$

$$= \left[ \frac{\theta}{2} - \cos\theta \right]_0^{2\pi} + \int_0^{2\pi} \frac{(\sin\theta)^2}{2}\, d\theta$$

$$= \pi + \frac{\pi}{2} = \frac{3\pi}{2}.$$

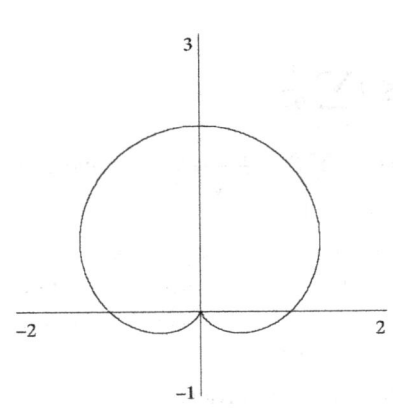

FIGURE FOR SOLUTION 4.21
A cardioid

**4.23** a. In spherical coordinates, $Q$ corresponds to

$$U = \left\{ \begin{pmatrix} r \\ \theta \\ \varphi \end{pmatrix} \middle| 0 < r \le 1,\ 0 \le \theta \le \frac{\pi}{2},\ 0 \le \varphi < \frac{\pi}{2} \right\}.$$

With this in mind, we can write our integral as an iterated integral:

$$\int_Q (x+y+z)|d^3\mathbf{x}| = \int_0^1 \left( \int_0^{\pi/2} \left( \int_0^{\pi/2} (r\cos\theta\cos\varphi + r\sin\theta\cos\varphi + r\sin\varphi)r^2\cos\varphi\,d\varphi \right) d\theta \right) dr.$$

b. We can use a table of integrals to evaluate the more unpleasant sine and cosine portions of the integral, but it is easier to argue that the $x$, $y$, and $z$ contributions to the total must be the same (symmetry), and then evaluate the $z$ contribution:

$$\int_0^1 \left( \int_0^{\pi/2} \left( \int_0^{\pi/2} r^3\sin\varphi\cos\varphi\,d\varphi \right) d\theta \right) dr = \int_0^1 \left( \int_0^{\pi/2} r^3 \left[ \frac{\sin^2\varphi}{2} \right]_0^{\pi/2} d\theta \right) dr$$

$$= \int_0^1 \left( \int_0^{\pi/2} \frac{r^3}{2} d\theta \right) dr$$

$$= \int_0^1 \frac{\pi r^3}{4} dr = \frac{\pi}{16}.$$

The value of our original integral is three times this: $3\pi/16$.

**4.25** By symmetry, the center of gravity is on the $z$-axis. The $z$-coordinate $\overline{z}$ of the center of gravity is

$$\overline{z} = \frac{\int_A z|dx\,dy\,dz|}{\int_A |dx\,dy\,dz|} = \frac{\int_0^1 z(\pi z)dz}{\int_0^1 (\pi z)dz} = \frac{1/3}{1/2} = \frac{2}{3}.$$

**4.27** a.  The function $1/\sqrt{|x-a|}$ is L-integrable over $[0,1]$ for every $a \in [0,1]$ (or even outside):

$$\int_0^1 \frac{1}{\sqrt{|x-a|}} dx \leq \int_{a-1}^{a+1} \frac{1}{\sqrt{|x-a|}} dx = \int_{-1}^1 \frac{1}{\sqrt{|x|}} dx = 4.$$

Then since

$$\sum_{k=1}^{\infty} \frac{1}{2^k} \int_0^1 \frac{1}{\sqrt{|x-a_k|}} dx \leq 4 \sum_{k=1}^{\infty} \frac{1}{2^k} \leq 4,$$

theorem 4.11.17 says that the series of functions $\sum_{k=1}^{\infty} \frac{1}{2^k} \frac{1}{\sqrt{|x-a_k|}}$ converges for almost all $x$, and that $f$ is L-integrable on $[0,1]$.

b. We did this in part a: "on a set of measure 0" is the same as "for almost all $x$". Note, however, that when $x$ is rational, the series does not converge, since we have a 0 in the denominator for one of the terms. So "almost all $x$" does not include any rational numbers.

c. Let us take the rationals in $[0,1]$ in the following order:

$$a_1 = 0, \ a_2 = 1, \ a_3 = \frac{1}{2}, \ a_4 = \frac{1}{3}, \ a_5 = \frac{2}{3}, \ a_6 = \frac{1}{4}, \ a_7 = \frac{3}{4}, \ a_8 = \frac{1}{5}, \ a_9 = \frac{2}{5}, \dots,$$

taking first all those with denominator 1, then those with denominator 2, then those with denominator 3, etc. Note that if $a_k = p/q$ for $p, q$ integers,

---

**Solution 4.27, part c:** This argument may seem opaque! Why take $x = 1/\sqrt{2}$? Since we were trying to find a point in the complement of a set of measure 0, why did we need a number with special properties?

These are hard questions to answer. It should be clear that the convergence of the series depends on $x$ being poorly approximated by rational numbers, in the sense that if you want to get close to $x$ you will have to use a large denominator. In fact, part b says that most real numbers are like that.

But actually finding one is a different matter. (If you are asked to "pick a number, any number," you will most likely come up with a rational number, which has probability 0.) It often happens that we can prove that some property of numbers (like being transcendental) is shared by almost all numbers, without being able to give a single example, at least easily (see section 0.6). The approximation of numbers by rationals is called *diophantine analysis*, and one of its first results is that quadratic irrationals, like $1/\sqrt{2}$, are poorly approximable; in part c we present a proof.

then $k \geq 2q - 3$, since there are at least two numbers for any denominator except 2 (accounting for the $-3$).

Now take $x = 1/\sqrt{2}$. Notice (this is the clever step) that

$$2q^2 \left| \frac{1}{\sqrt{2}} - \frac{p}{q} \right| \left| \frac{1}{\sqrt{2}} + \frac{p}{q} \right| = |q^2 - 2p^2| \geq 1$$

since it is a nonzero integer. Hence for any coprime integers $p, q$ with $p/q \in [0, 1]$, we have

> The second inequality in equation (1): since $1/\sqrt{2} < 1$ and $p/q \leq 1$, their sum is $\leq 2$.

$$\left| x - \frac{p}{q} \right| = \left| \frac{1}{\sqrt{2}} - \frac{p}{q} \right| \geq \frac{1}{2q^2 \left| \frac{1}{\sqrt{2}} + \frac{p}{q} \right|} \geq \frac{1}{4q^2}, \tag{1}$$

so

$$\frac{1}{\sqrt{|x - a_k|}} \leq 2q.$$

This leads to

> Equation (2): the sum $\sum_{p=0}^{q}$ accounts for all the rational numbers with denominator $q$; for all $q$ except $q = 1$, there are at most $q - 1$ of them. There are exactly $q - 1$ of them if $q$ is prime. This sum becomes multiplication by $q+1$ in the next step. The case where a quantity being summed has no index matching the index of the sum is discussed in a margin note in section 0.1.

$$\sum_{k=1}^{\infty} \frac{1}{2^k} \frac{1}{\sqrt{|x - a_k|}} \leq \sum_{q=1}^{\infty} \sum_{p=0}^{q} \frac{1}{2^{2q-3}} \, 2q = \sum_{q=1}^{\infty} \frac{2q(q+1)}{2^{2q-3}}. \tag{2}$$

This series is easily seen to be convergent by the ratio test.

**4.29** Write $a = a_1 + ia_2$ and $b = b_1 + ib_2$, so that $T(u) = au + b\bar{u}$ can be written

$$\begin{aligned} T(x + iy) &= (a_1 + ia_2)(x + iy) + (b_1 + ib_2)(x - iy) \\ &= a_1 x - a_2 y + b_1 x + b_2 y + i(a_1 y + a_2 x - b_1 y + b_2 x). \end{aligned} \tag{3}$$

we are identifying $\mathbb{C}$ with $\mathbb{R}^2$ in the standard way, with $x + iy$ written $\begin{pmatrix} x \\ y \end{pmatrix}$, so equation (3) is equivalent to the matrix multiplication

$$T \begin{bmatrix} x \\ y \end{bmatrix} = \begin{bmatrix} a_1 + b_1 & -a_2 + b_2 \\ a_2 + b_2 & a_1 - b_1 \end{bmatrix} \begin{bmatrix} x \\ y \end{bmatrix}.$$

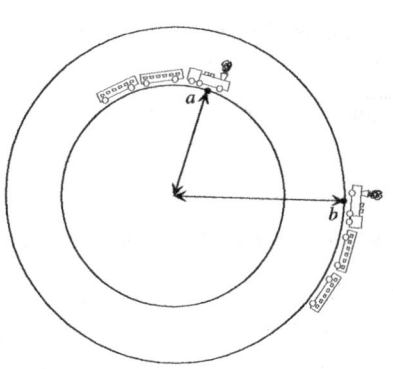

FIGURE FOR SOLUTION 4.29

One train represents $ae^{i\theta}$, the other $be^{-i\theta}$. There exists a value of $\theta$ for which the trains line up, i.e., for which $ae^{i\theta}$ and $be^{-i\theta}$ are multiples of each other by a positive real number.

Thus

$$\det T = (a_1 + b_1)(a_1 - b_1) - (a_2 + b_2)(-a_2 + b_2) = a_1^2 - b_1^2 + a_2^2 - b_2^2 = |a|^2 - |b|^2.$$

Now we show the result for the norm. Any complex number of absolute value 1 can be written $|e^{i\theta}| = 1$. This, with the triangle inequality, gives

$$\|T\| = \sup_\theta \left| ae^{i\theta} + be^{-i\theta} \right| = |a + b| \leq |a| + |b|.$$

Think of two circles centered at the origin, one with radius $|a|$, the other with radius $|b|$: $ae^{i\theta}$ travels around its circle in one direction, and $be^{-i\theta}$ travels around its circle in the other direction, as illustrated in the figure at left.

Clearly there is a value of $\theta$ where the two are on the same halfline from the origin through the circles; for this value of $\theta$, we see that $ae^{i\theta}$ and $be^{-i\theta}$ are multiples of each other by a positive real number, giving

$$\|T\| = |a + b| = |a| + |b|.$$

**5.1.1** Set $T = [\vec{\mathbf{v}}_1, \vec{\mathbf{v}}_2, \vec{\mathbf{v}}_3]$; then $T^\top T = \begin{bmatrix} 3 & 2 & 3 \\ 2 & 6 & 4 \\ 3 & 4 & 6 \end{bmatrix}$ and $\det(T^\top T) = 30$,

so the 3-dimensional volume of the parallelogram is $\sqrt{30}$.

**5.1.3** Here are two solutions.

*First solution*

Set $T = [\vec{\mathbf{v}}_1, \ldots, \vec{\mathbf{v}}_k]$. Since the vectors $\vec{\mathbf{v}}_1, \ldots, \vec{\mathbf{v}}_k$ are linearly dependent, $\operatorname{rank} T < k$. Further, $\operatorname{img} T^\top T \subset \operatorname{img} T^\top$, so

$$\operatorname{rank} T^\top T \leq \underbrace{\operatorname{rank} T^\top}_{\text{prop. 2.5.11}} = \operatorname{rank} T < k.$$

Since $T^\top T$ is a $k \times k$ matrix with rank $< k$, it is not invertible, hence its determinant is 0, so

$$\operatorname{vol}_k P(\vec{\mathbf{v}}_1, \ldots, \vec{\mathbf{v}}_k) = \sqrt{\det T^\top T} = 0.$$

*Second solution*

Since the vectors $\vec{\mathbf{v}}_1, \ldots, \vec{\mathbf{v}}_k$ are linearly dependent, there exist $a_1, \ldots, a_k$ not all 0 such that $a_1 \vec{\mathbf{v}}_1 + \cdots + a_k \vec{\mathbf{v}}_k = \mathbf{0}$, i.e.,

$$[\vec{\mathbf{v}}_1, \ldots, \vec{\mathbf{v}}_k] \begin{bmatrix} a_1 \\ \vdots \\ a_k \end{bmatrix} = \mathbf{0}.$$

Set $T = [\vec{\mathbf{v}}_1, \ldots, \vec{\mathbf{v}}_k]$. Since $T^\top T \mathbf{a} = \mathbf{0}$, it follows that $\ker (T^\top T) \neq \mathbf{0}$, so $T^\top T$ is not invertible, and $\det(T^\top T) = 0$, so

$$\operatorname{vol}_k P(\vec{\mathbf{v}}_1, \ldots, \vec{\mathbf{v}}_k) = \sqrt{\det T^\top T} = 0.$$

Solution 5.1.5: Here is another approach. By equation 5.1.9, the matrix $T^\top T$ is symmetric, so equation (1) shows that $T^\top T$ is a symmetric matrix representing a quadratic form taking only non-negative values, i.e., its signature is $(l, 0)$ for some $l \leq k$. So $T^\top T$ has $l$ positive eigenvalues, and the others are zero, by theorem 3.7.15. Since the determinant of a triangular matrix is the product of the eigenvalues (theorem 4.8.8), and a symmetric matrix is diagonalizable (the spectral theorem), and the determinant is basis independent (theorem 4.8.6), it follows that $\det T^\top T \geq 0$.

**5.1.5** Note first that if $\vec{\mathbf{v}}$ is a nonzero vector in $\mathbb{R}^k$,

$$(T^\top T \vec{\mathbf{v}}) \cdot \vec{\mathbf{v}} = \left(T^\top (T\vec{\mathbf{v}})\right)^\top \vec{\mathbf{v}} = (T\vec{\mathbf{v}})^\top T\vec{\mathbf{v}} = T\vec{\mathbf{v}} \cdot T\vec{\mathbf{v}} \geq 0. \qquad (1)$$

Denote by $A$ the $k \times k$ matrix $T^\top T$ and let $I$ be the $k \times k$ identity matrix. Consider the matrix $tA + (1 - t)I$ for $t \in [0, 1]$, which we can think of as $A$ (when $t = 1$) being transformed to $I$ (when $t = 0$). For $\vec{\mathbf{v}} \neq \mathbf{0}$ and $0 \leq t < 1$, we have

$$\left(tA + (1-t)I\right)\vec{\mathbf{v}} \cdot \vec{\mathbf{v}} = t\underbrace{A\vec{\mathbf{v}} \cdot \vec{\mathbf{v}}}_{\geq 0} + \underbrace{(1-t)}_{>0}\underbrace{\vec{\mathbf{v}} \cdot \vec{\mathbf{v}}}_{>0} > 0.$$

This implies that, for $0 < t \leq 1$, $\ker\left(tA + (1-t)I = \mathbf{0}\right.$ and thus that $\det\left(tA + (1-t)I\right.$ is never 0 when $0 \leq t < 1$. Since when $t = 0$, we have $\det\left(tA + (1-t)I = 1\right.$, and when $t = 1$, $\det\left(tA + (1-t)I = \det A\right.$, it follows that $\det A \geq 0$. (This uses the continuity of determinant, which is a polynomial function.)

### Note on parametrizations

A number of exercises in sections 5.2–5.4 involve finding parametrizations. The "small catalog of parametrizations" in section 5.2 should help. In many cases we adapt one of the standard parametrizations (spherical, polar, cylindrical) to the problem at hand. Note also the advice given in solution 5.1: *To find a parametrization for some region, we ask, "how should we describe where we are at some point in the region?"* Thus a first step is often to draw a picture.

If you memorized the spherical, cylindrical, and polar coordinates maps without thinking about why those are good descriptions of "where we are at some point", this is a good time to think about how these standard change of coordinates maps follow from high school trigonometry. Consider, for instance, the spherical coordinates map

$$S : \begin{pmatrix} r \\ \theta \\ \varphi \end{pmatrix} \mapsto \begin{pmatrix} r \cos\theta \cos\varphi \\ r \sin\theta \cos\varphi \\ r \sin\varphi \end{pmatrix}$$

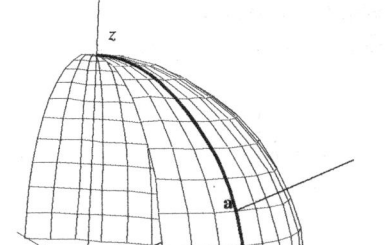

where $r$ is the radius of the sphere, $\theta$ is longitude, and $\varphi$ is latitude, as shown in the top figure at left (which you already saw as figure 4.10.4).

The middle figure at left shows the triangle whose vertices are a point **a** on the sphere, the origin, and the projection **p** of **a** onto the $(x, y)$-plane. Since $\cos\varphi$ is adjacent side over hypotenuse, the length of the side between **0** and **p** is $r \cos\varphi$. In the bottom triangle, $\cos\theta$ is adjacent side over hypotenuse $r\cos\varphi$, so the adjacent side (the side lying on the $x$-axis) is $r \cos\theta \cos\varphi$. This gives the first entry in the spherical coordinates map; it is the "$x$-coordinate" of the point **p**, which is the same as the $x$-coordinate of the point **a**. Since $\sin\theta$ is opposite side over hypotenuse, the "$y$-coordinate" of **p** and of **a** is $r \sin\theta \cos\varphi$. The $z$-coordinate of **a** should be the distance between **a** and **p**, as shown in the middle figure; that distance is indeed $r \sin\varphi$.

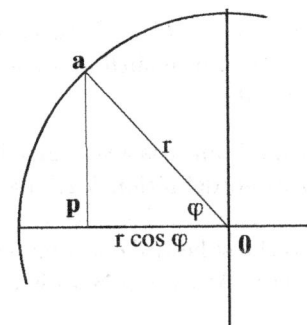

Trigonometry was invented by the ancient Greeks for the purpose of doing these computations.

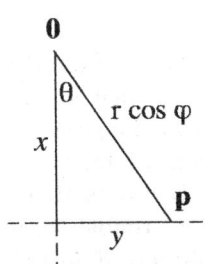

TOP: In spherical coordinates, a point **a** is specified by its distance from the origin ($r$), its longitude ($\theta$), and its latitude ($\varphi$). MIDDLE: This triangle lives in a plane perpendicular to the $(x, y)$-plane; you might think of slicing an orange in half vertically. BOTTOM: This triangle lives in the $(x, y)$-plane.

**5.2.1** Condition 3 of definition 5.2.3 requires that $\gamma : (U - X) \to M$ be one to one, of class $C^1$, with locally Lipschitz derivative. To show that it is one to one, we must show that if $\gamma \begin{pmatrix} r_1 \\ \theta_1 \end{pmatrix} = \gamma \begin{pmatrix} r_2 \\ \theta_2 \end{pmatrix}$, then $\begin{pmatrix} r_1 \\ \theta_1 \end{pmatrix} = \begin{pmatrix} r_2 \\ \theta_2 \end{pmatrix}$. Set

$$r_1 \cos\theta_1 = r_2 \cos\theta_2$$
$$r_1 \sin\theta_1 = r_2 \sin\theta_2$$
$$r_1 = r_2.$$

Since $r_1 = r_2$ by the third equation, $\cos\theta_1 = \cos\theta_2$ and $\sin\theta_1 = \sin\theta_2$. Since we are restricting $\theta$ to $0 < \theta < 2\pi$, this implies that $\theta_1 = \theta_2$. (But $\gamma$ is not one to one on $U$.)

The mapping $\gamma$ is clearly not just $C^1$ but $C^\infty$; we can differentiate it as many times as we like. Proposition 2.8.9 says that the derivative of a $C^2$ mapping is Lipschitz, so $\gamma$ has locally Lipschitz derivative.

For condition 4, we must show that $[\mathbf{D}\gamma(\mathbf{u})]$ is one to one for all $\mathbf{u}$ in $U - X$. We have

$$\left[\mathbf{D}\gamma\begin{pmatrix} r \\ \theta \end{pmatrix}\right] = \begin{bmatrix} \cos\theta & -r\sin\theta \\ \sin\theta & r\cos\theta \\ 1 & 0 \end{bmatrix}.$$

It is one to one if the only solution to

$$\left[\mathbf{D}\gamma\begin{pmatrix} r \\ \theta \end{pmatrix}\right]\begin{bmatrix} x \\ y \end{bmatrix} = \begin{bmatrix} x\cos\theta - yr\sin\theta \\ x\sin\theta + yr\cos\theta \\ x \end{bmatrix} = \begin{bmatrix} 0 \\ 0 \\ 0 \end{bmatrix}$$

is $x = 0$, $y = 0$.

Since $x = 0$, we have $yr\cos\theta = yr\sin\theta = 0$; since in $U - X$ we have $r > 0$, we can divide by $r$ to get $y\cos\theta = y\sin\theta = 0$. Since $\sin\theta$ and $\cos\theta$ do not simultaneously vanish, this implies that $y = 0$.

**5.2.3** By definition 3.1.18, $S$ is not a parametrization, but it is a parametrization by the relaxed definition of a parametrization, definition 5.2.3 (see the discussion at the beginning of section 5.2).

It is not difficult to show that $S$ is equivalent to the spherical coordinates map using latitude and longitude (definition 4.10.6). Any point $S(\mathbf{x})$ is on a sphere of radius $r$, since

$$r^2\sin^2\varphi\cos^2\theta + r^2\sin^2\varphi\sin^2\theta + r^2\cos^2\varphi = r^2.$$

Since $0 \leq r \leq \infty$, the mapping $S$ is onto.

The angle $\varphi$ tells what latitude a point is on. Going from 0 to $\pi$, it covers every possible latitude. The polar angle $\theta$ tells what longitude the point is on; going from 0 to $2\pi$, it covers every possible longitude.

However, $S$ is not a parametrization by definition 3.1.18, because it is not one to one. Trouble occurs in several places. If $r = 0$, then $S(\mathbf{x})$ is the origin, regardless of the values of $\theta$ and $\varphi$. For $\theta$, trouble occurs when $\theta = 0$ and $\theta = 2\pi$; for fixed $r$ and $\varphi$, these two values of $\theta$ give the same point. For $\varphi$, trouble occurs at 0 and $\pi$; if $\varphi = \pi$, the point is the south pole of a sphere of radius $r$, regardless of $\theta$, and if $\varphi = 0$, the point is the north pole of a sphere of radius $r$, regardless of $\theta$.

By definition 5.2.3, $S$ is a parametrization, since the trouble occurs only on a set of 3-dimensional volume 0. In the language of definition 5.2.3,

$$M = \mathbb{R}^3, \qquad U = \overbrace{[0, \infty)}^{\text{for } r} \times \overbrace{[0, 2\pi]}^{\text{for } \theta} \times \overbrace{[0, \pi]}^{\text{for } \varphi},$$

$$X = \underbrace{\left(\{0\} \times [0, 2\pi] \times [0, \pi]\right)}_{\text{trouble when } r=0} \cup \underbrace{\left([0, \infty) \times \{0, 2\pi\} \times [0, \pi]\right)}_{\text{trouble when } \theta=0 \text{ or } \theta=2\pi}$$

$$\cup \underbrace{\left([0, \infty] \times [0, 2\pi] \times \{0, \pi\}\right)}_{\text{trouble when } \varphi=0 \text{ or } \varphi=\pi}.$$

Trouble occurs on a union of five surfaces, which you can think of as five sides of a box whose sixth side is at infinity. The base of the box lies in the plane where $r = 0$; the sides of the base have length $\pi$ and $2\pi$. The base represents the set labeled "trouble when $r = 0$." Two parallel sides of the box stretching to infinity represent the set "trouble when $\theta = 0$ or $\theta = 2\pi$." The other set of parallel sides represents "trouble when $\varphi = 0$ or $\varphi = \pi$." By proposition 4.3.7 and definition 5.2.1, these surfaces have $\text{vol}_3 = 0$.

By the dimension formula, the derivative is invertible if it is one to one.

Next let us check (condition 4 of definition 5.2.3), that the derivative is one to one for all $\mathbf{u}$ in $U - X$. This will be true (theorem 4.8.3) if and only if the determinant of the derivative is not 0. The determinant is

$$\det\left[\mathbf{D}S\begin{pmatrix} r \\ \theta \\ \varphi \end{pmatrix}\right] = \det\begin{bmatrix} \sin\varphi\cos\theta & -r\sin\varphi\sin\theta & r\cos\varphi\cos\theta \\ \sin\varphi\sin\theta & r\sin\varphi\cos\theta & r\cos\varphi\sin\theta \\ \cos\varphi & 0 & -r\sin\varphi \end{bmatrix} = -2r^2\sin\varphi,$$

which is 0 only if $r = 0$, or $\varphi = 0$, or $\varphi = \pi$, which are not in $U - X$.

Polynomials are continuous by corollary 1.5.30. Theorem 1.5.28 discusses combining continuous functions

Next, we will show that $S : (U - X) \to \mathbb{R}^3$ is of class $C^1$ with locally Lipschitz derivative. We know the first derivatives exist, since we just computed the derivative. They are continuous, since they are all polynomials in $r$, sine, and cosine, which are continuous. In fact, $S : (U - X) \to \mathbb{R}^3$ is $C^\infty$, since its derivatives of all order are polynomials in $r$, sine, and cosine, and are thus continuous. We do not need to check anything about Lipschitz conditions because proposition 2.8.9 says that the derivative of a $C^2$ function is Lipschitz.

Finally, we need to show that $S(X)$ has 3-dimensional volume 0. If $r = 0$, then $S\begin{pmatrix} r \\ \theta \\ \varphi \end{pmatrix}$ is the origin, whatever the values of $\theta$ and $\varphi$. If $\theta = 0$ we have $S\begin{pmatrix} r \\ 0 \\ \varphi \end{pmatrix} = \begin{pmatrix} r\sin\varphi \\ 0 \\ r\cos\varphi \end{pmatrix}$, which is a surface in the $(x, z)$-plane. If $\theta = 2\pi$, we get the same surface. If $\varphi = 0$, we get $\begin{pmatrix} 0 \\ 0 \\ r \end{pmatrix}$, the $z$-axis going from 0 to $\infty$, and if $\varphi = \pi$ $\begin{pmatrix} 0 \\ 0 \\ -r \end{pmatrix}$, we get the $z$-axis going from 0 to $-\infty$.

**5.2.5**  The set
$$U = \bigcup_{i=1}^{\infty}\left(a_i - \frac{1}{2^{i+k}}, a_i - \frac{1}{2^{i+k}}\right)$$

is an open subset of $\mathbb{R}$, hence indeed a 1-dimensional manifold. The pathological property of $U$ is that its boundary has positive length, so that its characteristic function is not integrable.[5]

[5]The boundary of $U$ is the complement of $U$ in $[0, 1]$: by definition 1.5.10, a point $x$ is in the boundary if each neighborhood of $x$ intersects both $U$ and its complement. Since the length of $[0, 1]$ is 1, and the sum of the lengths of the intervals making up $U$ is $\epsilon$, the boundary $[0, 1] - U$ has positive length.

But $U$ is simply a countable union of disjoint open intervals. (In fact, every open subset of $\mathbb{R}$ is a countable union of open intervals.) List these intervals $I_1, I_2, \ldots$ in order of length (which is *not* the order in which they appear in $\mathbb{R}$). Let the length of $I_i$ be $l_i$, and let

$$J_i = \left( \sum_{j=1}^{i-1} l_i, \sum_{j=1}^{i} l_i \right),$$

so that $J_i$ is another interval of length $l_i$. Now let $J = \cup_{i=1}^{\infty} J_i$, and define $\gamma : J \to U$ to be the map that translates $J_i$ onto $I_i$. This is a parametrization according to definition 5.2.3. Indeed, let $X$ be empty, and check the conditions:

1. $\gamma(J) = U$, by definition;
2. $\gamma(J - X) = U$, since $X$ is empty;
3. $\gamma : J - X \to U$ is one to one, of class $C^1$, with Lipschitz derivative. These are local conditions, and $\gamma$ is locally a translation.
4. $\gamma'(x) \neq 0$ for all $x \in J$, since $\gamma'(x) = 1$ everywhere.
5. $\gamma(X)$ has 1-dimensional volume 0, since $X$ is empty.

**Solution 5.2.5:** The point of showing that $U$ can be parametrized is that we will be able to compute integrals over $U$ using the parametrization. We can't integrate a function $f$ over $U$, since the upper and lower sums

$$U_N(f\mathbf{1}_U) \text{ and } L_N(f\mathbf{1}_U)$$

do not have a common limit. But we can integrate $f \circ \gamma$ over $J$, since the boundary of $J$ is the increasing sequence of numbers (together with its limit)

$$a_i = \sum_{j=1}^{i} l_j, \ i = 0, 1, \ldots, \infty,$$

and certainly has volume 0.

**5.3.1**  a. Going back to Cartesian coordinates, set $x(t) = r(t) \cos \theta(t)$ and $y(t) = r(t) \sin \theta(t)$:

$$\gamma(t) = \begin{pmatrix} x(t) \\ y(t) \end{pmatrix} = \begin{pmatrix} r(t) \cos \theta(t) \\ r(t) \sin \theta(t) \end{pmatrix}.$$

So the formula $l = \int_a^b |\vec{\gamma}'(t)| \, dt$ (equation 5.3.4) gives

$$l = \int_a^b \sqrt{\left( (r\cos\theta)'(t) \right)^2 + \left( (r\sin\theta)'(t) \right)^2} \, dt$$

$$= \int_a^b \sqrt{\left( r'(t)\cos(\theta(t)) - r(t)\sin(\theta(t))\theta'(t) \right)^2 + \left( r'(t)\sin(\theta(t)) + r\cos(\theta(t))\theta'(t) \right)^2} \, dt$$

$$= \int_a^b \sqrt{\left( r'(t) \right)^2 + \left( r(t) \right)^2 \left( \theta'(t) \right)^2} \, dt.$$

When computing the first entry of $\gamma'(t)$, remember that the derivative of $\cos \theta(t)$ is the derivative at $t$ of the composition $\cos \circ \theta$, computed using the chain rule: $(\cos \circ \theta)'(t) = \cos'(\theta(t))\theta'(t) = -\sin(\theta(t))\theta'(t)$. Similarly, for the second entry, the derivative of $\sin \theta(t)$ is the derivative at $t$ of the composition $\sin \circ \theta$, giving $(\sin \circ \theta)'(t) = \sin'(\theta(t))\theta'(t) = \cos(\theta(t))\theta'(t)$.

b. The spiral is the curve parametrized by $\gamma(t) = \begin{pmatrix} e^{-\alpha t} \cos t \\ e^{-\alpha t} \sin t \end{pmatrix}$, whose derivative is $\vec{\gamma}'(t) = \begin{bmatrix} -e^{-\alpha t} \sin t - \alpha e^{-\alpha t} \cos t \\ e^{-\alpha t} \cos t - \alpha e^{-\alpha t} \sin t \end{bmatrix}$, so the length between $t = 0$ and $t = a$ is

$$\int_0^a |\vec{\gamma}'(t)| \, dt = \int_0^a \sqrt{\alpha^2 e^{-2\alpha t} + e^{-2\alpha t}} \, dt = \int_0^a \sqrt{1 + \alpha^2} e^{-\alpha t} \, dt$$

$$= \sqrt{\alpha^2 + 1} \, \frac{(e^{-\alpha a} - 1)}{-\alpha}.$$

The limit of this length as $\alpha \to 0$ is $a$: expressing $e^{-\alpha a}$ as its Taylor polynomial (equation 3.4.5) gives

$$\lim_{\alpha \to 0} \sqrt{\alpha^2+1}\,\frac{(e^{-\alpha a}-1)}{-\alpha} = \lim_{\alpha \to 0} \sqrt{\alpha^2+1}\,\frac{\left((1-a\alpha+\frac{a^2\alpha^2}{2}-\cdots)-1\right)}{-\alpha}$$

$$= \lim_{\alpha \to 0} \sqrt{\alpha^2+1}\left(a+\alpha(\text{higher degree terms})\right)$$

$$= a.$$

At $\alpha = 0$, the distance from the origin is 1, so the curve is a circle of radius 1 rather than a narrowing spiral. $\triangle$

Solution 5.3.1, part b: How do we know this curve is a spiral? Each point of the curve is known by polar coordinates; at time $t$ the distance from the point to the origin is $e^{\cdot\,\alpha t}$ and the polar angle is $t$. The beginning point, at $t = 0$, is the point $\begin{pmatrix} 1 \\ 0 \end{pmatrix}$ in the $(x,y)$-plane – the point with coordinates

$$x = e^0 \cos 0, \quad y = e^0 \sin 0.$$

As $t$ increases, the polar angle increases and the distance from the origin decreases; as $t \to \infty$, we have $e^{\cdot\,\alpha t} \to 0$.

c. For every $2\pi$ increment in $t$ the spiral turns around the origin once: $\theta(t) = \theta(t+2\pi)$. So the spiral turns infinitely many times about the origin as $t \to \infty$. The length does not tend to infinity, since

$$\lim_{t\to\infty} l = \lim_{t\to\infty} \frac{\sqrt{\alpha^2+1}}{\alpha}(1-e^{-\alpha t}) = \frac{\sqrt{\alpha^2+1}}{\alpha}.$$

**5.3.3** a. Since

$$x(t) = r(t)\cos\varphi(t)\cos\theta(t),\ \ y(t) = r(t)\cos\varphi(t)\sin\theta(t),\ \ z(t) = r(t)\sin\varphi(t),$$

we have

$$l = \int \Big((r'\cos\varphi\cos\theta - r\sin\varphi\cos\theta\,\varphi' - r'\cos\varphi\sin\theta\,\theta')^2$$

$$+ (r'\cos\varphi\sin\theta - r\sin\varphi\sin\theta\,\varphi' + r'\cos\varphi\cos\theta\,\theta')^2$$

$$+ (r'\sin\varphi - r\cos\varphi\,\varphi')^2\Big)^{1/2} dt$$

$$= \int \sqrt{(r')^2 + r^2\,(\varphi')^2 + r^2\cos^2\varphi\,(\theta')^2}\ dt.$$

Solution 5.3.3: In the exercise, the order in which we listed the variables $\theta$ and $\varphi$ in the domain is reversed, compared to definition 4.10.6. You should think that here we have

$$S : \begin{pmatrix} r \\ \varphi \\ \theta \end{pmatrix} \mapsto \begin{pmatrix} r\cos\theta\cos\varphi \\ r\sin\theta\cos\varphi \\ r\sin\varphi \end{pmatrix}.$$

Solution 5.3.3, part b: Even though $\theta(t) = \tan t$ tends to infinity, the curve does not become long.

b. We get the integral

$$\int_0^a \sqrt{(-\sin t)^2 + (\cos t)^2 + (\cos t)^2(\cos t)^2\frac{1}{(\cos t)^4}}\ dt = \int_0^a \sqrt{2}\ dt = a\sqrt{2}.$$

**5.3.5** a. The map

$$\gamma : \begin{pmatrix} r \\ \theta \end{pmatrix} \mapsto \begin{pmatrix} 2r\cos\theta \\ 3r\sin\theta \\ r^2 \end{pmatrix}$$

parametrizes the locus given by the equation $z = \frac{x^2}{4}+\frac{y^2}{9}$ (which is an elliptic paraboloid)[6]; it maps the part of the $(r,\theta)$-plane where $0 \le \theta \le 2\pi$, $r \le a$

---

[6]How did we find this parametrization? The surface cut horizontally at height $z$ is an ellipse, so we were guided by the parametrization $t \mapsto \begin{pmatrix} a\cos t \\ b\sin t \end{pmatrix}$ for the ellipse $\frac{x^2}{a^2} + \frac{y^2}{b^2} = 1$; see example 4.10.19. We got the $r^2$ by noting that then $t \mapsto \begin{pmatrix} ra\cos t \\ rb\sin t \end{pmatrix}$ parametrizes the ellipse $\frac{x^2}{a^2} + \frac{y^2}{b^2} = r^2$.

to the region of the paraboloid where $z \le a^2$. Thus the surface area is given by the integral

$$\int_0^{2\pi}\int_0^a \sqrt{\det\left(\left[\mathbf{D}\gamma\begin{pmatrix}r\\\theta\end{pmatrix}\right]^\top [\mathbf{D}\gamma(\begin{smallmatrix}r\\\theta\end{smallmatrix})]\right)}\,dr\,d\theta = \int_0^{2\pi}\int_0^a \sqrt{\det\begin{bmatrix}2\cos\theta & 3\sin\theta & 2r\\-2r\sin\theta & 3r\cos\theta & 0\end{bmatrix}\begin{bmatrix}2\cos\theta & -2r\sin\theta\\3\sin\theta & 3r\cos\theta\\2r & 0\end{bmatrix}}\,dr\,d\theta$$

$$= \int_0^{2\pi}\int_0^a 2r\sqrt{9 + r^2(4\sin^2\theta + 9\cos^2\theta)}\,dr\,d\theta.$$

This integral cannot be evaluated in terms of elementary functions, but it can be evaluated in terms of elliptic functions, using MAPLE (at least, version 7 or higher):

$$\int_0^{2\pi}\int_0^a 2r\sqrt{9 + r^2(4\sin^2\theta + 9\cos^2\theta)}\,dr\,d\theta = \frac{2}{3\sqrt{1+a^2}}\left(36\sqrt{\frac{1+a^2}{9+4a^2}}a^2\,\text{ellipticK}\sqrt{-5\frac{a^2}{9+4a^2}}\right.$$

$$+81\sqrt{\frac{1+a^2}{9+4a^2}}\text{elliptic }\pi\left(\frac{-5}{4},\sqrt{-5\frac{a^2}{9+4a^2}}\right) + 16\sqrt{\frac{1+a^2}{9+4a^2}}a^4\text{ellipticE}\sqrt{-5\frac{a^2}{9+4a^2}}$$

$$\left.+36\sqrt{\frac{1+a^2}{9+4a^2}}a^2\text{ellipticE}\sqrt{-5\frac{a^2}{9+4a^2}} - 9\pi\sqrt{1+a^2}\right).$$

The graph of the area as a function of $a$ is shown below.

The elliptic functions "elliptic K," "elliptic E," and "elliptic $\pi$" are tabulated functions; tables with these functions can be found, for example, in *Handbook of Mathematical Functions*, edited by Milton Abramowitz and Irene Stegun (Dover Publications, Inc.)

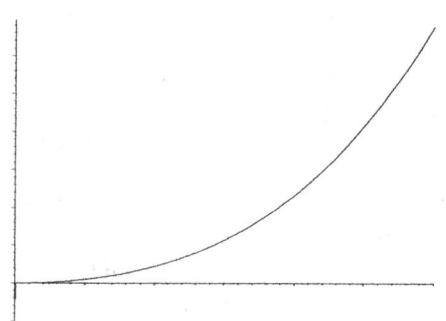

b. It is easier to begin by finding the volume of the region outside the elliptic paraboloid, and then subtract it from the volume of the elliptic cylinder. Call $U_a$ the ellipse

$$\frac{x^2}{4} + \frac{y^2}{9} \le a^2.$$

Then the volume of the region above this ellipse in the plane $z = 0$ and beneath the paraboloid is

$$\int_{U_a}\left(\frac{x^2}{4} + \frac{y^2}{9}\right)|dx\,dy| = \int_0^{2\pi}\int_0^a \det\begin{bmatrix}2\cos\theta & -2r\sin\theta\\3\sin\theta & 3r\cos\theta\end{bmatrix}r^2\,dr = 3a^4\pi.$$

The elliptic cylinder has total volume $\pi(6a^2)a^2 = 6\pi a^4$, and the difference of the two is $3\pi a^4$.

**5.3.7** a. The obvious way to parametrize the surface of the ellipsoid is

Solution 5.3.7: This parametrization is inspired by the spherical change of coordinates for the unit sphere:

$$\gamma\begin{pmatrix}\varphi \\ \theta\end{pmatrix} = \begin{pmatrix} a\cos\varphi\cos\theta \\ b\cos\varphi\sin\theta \\ c\sin\varphi \end{pmatrix}, \quad -\frac{\pi}{2} < \varphi < \frac{\pi}{2},\ 0 \le \theta < 2\pi.$$

$$\begin{pmatrix}\theta \\ \varphi\end{pmatrix} \mapsto \begin{pmatrix}\cos\theta\cos\varphi \\ \sin\theta\cos\varphi \\ \sin\varphi\end{pmatrix}.$$

Scaling a sphere in three directions with three different scaling factors $a, b, c$ gives an ellipsoid.

In part b we are using equation 5.3.35.

b. The area is then given by

$$\int_0^{2\pi}\int_{-\pi/2}^{\pi/2}\left|\left(\overrightarrow{D_1\gamma}\begin{pmatrix}\varphi\\\theta\end{pmatrix}\right)\times\left(\overrightarrow{D_2\gamma}\begin{pmatrix}\varphi\\\theta\end{pmatrix}\right)\right|d\varphi\,d\theta$$

$$=\int_0^{2\pi}\int_{-\pi/2}^{\pi/2}\left|\begin{bmatrix}-a\sin\varphi\cos\theta\\-b\sin\varphi\sin\theta\\c\cos\varphi\end{bmatrix}\times\begin{bmatrix}-a\cos\varphi\sin\theta\\b\cos\varphi\cos\theta\\0\end{bmatrix}\right|d\varphi\,d\theta$$

$$=\int_0^{2\pi}\int_{-\pi/2}^{\pi/2}\sqrt{b^2c^2\cos^4\varphi\cos^2\theta+a^2c^2\cos^4\varphi\sin^2\theta+a^2b^2\cos^2\varphi\sin^2\varphi}\,d\varphi\,d\theta.$$

**5.3.9** a. We parametrize $S$ by $\gamma\begin{pmatrix}x\\y\end{pmatrix}=\begin{pmatrix}x\\y\\x^3+y^4\end{pmatrix}$. We could use our standard approach and set up the integral as

$$\int_S (x+y+\overbrace{x^3+y^4}^{z})\sqrt{\det[\mathbf{D}\gamma(\mathbf{x})]^\top[\mathbf{D}\gamma(\mathbf{x})]}|dx\,dy|$$

$$=\int_{-1}^{1}\int_{-\sqrt{1-y^2}}^{\sqrt{1-y^2}}(x+y+x^3+y^4)\sqrt{\det\left(\begin{bmatrix}1&0&3x^2\\0&1&4y^3\end{bmatrix}\begin{bmatrix}1&0\\0&1\\3x^2&4y^3\end{bmatrix}\right)}\,dx\,dy,$$

Solution 5.3.9: In part b, we use polar coordinates, but the result will not be computable in elementary terms in any case.

but since we are in $\mathbb{R}^3$ it is probably easier computationally to use the length of the cross product to give the area of the parallelogram spanned by the partial derivatives $\overrightarrow{D_1\gamma}$ and $\overrightarrow{D_2\gamma}$ (see proposition 1.4.19 or equation 5.3.35). Thus we have

$$\int_{-1}^{1}\int_{-\sqrt{1-y^2}}^{\sqrt{1-y^2}}(x+y+\overbrace{x^3+y^4}^{z})\left|\begin{bmatrix}1\\0\\3x^2\end{bmatrix}\times\begin{bmatrix}0\\1\\4y^3\end{bmatrix}\right|dx\,dy$$

$$=\int_{-1}^{1}\int_{-\sqrt{1-y^2}}^{\sqrt{1-y^2}}(x+y+x^3+y^4)\sqrt{(9x^4+16y^6+1}\,dx\,dy.$$

b. It is easier to pass to polar coordinates, so that the integral will be over a rectangle. We find

$$\int_0^{2\pi}\int_0^1(r\cos\theta+r\sin\theta+r^3\cos^3\theta+r^4\sin^4\theta)\sqrt{9r^4\cos^4\theta+16r^6\sin^6\theta+1}\,r\,dr\,d\theta.$$

We have no idea how to compute this integral symbolically. To evaluate it numerically, we need to choose a numerical method. The following sequence of MATLAB instructions performs a midpoint Riemann sum, breaking the domain into 100 intervals on each side, and evaluating the function at the center of each rectangle. (MATLAB has a command "dblquad" that should compute double integrals, but we could not get it to work, even on MATLAB's own example.)

First, define the function:

function prob545=integrand (x,y)
prob545 = x$^2$*(cos(y)+sin(y)+x$^2$* (cos(y))$^3$ +x$^3$* (sin(y))$^4$)* sqrt(9*(x*cos(y))$^4$
  +16*(x*sin(y))$^6$+1);

Then create a matrix $A$ whose values are the values of this function at the points

$$\begin{pmatrix} (2i-1)/200 \\ 2\pi(2j-1)/200 \end{pmatrix}, \quad i,j = 1, \ldots, 200.$$

This is done by the following commands:

$N = 100$
for $I = 1 : N$,
        for $J = 1 : N$,
                $A(I, J) = $ integrand((2*I-1)/(2*N),2*pi*(2*J-1)/(2*N));
        end,
end;

Now sum the elements of the matrix, and multiply by $2\pi/(100^2)$, the area of the small rectangles:

$$\text{sum(sum(A(1:N,1:N)))*2*pi/(N}^2)$$

The answer the computer comes up with is 0.96117719450855.

If you want a more precise answer, you might try the 2-dimensional Simpson's method. The following MATLAB program will do this.

```
            N=101
            S(1)=1;
            S(N)=1;
            for i=1:(N-1)/2, S(2*i)=4; end
            for i=1:(N-3)/2, S(2*i+1)=2; end
            for i=1:(N-3)/2, S(2*i+1)=2; end
            for i=1:N,
              for j=1:N,
                    SS(i,j) = S(i)*S(j);
              end
            end

            for I = 1:N,
                for J = 1:N,
                   B(I,J) = integrand((I-1)/(N-1),2*pi*(J-1)/(N-1));
                end,
            end;
            (sum(sum(SS(1:N,1:N).*B(1:N,1:N))))*2*pi/(9*(N-1)^2)
```

This yields the estimate 0.96143451150993. If you repeat the procedure with $N = 201$, you will find 0.96143438060954. It seems likely that the digits 0.9614343 are accurate; this is in agreement with the estimate (from theorem 4.6.2) that says the error should be a constant times $1/N^4$, where the constant is computed from the partials of $f$ up to order 4.

**5.3.11**   a.   The surface $X$ is parametrized by

$$\begin{pmatrix} r \\ \theta \end{pmatrix} \mapsto \begin{pmatrix} r\cos\theta \\ r\sin\theta \\ r^k\cos k\theta \\ r^k\sin k\theta \end{pmatrix}, \quad 0 \le \theta < 2\pi, \ r \ge 0.$$

b.   We need to integrate over $0 \le \theta < 2\pi$, $0 \le r \le R$ the area of the parallelogram in $\mathbb{R}^4$ spanned by the partial derivatives of the parametrizing map above. As in example 5.3.10, we compute the area of the parallelogram using the formula

$$\int_U \sqrt{\det\left(\left[\mathbf{D}\gamma\begin{pmatrix} u \\ v \end{pmatrix}\right]^\top \left[\mathbf{D}\gamma\begin{pmatrix} u \\ v \end{pmatrix}\right]\right)} \, |du\,dv|,$$

where $U$ is the domain of integration. Since

$$\left[\mathbf{D}\gamma\begin{pmatrix} r \\ \theta \end{pmatrix}\right] = \begin{bmatrix} \cos\theta & -r\sin\theta \\ \sin\theta & r\cos\theta \\ kr^{k-1}\cos k\theta & -kr^k\sin k\theta \\ kr^{k-1}\sin k\theta & kr^k\cos k\theta \end{bmatrix},$$

we have

$$\sqrt{\det\left(\left[\mathbf{D}\gamma\begin{pmatrix} r \\ \theta \end{pmatrix}\right]^\top \left[\mathbf{D}\gamma\begin{pmatrix} r \\ \theta \end{pmatrix}\right]\right)} = \sqrt{\left(1 + k^2 r^{2k-2}\right)\left(r^2 + k^2 r^{2k}\right)}$$

$$= r + k^2 r^{2k-1}.$$

So we find

$$\int_0^{2\pi}\int_0^R (r + k^2 r^{2k-1}) \, dr \, d\theta = 2\pi\left[\frac{r^2}{2} + k^2\frac{r^{2k}}{2k}\right]_0^R = 2\pi\left(\frac{R^2}{2} + k^2\frac{R^{2k}}{2k}\right).$$

Solution 5.3.13, part a: More generally, the same sort of computation shows that when a surface is scaled by $a$, its Gaussian curvature is scaled by $1/a^2$. The mean curvature is scaled by $1/a$.

**5.3.13**   a.   The Gaussian curvature $K$ of the sphere $S$ of radius $R$ is $1/R^2$. This computation is very similar to that in solution 3.8.3. Any point can be rotated to the point $\mathbf{a} = \begin{pmatrix} 0 \\ 0 \\ R \end{pmatrix}$, so it is enough to compute the curvatures there. Near $\mathbf{a}$ the "best coordinates" are $X = x$, $Y = y$, $Z = z - R$, so that

$$Z = \overbrace{\sqrt{R^2 - X^2 - Y^2}}^{z} - R = R\left(\sqrt{1 - \frac{X^2}{R^2} - \frac{Y^2}{R^2}} - 1\right).$$

As in solution 3.8.3, we use equation 3.4.9 to compute

$$Z \approx R\left(1 + \frac{1}{2}\left(-\frac{X^2}{R^2} - \frac{Y^2}{R^2}\right) + \cdots - 1\right).$$

By definition 3.8.8, the Gaussian curvature $K$ of a surface at a point $\mathbf{a}$ is

$$K(\mathbf{a}) = A_{2,0}A_{0,2} - A_{1,1}^2.$$

The quadratic terms of the Taylor polynomial are $-\dfrac{1}{2}\left(\dfrac{X^2}{R} + \dfrac{Y^2}{R}\right)$, so

$$A_{1,1} = 0, \quad A_{2,0} = A_{0,2} = 1/R, \quad \text{giving} \quad K = 1/R^2.$$

The area of the sphere of radius $R$ is $4\pi R^2$, so its total curvature $\mathbf{K}(S)$ is $4\pi$, independent of $R$:

$$\mathbf{K}(S) = \int_S |K(\mathbf{x})|\,|d^2\mathbf{x}| = \frac{1}{R^2}4\pi R^2 = 4\pi.$$

This parametrization is an example of parametrizing as a graph:

$$\gamma\begin{pmatrix} x \\ y \end{pmatrix} = \begin{pmatrix} x \\ y \\ x^2 - y^2 \end{pmatrix}.$$

b. The hyperboloid of equation $z = x^2 - y^2$ is parametrized by $x$ and $y$. We computed the Gaussian curvature of this surface in example 3.8.11, and found it to be

$$K\begin{pmatrix} x \\ y \\ x^2 - y^2 \end{pmatrix} = \frac{-4}{(1 + 4x^2 + 4y^2)^2}.$$

Thus our total curvature is

$$\int_{\mathbb{R}^2} \frac{-4}{(1 + 4x^2 + 4y^2)^2} \left\| \begin{bmatrix} 1 \\ 0 \\ 2x \end{bmatrix} \times \begin{bmatrix} 0 \\ 1 \\ -2y \end{bmatrix} \right\| |dx\,dy| = 4\int_{\mathbb{R}^2} \frac{|dx\,dy|}{(1 + 4x^2 + 4y^2)^{3/2}}.$$

This integral exists (as a Lebesgue integral), and can be computed in polar coordinates (this uses theorems 4.11.20 and 4.11.21), to yield

$$4\int_0^{2\pi}\left(\int_0^\infty \frac{r}{(1 + 4r^2)^{3/2}}\,dr\right)d\theta = \pi\int_0^\infty \frac{8r}{(1 + 4r^2)^{3/2}}\,dr = \pi.$$

**5.3.15**   a. For the point $\begin{pmatrix} a \\ b \\ c \\ d \end{pmatrix} \in \mathbb{R}^4$ to be on the unit sphere, we must have $a^2 + b^2 + c^2 + d^2 = 1$, so a first step in showing that $\gamma$ parametrizes $S^3$ is to show that the sum of the squares of the coordinates is 1:

$$\cos^2\psi\cos^2\varphi\cos^2\theta + \cos^2\psi\cos^2\varphi\sin^2\theta + \cos^2\psi\sin^2\varphi + \sin^2\psi$$
$$= \cos^2\psi\cos^2\varphi + \cos^2\psi\sin^2\varphi + \sin^2\psi = \cos^2\psi + \sin^2\psi = 1.$$

Next we must check the domains. Except possibly on sets of 3-dimensional volume 0 (individual points, for example), is $\gamma$ one to one and onto? Think of the fourth coordinate, $\sin\psi$, as being the "height" of $S^3$, analogous to the $z$-coordinate for the unit sphere in $\mathbb{R}^3$. Clearly $\sin\psi$ must run through $[-1, 1]$, and does run through these values exactly once as $\psi$ varies from $-\pi/2$ to $\pi/2$.

For any fixed value of $\psi \in (-\pi/2, \pi/2)$, the other coordinates satisfy

$$a^2 + b^2 + c^2 + (\sin\psi)^2 = 1, \quad \text{i.e.,} \quad a^2 + b^2 + c^2 = (\cos\psi)^2,$$

so $\begin{pmatrix} a \\ b \\ c \end{pmatrix}$ runs through the 2-sphere in $\mathbb{R}^3$ of radius $\cos\psi$. For fixed $\sin\psi$, the first three coordinates are:

$$\begin{pmatrix} \cos\psi\cos\theta\cos\varphi \\ \cos\psi\sin\theta\cos\varphi \\ \cos\psi\sin\varphi \end{pmatrix},$$

which is the parametrization of the sphere in $\mathbb{R}^3$ with radius $\cos\psi$ by spherical coordinates (definition 4.10.6). Thus $\gamma$ is a parametrization of the 3-sphere in $\mathbb{R}^4$.

b. To compute $\mathrm{vol}_3(S^3)$ we use equation 5.3.3. First we compute

$$\left[\mathbf{D}\gamma\begin{pmatrix} \theta \\ \varphi \\ \psi \end{pmatrix}\right] \quad\text{and}\quad \left[\mathbf{D}\gamma\begin{pmatrix} \theta \\ \varphi \\ \psi \end{pmatrix}\right]^\top \left[\mathbf{D}\gamma\begin{pmatrix} \theta \\ \varphi \\ \psi \end{pmatrix}\right]:$$

$$\left[\mathbf{D}\gamma\begin{pmatrix} \theta \\ \varphi \\ \psi \end{pmatrix}\right] = \begin{bmatrix} -\cos\psi\cos\varphi\sin\theta & -\cos\psi\sin\varphi\cos\theta & -\sin\psi\cos\varphi\cos\theta \\ \cos\psi\cos\varphi\cos\theta & -\cos\psi\sin\varphi\sin\theta & -\sin\psi\cos\varphi\sin\theta \\ 0 & \cos\psi\cos\varphi & -\sin\psi\sin\varphi \\ 0 & 0 & \cos\psi \end{bmatrix}.$$

$$\left[\mathbf{D}\gamma\begin{pmatrix} \theta \\ \varphi \\ \psi \end{pmatrix}\right]^\top = \begin{bmatrix} -\cos\psi\cos\varphi\sin\theta & \cos\psi\cos\varphi\cos\theta & 0 & 0 \\ -\cos\psi\sin\varphi\cos\theta & -\cos\psi\sin\varphi\sin\theta & \cos\psi\cos\varphi & 0 \\ -\sin\psi\cos\varphi\cos\theta & -\sin\psi\cos\varphi\sin\theta & -\sin\psi\sin\varphi & \cos\psi \end{bmatrix},$$

giving

Computing this product is a little easier if you remember that $A^\top A$ is always symmetric (see exercise 1.2.16 and equation 5.1.9).

$$\left[\mathbf{D}\gamma\begin{pmatrix} \theta \\ \varphi \\ \psi \end{pmatrix}\right]^\top \left[\mathbf{D}\gamma\begin{pmatrix} \theta \\ \varphi \\ \psi \end{pmatrix}\right] = \begin{bmatrix} \cos^2\psi\cos^2\varphi & 0 & 0 \\ 0 & \cos^2\psi & 0 \\ 0 & 0 & 1 \end{bmatrix}.$$

The determinant of this matrix is $\cos^4\psi\cos^2\varphi$, so applying equation 5.3.3 to our problem gives

$$\mathrm{vol}_3 S^3 = \int_{-\pi/2}^{\pi/2}\int_{-\pi/2}^{\pi/2}\int_0^{2\pi} \cos^2\psi\cos\varphi\,d\theta\,d\varphi\,d\psi$$

$$= 2\pi\int_{-\pi/2}^{\pi/2}\int_{-\pi/2}^{\pi/2} \cos^2\psi\cos\varphi\,d\varphi\,d\psi$$

To integrate $\cos^2\psi$ we used the fact that $\cos^2\psi$ has average $1/2$ over one full period, which for $\cos^2$ is $\pi$.

You may notice that this result is the same as the value for $\mathrm{vol}_3 S^3$ given in table 5.3.3.

$$= 2\pi\int_{-\pi/2}^{\pi/2} [\sin\varphi]_{-\pi/2}^{\pi/2}\cos^2\psi\,d\psi = 4\pi\int_{-\pi/2}^{\pi/2}\cos^2\psi\,d\psi = 4\pi\frac{\pi}{2}$$

$$= 2\pi^2.$$

**5.3.17** It's pretty easy to write down the appropriate integrals, but one of them is easy to evaluate in elementary terms, one is hard, and one is very hard.

Denote by $\bar{x}$ and $\bar{y}$ the $x$ and $y$ components of the center of gravity. Then (by definition 4.2.1)

$$\bar{x} = \frac{\int_C x\,|d\mathbf{x}|}{\int_C |d\mathbf{x}|} \quad \text{and} \quad \bar{y} = \frac{\int_C y\,|d\mathbf{x}|}{\int_C |d\mathbf{x}|}.$$

The curve is parametrized by $x \mapsto \begin{pmatrix} x \\ x^2 \end{pmatrix}$, so the integrals become

In equation (1), the element of length $|d^1\mathbf{x}|$ is being evaluated on the vector $\begin{bmatrix} 1 \\ 2x \end{bmatrix}$. We could also write this as $\left|\left[\begin{smallmatrix} 1 \\ 2x \end{smallmatrix}\right]\right|$.

$$\int_0^a x|d^1\mathbf{x}|\begin{bmatrix} 1 \\ 2x \end{bmatrix}\,dx = \int_0^a x\sqrt{1+4x^2}\,dx$$

$$\int_0^a y|d^1\mathbf{x}|\begin{bmatrix} 1 \\ 2x \end{bmatrix}\,dx = \int_0^a x^2\sqrt{1+4x^2}\,dx \qquad (1)$$

$$\int_0^a |d^1\mathbf{x}|\begin{bmatrix} 1 \\ 2x \end{bmatrix}\,dx = \int_0^a \sqrt{1+4x^2}\,dx.$$

The first of these is easy, setting $4x^2 = u$, so $8x\,dx = du$. This leads to

$$\int_0^a x\sqrt{1+4x^2}\,dx = \frac{1}{12}\left((1+4a^2)^{3/2} - 1\right).$$

The third is a good bit harder. Substitute $2x = \tan\theta$, so that $2\,dx = d\theta/\cos^2\theta$. Our integral becomes

$$\int \sqrt{1+4x^2}\,dx = \frac{1}{2}\int \frac{1}{\cos^3\theta}\,d\theta = \frac{1}{2}\int \frac{\cos\theta}{(1-\sin^2\theta)^2}\,d\theta.$$

Now substitute $\sin\theta = t$, so that $\cos\theta\,d\theta = dt$, and the integral becomes

$$\frac{1}{2}\int \frac{dt}{(1-t^2)^2} = \frac{1}{8}\left(\int \frac{dt}{1-t} + \int \frac{dt}{1+t} + \int \frac{dt}{(1-t)^2} + \int \frac{dt}{(1+t)^2}\right),$$

using partial fractions. This is easy to integrate, giving

$$\frac{1}{4}\frac{t}{1-t^2} + \frac{1}{8}\ln\left|\frac{1+t}{1-t}\right|.$$

Substituting back leads to

$$\int_0^a \sqrt{1+4x^2}\,dx = \frac{1}{2}\left(a\sqrt{1+4a^2} + \frac{1}{2}\ln\left|2a + \sqrt{1+4a^2}\right|\right). \qquad (2)$$

The second integral is trickier yet. Begin by integrating by parts:

$$\underbrace{\int x^2\sqrt{1+4x^2}\,dx}_{A} = \int \frac{x}{8}\cdot 8x\sqrt{1+4x^2}\,dx = \frac{x}{12}(1+4x^2)^{3/2} - \frac{1}{12}\int (1+4x^2)\sqrt{1+4x^2}\,dx$$

$$= \frac{x}{12}(1+4x^2)^{3/2} - \frac{1}{12}\underbrace{\int \sqrt{1+4x^2}\,dx}_{\text{see equation (2)}} - \frac{4}{12}\underbrace{\int x^2\sqrt{1+4x^2}\,dx}_{A}.$$

Now insert the value computed in equation (2), and add the term on the far right to both sides of the equation, to get

$$\frac{4}{3}\int x^2\sqrt{1+4x^2}\,dx = \frac{x}{12}(1+4x^2)^{3/2} - \frac{1}{24}\left(x\sqrt{1+4x^2} + \frac{1}{2}\ln\left|2x + \sqrt{1+4x^2}\right|\right).$$

Finally, we find

$$\int_0^a x^2\sqrt{1+4x^2}\,dx = \frac{a}{16}(1+4a^2)^{3/2} - \frac{1}{32}\left(a\sqrt{1+4a^2} + \frac{1}{2}\ln\left|2a+\sqrt{1+4a^2}\right|\right).$$

We do not consider an ability to compute such complicated integrals an important mathematical skill. In general, it is far better to use MAPLE or similar software to carry out such computations.

This gives

$$\overline{x} = \frac{1}{6}\frac{\left((1+4a^2)^{3/2}-1\right)}{a\sqrt{1+4a^2}+\frac{1}{2}\ln\left|2a+\sqrt{1+4a^2}\right|}$$

and

$$\overline{y} = \frac{\frac{a}{8}(1+4a^2)^{3/2} - \frac{1}{16}\left(a\sqrt{1+4a^2}+\frac{1}{2}\ln\left|2a+\sqrt{1+4a^2}\right|\right)}{a\sqrt{1+4a^2}+\frac{1}{2}\ln\left|2a+\sqrt{1+4a^2}\right|}.$$

**5.3.19** To go from the third to the last line of equation 5.3.53, first we do an integration by parts, with

$$du = \sqrt{1+r^2}\,r\,dr \text{ and } v = r, \quad \text{so that} \quad u = \frac{(1+r^2)^{3/2}}{3} \text{ and } dv = dr.$$

This gives

$$4\pi\int_0^R\sqrt{1+r^2}\,r^2\,dr = \frac{4\pi}{3}\left[(1+r^2)^{3/2}r\right]_0^R - \frac{4\pi}{3}\int_0^R(1+r^2)^{3/2}\,dr$$

$$= \frac{4\pi}{3}R(1+R^2)^{3/2} - \frac{4\pi}{3}\left(\int_0^R\sqrt{1+r^2}+\sqrt{1+r^2}\,r^2\,dr\right).$$

Multiplying each side by $3/(4\pi)$ gives

$$3\int_0^R\sqrt{1+r^2}\,r^2\,dr = R(1+R^2)^{3/2} - \int_0^R\sqrt{1+r^2}\,dr - \int_0^R\sqrt{1+r^2}\,r^2\,dr,$$

and adding the last term on the right to both sides gives

$$4\int_0^R\sqrt{1+r^2}\,r^2\,dr = R(1+R^2)^{3/2} - \int_0^R\sqrt{1+r^2}\,dr. \tag{1}$$

Now we fiddle with the last term on the right side. To go from the first to the second line below, we set $\tan\theta = r$, so that $dr = \frac{d\theta}{\cos\theta}$, and $\sin\theta = \tan\theta\sqrt{\frac{1}{1+\tan^2\theta}} = r\sqrt{\frac{1}{1+r^2}}$:

$$-\int_0^R\sqrt{1+r^2}\,dr = -\int_0^R\frac{1+r^2}{\sqrt{1+r^2}}\,dr = -\overbrace{\int_0^R\frac{1}{\sqrt{1+r^2}}\,dr}^{\text{1st}} - \overbrace{\int_0^R\frac{r^2}{\sqrt{1+r^2}}\,dr}^{\text{2nd}}$$

$$= \underbrace{-\int_0^{\arctan R}\frac{\cos\theta}{\cos^2\theta}\,d\theta}_{\substack{\text{from 1st term in braces, using}\\\text{trig. as described above}}} \underbrace{-\left[r\sqrt{1+r^2}\right]_0^R + \int_0^R\sqrt{1+r^2}\,dr}_{\substack{\text{from 2nd term in braces,}\\\text{by integration by parts}}}$$

So, subtracting from both sides the last term on the second line, we get

$$-2\int_0^R \sqrt{1+r^2}\,dr = -\int_0^{\arctan R} \frac{\cos\theta}{\cos^2\theta}\,d\theta - R\sqrt{1+R^2}$$

$$= -\frac{1}{2}\int_0^{\arctan R} \frac{2\cos\theta}{1-\sin^2\theta}\,d\theta - R\sqrt{1+R^2}$$

$$= -\frac{1}{2}\int_0^{\arctan R} \left(\frac{\cos\theta}{1+\sin\theta} + \frac{\cos\theta}{1-\sin\theta}\right)d\theta - R\sqrt{1+R^2}$$

$$= -\frac{1}{2}\left[\ln\left(\frac{1+\sin\theta}{1-\sin\theta}\right)\right]_0^{\arctan R} - R\sqrt{1+R^2}$$

$$= -\frac{1}{2}\left[\ln\frac{\dfrac{r+\sqrt{1+r^2}}{\sqrt{1+r^2}}}{\dfrac{\sqrt{1+r^2}-r}{\sqrt{1+r^2}}}\right]_0^R - R\sqrt{1+R^2}$$

$$= -\frac{1}{2}\ln\frac{(R+\sqrt{1+R^2})^2}{1+R^2-R^2} - R\sqrt{1+R^2}$$

$$= -\ln(R+\sqrt{1+R^2}) - R\sqrt{1+R^2}.$$

Thus

$$-\int_0^R \sqrt{1+r^2}\,dr = -\frac{1}{2}\ln\left(R+\sqrt{1+R^2}\right) - \frac{R\sqrt{1+R^2}}{2}. \tag{2}$$

Now we multiply equation (1) by $\pi$, replacing the second term on the right side by its equivalent value in equation (2), to get

$$4\pi\int_0^R \sqrt{1+r^2}\,r^2\,dr = \pi\left(R(1+R^2)^{3/2} - \frac{1}{2}\ln\left(R+\sqrt{1+R^2}\right) - \frac{R\sqrt{1+R^2}}{2}\right).$$

**5.3.21** We will want to use equation 5.3.3 for computing the $k$-dimensional volume of a $k$-dimensional manifold in $\mathbb{R}^n$, in this case a 2-dimensional manifold in $\mathbb{C}^3$, which we can think of as being $\mathbb{R}^6$.

Let us write $z = s(\cos\theta + i\sin\theta)$, so that $s = |z|$. Then using de Moivre's formula

$$\big(s(\cos\theta + i\sin\theta)\big)^p = s^p(\cos p\theta + i\sin p\theta),$$

This uses equations 0.7.9 and 0.7.10, which give

$$z = |z|(\cos\theta + i\sin\theta).$$

we have

$$\gamma\binom{s}{\theta} = \begin{pmatrix} s^p\cos p\theta \\ s^p\sin p\theta \\ s^q\cos q\theta \\ s^q\sin q\theta \\ s^r\cos r\theta \\ s^r\sin r\theta \end{pmatrix} \quad\text{and}\quad \left[\mathbf{D}\gamma\binom{s}{\theta}\right] = \begin{bmatrix} ps^{p-1}\cos p\theta & -ps^p\sin p\theta \\ ps^{p-1}\sin p\theta & ps^p\cos p\theta \\ qs^{q-1}\cos q\theta & -qs^q\sin q\theta \\ qs^{q-1}\sin q\theta & qs^q\cos q\theta \\ rs^{r-1}\cos r\theta & -rs^r\sin r\theta \\ rs^{r-1}\sin r\theta & rs^r\cos r\theta \end{bmatrix}.$$

Computing $\left[\mathbf{D}\gamma\left(\begin{smallmatrix}s\\\theta\end{smallmatrix}\right)\right]^{\top}\left[\mathbf{D}\gamma\left(\begin{smallmatrix}s\\\theta\end{smallmatrix}\right)\right]$ and taking the determinant gives

$$\overbrace{\det\begin{bmatrix} p^2s^{2p-2}+q^2s^{2q-2}+r^2s^{2r-2} & 0 \\ 0 & p^2s^{2p}+q^2s^{2q}+r^2s^{2r} \end{bmatrix}}^{\left[\mathbf{D}\gamma\left(\begin{smallmatrix}s\\\theta\end{smallmatrix}\right)\right]^{\top}\left[\mathbf{D}\gamma\left(\begin{smallmatrix}s\\\theta\end{smallmatrix}\right)\right]}$$
$$= \left(p^2s^{2p-1}+q^2s^{2q-1}+r^2s^{2r-1}\right)^2,$$

so taking the square root of the determinant is easy. (This is the kind of thing that happens with complex numbers.) Thus if we call our surface $S$, equation 5.3.3 gives

$$\operatorname{vol}_k S = \int_{|z|\le 1} \sqrt{\det\left[\mathbf{D}\gamma\left(\begin{smallmatrix}s\\\theta\end{smallmatrix}\right)\right]^{\top}\left[\mathbf{D}\gamma\left(\begin{smallmatrix}s\\\theta\end{smallmatrix}\right)\right]}\,ds\,d\theta$$
$$= \int_0^{2\pi}\int_0^1 \left(p^2s^{2p-1}+q^2s^{2q-1}+r^2s^{2r-1}\right)\,ds\,d\theta$$
$$= 2\pi\left(\left[\frac{p^2s^{2p}}{2p}\right]_0^1 + \left[\frac{q^2s^{2q}}{2q}\right]_0^1 + \left[\frac{r^2s^{2r}}{2r}\right]_0^1\right)$$
$$= 2\pi\left(\frac{p}{2}+\frac{q}{2}+\frac{r}{2}\right) = \pi(p+q+r).$$

**5.4.1** a. The image of the hyperboloid under the Gauss map is the part of the sphere where $|\cos\varphi| \ge \dfrac{1}{\sqrt{1+a^2}}$. Set $\varphi_a = \arccos 1/\sqrt{1+a^2}$. Using spherical coordinates, we find the area of this region to be

$$\int_0^{2\pi}\int_{-\varphi_a}^{\varphi_a} \cos\varphi\,|d\varphi|\,|d\theta| = 2\pi(2\sin\varphi_a) = 4\pi\frac{a}{\sqrt{1+a^2}}.$$

FIGURE FOR SOLUTION 5.4.1, part a: At right, we see the branch of the hyperbola of equation

$$x^2 + a^2z^2 = 1$$

where $x > 0$, together with representative normals, and the same normals anchored at the origin. These normals fill the arc

$$\pi - \frac{a}{\sqrt{a^2+1}} < \varphi < \pi + \frac{a}{\sqrt{a^2+1}}.$$

The image of the hyperboloid under the Gauss map is obtained by rotating this arc around the $z$-axis.

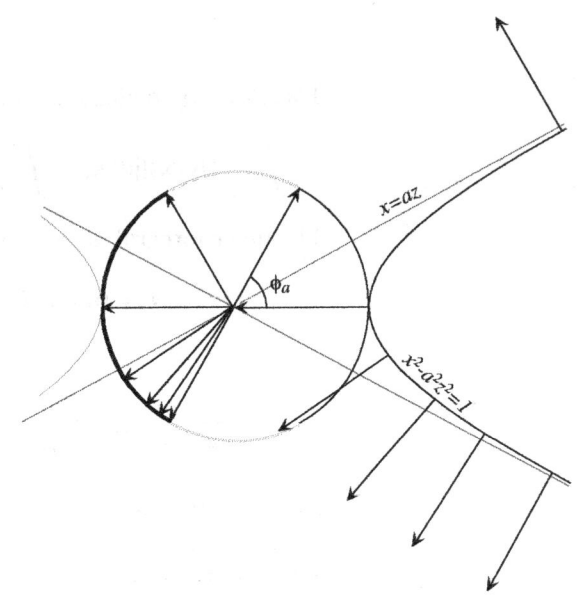

b. In this case, we are rotating the branch of the hyperbola of equation $a^2z^2 = x^2 + 1$ where $z > 0$ around the $z$ axis. The image of the Gauss map is also obtained by revolving the set of normals to the hyperbola around the $z$ axis. The normals (choosing the upward-pointing normal) satisfy $\arcsin \frac{a}{\sqrt{1+a^2}} \leq \varphi \leq \pi/2$. Thus the image of the Gauss map has area

$$\int_{\arcsin \frac{a}{\sqrt{1+a^2}}}^{\frac{\pi}{2}} \cos \varphi |d\varphi| \, |d\theta| = 2\pi \left[\sin \phi\right]_{\arcsin \frac{a}{\sqrt{1+a^2}}}^{\frac{\pi}{2}} = 2\pi \left(1 - \frac{a}{\sqrt{1+a^2}}\right).$$

**5.4.3** The map

$$\gamma : \begin{pmatrix} r \\ \theta \end{pmatrix} \mapsto \begin{pmatrix} -r \sin \theta \\ r \cos \theta \\ \theta \end{pmatrix}, \quad 0 \leq r \leq R, \ 0 \leq \theta < a,$$

parametrizes our surface: indeed

$$x \cos z + y \sin z = -r \sin \theta \cos \theta + r \sin \theta \cos \theta = 0,$$

so the image of $\gamma$ is indeed part of the surface, and moreover $\gamma$ is invertible:

$$\gamma^{-1} \begin{pmatrix} x \\ y \\ z \end{pmatrix} = \begin{pmatrix} \sqrt{x^2 + y^2} \\ z \end{pmatrix}.$$

We have

$$\det \left(\left[\mathbf{D}\gamma\begin{pmatrix} r \\ \theta \end{pmatrix}\right]\right)^{\top} \left(\left[\mathbf{D}\gamma\begin{pmatrix} r \\ \theta \end{pmatrix}\right]\right) = \det \begin{bmatrix} -\sin \theta & \cos \theta & 0 \\ -r \cos \theta & -r \sin \theta & 1 \end{bmatrix} \begin{bmatrix} -\sin \theta & -r \cos \theta \\ \cos \theta & -r \sin \theta \\ 0 & 1 \end{bmatrix} = 1 + r^2.$$

Moreover, we saw in example 3.8.12 that the curvature depends only on $r$, and we have

$$K(r) = \frac{-1}{(1+r^2)^2}.$$

Therefore, by definition 5.3.2,

$$\int_{X_{a,R}} |K(\mathbf{x})| |d^2\mathbf{x}| = \int_0^a \left(\int_0^R \frac{1}{(1+r^2)^2} \sqrt{1+r^2} dr\right) d\theta = \frac{aR}{\sqrt{1+R^2}}.$$

The inner integral is calculated by setting

$$r = \tan t, \ dr = \frac{dt}{\cos^2 t} \quad \text{and} \quad 1 + r^2 = \frac{1}{\cos^2 t},$$

leading to

$$\int_0^R \frac{dr}{(1+r^2)^{3/2}} = \int_0^{\arctan R} \frac{\cos^3 t \, dt}{\cos^2 t} = \left[\sin t\right]_0^{\arctan R} = \frac{R}{\sqrt{1+R^2}}.$$

**5.5.1** a. In base 3, the first digit after the decimal point corresponds to thirds, the second to ninths, the third to twenty-sevenths, and so on. So removing the open middle third of $[0, 1]$ removes all numbers greater than

Solution 5.5.1: Although the rational numbers 1/3, 2/3, 1/9, 2/9, 4/9, 5/9, 7/9, 8/9, 1/27,... are elements of $C$, there are elements of $C$ not in this list.

We could also use the same procedure as we used to show that the set of real numbers is uncountable (equation 0.6.2). Make any list of elements of $C$, for example

$$0.0000000200\ldots$$
$$0.2000000200\ldots$$
$$0.0000000200\ldots$$
$$0.0000200200\ldots$$
$$0.0000020200\ldots$$

$$\ldots$$

Now change every digit along the diagonal so that 2 becomes 0 and 0 becomes 2, and use those digits to form the number $222222\ldots$. This number cannot be the $n$th number on the list, because it doesn't have the same $n$th decimal.

1/3 and less than 2/3, which are all numbers that in base 3 start with 0.1. The number 1/3 itself remains, but it can be written $0.0222\ldots$.

At the next step, removing the open middle third of $[0, 1/3]$ corresponds to removing all $x$ such that $1/9 < x < 2/9$, which correspond to numbers that in base 3 start with 0.01. Again, the number 1/9 itself remains, but can be written $0.00222\ldots$. Similarly, removing the open middle third of $[2/3, 1]$ corresponds to removing all $x$ such that $7/9 < x < 8/9$, which corresponds to numbers that in base 3 start with 0.21. The number 7/9 remains, but it can be written $0.20222\ldots$.

Thus going one level deeper into the procedure corresponds to looking at one more digit after the decimal, when numbers are written in base 3. Removing the second third of $[0, 1]$ corresponds to removing numbers that begin with 0.1, and removing the second and eighth ninth corresponds to removing numbers that begin with 0.01 and 0.21 respectively.

We can show that $C$ is an uncountable set, i.e., that the elements of $C$ cannot be put in one-to-one correspondence with the positive integers, by noting that by changing every 2 to 1, any element of $C$ becomes a number written in base 2; conversely, any number written in base 2 can be turned into an element of $C$ by changing 1 to 2. Since any real number can be written in base 2, and the reals are uncountable (see section 0.6), the elements of $C$ are uncountable.

b. For $C$ to be pavable, its characteristic function $\mathbf{1}_C$ must be integrable (definition 4.1.17). Theorem 4.3.10 says that a bounded function with bounded support is integrable if it is continuous except on a set of volume 0. Certainly $\mathbf{1}_C$ is bounded, since $\mathbf{1}_C(x)$ can only equal 0 or 1, and it has bounded support, since its support is $[0, 1]$. Since $\mathbf{1}_C$ equals 0 everywhere except on $C$, and $C$ is closed, $\mathbf{1}_C(x)$ is continuous except on $C$, so if we can show that $C$ has length 0, we will have also shown that $C$ is pavable.

By proposition 4.1.22, we can show that $C$ has length 0 by showing that for every $\epsilon > 0$, we can put every element of $C$ in a 1-dimensional dyadic cube such that the total length of the cubes is $\leq \epsilon$. We will actually show that we can do this using non-dyadic cubes, and will not bother to spell out how we could convert that argument to dyadic cubes.

Consider the sequence $C_1, C_2, C_3, \ldots$, where $C_1$ is the interval $[0, 1]$ with the open middle third removed, $C_2$ is the interval with the middle third, second ninth, and eighth ninth removed, and so on. Thus $\cap_{j \to \infty} C_j = C$.[7] Further, $C_1$ has length 2/3, $C_2$ has length $4/9 = (2/3)^2$, etc. These sets include every element of $C$, and

$$\mathrm{vol}_1\, C = \lim_{j \to \infty} \mathrm{vol}_1\, C_j = \lim_{j \to \infty} \left(\frac{2}{3}\right)^j = 0.$$

---

[7]Rather than speak of the intersection of the $C_j$ as $j \to \infty$ it might seem more natural to think of this as the limit of the $C_j$ as $j \to \infty$, but we have not defined the limit of a sequence of sets.

The volume $\mathrm{vol}_k\, C$ is $3^k\, \mathrm{vol}_k\, A$, not $3\, \mathrm{vol}_k\, A$ because we are assuming that $C$ is a $k$-dimensional object, and multiplying one dimension of a $k$-dimensional object while leave other dimensions unchanged will not produce a self-similar object. To enlarge a square while keeping it square, one must multiply side and length by the same factor.

Here we use the formula

$$\ln(a^k) = k \ln a.$$

In addition to $\ln(a^k) = k \ln a$, equation (2) uses the formula

$$\ln(ab) = \ln a + \ln b.$$

c. To determine for what if any $k$ the set $C$ has $k$-dimensional volume that is not 0 or infinity, we can use the procedure of example 5.5.1. Note that $C$ consists of two equal parts, the first third and last third of $[0,1]$. Denote by $A$ the first third. Then $\mathrm{vol}_k\, C = 2\, \mathrm{vol}_k\, A$. But imagine that $A$ is a rubber band; if you stretch it by a factor of 3 so that it fills the interval $[0,1]$, it will be identical to $C$. Thus $\mathrm{vol}_k\, C = 3^k\, \mathrm{vol}_k\, A$.

The only value of $k$ (other than 0 or infinity) that satisfies

$$\mathrm{vol}_k\, C = 2\, \mathrm{vol}_k\, A$$
$$\mathrm{vol}_k\, C = 3^k\, \mathrm{vol}_k\, A$$

is $2 = 3^k$, i.e., $\ln 2 = k \ln 3$, or $k = \dfrac{\ln 2}{\ln 3} \approx 0.63093$.

You might want to compute the $\frac{\ln 2}{\ln 3}$-dimensional volume of $C$:

$$\mathrm{vol}_{\frac{\ln 2}{\ln 3}}\, C = \lim_{j \to \infty} 2^j \left(\frac{1}{3}\right)^{j \frac{\ln 2}{\ln 3}}. \tag{1}$$

The 2 gets exponent $j$ instead of $j\frac{\ln 2}{\ln 3}$ because it is the *number* of pieces contained in $C$: $C_1$ has two pieces, $C_2$ has four pieces, etc., so $C_j$ has $2^j$ pieces. Now take the log of each side of equation (1):

$$\ln\left(\mathrm{vol}_{\frac{\ln 2}{\ln 3}}\, C\right) = \ln\left(\lim_{j \to \infty} 2^j \left(\frac{1}{3}\right)^{j \frac{\ln 2}{\ln 3}}\right) \tag{2}$$

$$= \lim_{j \to \infty} j \ln 2 + j\frac{\ln 2}{\ln 3}(-\ln 3) = \lim_{j \to \infty} \left(j \ln 2 - j \ln 2\right) = 0,$$

so $\mathrm{vol}_{\frac{\ln 2}{\ln 3}}\, C = 1$.

**5.1** To find a parametrization for some region, we ask, "how should we describe where we are at some point in the region?" Using the trigonometric formulas discussed in "Note on parametrizations" (immediately before solution 5.2.1), we see that the map

$$f : u \mapsto \begin{pmatrix} x = R + r \cos u \\ y = 0 \\ z = r \sin u \end{pmatrix}$$

parametrizes the circle in the $(x, z)$-plane of radius $r$ centered at $\begin{pmatrix} R \\ 0 \\ 0 \end{pmatrix}$. Imagine that circle (the small circle at the top of the figure in the margin) traveling in a perfect circle around the $z$-axis, to form a torus. The shaded annulus at the bottom of the figure is the projection of this torus onto the $(x, y)$-plane.

As the point

$$\begin{pmatrix} R + r \cos u \\ r \sin u \end{pmatrix}$$

in the little circle turns by angle $v$ around the $z$ axis, its position is given by

$$\begin{pmatrix} (R + r \cos u) \cos v \\ (R + r \cos u) \sin v \\ r \sin u \end{pmatrix}.$$

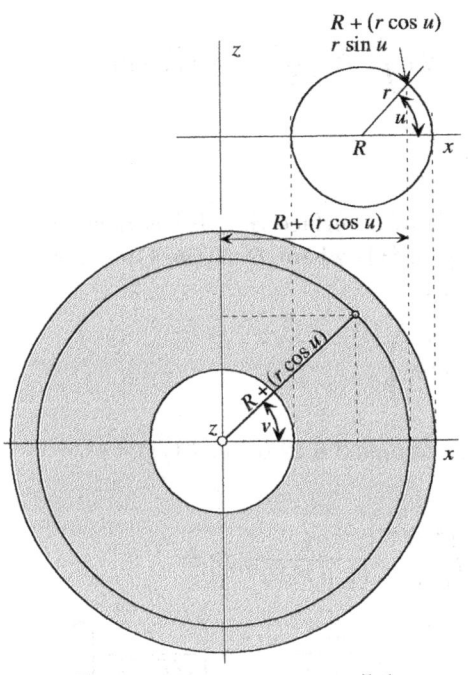

FIGURE FOR SOLUTION 5.1.

TOP: In the $(x, z)$-plane, we travel on a circle of radius $r$ centered at $R$. Polar coordinates describe where we are on that circle. As this small circle orbits the $z$-axis, it forms a torus. BOTTOM: The shaded annulus is the projection of that torus onto the $(x, y)$-plane.

**5.3** As we did in solution 5.3.1, we return to Cartesian coordinates, setting $x = r(t) \cos \theta(t)$ and $y = r(t) \sin \theta(t)$; since $r(t) = 1/t^\alpha$ and $\theta(t) = t$, this gives

$$\gamma(t) = \begin{pmatrix} x(t) \\ y(t) \end{pmatrix} = \begin{pmatrix} \frac{1}{t^\alpha} \cos t \\ \frac{1}{t^\alpha} \sin t \end{pmatrix}.$$

So the length of the spiral is

$$\int_1^\infty |\gamma'(t)|\, dt = \int_1^\infty \left| \begin{bmatrix} -t^{-\alpha} \sin t - \alpha t^{-\alpha-1} \cos t \\ t^{-\alpha} \cos t - \alpha t^{-\alpha-1} \sin t \end{bmatrix} \right| dt$$

$$= \int_1^\infty \sqrt{t^{-2\alpha} + \alpha^2 t^{-2\alpha-2}}\, dt$$

$$= \int_1^\infty \frac{1}{t^{\alpha+1}} \sqrt{\alpha^2 + t^2}\, dt.$$

For $t$ sufficiently large,

$$\sqrt{\alpha^2 + t^2} = t \sqrt{1 + \frac{\alpha^2}{t^2}} \quad \text{satisfies} \quad \frac{1}{2} t < t \sqrt{1 + \frac{\alpha^2}{t^2}} < 2t;$$

i.e., it is approximately $t$. Thus for $t$ sufficiently large, the integrand $\frac{1}{t^{\alpha+1}}\sqrt{a^2+t^2}$ behaves like $\frac{t}{t^{\alpha+1}} = \frac{1}{t^{\alpha}}$, so if $\alpha > 1$, the length is finite; if $\alpha \le 1$, it is not.

**5.5**  a.   The surface is parametrized by $\gamma : \begin{pmatrix} x \\ \theta \end{pmatrix} \mapsto \begin{pmatrix} x \\ f(x)\cos\theta \\ f(x)\sin\theta \end{pmatrix}$ (see equation 5.2.10).

b. The surface area is given by

$$\int_0^{2\pi}\int_a^b \underbrace{\left| \begin{bmatrix} 1 \\ f'(x)\cos\theta \\ f'(x)\sin\theta \end{bmatrix} \times \begin{bmatrix} 0 \\ -f(x)\sin\theta \\ f(x)\cos\theta \end{bmatrix} \right|}_{\sqrt{|[\mathbf{D}\gamma(\mathbf{u})]^\top[\mathbf{D}\gamma(\mathbf{u})]|}\ (\text{equation 5.3.35})}\, dx\, d\theta = 2\pi\int_a^b f(x)\sqrt{1+(f'(x))^2}\, dx.$$

**5.7** Recall that the total curvature $\mathbf{K}(S)$ of a surface $S \subset \mathbb{R}^3$ is defined in exercise 5.3.13 to be $\mathbf{K}(S) = \int_S |K(\mathbf{x})||d^2\mathbf{x}|$, where $K$ is the Gaussian curvature. The helicoid is parametrized by

$$\begin{pmatrix} r \\ z \end{pmatrix} \mapsto \begin{pmatrix} r\cos z \\ r\sin z \\ z \end{pmatrix}.$$

Its Gaussian curvature was computed in example 3.8.12 to be $-1/(1+r^2)^2$. Thus our total curvature is

$$\int_0^a\int_0^\infty \frac{1}{(1+r^2)^2}\left| \begin{bmatrix} \cos z \\ \sin z \\ 0 \end{bmatrix} \times \begin{bmatrix} -r\sin z \\ r\cos z \\ 1 \end{bmatrix} \right| |dr\, dz| = \int_0^a\int_0^\infty \frac{1}{(1+r^2)^{3/2}}\, dr\, dz$$

$$= a\int_0^\infty \frac{dr}{(1+r^2)^{3/2}} = a\left[\frac{r}{\sqrt{1+r^2}}\right]_0^\infty$$

$$= a.$$

**5.9**  a. In base 5, the number $1/5$ is written $0.1$, $2/5$ is written $0.2$, and so on, so removing the open middle fifth is equivalent to removing the numbers that start with $0.2$ in base 5, except for $0.2$ itself, which can be written as the limit of the sequence $0.04, 0.044, 0.0444\ldots$. Removing the open middle fifths of the four remaining fifths is equivalent to removing the numbers that can only be written beginning $0.02, 0.12, 0.32,$ or $0.42$, and so on. One way to show that $C$ is an uncountable set is to throw out all the numbers in $C$ containing the digits 3 or 4. What remains is all the real numbers, written in base 2. Since the reals are uncountable (see section 0.6), the elements of $C$ are uncountable.

b.   For $C$ to be pavable, its characteristic function $\mathbf{1}_C$ must be integrable (definition 4.1.17).  Theorem 4.3.10 says that a bounded function with bounded support is integrable if it is continuous except on a set of volume 0. Certainly $\mathbf{1}_C$ is bounded, since $\mathbf{1}_C(x)$ can only equal 0 or 1, and its support is $[0,1]$. Since $\mathbf{1}_C$ equals 0 everywhere except on $C$, and $C$ is closed, $\mathbf{1}_C(x)$ is continuous except on $C$, so if we can show that $C$ has length 0, we will have also shown that $C$ is pavable.

By proposition 4.1.22, we can do this by showing that for every $\epsilon > 0$, we can put every element of $C$ in a 1-dimensional dyadic cube such that the total length of the cubes is $\leq \epsilon$. We will actually show that we can do this using non-dyadic cubes, and will not bother to spell out how we could convert that argument to dyadic cubes.

Consider the sequence $C_1, C_2, C_3, \ldots$, where $C_1$ is the interval $[0,1]$ with the open middle fifth removed, $C_2$ is the interval at the next level, and so on. Thus $\cap_{j \to \infty} C_j = C$. Further, $C_1$ has length $4/5$, $C_2$ has length $16/25 = (4/5)^2$, etc. These sets include every element of $C$, and

$$\text{vol}_1\, C = \lim_{j \to \infty} \text{vol}_1\, C_j = \lim_{j \to \infty} \left(\frac{4}{5}\right)^j = 0.$$

c. Note that $C$ is made up of two identical segments, from 0 to 2/5 and from 3/5 to 1. But stretching the first segment by a factor of 5/2 would make it identical to $C$. So

$$\text{vol}_k\, C = 2\,\text{vol}_k\, A$$
$$\text{vol}_k\, C = (5/2)^k\,\text{vol}_k\, A.$$

Therefore,

Note that this dimension is a little bit bigger than the dimension of the triadic Cantor set in exercise 5.5.1, which was about .63; this set is a little bit "thicker," more like a line.

$$\left(\frac{5}{2}\right)^k = 2, \quad \text{i.e.,} \quad k = \frac{\log 2}{\log 5/2} = \frac{\log 2}{\log 5 - \log 2} \approx 0.756471.$$

**6.1.1 a.** On $\mathbb{R}^3$ there are three elementary 1-forms, $dx, dy,$ and $dz$; three elementary 2-forms, $dx \wedge dy$, $dx \wedge dz$, and $dy \wedge dz$; one elementary 0-form, the number 1; and one elementary 3-form, $dx \wedge dy \wedge dz$.

**b.** On $\mathbb{R}^4$ there are

1 elementary 0-form: the number 1.

4 elementary 1-forms: $dx_1, dx_2, dx_3,$ and $dx_4$.

6 elementary 2-forms: $dx_1 \wedge dx_2$, $dx_1 \wedge dx_3$, $dx_1 \wedge dx_4$, $dx_2 \wedge dx_3$, $dx_2 \wedge dx_4$, and $dx_3 \wedge dx_4$.

4 elementary 3-forms: $dx_1 \wedge dx_2 \wedge dx_3$, $dx_1 \wedge dx_2 \wedge dx_4$, $dx_1 \wedge dx_3 \wedge dx_4$, and $dx_2 \wedge dx_3 \wedge dx_4$.

1 elementary 4-form: $dx_1 \wedge dx_2 \wedge dx_3 \wedge dx_4$.

**c.** Since $\binom{m}{k} = \binom{5}{4} = \frac{5!}{4!1!} = 5$, there are five elementary 4-forms on $\mathbb{R}^5$:

$$dx_1 \wedge dx_2 \wedge dx_3 \wedge dx_4 \qquad dx_1 \wedge dx_2 \wedge dx_3 \wedge dx_5 \qquad dx_1 \wedge dx_2 \wedge dx_4 \wedge dx_5$$
$$dx_1 \wedge dx_3 \wedge dx_4 \wedge dx_5 \qquad dx_2 \wedge dx_3 \wedge dx_4 \wedge dx_5$$

**6.1.3 a.** $\det \begin{bmatrix} 1 & 1 \\ 2 & 2 \end{bmatrix} = 0$

**b.** $\det \begin{bmatrix} 1 & -2 \\ 0 & 1 \end{bmatrix} + 2 \det \begin{bmatrix} 0 & 1 \\ 1 & 0 \end{bmatrix} = 1 + 2 \cdot (-1) = -1.$

**c.** $\det \begin{bmatrix} 2 & 2 \\ 0 & -3 \end{bmatrix} = -6.$

**d.** The answer is 0; no need to compute $\det \begin{bmatrix} 1 & -2 & 2 \\ 3 & 1 & 2 \\ 3 & 1 & 2 \end{bmatrix} = 0$; exchanging the two $dx_2$ changes the sign of the answer while leaving it unchanged.

**6.1.5 a.** The forms (ii) and (iii) could be forms on $\mathbb{R}^3$; the forms (ii), (iii), and (iv) could be forms on $\mathbb{R}^4$. The forms (ii) and (iii) are elementary forms; the others are not because the 1-forms are not in ascending order.

**b.** The $y$-component is $-3$.

**c.** The $x_2, x_4$-component of signed volume is

$$dx_2 \wedge dx_4 \begin{bmatrix} 2 & -2 \\ 1 & 1 \\ 0 & 2 \\ 4 & -3 \end{bmatrix} = \det \begin{bmatrix} 1 & 1 \\ 4 & -3 \end{bmatrix} = -7.$$

**6.1.7 a.** On $\mathbb{R}^4$ there are

1 elementary 0-form, since $\begin{pmatrix} 4 \\ 0 \end{pmatrix} = \dfrac{4!}{0!4!} = 1;$    4 elementary 1-forms, since $\begin{pmatrix} 4 \\ 1 \end{pmatrix} = \dfrac{4!}{1!3!} = 4;$

6 elementary 2-forms, since $\begin{pmatrix} 4 \\ 2 \end{pmatrix} = \dfrac{4!}{2!2!} = 6;$    4 elementary 3-forms, since $\begin{pmatrix} 4 \\ 3 \end{pmatrix} = \dfrac{4!}{3!1!} = 4;$

1 elementary 4-form, since $\begin{pmatrix} 4 \\ 4 \end{pmatrix} = \dfrac{4!}{4!0!} = 1.$

**b.** On $\mathbb{R}^5$ there are

5 elementary 1-forms, since $\begin{pmatrix} 5 \\ 1 \end{pmatrix} = \dfrac{5!}{1!4!} = 5$

5 elementary 4-forms, since $\begin{pmatrix} 5 \\ 4 \end{pmatrix} = \dfrac{5!}{4!1!} = 5$

10 elementary 2-forms, since $\begin{pmatrix} 5 \\ 2 \end{pmatrix} = \dfrac{5!}{2!3!} = 10$

10 elementary 3-forms, since $\begin{pmatrix} 5 \\ 3 \end{pmatrix} = \dfrac{5!}{3!2!} = 10$

**c.** If $E$ is a vector space of dimension 7, then 35 elementary 3-forms form a basis of $A^3(E)$, since $\begin{pmatrix} 7 \\ 3 \end{pmatrix} = \dfrac{7!}{3!4!} = 35.$

**d.** True, since $\begin{pmatrix} 3 \\ 3 \end{pmatrix} = \dfrac{3!}{3!0!} = 1.$

**6.1.9 a.** $\sin(x_4) \det \begin{bmatrix} 3 & -1 \\ 2x_1 & x_4 \end{bmatrix} = \sin x_4 (3x_4 + 2x_1)$

**b.** $e^x \det[2] = 2e^x$    **c.** $(-x_1)^2 e^{x_2} \det \begin{bmatrix} 3 & -1 & -1 \\ 2 & 1 & -1 \\ 1 & 0 & 1 \end{bmatrix} = 7x_1^2 e^{x_2}$

**6.1.11** The 2-form $\varphi$ on $\mathbb{R}^3$ is a linear combination of the three elementary 2-forms on $\mathbb{R}^3$:

$$\varphi = \alpha_{1,2} dx_1 \wedge dx_2 + \alpha_{1,3} dx_1 \wedge dx_3 + \alpha_{2,3} dx_2 \wedge dx_3.$$

To find the coefficients, we evaluate $\varphi$ on the standard basis vectors; see theorem 6.1.8, part b. Set $\vec{a} = \begin{bmatrix} a_1 \\ a_2 \\ a_3 \end{bmatrix}$. Then

$$\alpha_{1,2} = \varphi(\vec{e}_1, \vec{e}_2) = \det \begin{bmatrix} a_1 & 1 & 0 \\ a_2 & 0 & 1 \\ a_3 & 0 & 0 \end{bmatrix} = a_3$$

$$\alpha_{1,3} = \varphi(\vec{e}_1, \vec{e}_3) = \det \begin{bmatrix} a_1 & 1 & 0 \\ a_2 & 0 & 0 \\ a_3 & 0 & 1 \end{bmatrix} = -a_2$$

$$\alpha_{2,3} = \varphi(\vec{e}_2, \vec{e}_3) = \det \begin{bmatrix} a_1 & 0 & 0 \\ a_2 & 1 & 0 \\ a_3 & 0 & 1 \end{bmatrix} = a_1$$

So
$$\varphi = a_3 dx_1 \wedge dx_2 - a_2 dx_1 \wedge dx_3 + a_1 dx_2 \wedge dx_3.$$

**6.1.13** If $\varphi$ and $\psi$ are 2-forms, then
$$\begin{aligned}
\varphi \wedge \psi(\vec{\mathbf{v}}_1, \vec{\mathbf{v}}_2, \vec{\mathbf{v}}_3, \vec{\mathbf{v}}_4) &= \varphi(\vec{\mathbf{v}}_1, \vec{\mathbf{v}}_2)\psi(\vec{\mathbf{v}}_3, \vec{\mathbf{v}}_4) - \varphi(\vec{\mathbf{v}}_1, \vec{\mathbf{v}}_3)\psi(\vec{\mathbf{v}}_2, \vec{\mathbf{v}}_4) \\
&\quad + \varphi(\vec{\mathbf{v}}_1, \vec{\mathbf{v}}_4)\psi(\vec{\mathbf{v}}_2, \vec{\mathbf{v}}_3) + \varphi(\vec{\mathbf{v}}_2, \vec{\mathbf{v}}_3)\psi(\vec{\mathbf{v}}_1, \vec{\mathbf{v}}_4) \\
&\quad - \varphi(\vec{\mathbf{v}}_2, \vec{\mathbf{v}}_4)\psi(\vec{\mathbf{v}}_1, \vec{\mathbf{v}}_3) + \varphi(\vec{\mathbf{v}}_3, \vec{\mathbf{v}}_4)\psi(\vec{\mathbf{v}}_1, \vec{\mathbf{v}}_2).
\end{aligned}$$

**6.1.15** In example 6.1.20, $\varphi$ and $\psi$ are 1-forms. Then
$$\varphi \wedge \psi(\vec{\mathbf{v}}_1, \vec{\mathbf{v}}_2) = \varphi(\vec{\mathbf{v}}_1)\psi(\vec{\mathbf{v}}_2) - \varphi(\vec{\mathbf{v}}_2)\psi(\vec{\mathbf{v}}_1) = -\varphi \wedge \psi(\vec{\mathbf{v}}_2, \vec{\mathbf{v}}_1).$$

In example 6.1.21, $\varphi$ is a 2-form, and $\psi$ is still a 1-form. Recall from the example that in this case there are three shuffles, two with positive signature and one with negative signature. This gives
$$\psi \wedge \varphi(\vec{\mathbf{v}}_1, \vec{\mathbf{v}}_2, \vec{\mathbf{v}}_3) = \psi(\vec{\mathbf{v}}_1)\varphi(\vec{\mathbf{v}}_2, \vec{\mathbf{v}}_3) - \psi(\vec{\mathbf{v}}_2)\varphi(\vec{\mathbf{v}}_1, \vec{\mathbf{v}}_3) + \psi(\vec{\mathbf{v}}_3)\varphi(\vec{\mathbf{v}}_1, \vec{\mathbf{v}}_2),$$

whereas
$$\varphi \wedge \psi(\vec{\mathbf{v}}_1, \vec{\mathbf{v}}_2, \vec{\mathbf{v}}_3) = \varphi(\vec{\mathbf{v}}_1, \vec{\mathbf{v}}_2)\psi(\vec{\mathbf{v}}_3) - \varphi(\vec{\mathbf{v}}_1, \vec{\mathbf{v}}_3)\psi(\vec{\mathbf{v}}_2) + \varphi(\vec{\mathbf{v}}_2, \vec{\mathbf{v}}_3)\psi(\vec{\mathbf{v}}_1),$$

and indeed these are equal.

Solution 6.1.17: Recall from section 4.8 that $\mathrm{Perm}_n$ is the set of permutations of the set $\{1, \ldots, n\}$. Thus $\mathrm{Perm}_{(k+l)}$ is the set of permutations of the set $\{1, \ldots, n\}$, where $n = k + l$.

**6.1.17** 1. *Distributivity.* This is just busy work. Let $S(k, l) \subset \mathrm{Perm}_{(k+l)}$ be the set of $(k, l)$-shuffles, i.e., the set of permutations $\sigma$ of $1, \ldots, k+l$ such that
$$\sigma(1) < \cdots < \sigma(k), \quad \sigma(k+1) < \cdots < \sigma(k+l).$$

If $\varphi \in A^k(E)$, $\psi_1, \psi_2 \in A^l(E)$, then
$$\varphi \wedge (\psi_1 + \psi_2)(\mathbf{v}_1, \ldots, \mathbf{v}_{k+l}) = \sum_{\sigma \in S(k,l)} \mathrm{sgn}(\sigma)\varphi(\mathbf{v}_{\sigma(1)}, \ldots, \mathbf{v}_{\sigma(k)})(\psi_1 + \psi_2)(\mathbf{v}_{\sigma(k+1)}, \ldots, \mathbf{v}_{\sigma(k+l)})$$

$$= \sum_{\sigma \in S(k,l)} \mathrm{sgn}(\sigma)\varphi(\mathbf{v}_{\sigma(1)}, \ldots, \mathbf{v}_{\sigma(k)})\Big(\psi_1(\mathbf{v}_{\sigma(k+1)}, \ldots, \mathbf{v}_{\sigma(k+l)}) + \psi_2(\mathbf{v}_{\sigma(k+1)}, \ldots, \mathbf{v}_{\sigma(k+l)})\Big)$$

$$= \sum_{\sigma \in S(k,l)} \mathrm{sgn}(\sigma)\varphi(\mathbf{v}_{\sigma(1)}, \ldots, \mathbf{v}_{\sigma(k)})\psi_1(\mathbf{v}_{\sigma(k+1)}, \ldots, \mathbf{v}_{\sigma(k+l)})$$

$$\qquad + \sum_{\sigma \in S(k,l)} \mathrm{sgn}(\sigma)\varphi(\mathbf{v}_{\sigma(1)}, \ldots, \mathbf{v}_{\sigma(k)})\psi_2(\mathbf{v}_{\sigma(k+1)}, \ldots, \mathbf{v}_{\sigma(k+l)})$$

$$= (\varphi \wedge \psi_1)(\mathbf{v}_1, \ldots, \mathbf{v}_{k+l}) + (\varphi \wedge \psi_2)(\mathbf{v}_1, \ldots, \mathbf{v}_{k+l}).$$

2. *Associativity.* The proof of associativity is not quite so easy. Let us introduce the notation $S(k_1, k_2, k_3) \subset \mathrm{Perm}_{(k_1+k_2+k_3)}$ for the triple shuffles, i.e., permutations $\sigma \in \mathrm{Perm}_{(k_1+k_2+k_3)}$ such that $\sigma(i) < \sigma(j)$ whenever
$$1 \le i < j \le k_1, \quad \text{or} \quad k_1 \le i < j \le k_2, \quad \text{or} \quad k_2 \le i < j \le k_3.$$

Imagine you have a pack of 39 cards, say the spades, hearts, and diamonds of a standard deck, originally arranged with first the diamonds, then the hearts, then the spades, each arranged from lowest to highest. Then the set of shuffles $S(13, 13, 13)$ consists of the arrangements of the deck such that a lower spade comes sooner than a higher spade, and similarly for the hearts and diamonds, but nothing is said about the position of a heart relative to a spade.

The key to associativity is the observation that any such arrangement can be achieved uniquely by first shuffling together the hearts and spades, and then shuffling that pack with the diamonds, or equivalently, first shuffling the diamonds and hearts, and then shuffling that pack with the pack of spades.

To help translate this into mathematics, let us further invent the notation

$$S(\hat{k}_1, k_2, k_3) \subset S(k_1, k_2, k_3)$$

to denote the subset of the triple shuffles $\sigma$ such that $\sigma(i) = i$ when $i \leq k_1$, and similarly $S(k_1, k_2, \hat{k}_3) \subset S(k_1, k_2, k_3)$ the triple shuffles such that $\sigma(i) = i$ when $i > k_2$, like the shuffles of the diamonds and hearts which don't affect the spades, and the shuffles of hearts and spades that don't affect the diamonds.

The principle above can be restated in mathematical terms:

**Lemma 1.** *Every element $\sigma \in S(k_1, k_2, k_3)$ can be written uniquely as*

$$\sigma = \tau_2 \circ \tau_1$$

*with $\tau_1 \in S(\hat{k}_1, k_2, k_3)$ and $\tau_2 \in S(k_1, k_2 + k_3)$, and also as*

$$\sigma = \tau'_2 \circ \tau'_1$$

*with $\tau'_1 \in S(k_1, k_2, \hat{k}_3)$ and $\tau_2 \in S(k_1 + k_2, k_3)$.*

Using this notation and lemma 1, associativity is the following computation:

$$\varphi_1 \wedge (\varphi_2 \wedge \varphi_3)(\mathbf{v}_1, \ldots, \mathbf{v}_{k_1+k_2+k_3})$$

$$= \sum_{\sigma \in S(k_1, k_2+k_3)} \mathrm{sgn}(\sigma) \varphi_1(\mathbf{v}_{\sigma(1)}, \ldots, \mathbf{v}_{\sigma(k_1)})$$

$$\left( \sum_{\tau \in S(\hat{k}_1, k_2, k_3)} \mathrm{sgn}(\tau) \Big( \varphi_2(\mathbf{v}_{\sigma(\tau(k_1+1))}, \ldots, \mathbf{v}_{\sigma(\tau(k_1+k_2))}) \Big) \Big( \varphi_3(\mathbf{v}_{\sigma(\tau(k_1+k_2+1))}, \ldots, \mathbf{v}_{\sigma(\tau(k_1+k_2+k_3))}) \Big) \right)$$

$$= \sum_{\sigma \in S(k_1, k_2+k_3)} \sum_{\tau \in S(\hat{k}_1, k_2, k_3)} \mathrm{sgn}(\tau) \mathrm{sgn}(\sigma) \varphi_1(\mathbf{v}_{\sigma(\tau(1))}, \ldots, \mathbf{v}_{\sigma(\tau(k_1))})$$

$$\left( \Big( \varphi_2(\mathbf{v}_{\sigma(\tau(k_1+1))}, \ldots, \mathbf{v}_{\sigma(\tau(k_1+k_2))}) \Big) \Big( \varphi_3(\mathbf{v}_{\sigma(\tau(k_1+k_2+1))}, \ldots, \mathbf{v}_{\sigma(\tau(k_1+k_2+k_3))}) \Big) \right)$$

$$= \sum_{\alpha \in S(k_1, k_2, k_3)} \mathrm{sgn}(\alpha) \varphi_1(\mathbf{v}_{\alpha(1)}, \ldots, \mathbf{v}_{\alpha(k_1)}) \varphi_2(\mathbf{v}_{\alpha(k_1+1)}, \ldots, \mathbf{v}_{\alpha(k_1+k_2)}) \varphi_3(\mathbf{v}_{\alpha(k_1+k_2+1)}, \ldots, \mathbf{v}_{\alpha(k_1+k_2+k_3)})$$

The second equality uses the fact that $\sigma(\tau(i)) = \sigma(i)$ when $1 \le i \le k_1$, since $\tau \in S(\hat{k}_1, k_2, k_3)$. The third uses lemma 1, i.e., that the $\sigma \circ \tau$ are exactly running through the elements of $S(k_1, k_2, k_3)$, and the fact that $\operatorname{sgn}(\sigma)\operatorname{sgn}(\tau) = \operatorname{sgn}(\sigma \circ \tau)$.

An exactly parallel development shows that

$$\big((\varphi_1 \wedge \varphi_2) \wedge \varphi_3\big)(\mathbf{v}_1, \ldots, \mathbf{v}_{k_1+k_2+k_3})$$

is also the same sum over the triple shuffles. This completes the discussion of associativity.

3. *Skew commutativity.* For skew commutativity, we want again to look carefully at permutations. Let us define $\alpha \in \operatorname{Perm}_{(k_1+k_2)}$ to be the permutation that puts the last indices at the beginning and the first indices at the end, i.e.,

$$\alpha(i) = \begin{cases} i + k_1 & \text{if } i \le k_2 \\ i - k_2 & \text{if } i > k_2. \end{cases}$$

Then every permutation in $\tau \in S(k_2, k_1)$ can be written uniquely $\tau = \sigma \circ \alpha$ with $\sigma \in S(k_1, k_2)$.

Using this, we see that

$$(\varphi \wedge \psi)(\mathbf{v}_1, \ldots, \mathbf{v}_{k_1+k_2}) = \sum_{\sigma \in S(k_1,k_2)} \operatorname{sgn}(\sigma)\varphi(\mathbf{v}_{\sigma(1)}, \ldots, \mathbf{v}_{\sigma(k_1)})\psi(\mathbf{v}_{\sigma(k_1+1)}, \ldots, \mathbf{v}_{\sigma(k_1+k_2)}),$$

whereas

$$(\psi \wedge \varphi)(\mathbf{v}_1, \ldots, \mathbf{v}_{k_1+k_2}) = \sum_{\tau \in S(k_2,k_1)} \psi(\mathbf{v}_{\tau(1)}, \ldots, \mathbf{v}_{\tau(k_2)})\varphi(\mathbf{v}_{\tau(k_2+1)}, \ldots, \mathbf{v}_{\tau(k_1+k_2)})$$

$$= \sum_{\sigma \in S(k_1,k_2)} \operatorname{sgn}(\sigma \circ \alpha)\psi(\mathbf{v}_{\sigma\circ\alpha(1)}, \ldots, \mathbf{v}_{\sigma\circ\alpha(k_2)})\psi(\mathbf{v}_{\sigma\circ\alpha(k_2+1)}, \ldots, \mathbf{v}_{\sigma\circ\alpha(k_1+k_2)})$$

$$= \operatorname{sgn}(\alpha) \sum_{\sigma \in S(k_1,k_2)} \operatorname{sgn}(\sigma)\psi(\mathbf{v}_{\sigma(k_1+1)}, \ldots, \mathbf{v}_{\sigma(k_1+k_2)})\psi(\mathbf{v}_{\sigma(1)}, \ldots, \mathbf{v}_{\sigma(k_1)}).$$

So we need to show that $\operatorname{sgn}(\alpha) = (-1)^{k_1 k_2}$. This can be done by counting transpositions: imagine switching $k_1 + 1$ first with $k_1$, then with $k_1 - 1$, and so forth, until it arrives in the first position. We will have made $k_1$ transpositions. Now move $k_1 + 2$ into second position; this will require another $k_1$ transpositions. Now move $k_1 + 3$ into third position, $\ldots$, and finally $k_1 + k_2$ into position $k_2$. This will realize $\alpha$ as $k_2 k_1$ transpositions.

**6.2.1 a.**

$$\int_{\gamma(I)} x\,dy + y\,dz = \int_I (x\,dy + y\,dz)\left(P_{\begin{pmatrix}\sin t\\\cos t\\t\end{pmatrix}}\left(\begin{bmatrix}\cos t\\-\sin t\\1\end{bmatrix}\right)\right)\,dt$$

$$= \int_{-1}^1 (-(\sin t)^2 + \cos t)\,dt = \int_{-1}^1 \left(\frac{1}{2}\cos 2t - \frac{1}{2} + \cos t\right)dt$$

To get the second equality in the second line, we use

$$\cos 2t = \cos^2 t - \sin^2 t$$
$$= 2\cos^2 t - 1$$
$$= 1 - 2\sin^2 t.$$

$$= \left[\frac{\sin 2t}{4} - \frac{1}{2}t + \sin t\right]_{-1}^1 = \frac{2\sin 2}{4} - 1 + 2\sin 1$$

$$= \frac{\sin 2}{2} + 2\sin 1 - 1.$$

In the second line, the variable $x$ is replaced by $\sin t$, and $y$ is replaced by $\cos t$, since the parallelogram is anchored at $\begin{pmatrix}\sin t\\\cos t\\t\end{pmatrix}$.

b. To get from the last equality below, we used MAPLE. It is possible to compute the integral by hand, but it requires a lot of computing, the kind of thing that computers are more reliable at than people.

$$\int_{\gamma(U)} x_1\,dx_2 \wedge dx_3 + x_2\,dx_3 \wedge dx_4$$

$$= \int_U x_1\,dx_2 \wedge dx_3 + x_2\,dx_3 \wedge dx_4\left(P_{\begin{pmatrix}uv\\u^2+v^2\\u-v\\\ln(u+v+1)\end{pmatrix}}\left(\begin{bmatrix}v\\2u\\1\\1/(u+v+1)\end{bmatrix},\begin{bmatrix}u\\2v\\-1\\1/(u+v+1)\end{bmatrix}\right)\right)\,du\,dv$$

$$= \int_U \left(uv\det\begin{bmatrix}2u & 2v\\1 & -1\end{bmatrix} + (u^2+v^2)\det\begin{bmatrix}1 & -1\\1/(u+v+1) & 1/(u+v+1)\end{bmatrix}\right)|du\,dv|$$

$$= \int_U \left(-uv(2u+2v) + 2\frac{u^2+v^2}{1+u+v}\right)du\,dv$$

$$= \int_0^2\left(\int_0^{2-u}(-2u^2v - 2uv^2)\,dv\right)du + 2\int_0^2\int_0^{2-u}\left(\frac{u^2+v^2}{1+u+v}\,dv\right)du = \frac{64}{45} + \frac{4}{3}\ln 3$$

**6.2.3**

a. $$\int_{\gamma(U)}(x_1 + x_4)\,dx_2 \wedge dx_3 = \int_U (e^u + \sin v)\det\begin{bmatrix}0 & -e^{-v}\\-\sin u & 0\end{bmatrix}|du\,dv|$$

$$= -\int_0^1\int_{-u}^u (e^u + \sin v)(e^v \sin u)\,dv\,du$$

b. $\displaystyle\int_U (u-v)(w-v)\,dx_1 \wedge dx_3 \wedge dx_4 \left( \begin{bmatrix} 1 \\ 1 \\ 0 \\ 0 \end{bmatrix} \begin{bmatrix} 1 \\ -1 \\ 1 \\ -1 \end{bmatrix} \begin{bmatrix} 0 \\ 0 \\ 1 \\ 1 \end{bmatrix} \right) |du\,dv\,dw|$

$$= \int_U (u-v)(w-v) \det \begin{bmatrix} 1 & 1 & 0 \\ 0 & 1 & 1 \\ 0 & -1 & 1 \end{bmatrix} |du\,dv\,dw|$$

$$= \int_U (u-v)(w-v)2\,|du\,dv\,dw|$$

$$= \int_0^1 \left( \int_{-(1-w)}^{1-w} \left( \int_{-\sqrt{(w-1)^2-v^2}}^{\sqrt{(w-1)^2-v^2}} 2(u-v)(w-v)\,du \right) dv \right) dw.$$

**Solution 6.2.3:** Actually, the problem only asks you to go to the next-to-last line of both computations; in part b, the final step is better done passing to cylindrical coordinates.

**6.2.5** We will identify a point in $\mathbb{C}^3$ with the point $\begin{pmatrix} x_1 \\ y_1 \\ x_2 \\ y_2 \\ x_3 \\ y_3 \end{pmatrix} \in \mathbb{R}^6$. Then

the point $\begin{pmatrix} z \\ z^2 \\ z^3 \end{pmatrix} \in \mathbb{C}^3$ is identified to $\begin{pmatrix} x \\ y \\ x^2 - y^2 \\ 2xy \\ x^3 - 3xy^2 \\ 3x^2y - y^3 \end{pmatrix}$, since the real and

imaginary parts of $z^2$ are $x^2 - y^2$ and $2xy$ respectively, and the real and imaginary parts of $z^3$ are $x^3 - 3xy^2$ and $3x^2y - y^3$ respectively. Then

$$\vec{D_x\gamma}(z) = \begin{bmatrix} 1 \\ 0 \\ 2x \\ 2y \\ 3x^2 - 3y^2 \\ 6xy \end{bmatrix} \quad \text{and} \quad \vec{D_y\gamma}(z) = \begin{bmatrix} 0 \\ 1 \\ -2y \\ 2x \\ -6xy \\ 3x^2 - 3y^2 \end{bmatrix}.$$

If we denote $dx_1 \wedge dy_1 + dx_2 \wedge dy_2 + dx_3 \wedge dy_3$ by $\varphi$, definition 6.2.1 then gives

$$\int_{[\gamma(S)]} \varphi = \int_S \varphi \left( P_{\gamma(z)}(\vec{D_x\gamma}(z), \vec{D_y\gamma}(z)) \right) |dx|\,|dy|$$

$$= \int_S \det \begin{bmatrix} 1 & 0 \\ 0 & 1 \end{bmatrix} + \det \begin{bmatrix} 2x & -2y \\ 2y & 2x \end{bmatrix} + \det \begin{bmatrix} 3x^2 - 3y^2 & -6xy \\ 6xy & 3x^2 - 3y^2 \end{bmatrix} |dx|\,|dy|$$

$$= \int_S (1 + 4x^2 + 4y^2 + 9x^4 + 9y^4 + 18x^2y^2)|dx|\,|dy|.$$

Since $S$ is a disc, we now pass to polar coordinates. This gives

$$\int_S \left(1+4(x^2+y^2)+9(x^2+y^2)^2\right)|dx|\,|dy|$$

$$= \int_0^{2\pi} \left(\int_0^1 \left(1 + 4r^2(\cos^2\theta + \sin^2\theta) + 9r^4(\cos^2\theta + \sin^2\theta)\right)r\,dr\right)d\theta$$

In going from the second to third lines, remember the $r$ in $r\,dr$.

$$= \int_0^{2\pi} \left[\frac{r^2}{2} + r^4 + \frac{9r^6}{6}\right]_0^1 d\theta$$

$$= \int_0^{2\pi} \left(\frac{1}{2} + 1 + \frac{9}{6}\right)d\theta = 6\pi.$$

Solution 6.3.1: Recall that definition 6.3.1 requires the orienting form to be nonzero.

**6.3.1** It does not define an orientation because any vector on the line can be written $\begin{bmatrix} a \\ -a \end{bmatrix}$, and $(dx+dy)\begin{bmatrix} a \\ -a \end{bmatrix} = 0$, so $dx + dy$ corresponds to $0 \in A^1(V)$, where $V$ denotes the line.

**6.3.3** On the unit circle, $dx + dy$ vanishes at $\begin{pmatrix} 1/\sqrt{2} \\ 1/\sqrt{2} \end{pmatrix}$ and $\begin{pmatrix} -1/\sqrt{2} \\ -1/\sqrt{2} \end{pmatrix}$.

Solution 6.3.5: $\vec{v}, \vec{w}$ are vectors tangent to the surface $P$ – in fact, in $P$, since $P$ is a plane.

**6.3.5** Treating the plane $P$ as a surface, we have

$$\omega_\mathbf{x}(\vec{v}, \vec{w}) = \det\begin{bmatrix} 1 & v_1 & w_1 \\ 1 & v_2 & w_2 \\ 1 & v_3 & w_3 \end{bmatrix},$$

so $\omega = dy \wedge dz - dx \wedge dz + dx \wedge dy$. But as we saw in example 6.1.14, when these forms are restricted to the plane, $[dx\wedge dy]_P = [dy\wedge dz]_P = -[dx\wedge dz]_P$, so $\omega = 3\,dx \wedge dy$. Of course, its positive multiple $dx \wedge dy$ gives the same orientation.

**6.3.7 a.** The basis $\vec{v}_1, \vec{v}_2$ is direct for $V$ oriented by $dx \wedge dz$ since

$$dx \wedge dz \left(\begin{bmatrix} 1 \\ 0 \\ 1 \end{bmatrix}, \begin{bmatrix} 0 \\ 1 \\ 2 \end{bmatrix}\right) = \det\begin{bmatrix} 1 & 0 \\ 1 & 2 \end{bmatrix} = 2.$$

We could do part c by direct computation: Since

$$\vec{w}_1 = 2\vec{v}_1 - 3\vec{v}_2$$

and

$$\vec{w}_2 = \vec{v}_1 + 2\vec{v}_2,$$

the change of basis matrix is

$$\begin{bmatrix} 2 & 1 \\ -3 & 2 \end{bmatrix},$$

with determinant $+7$. But this is unnecessary work.

**b.** The basis $\vec{w}_1, \vec{w}_2$ is also direct since

$$dx \wedge dz \left(\begin{bmatrix} 2 \\ -3 \\ -4 \end{bmatrix}, \begin{bmatrix} 1 \\ 2 \\ 5 \end{bmatrix},\right) = \det\begin{bmatrix} 2 & 1 \\ -4 & 5 \end{bmatrix} = 14.$$

Since

$$\vec{v}_1 = (2/7)\vec{w}_1 + (3/7)\vec{w}_2 \quad \text{and} \quad \vec{v}_2 = -(1/7)\vec{w}_1 + (2/7)\vec{w}_2,$$

the change of basis matrix is $\begin{bmatrix} 2/7 & -1/7 \\ 3/7 & 2/7 \end{bmatrix}$, with determinant $1/7$.

**c.** By theorem 4.8.5, the determinant is 7, the inverse of the determinant in part b.

**6.3.9**   a.  The vector field $\vec{F}\begin{pmatrix} x \\ y \end{pmatrix} = \begin{bmatrix} -y \\ x-1 \end{bmatrix}$ describes the orientation "increasing polar angle" for the circle of equation $(x-1)^2 + y^2 = 4$, shown at left below.  Again, we can confirm this by taking the dot product of this vector and a vector going from the center of the circle to a point on the circle: $\begin{bmatrix} -y \\ x-1 \end{bmatrix} \cdot \begin{bmatrix} x-1 \\ y \end{bmatrix} = 0$. To get the desired unit vector field, we divide $\vec{F}\begin{pmatrix} x \\ y \end{pmatrix} = \begin{bmatrix} -y \\ x-1 \end{bmatrix}$ by its length, which is $\sqrt{(x-1)^2 + y^2} = 2$, so the answer is $\frac{1}{2}\begin{bmatrix} -y \\ x-1 \end{bmatrix}$.

b.  Polar angle is defined from the origin, and every point of the circle of equation $(x-2)^2 + y^2 = 1$ is to the right of the origin. As shown at right in the figure below, as the polar angle increases, we move simultaneously counterclockwise on part of the circle and clockwise on another part of the circle. Polar angle defined from the origin orients the circle if the origin is inside the circle but does not orient it if the origin is outside the circle.

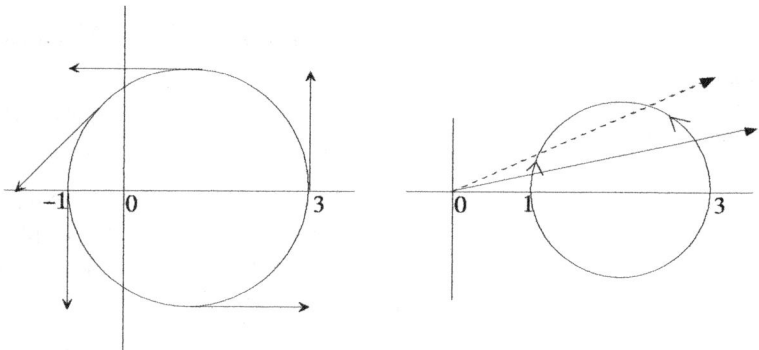

FIGURE FOR SOLUTION 6.3.9. LEFT: This illustrates part a. The vector field $\vec{F}\begin{pmatrix} x \\ y \end{pmatrix} = \begin{bmatrix} -y \\ x-1 \end{bmatrix}$ describes the orientation "increasing polar angle" for the circle of equation $(x-1)^2 + y^2 = 4$. RIGHT: Part b. The circle of equation $(x-2)^2 + y^2 = 1$. As the polar angle increases (from the solid arrow to the dotted arrow), we move simultaneously counterclockwise on part of the circle and clockwise on another part of the circle.

**6.3.11**   a.  The surface $S \subset \mathbb{R}^4$ is given by the equation $\mathbf{f}(\mathbf{x}) = \mathbf{0}$, where $\mathbf{f}\begin{pmatrix} x_1 \\ x_2 \\ x_3 \\ x_4 \end{pmatrix} = \begin{pmatrix} x_1^2 - x_2^2 - x_3 \\ 2x_1 x_2 - x_4 \end{pmatrix}$. The derivatives are

$$\mathbf{D}f_1 = [2x_1 \quad -2x_2 \quad -1 \quad 0] \quad \text{and} \quad \mathbf{D}f_2 = [2x_2 \quad 2x_1 \quad 0 \quad -1],$$

so we are looking for a 2-form $\omega$ that satisfies

$$\omega_{\mathbf{x}}(\vec{\mathbf{v}}, \vec{\mathbf{w}}) = \det \begin{bmatrix} 2x_1 & 2x_2 & v_1 & w_1 \\ -2x_2 & 2x_1 & v_2 & w_2 \\ -1 & 0 & v_3 & w_3 \\ 0 & -1 & v_4 & w_4 \end{bmatrix},$$

i.e.,

$$\omega_{\mathbf{x}} = dx_1 \wedge dx_2 - 2x_2\, dx_1 \wedge dx_3 + 2x_1\, dx_1 \wedge dx_4 - 2x_1\, dx_2 \wedge dx_3$$
$$- 2x_2\, dx_2 \wedge dx_4 + 4(x_1^2 + x_2^2)\, dx_3 \wedge dx_4.$$

b. There is no simple way to determine whether an elementary 2-form on $\mathbb{R}^4$ is a multiple of this rather complicated form (which it must be if it is to determine an orientation of $S$). Instead we approach the problem directly. We need to see whether there exists an elementary 2-form on $\mathbb{R}^4$ that does not vanish on two linearly independent tangent vectors to the surface and that varies continually with $\mathbf{x} \in S$. Any tangent vector to $S$ is in the kernel of the derivative of $\mathbf{f}$ (theorem 3.2.4). Thus it is any

vector $\begin{bmatrix} a \\ b \\ c \\ d \end{bmatrix}$ satisfying $\begin{bmatrix} 2x_1 & -2x_2 & -1 & 0 \\ 2x_2 & 2x_1 & 0 & -1 \end{bmatrix} \begin{bmatrix} a \\ b \\ c \\ d \end{bmatrix} \begin{bmatrix} 0 \\ 0 \end{bmatrix}$. The two vectors

$\begin{bmatrix} 0 \\ 1 \\ -2x_2 \\ 2x_1 \end{bmatrix}$, $\begin{bmatrix} 1 \\ 0 \\ 2x_1 \\ 2x_2 \end{bmatrix}$ are linearly independent and satisfy that equation, so we can take them as a basis for the tangent space $T_{\mathbf{x}}S$. Now we evaluate the six elementary 2-forms on $\mathbb{R}^4$ on these two vectors. The 2-form $dx_1 \wedge dx_3$ gives $2x_2$, which certainly vanishes at the origin; the 2-form $dx_1 \wedge dx_4$ gives $-2x_1$, which also vanishes at the origin. But the 2-form $dx_1 \wedge dx_2$ gives $-1$:

$$dx_1 \wedge dx_2 \left( \begin{bmatrix} 0 \\ 1 \\ -2x_2 \\ 2x_1 \end{bmatrix}, \begin{bmatrix} 1 \\ 0 \\ 2x_1 \\ 2x_2 \end{bmatrix} \right) = -1.$$

Thus it orients $S$. It is nonzero at every point $\mathbf{x} \in S$, and it varies continually with $\mathbf{x} \in S$ (in the easiest way, by not varying at all).

**6.3.13** a. The first part of the question corresponds to: "does $dx_1 \wedge dy_2$ vanish on the tangent space to $S$ at any point of $S$?" We will use the parametrization $\gamma$ in equation 6.4.13. Clearly

$$\vec{D_1\gamma} = \begin{bmatrix} 1 \\ 0 \\ 2x \\ 2y \\ 3x^2 - 3y^2 \\ 6xy \end{bmatrix}, \qquad \vec{D_2\gamma} = \begin{bmatrix} 0 \\ 1 \\ -2y \\ 2x \\ -6xy \\ 3x^2 - 3y^2 \end{bmatrix}$$

form a basis of the tangent space to $S$ at the point $\gamma\begin{pmatrix} x \\ y \end{pmatrix}$, so it is enough to compute $dx_1 \wedge dy_2$ on these two vectors. This gives

$$dx_1 \wedge dy_2 \left( \overrightarrow{D_1\gamma}, \overrightarrow{D_2\gamma} \right) = \det \begin{bmatrix} 1 & 0 \\ 2y & 2x \end{bmatrix} = 2x.$$

In particular, we see that this 2-form vanishes on the tangent space to $S$ at the points $\gamma\begin{pmatrix} 0 \\ y \end{pmatrix}$, so it does not define an orientation on the whole surface; it defines an orientation everywhere except at those points.

b. Since

$$dx_1 \wedge dy_1 \left( \overrightarrow{D_1\gamma}, \overrightarrow{D_2\gamma} \right) = \det \begin{bmatrix} 1 & 0 \\ 0 & 1 \end{bmatrix} = 1,$$

$dx_1 \wedge dy_1$ defines an orientation everywhere.

c. Since

$$dx_1 \wedge dy_2 = 2x\, dx_1 \wedge dy_1,$$

$dx_1 \wedge dy_2$ defines the same orientation as $dx_1 \wedge dy_1$ on the part of $S$ parametrized by the halfplane $x > 0$, and the opposite orientation on the part parametrized by $x < 0$.

Solution 6.3.15: When speaking of change of basis matrices, speaking of "from" and "to" can be confusing. In the exercise we asked for the change of basis matrix going from the $\mathbf{v}_i$ to the $\mathbf{w}_i$, but what is actually written expresses the basis vectors $\mathbf{w}_i$ in terms of the $\mathbf{v}_j$: the columns of the change of basis matrix $C$ are the coordinates of the $\mathbf{w}_i$ with respect to the $\mathbf{v}_j$.

**6.3.15** a. Any vector $\mathbf{v}$ can be written uniquely as

$$\mathbf{v} = c_1\mathbf{v}_1 + \cdots + c_n\mathbf{v}_n = (a_1 + ib_1)\mathbf{v}_1 + \cdots + (a_n + ib_n)\mathbf{v}_n$$
$$= a_1\mathbf{v}_1 + b_1(i\mathbf{v}_1) + \cdots + a_n\mathbf{v}_n + b_n(i\mathbf{v}_n).$$

b. Let us do the $2 \times 2$ case first. Let $C$ be the change of basis matrix

$$C = \begin{bmatrix} c_{1,1} & c_{1,2} \\ c_{2,1} & c_{2,2} \end{bmatrix}.$$

We know that $\mathbf{w}_1 = c_{1,1}\mathbf{v}_1 + c_{2,1}\mathbf{v}_2$. If we write $c_{k,l} = a_{k,l} + ib_{k,l}$, this leads to

$$\mathbf{w}_1 = (a_{1,1} + ib_{1,1})\mathbf{v}_1 + (a_{2,1} + ib_{2,1})\mathbf{v}_2 = a_{1,1}\mathbf{v}_1 + b_{1,1}(i\mathbf{v}_1) + a_{2,1}\mathbf{v}_2 + b_{2,1}(i\mathbf{v}_2)$$
$$i\mathbf{w}_1 = i(a_{1,1} + ib_{1,1})\mathbf{v}_1 + i(a_{2,1} + ib_{2,1})\mathbf{v}_2 = -b_{1,1}\mathbf{v}_1 + a_{1,1}(i\mathbf{v}_1) - b_{2,1}\mathbf{v}_2 + a_{2,1}(i\mathbf{v}_2).$$

A similar computation for $\mathbf{w}_2$ and $i\mathbf{w}_2$ leads to the change of basis matrix $\widetilde{C}: \mathbb{R}^4 \to \mathbb{R}^4$

$$\widetilde{C} = \begin{bmatrix} a_{1,1} & -b_{1,1} & a_{1,2} & -b_{1,2} \\ b_{1,1} & a_{1,1} & b_{1,2} & a_{1,2} \\ a_{2,1} & -b_{2,1} & a_{2,2} & -b_{2,2} \\ b_{2,1} & a_{2,1} & b_{2,2} & a_{2,2,} \end{bmatrix}$$

where the first column corresponds to the coefficients with which we have written $\mathbf{w}_i$, the second to the coefficients with which we have written $i\mathbf{w}_i$, and so on.

In the general case, the change of matrix $\widetilde{C} : \mathbb{R}^{2n} \to \mathbb{R}^{2n}$ is

$$\widetilde{C} = \begin{bmatrix} a_{1,1} & -b_{1,1} & \cdots & a_{1,n} & -b_{1,n} \\ b_{1,1} & a_{1,1} & \cdots & b_{1,n} & a_{1,n} \\ \vdots & \vdots & \ddots & \vdots & \vdots \\ a_{n,1} & -b_{n,1} & \cdots & a_{n,n} & -b_{n,n} \\ b_{n,1} & a_{n,1} & \cdots & b_{n,n} & a_{n,n}. \end{bmatrix}.$$

Solution 6.3.15 uses the fact that if $c = a + ib \in \mathbb{C}$, and we define $M_c = \begin{bmatrix} a & -b \\ b & a \end{bmatrix}$, then

$$M_{c_1 + c_2} = M_{c_1} + M_{c_2}$$

and

$$M_{c_1 c_2} = M_{c_1} M_{c_2}.$$

(If you change $c = a + ib$ to $\bar{c} = a - ib$, you would then replace $\begin{bmatrix} a & -b \\ b & a \end{bmatrix}$ by $\begin{bmatrix} a & b \\ -b & a \end{bmatrix}$, and the statement is the object of exercise 1.2.20.)

Note that each complex entry $c = a + ib$ has been replaced by the $2 \times 2$ real matrix $\begin{bmatrix} a & -b \\ b & a \end{bmatrix}$.

c. The easiest way to see this is by row reduction, and this is easiest in terms of multiplication by elementary matrices. Suppose $C = E_1 \ldots E_N$, where the $E_k$ are elementary matrices. Then $\widetilde{C}$ can be written

$$\widetilde{C} = \widetilde{E}_1 \ldots \widetilde{E}_n,$$

where (as in part b) each $\widetilde{E}_k$ is obtained from $E_k$ by replacing each complex entry $c = a + ib$ by the $2 \times 2$ real matrix $\begin{bmatrix} a & -b \\ b & a \end{bmatrix}$.

It is still true that if $E_k$ corresponds to adding a multiple of a row onto another, $\det E_k = \det \widetilde{E}_k = 1$. If $E_k$ corresponds to multiplying a row through by $c = a + ib$, then

$$\det E_k = c, \quad \text{but} \quad \det \widetilde{E}_k = \det \begin{bmatrix} a & -b \\ b & a \end{bmatrix} = a^2 + b^2 = |c|^2.$$

Finally, if $E_k$ corresponds to exchanging two rows, then $\det E_k = -1$, but $\det \widetilde{E}_k = 1$. Thus in all cases, $\det \widetilde{E}_k = |\det E_k|^2$, and we see that

$$\det \widetilde{C} = \det \widetilde{E}_1 \cdot \ldots \cdot \det \widetilde{E}_N = |\det E_1|^2 \cdot \ldots \cdot |\det E_N|^2$$
$$= |\det E_1 \cdot \ldots \cdot \det E_N|^2 = |\det C|^2.$$

d. The criterion for the basis $\mathbf{w}_1, i\mathbf{w}_1, \ldots, \mathbf{w}_n, i\mathbf{w}_n$ to be a direct basis is that the change of basis matrix have positive determinant. The change of basis matrix is the matrix $\widetilde{C}$ above, with determinant $|\det C|^2$, which is certainly positive.

**6.3.17** a. The form $\omega$ does not orient $M$. Here are two solutions.

*Computational solution:*

Parametrize the circle centered at $\begin{pmatrix} 3 \\ 0 \end{pmatrix}$ by $\theta \mapsto \begin{pmatrix} 3 + \cos\theta \\ \sin\theta \end{pmatrix}$. A vector in the tangent space to this circle can be written $\begin{bmatrix} -\sin\theta \\ \cos\theta \end{bmatrix}$, and

$$(x\,dy - y\,dx)\begin{bmatrix} -\sin\theta \\ \cos\theta \end{bmatrix} = (3 + \cos\theta)\cos\theta - \sin\theta(-\sin\theta) = 3\cos\theta + 1.$$

So $\omega$ vanishes on the tangent space to the circle at $\theta = \arccos(-1/3)$.

*More elegant solution:*

As seen in the figure below, at two points $\begin{pmatrix} x \\ y \end{pmatrix}$ on $C_2$, the tangent space to $C_2$ contains the vector $\begin{bmatrix} x \\ y \end{bmatrix}$. At those points, $\omega$ does not orient $C_2$, since

$$x\,dy - y\,dx \begin{bmatrix} x \\ y \end{bmatrix} = xy - xy = 0.$$

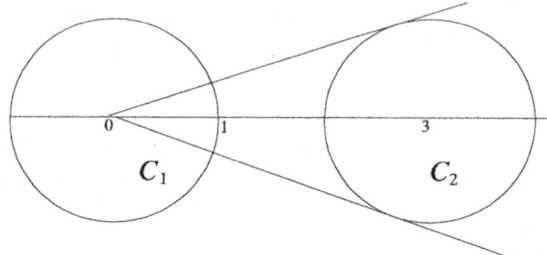

FIGURE FOR SOLUTION 6.3.17, part a. At two points on $C_2$, the tangent line to $C_2$ passes through the origin. The form $\omega = x\,dy - y\,dx$ vanishes on those tangent spaces, so $\omega$ does not orient $C_2$.

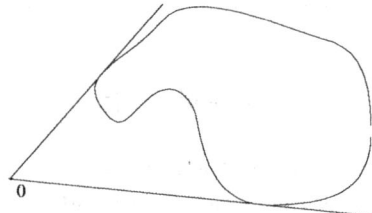

FIGURE FOR SOLUTION 6.3.17, part b: Any smooth simple closed curve in $\mathbb{R}^2$ that does not enclose the origin has at least two points at which the tangent space is not oriented by $\omega$.

b. As shown by the figure in the margin, any smooth simple closed curve in $\mathbb{R}^2$ that does not enclose the origin will have at least two points at which the tangent space is not oriented by $\omega$: there will exist at least two radial vectors from the origin tangent to the curve. These vectors can be written $\begin{bmatrix} x \\ y \end{bmatrix}$, and $\omega = x\,dy - y\,dx$ vanishes on them:

$$x\,dy - y\,dx \left( \begin{bmatrix} x \\ y \end{bmatrix} \right) = xy - yx = 0.$$

We can argue more formally as follows. Let the curve be parametrized by $\gamma$ and let the polar angle measured from the origin be denoted $\theta$. Since the curve does not enclose the origin, the polar angle function $\theta$ restricted to the curve has a maximum. (It can have several local maxima and minima.) At that maximum, the derivative of $\theta$ vanishes on the tangent space to the curve (theorem 3.7.1); i.e., if $\vec{v}$ is tangent to the curve at $\mathbf{c}$, then $[\mathbf{D}\theta(\mathbf{c})]\vec{v} = 0$. The derivative of $\theta$ is

$$[\mathbf{D}\theta] = d\left( \arctan \frac{y}{x} \right) = \begin{bmatrix} \dfrac{-y}{x^2 + y^2} & \dfrac{x}{x^2 + y^2} \end{bmatrix}.$$

Thus if $\vec{v}$ is tangent to the curve at a local minimum or maximum of $\theta$, then $\omega(\vec{v}) = 0$:

$$\begin{bmatrix} \dfrac{-y}{x^2 + y^2} & \dfrac{x}{x^2 + y^2} \end{bmatrix} \begin{bmatrix} v_1 \\ v_2 \end{bmatrix} = \frac{-yv_1}{x^2 + y^2} + \frac{xv_2}{x^2 + y^2} = 0$$

implies $-yv_1 + xv_2 = 0$, which is the same as $x\,dy - y\,dx \begin{bmatrix} v_1 \\ v_2 \end{bmatrix} = 0.$

**6.4.1** No, it does not preserve orientation. The derivative of the function

$$f(\mathbf{x}) = x^2 + y^2 - z^2 \text{ is } [2x,\ 2y,\ -2z], \text{ so } \vec{\mathbf{n}}(\mathbf{x}) = \begin{bmatrix} x \\ y \\ -z \end{bmatrix} \text{ is a normal vector}$$

field. This vector is outward pointing, since (for example) at the point

$$\begin{pmatrix} 1/2 \\ 0 \\ 1/2 \end{pmatrix}, \text{ it is } \begin{bmatrix} 1/2 \\ 0 \\ -1/2 \end{bmatrix}; \text{ i.e., it increases in the } x \text{ direction and decreases in}$$

the $z$ direction.

This normal vector field $\vec{\mathbf{n}}$ corresponds to

$$\omega_{\mathbf{x}} = x\, dy \wedge dz - y\, dx \wedge dz - z\, dx \wedge dy,$$

since

$$x\, dy \wedge dz - y\, dx \wedge dz - z\, dx \wedge dy(\vec{\mathbf{v}}, \vec{\mathbf{w}}) = \det[\vec{\mathbf{n}}, \vec{\mathbf{v}}, \vec{\mathbf{w}}].$$

Now compute

$$x\, dy \wedge dz - y\, dx \wedge dz - z\, dx \wedge dy \left( P_{\gamma\binom{r}{\theta}} \left( \underbrace{\begin{bmatrix} \cos\theta \\ \sin\theta \\ 1 \end{bmatrix}}_{\overrightarrow{D_1\gamma}}, \underbrace{\begin{bmatrix} -r\sin\theta \\ r\cos\theta \\ 0 \end{bmatrix}}_{\overrightarrow{D_2\gamma}} \right) \right) = -2r^2 < 0.$$

We know that $r \neq 0$ since it corresponds to the $z$ coordinate, and $0 < z < 1$ (see equation 5.2.3); if we did not exclude the vertex $z = 0$ from the cone it would not be a manifold.

**6.4.3** If we take $0 < \theta < 2\pi$ and $-\pi/2 < \varphi < \pi/2$, then

$$\det\left[ \vec{\mathbf{n}}, \overrightarrow{D_1\gamma}, \overrightarrow{D_2\gamma} \right] = -\cos\varphi$$

will always be negative; the mapping will be orientation reversing.

**6.4.5** Parametrize the domain of integration by $\binom{r}{\theta} \mapsto \begin{pmatrix} r\cos\theta \\ r\sin\theta \\ r^k\cos k\theta \\ r^k\sin k\theta \end{pmatrix}$, for

$0 \le r \le 1$ and $0 \le \theta \le 2\pi$. The integral becomes

$$\int_0^{2\pi}\int_0^1 (dx_1 \wedge dy_1 + dy_1 \wedge dx_2)\left( \begin{bmatrix} \cos\theta \\ \sin\theta \\ kr^{k-1}\cos k\theta \\ kr^{k-1}\sin k\theta \end{bmatrix}, \begin{bmatrix} -r\sin\theta \\ r\cos\theta \\ -kr^k\sin k\theta \\ kr^k\cos k\theta \end{bmatrix} \right) dr\, d\theta$$

$$= \int_0^{2\pi}\int_0^1 \left( \det\begin{bmatrix} \cos\theta & -r\sin\theta \\ \sin\theta & r\cos\theta \end{bmatrix} + \det\begin{bmatrix} \sin\theta & r\cos\theta \\ kr^{k-1}\cos k\theta & -kr^k\sin k\theta \end{bmatrix} \right)$$

$$= \int_0^{2\pi}\int_0^1 r - kr^k\left( \sin\theta\sin k\theta + \cos\theta\cos k\theta \right)$$

$$= \int_0^{2\pi}\int_0^1 \left( r - kr^k\cos(k-1)\theta \right) dr\, d\theta = \begin{cases} \pi & \text{if } k \neq 1 \\ 0 & \text{if } k = 1 \end{cases}.$$

---

Solution 6.4.1: The transpose of the derivative is a normal vector field to the surface; see example 6.3.6. Of course, a multiple of a normal vector field is also a normal vector field.

This parametrization uses de Moivre's formula:

$$z^k = r^k(\cos k\theta + i\sin k\theta).$$

**6.4.7 a.** Any $\omega \in A^n(\mathbb{R}^n)$ is a multiple of the determinant. Since $T$ is not invertible, its columns $T\vec{e}_1, \ldots, T\vec{e}_n$ are linearly dependent, and therefore

$$\omega(T\vec{e}_1, \ldots, T\vec{e}_n) = 0.$$

**b.** Suppose that $\omega \in A^k(V)$ is nonzero. What is to be shown is that if $T : V \to W$ is a linear transformation that is not 1–1, and $\vec{v}_1, \ldots \vec{v}_k$ is basis for $V$, then $\omega(T(\vec{v}_1), \ldots, T(\vec{v}_k))$ is neither positive or negative; i.e., it is zero.

If $T$ is not 1–1, the vectors $T(\vec{v}_1), \ldots, T(\vec{v}_k)$ are linearly dependent, so one of them is a linear combination of the others. We may assume that it is the last, say

$$T(\vec{v}_k) = a_1 T(\vec{v}_1) + \cdots + a_{k-1} T(\vec{v}_{k-1}).$$

Then

$$\omega\Big(T(\vec{v}_1), \ldots, T(\vec{v}_k)\Big) = \omega\Big(T(\vec{v}_1), T(\vec{v}_2), \ldots, T(\vec{v}_{k-1}), a_1 T(\vec{v}_1) + \cdots + a_{k-1} T(\vec{v}_{k-1})\Big)$$

$$= a_1 \omega\Big(T(\vec{v}_1), T(\vec{v}_2), \ldots, T(\vec{v}_1)\Big) + \cdots + a_{k-1}\omega\Big((T(\vec{v}_1), \ldots, T(\vec{v}_{k-1}), T(\vec{v}_{k-1})\Big).$$

In each of the $k - 1$ terms of this last sum, the same vector appears twice. Since $\omega$ is antisymmetric, this forces all the terms to vanish.

**6.4.9 a.** The space $M_1(2, 2)$ contains all $2 \times 2$ matrices with determinant 0 except the zero matrix; i.e., it is the complement of the zero matrix in the locus defined by the equation $\det(A) = 0$. (You might think of it as a "double" cone like the one in figure 5.2.1, the vertex being the zero matrix.)

It separates $2 \times 2$ matrices with positive determinant from those with negative determinant. At all points $A$ of $M_1(2, 2)$, we have $[\mathbf{D}\det(A)] \neq 0$ (by exercise 3.1.19), so $M_1(2, 2)$ is a manifold and does have a tangent space. Therefore we can choose a vector field orthogonal to the tangent space, pointing towards matrices with positive determinant. This vector field orients the manifold. The normal vector field pointing towards matrices with negative determinant gives the opposite orientation.

**b.** This is rather similar to example 6.4.9, though the computation comes out differently. We will parametrize three open subsets

$$U_1, U_2, U_3 \subset M_1(3, 3),$$

whose union is all of $M_1(3, 3)$. We will see that the orientations induced by these parametrizations are all compatible. The subsets $U_i$ are the subsets where the $i$th column is not $\mathbf{0}$, and the parametrizations are

$$\gamma_1 \begin{pmatrix} a_1 \\ b_1 \\ c_1 \\ u_1 \\ v_1 \end{pmatrix} = \underbrace{\begin{bmatrix} a_1 & u_1 a_1 & v_1 a_1 \\ b_1 & u_1 b_1 & v_1 b_1 \\ c_1 & u_1 c_1 & v_1 c_1 \end{bmatrix}}_{U_1}, \qquad \gamma_2 \begin{pmatrix} a_2 \\ b_2 \\ c_2 \\ u_2 \\ v_2 \end{pmatrix} = \underbrace{\begin{bmatrix} v_2 a_2 & a_2 & u_2 a_2 \\ v_2 b_2 & b_2 & u_2 b_2 \\ v_2 c_2 & c_2 & u_2 c_2 \end{bmatrix}}_{U_2},$$

$$\gamma_3 \underbrace{\begin{pmatrix} a_3 \\ b_3 \\ c_3 \\ u_3 \\ v_3 \end{pmatrix}}_{U_3} = \begin{bmatrix} u_3 a_3 & v_3 a_3 & a_3 \\ u_3 b_3 & v_3 b_3 & b_3 \\ u_3 c_3 & v_3 c_3 & c_3 \end{bmatrix}$$

On $U_1 \cap U_2$, the parametrizing variables are related by

For example, we get $a_2 = u_1 a_1$ by comparing the first row, second entry in the matrices for $U_1$ and $U_2$. Then substituting $u_1 a_1$ for $a_2$ in the first row, third entry of the second matrix and comparing it with the corresponding entry of the first matrix gives $u_2 u_1 = v_1$.

$$a_2 = u_1 a_1, \ b_2 = u_1 b_1, \ c_2 = u_1 c_1, \ u_2 = \frac{v_1}{u_1}, \ v_2 = \frac{1}{u_1}.$$

The mapping $\gamma_2^{-1} \circ \gamma_1$ is given by

$$\gamma_2^{-1} \circ \gamma_1 \begin{pmatrix} a_1 \\ b_1 \\ c_1 \\ u_1 \\ v_1 \end{pmatrix} = \begin{pmatrix} u_1 a_1 \\ u_1 b_1 \\ u_1 c_1 \\ v_1/u_1 \\ 1/u_1 \end{pmatrix}, \text{ with derivative } \begin{bmatrix} u_1 & 0 & 0 & a_1 & 0 \\ 0 & u_1 & 0 & b_1 & 0 \\ 0 & 0 & u_1 & c_1 & 0 \\ 0 & 0 & 0 & -\frac{v_1}{u_1^2} & \frac{1}{u_1} \\ 0 & 0 & 0 & -\frac{1}{u_1^2} & 0 \end{bmatrix}.$$

The determinant of this matrix is $u_1^3/u_1^3 = 1$. Thus, by proposition 6.4.8, $\gamma_1$ and $\gamma_2$ define orientations on $U_1$ and $U_2$ that are compatible on $U_1 \cap U_2$.

Exactly the same computation shows that $\gamma_1$ and $\gamma_3$ define compatible orientations on $U_1 \cap U_3$, so together they define an orientation of $M_1(3,3)$.

**6.5.1** The work form (a) is identical to (j) and (l). The work form evaluated on $P_\mathbf{x}(\vec{v})$ in (b) is identical to (i).

The flux form in (k) is identical to (d) and (h). The flux form evaluated on $P_\mathbf{x}(\vec{v}, \vec{w})$, given in (c), is identical to (e) and (f).

The mass form in (g) has no equivalents.

**6.5.3 a.** To correct this expression, one can either change the flux form to a mass form, getting $M_f\big(P_\mathbf{x}(\vec{v}_1, \vec{v}_2, \vec{v}_3)\big)$, or change the 3-parallelogram to a 2-parallelogram, getting $\Phi_{\vec{F}}\big(P_\mathbf{x}(\vec{v}_1, \vec{v}_2)\big)$.

b. The work form field is associated to a vector field, not a function; this should be $W_{\vec{F}}$.

c. The mass form field is a function of a 3-parallelogram; this should be $M_f\big(P_\mathbf{x}(\vec{u}, \vec{v}, \vec{w})\big)$ (or $M_f\big(P_\mathbf{x}(\vec{v}_1, \vec{v}_2, \vec{v}_3)\big)$ or the equivalent).

d. The cross product takes two vectors and gives a vector, while the dot product takes two vectors and gives a number. This should be $\vec{v}_1 \cdot (\vec{v}_2 \times \vec{v}_3)$.

e. This correct; the flux form $\Phi$ is associated to a vector field $\vec{F}$.

f. This expression is meaningful; it is the dot product of two vectors. It could also be written $\det[\vec{F}, \vec{v}, \vec{w}]$.

g. $W_{\vec{F}}\big(P_\mathbf{x}(\vec{v}_1)\big)$ or $\Phi_{\vec{F}}\big(P_\mathbf{x}(\vec{v}_1, \vec{v}_2)\big)$.

h. The mass form is associated to a function, not a vector field; this should be $M_f$.

i. This is correct as is.

**6.5.5** We will first show that $W_{\vec{F}} \wedge \Phi_{\vec{G}} = M_{\vec{F} \cdot \vec{G}}$. By proposition 6.1.22,

$$\underbrace{(F_1 dx + F_2 dy + F_3 dz)}_{W_{\vec{F}}} \wedge \underbrace{(G_1 dy \wedge dz + G_2 dz \wedge dx + G_x dz \wedge dy)}_{\Phi_{\vec{G}}}$$

$$= F_1 G_1\, dx \wedge dy \wedge dz + F_2 G_2\, dy \wedge dz \wedge dx + F_3 G_3\, dz \wedge dx \wedge dy$$

$$= (F_1 G_1 + F_2 G_2 + F_3 G_3)\, dx \wedge dy \wedge dz$$

$$= \vec{F} \cdot \vec{G}\, dx \wedge dy \wedge z = M_{\vec{F} \cdot \vec{G}}.$$

It follows that $M_{\vec{F} \cdot \vec{G}} = M_{\vec{G} \cdot \vec{F}} = W_{\vec{G}} \wedge \Phi_{\vec{F}}$.

**Solution 6.5.5:** You do not want to evaluate $W_{\vec{F}} \wedge \Phi_{\vec{G}}$ on vectors, as we did in example 6.1.21; that leads to horrendous computations.

In the computation at right, remember to discard terms containing $dx_i \wedge dx_i$ (for example, $F_1 G_2\, dx \wedge dz \wedge dx$), since such terms equal 0.

**6.5.7** $\vec{F} \begin{pmatrix} 0 \\ 1 \\ 2 \end{pmatrix} = \begin{bmatrix} 0 \\ -1 \\ -2 \end{bmatrix}$, so $W_{\vec{F}}\big(P_{\mathbf{a}}(\vec{\mathbf{u}})\big) = \begin{bmatrix} 0 \\ -1 \\ -2 \end{bmatrix} \cdot \begin{bmatrix} 1 \\ -1 \\ 1 \end{bmatrix} = -1.$

**6.5.9 a.** One such parallelogram is anchored at $\begin{pmatrix} 1 \\ 1 \\ 0 \end{pmatrix}$ and spanned by $\begin{bmatrix} 1 \\ 1 \\ 0 \end{bmatrix}, \begin{bmatrix} 1 \\ 1 \\ 1 \end{bmatrix}$, in that order:

$$y\, dy \wedge dz + x\, dx \wedge dz - z\, dx \wedge dy \left( P_{\begin{pmatrix} 1 \\ 1 \\ 0 \end{pmatrix}}\left( \begin{bmatrix} 1 \\ 1 \\ 0 \end{bmatrix}, \begin{bmatrix} 1 \\ 1 \\ 1 \end{bmatrix} \right) \right) = dy \wedge dz + dx \wedge dz \left( \begin{bmatrix} 1 \\ 1 \\ 0 \end{bmatrix}, \begin{bmatrix} 1 \\ 1 \\ 1 \end{bmatrix} \right)$$

$$= \det \begin{bmatrix} 1 & 1 \\ 0 & 1 \end{bmatrix} + \det \begin{bmatrix} 1 & 1 \\ 0 & 1 \end{bmatrix} = 2.$$

This means that the flux of the vector field $\vec{F} \begin{pmatrix} x \\ y \\ z \end{pmatrix} = \begin{bmatrix} y \\ -x \\ -z \end{bmatrix}$ through the parallelogram is 2.

One way to do this problem is to choose two vectors at random and do the computation. Almost any two vectors will give either positive or negative flux; if you get negative flux, just switch the order in which the two vectors are listed.

A more algebraic approach is to note that since the parallelogram is anchored at $\begin{pmatrix} 1 \\ 1 \\ 0 \end{pmatrix}$, the 2-form $y\, dy \wedge dz + x\, dx \wedge dz - z\, dx \wedge dy$ becomes $dy \wedge dz + dx \wedge dz$; construct your vectors $\vec{v}$ and $\vec{w}$ so that

$$\det \begin{bmatrix} v_2 & w_2 \\ v_3 & w_3 \end{bmatrix} + \det \begin{bmatrix} v_1 & w_1 \\ v_3 & w_3 \end{bmatrix} > 0.$$

A more geometric approach is to note that figure 6.5.4 suggests that the flux through any parallelogram lying in the vertical plane where $x = y$ will be nonzero. How can one guess, without doing the computation, whether the flux will be positive or negative? Remember (proposition 1.4.20) that $\det[\vec{\mathbf{a}}, \vec{\mathbf{b}}, \vec{\mathbf{c}},]$ is positive if the vectors, in that order, satisfy the right-hand rule. So (definition 6.5.2), the flux will be positive if $\vec{F}(\mathbf{x}), \vec{\mathbf{v}}, \vec{\mathbf{w}}$ satisfy the right-hand rule.

b. If we anchor the parallelogram at $\begin{pmatrix} 0 \\ 0 \\ -1 \end{pmatrix}$, $\Phi$ gives a positive number (positive flux); if we anchor it at $\begin{pmatrix} 0 \\ 0 \\ 1 \end{pmatrix}$, $\Phi$ gives a negative number (negative flux):

$$y\,dy \wedge dz + x\,dx \wedge dz - z\,dx \wedge dy \left( P_{\begin{pmatrix} 0 \\ 0 \\ -1 \end{pmatrix}} \left( \begin{bmatrix} 1 \\ 0 \\ 0 \end{bmatrix}, \begin{bmatrix} 0 \\ 1 \\ 0 \end{bmatrix} \right) \right) = dx \wedge dy \left( \begin{bmatrix} 1 \\ 0 \\ 0 \end{bmatrix}, \begin{bmatrix} 0 \\ 1 \\ 0 \end{bmatrix} \right) = 1$$

$$y\,dy \wedge dz + x\,dx \wedge dz - z\,dx \wedge dy \left( P_{\begin{pmatrix} 0 \\ 0 \\ 1 \end{pmatrix}} \left( \begin{bmatrix} 1 \\ 0 \\ 0 \end{bmatrix}, \begin{bmatrix} 0 \\ 1 \\ 0 \end{bmatrix} \right) \right) = -dx \wedge dy \left( \begin{bmatrix} 1 \\ 0 \\ 0 \end{bmatrix}, \begin{bmatrix} 0 \\ 1 \\ 0 \end{bmatrix} \right) = -1.$$

Of course these aren't the only possibilities. How did we hit on these? The parallelogram is parallel to the $(x, y)$-plane, so figure 6.5.4 and the right-hand rule say that anchoring it "below" that plane will give positive flux and anchoring it "above" that plane will give negative flux.

**6.5.11**  a. $W_{\begin{bmatrix} x \\ y \end{bmatrix}} \left( P_{\begin{pmatrix} 1 \\ 1 \end{pmatrix}} \begin{bmatrix} 2 \\ 3 \end{bmatrix} \right) = \begin{bmatrix} 1 \\ 1 \end{bmatrix} \cdot \begin{bmatrix} 2 \\ 3 \end{bmatrix} = 5.$

b. $W_{\begin{bmatrix} x^2 \\ \sin xy \end{bmatrix}} \left( P_{\begin{pmatrix} -1 \\ -\pi \end{pmatrix}} \begin{bmatrix} e \\ \pi \end{bmatrix} \right) = \begin{bmatrix} 1 \\ 0 \end{bmatrix} \cdot \begin{bmatrix} e \\ \pi \end{bmatrix} = e.$

c. $\begin{bmatrix} 0 \\ 1 \\ 1 \end{bmatrix} \cdot \begin{bmatrix} 2 \\ 3 \\ -1 \end{bmatrix} = 2.$     d. $\begin{bmatrix} \sin 1 \\ \cos(-1) \\ e^0 \end{bmatrix} \cdot \begin{bmatrix} 0 \\ 1 \\ 0 \end{bmatrix} = \cos 1.$

**6.5.13**  We have $\vec{F}(\mathbf{x}) = \begin{bmatrix} 1 \\ 0 \\ -1 \end{bmatrix}$, and $f(\mathbf{x}) = -2$, so

$$W_{\vec{F}}\big(P_{\mathbf{x}}(\vec{v}_1)\big) = \begin{bmatrix} 1 \\ 0 \\ -1 \end{bmatrix} \cdot \begin{bmatrix} 0 \\ 1 \\ 1 \end{bmatrix} = -1$$

$$\Phi_{\vec{F}}\big(P_{\mathbf{x}}(\vec{v}_1, \vec{v}_2)\big) = \begin{bmatrix} 1 \\ 0 \\ -1 \end{bmatrix} \cdot \left( \begin{bmatrix} 0 \\ 1 \\ 1 \end{bmatrix} \times \begin{bmatrix} 1 \\ 1 \\ 0 \end{bmatrix} \right) = \begin{bmatrix} 1 \\ 0 \\ -1 \end{bmatrix} \cdot \begin{bmatrix} -1 \\ 1 \\ -1 \end{bmatrix} = 0,$$

$$M_f\big(P_{\mathbf{x}}(\vec{v}_1, \vec{v}_2, \vec{v}_3)\big) = -2\det \begin{bmatrix} 0 & 1 & -1 \\ 1 & 1 & 1 \\ 1 & 0 & 1 \end{bmatrix} = -2\big(-1(1) + 1(2)\big) = -2.$$

**6.5.15** We need to show that the formula

$$\Phi_{\vec{F}(\mathbf{x})}(\vec{v}_1, \ldots, \vec{v}_{n-1}) = \det[\vec{F}(\mathbf{x}), \vec{v}_1, \ldots, \vec{v}_{n-1}]$$

is indeed multilinear and antisymmetric as a function of the $\vec{v}_i$. This follows immediately from the corresponding statement about the determinant: it is linear as a function of each column, in particular as a function of the $\vec{v}_i$, and it is antisymmetric as a function of its columns; in particular, if you exchange two of the $\vec{v}_i$ you change the sign.

**6.5.17** We will need to compute four integrals. The first side of the square, which we will call $C_1$, is parametrized by $\gamma_1(t) = \begin{pmatrix} 0 \\ t \end{pmatrix}$, $0 \le t \le a$. This leads to

Solution 6.5.17: These computations use equation 6.5.14.

Since $\vec{F}\begin{pmatrix} x \\ y \end{pmatrix} = \begin{bmatrix} xy \\ ye^x \end{bmatrix}$, we have

$$\vec{F}\big(\gamma(t)\big) = \begin{bmatrix} 0 \\ t \end{bmatrix}.$$

$$\int_{C_1} W_{\vec{F}} = \int_0^a \vec{F}\big(\gamma(t)\big) \cdot \gamma_1'(t)\, dt = \int_0^a \begin{bmatrix} 0 \\ t \end{bmatrix} \cdot \begin{bmatrix} 0 \\ 1 \end{bmatrix}\, dt = \frac{a^2}{2}.$$

The second side, $C_2$, is parametrized by $\gamma_2(t) = \begin{pmatrix} t \\ a \end{pmatrix}$, $0 \le t \le b$. This leads to

$$\int_0^b \begin{bmatrix} ta \\ ae^t \end{bmatrix} \cdot \begin{bmatrix} 1 \\ 0 \end{bmatrix}\, dt = \frac{ab^2}{2}.$$

The third side, $C_3$, is parametrized by $\gamma_3(t) = \begin{pmatrix} b \\ t \end{pmatrix}$, $0 \le t \le a$. But this orientation reverses the orientation of $C_3$, so we find the integral

$$\int_a^0 \begin{bmatrix} bt \\ te^b \end{bmatrix} \cdot \begin{bmatrix} 0 \\ 1 \end{bmatrix}\, dt = -\frac{a^2e^b}{2}.$$

Finally, the last side $C_4$ is parametrized by $\gamma_4(t) = \begin{pmatrix} t \\ 0 \end{pmatrix}$, $0 \le t \le b$. (Again, this parametrization reverses orientation.) This gives the integral

$$\int_a^0 \begin{bmatrix} 0 \\ 0 \end{bmatrix} \cdot \begin{bmatrix} 1 \\ 0 \end{bmatrix}\, dt = 0.$$

So the total integral is

$$\frac{a^2}{2} + \frac{ab^2}{2} - \frac{a^2e^b}{2}.$$

**6.5.19** The obvious way to parametrize the sphere is with the spherical coordinates map

$$\gamma : \begin{pmatrix} \theta \\ \varphi \end{pmatrix} = \begin{pmatrix} R\cos\varphi\cos\theta \\ R\cos\varphi\sin\theta \\ R\sin\varphi \end{pmatrix}, \quad 0 \le \theta < 2\pi, \quad -\frac{\pi}{2} < \varphi < \frac{\pi}{2}.$$

This is compatible with the orientation given by an outward-pointing vector field, since

$$\det\left( \underbrace{\begin{bmatrix} R\cos\varphi\cos\theta \\ R\cos\varphi\sin\theta \\ R\sin\varphi \end{bmatrix}}_{\text{outward-pointing vector}}, \underbrace{\begin{bmatrix} -R\cos\varphi\sin\theta \\ R\cos\varphi\cos\theta \\ 0 \end{bmatrix}}_{\overrightarrow{D_\theta\gamma}\begin{pmatrix}\theta\\\varphi\end{pmatrix}}, \underbrace{\begin{bmatrix} -R\sin\varphi\cos\theta \\ -R\sin\varphi\sin\theta \\ R\cos\varphi \end{bmatrix}}_{\overrightarrow{D_\varphi\gamma}\begin{pmatrix}\theta\\\varphi\end{pmatrix}} \right) = R^2\cos\varphi,$$

and $\cos\varphi > 0$ for $|\varphi| < \pi/2$. Thus (note that on the sphere, $r = R$) the flux becomes

$$\int_0^{2\pi} \int_{-\pi/2}^{\pi/2} \underbrace{\begin{bmatrix} R^{a+1}\cos\varphi\cos\theta \\ R^{a+1}\cos\varphi\sin\theta \\ R^{a+1}\sin\varphi \end{bmatrix}}_{\vec{F}\gamma\begin{pmatrix}\theta\\\varphi\end{pmatrix}} \cdot \left( \overrightarrow{D_\theta\gamma}\begin{pmatrix}\theta\\\varphi\end{pmatrix} \times \overrightarrow{D_\varphi\gamma}\begin{pmatrix}\theta\\\varphi\end{pmatrix} \right) |d\varphi||d\theta|$$

$$= \int_0^{2\pi} \int_{-\pi/2}^{\pi/2} R^{a+1} \begin{bmatrix} \cos\varphi\cos\theta \\ \cos\varphi\sin\theta \\ \sin\varphi \end{bmatrix} \cdot \begin{bmatrix} R^2\cos^2\varphi\cos\theta \\ R^2\cos^2\varphi\sin\theta \\ R^2\cos\varphi\sin\varphi \end{bmatrix} |d\varphi||d\theta|$$

$$= 2\pi R^{a+3} \int_{-\pi/2}^{\pi/2} \cos\varphi |d\varphi| = 4\pi R^{a+3}.$$

**6.5.21 a.** No, it does not preserve orientation, since

Here we use definition 6.4.2.

$$\det\left[ \mathbf{D}\gamma\begin{pmatrix} u \\ v \\ w \end{pmatrix} \right] = \det \begin{bmatrix} \cos v\cos w & -u\cos w\sin v & -R\sin w - u\cos v\sin w \\ \cos v\sin w & -u\sin w\sin v & R\cos w + u\cos v\cos w \\ \sin v & u\cos v & 0 \end{bmatrix}$$

$$= -Ru - u^2\cos v.$$

This quantity is never positive (in fact, it is strictly negative except when $u = 0$, which happens on the core circle of the torus): by definition $R > 0$ and $u \ge 0$ and

$$Ru > |u^2\cos v|, \quad \text{since} \quad -u \le u\cos v \le u \text{ and } u \le r < R,$$

so that even when $v$ is between $\pi/2$ and $3\pi/2$ (so that $\cos v$ is negative), if $u > 0$, then $-Ru - u^2\cos v < 0$.

b. The integral becomes

$$\int_0^{2\pi} \int_0^{2\pi} \int_0^r f\left(\gamma\begin{pmatrix} u \\ v \\ w \end{pmatrix}\right) \det\left[\mathbf{D}\gamma\begin{pmatrix} u \\ v \\ w \end{pmatrix}\right] \, du \, dv \, dw$$

To compute $f(\gamma(\mathbf{u}))$, don't multiply out in detail; just note that this becomes

$$(R + u \cos v)^2 (\cos^2 w + \sin^2 w).$$

$$= \int_0^{2\pi} \int_0^{2\pi} \int_0^r \overbrace{-(R + u \cos v)^2}^{f(\gamma(\mathbf{u}))} \overbrace{u(R + u \cos v)}^{-\det[\mathbf{D}\gamma]} \, du \, dv \, dw$$

$$= 2\pi \int_0^{2\pi} \int_0^r (-R^3 u - 3R^2 u^2 \cos v - 3R u^3 \cos^2 v - u^4 \cos^3 v) \, du \, dv$$

$$= -2\pi \int_0^{2\pi} \left( \frac{R^3 r^2}{2} + R^2 r^3 \cos v + \frac{3 R r^4 \cos^2 v}{4} + \frac{r^5 \cos^3 v}{5} \right) dv$$

$$= -\pi^2 \left( 2 R^3 r^2 + \frac{3 R r^4}{2} \right)$$

Since the parametrization reverses orientation, we should multiply that integral by $-1$.

**6.6.1**  Any single point is in a single dyadic cube $C \in \mathcal{D}_N(\mathbb{R}^n)$ for any $N$. Equation 5.2.2 of definition 5.2.1 thus becomes

$$\lim_{N \to \infty} \left( \frac{1}{2^N} \right)^0 = 1 \neq 0.$$

**6.6.3 a.** Let $X$ be the subset of $\mathbb{R}^3$ where $xyz \leq 1$ and $x^2 + y^2 + z^2 \leq 4$. This set is sort of like an apple with four shallow bites taken out. The sets $Z_1$, $Z_2$ of equation

$$f\begin{pmatrix} x \\ y \\ z \end{pmatrix} = xyz = 1 \quad \text{and} \quad g\begin{pmatrix} x \\ y \\ z \end{pmatrix} = x^2 + y^2 + z^2 = 4$$

are both smooth 2-manifolds, since their derivatives do not vanish on $Z_1$ and $Z_2$ respectively. Therefore the points where both the equality $xyz = 1$ and the inequality $x^2 + y^2 + z^2 < 4$ are satisfied are smooth points of the boundary, as are the points where $xyz < 1$ and $x^2 + y^2 + z^2 - 4 = 0$.

Any nonsmooth points are necessarily points where both equalities

$$xyz = 1 \quad \text{and} \quad x^2 + y^2 + z^2 = 4$$

are satisfied. These points form a smooth curve $C$ (actually, a disjoint union of four curves), since the derivative

$$\left[ \mathbf{D}\left( \begin{smallmatrix} f \\ g \end{smallmatrix} \right) \begin{pmatrix} x \\ y \\ z \end{pmatrix} \right] = \begin{bmatrix} 2x & 2y & 2z \\ yz & xz & xy \end{bmatrix}$$

has rank 2 at all points of $C$.[8] So $C$ has 2-dimensional volume 0, satisfying part 1 of definition 6.6.6.

Part 2 is clear: the smooth part of the boundary is the union of the part of the sphere of radius 2 where $xyz < 1$, which certainly has finite area, and the part of the smooth surface of equation $xyz = 1$ where $x^2 + y^2 + z^2 < 4$, which is a 2-dimensional piece with boundary, and has finite area by theorem 6.6.11.

b. In this case we have sliced the bitten apple in two, and kept one of the pieces. (The pieces are not equal: one contains an entire bite and small pieces of the other three; the other has most of the other three. We kept the first.) Define $Z_3$ to be the plane of equation

$$h \begin{pmatrix} x \\ y \\ z \end{pmatrix} = x + y + z = 0.$$

We can repeat the above argument to show that any points where exactly one of the inequalities $f \leq 1$, $g \leq 4$, $h \geq 0$ is satisfied as an equality are smooth points of the boundary (either the peel of the apple, or exposed by the bite, or on the slice). Thus the nonsmooth points of the boundary are those where at least two of the inequalities are satisfied as equalities.

Any pair of equations defines a smooth curve, of two-dimensional volume 0, so the set of nonsmooth points, being a subset of a union of three such curves, has 2-dimensional volume 0. Moreover, as before, the projections of these curves onto the $(x, y)$-plane have area zero, so can be covered by dyadic squares of arbitrarily small area, and the parts of each of the surfaces $Z_1, Z_2, Z_2$ above these squares still have arbitrarily small area.

**6.6.5** a. The only thing to check is that $\omega$ is never 0, i.e., that at every $\mathbf{x} \in X$, we can find vectors $\vec{\mathbf{v}}_1, \ldots, \vec{\mathbf{v}}_{n-1} \in T_{\mathbf{x}}X$ such that

$$\det \left[ \vec{\boldsymbol{\nabla}} f(\mathbf{x}), \vec{\mathbf{v}}_1, \ldots, \vec{\mathbf{v}}_{n-1} \right] \neq 0.$$

That is the same thing as requiring that the vectors $\vec{\boldsymbol{\nabla}} f(\mathbf{x}), \vec{\mathbf{v}}_1, \ldots, \vec{\mathbf{v}}_{n-1}$ be linearly independent, and it is enough to take $\vec{\mathbf{v}}_1, \ldots, \vec{\mathbf{v}}_{n-1}$ to be a basis of $T_{\mathbf{x}}X$, since $\vec{\boldsymbol{\nabla}} f(\mathbf{x})$ is orthogonal to $T_{\mathbf{x}}X$.

b. Let us take $X$ to be the boundary of the region $Y = \{\mathbf{x} \mid f(\mathbf{x}) \leq 0\}$. We will give $Y$ the standard orientation of $\mathbb{R}^n$, given by det. Then $\omega$ defines

---

[8]You may wonder how we know the derivative has rank 2. Certainly it must be either 1 or 2. It can't be 0, since $x, y$ and $z$ are nonzero; it can't be greater than 2, because the codomain is $\mathbb{R}^2$. Setting the determinant of the first two columns to 0 gives $2x^2 z = 2y^2 z$, so the first two columns can be linearly dependent only if $x = \pm y$ or if $z = 0$ (which is impossible, since $xyz = 1$). Doing the same with the last two columns gives $2xy^2 = 2xz^2$, so those columns can be linearly dependent only if $y = \pm z$. Putting these together says that for $xyz = 1$ to be satisfied, we would have to have $x = y = z$. But that contradicts $x^2 + y^2 + z^2 = 4$. Therefore, two of the columns are linearly independent.

the boundary orientation of $\partial Y = X$. Indeed, $\vec{\nabla} f(\mathbf{x})$ is an outward-pointing vector, and by definition 6.6.16 of the boundary orientation is defined by

$$\omega^\partial(\vec{v}_1, \ldots, \vec{v}_{n-1}) = \omega(\vec{v}_{\text{out}}, \vec{v}_1, \ldots, \vec{v}_{n-1})$$

where $\omega$ is a form defining the orientation of the ambient space, in this case the orientation det of $\mathbb{R}^n$. Thus applied to our case, the formula above reads

Of course, if we had considered $X$ as the boundary of the set $\{\mathbf{x} \mid f(\mathbf{x}) \geq 0\}$, then the orientation given by $\omega$ would be the opposite of the boundary orientation.

$$\omega^\partial(\vec{v}_1, \ldots, \vec{v}_{n-1}) = \det\left[\vec{v}_{\text{out}}, \vec{v}_1, \ldots, \vec{v}_{n-1}\right],$$

and the result follows since $\vec{\nabla} f(\mathbf{x})$ is an outward-pointing vector.

**6.6.7**    a. All vectors in $T_{\mathbf{x}}P$ are of the form $\begin{bmatrix} a \\ b \\ -a-b \end{bmatrix}$ for some $a$ and $b$ in $\mathbb{R}$. The 2-form defined by the normal $\vec{N} = \begin{bmatrix} 1 \\ 1 \\ 1 \end{bmatrix}$ takes two vectors in

$$T_{\mathbf{x}}P, \quad \vec{v} = \begin{bmatrix} v_1 \\ v_2 \\ -v_1-v_2 \end{bmatrix} \text{ and } \vec{w} = \begin{bmatrix} w_1 \\ w_2 \\ -w_1-w_2 \end{bmatrix}, \text{ and returns the number}$$

$$\det\left[\vec{N}, \vec{v}, \vec{w}\right] = \det\begin{bmatrix} 1 & v_1 & w_1 \\ 1 & v_2 & w_2 \\ 1 & -v_1-v_2 & -w_1-w_2 \end{bmatrix} = 3(v_1 w_2 - v_2 w_1).$$

Compare the action of the listed forms on $\vec{v}$ and $\vec{w}$:

$$dx \wedge dy(\vec{v}, \vec{w}) = v_1 w_2 - v_2 w_1,$$
$$dx \wedge dz(\vec{v}, \vec{w}) = -(v_1 w_2 - v_2 w_1),$$
$$dy \wedge dz(\vec{v}, \vec{w}) = v_1 w_2 - v_2 w_1.$$

The first and last forms, $dx \wedge dy$ and $dy \wedge dz$, equal the form defined by the normal, multiplied by the positive real number $1/3$, so they define the same orientation of $P$. The second is minus the form defined by the normal, so it defines the opposite orientation.

b.  We will see that $X$ satisfies the conditions of definition 6.6.6, and that its nonsmooth boundary is empty, i.e., that all points in the boundary are smooth points (satisfy definition 6.6.2).

Indeed, we can take

$$f\begin{pmatrix} x \\ y \\ z \end{pmatrix} = x + y + z \quad \text{and} \quad g\begin{pmatrix} x \\ y \\ z \end{pmatrix} = 1 - (x^2 + y^2 + z^2).$$

Then

$$\left[\mathbf{D}\begin{pmatrix} f \\ g \end{pmatrix}\begin{pmatrix} x \\ y \\ z \end{pmatrix}\right] = \begin{bmatrix} 1 & 1 & 1 \\ -2x & -2y & -2z \end{bmatrix},$$

which has rank two (hence is surjective) unless $x = y = z$. In $P$, this can only happen at the origin, which is not in $\partial_P X$.

It is easy to check that our map takes its values in $\partial_P X$. It is also easy to see that it has injective derivative, and that it is injective on $(0, 2\pi)$. So the only problem is to see that it is surjective (with domain $[0, 2\pi]$). Given two coordinates $\binom{a}{b}$, we can solve

$$\begin{array}{ccc} u - v = a & & u = \dfrac{a-b}{2} \\ & \text{to find} & \\ -u - v = b, & & v = -\dfrac{a+b}{2} \end{array}.$$

If

$$2a^2 + 2b^2 + 2ab = \frac{3}{2}(a+b)^2 + \frac{1}{2}(a-b)^2 = 1,$$

we find $|u| \leq \sqrt{2}/2$ and $|v| \leq \sqrt{6}/6$. Thus we can set

$$u = \frac{\cos t}{\sqrt{2}} \quad \text{and} \quad v = \sin s \sqrt{6}$$

and since $2u^2 + 6v^2 = 1$, we can take $t = s$. This shows that the map is surjective.

c. The parametrization $\gamma$ is consistent with the orientation of $\partial X$ if $\vec{\gamma}'(t)$ is in the same direction as the unit tangent vector field to $\partial X$ that defines the orientation of $\partial X$. Definition 6.6.16 and example 6.6.19 tell us that this is equivalent to having

$$\det \left[ \vec{N}, \vec{v}_{\text{out}}(t), \vec{\gamma}'(t) \right] > 0$$

for $0 \leq t \leq 2\pi$. Note that $\gamma(t)$ is in $P$ and points out of $X$ at $\mathbf{x} = \vec{\gamma}(t)$, so we can set $\vec{v}_{\text{out}}(t) = \vec{\gamma}(t)$. From part a, $\det \left[ \vec{N}, \vec{v}, \vec{w} \right] = 3\, dy \wedge dz(\vec{v}, \vec{w})$ for two vectors $\vec{v}, \vec{w} \in P$. Both $\vec{\gamma}(t)$ and $\vec{\gamma}'(t)$ are in $P$, so

$$\det \left[ \vec{N}, \vec{\gamma}(t), \vec{\gamma}'(t) \right] = 3\, dy \wedge dz \left[ \vec{\gamma}(t), \vec{\gamma}'(t) \right].$$

It is easy to calculate the value of the latter expression:

$$3\, dy \wedge dz \left[ \vec{v}_{\text{out}}(t), \vec{\gamma}'(t) \right] = 3 \det \begin{bmatrix} -\dfrac{\cos t}{\sqrt{2}} - \dfrac{\sin t}{\sqrt{6}} & \dfrac{\sin t}{\sqrt{2}} - \dfrac{\cos t}{\sqrt{6}} \\ 2\dfrac{\sin t}{\sqrt{6}} & 2\dfrac{\cos t}{\sqrt{6}} \end{bmatrix} = -\sqrt{3}.$$

This is always negative, so the parametrization is not compatible with the boundary orientation of $\partial X$ and is, moreover, orientation reversing.

d. Since $\gamma$ from part c is orientation reversing, a form compatible with the orientation of $\partial X$ will yield a negative real number when it acts on $\vec{\gamma}'(t)$ at $\mathbf{x} = \gamma(t)$. Consider the actions of the following 1-forms on $\vec{\gamma}'(t)$, at any position $\mathbf{x} = \gamma(t)$:

$$dx\left( \vec{\gamma}'(t) \right) = -\frac{\sin t}{\sqrt{2}} - \frac{\cos t}{\sqrt{6}},$$

$$dy\left( \vec{\gamma}'(t) \right) = \frac{\sin t}{\sqrt{2}} - \frac{\cos t}{\sqrt{6}},$$

$$dz\left( \vec{\gamma}'(t) \right) = 2\frac{\cos t}{\sqrt{6}}.$$

Since these are not always less than 0 for $0 \le t < 2\pi$, they cannot define the orientation at every point.

e. Use the same strategy as in part d. The actions of the listed forms on the vector $\vec{\gamma}'(t)$ anchored at $\mathbf{x} = \gamma(t)$ are as follow:

$$
\begin{aligned}
x\,dy - y\,dx &\quad \text{yields} \quad -1/\sqrt{3}, \\
x\,dz - z\,dx &\quad \text{yields} \quad 1/\sqrt{3}, \\
y\,dz - z\,dy &\quad \text{yields} \quad -1/\sqrt{3}.
\end{aligned}
$$

The first and last forms, $x\,dy - y\,dx$ and $y\,dz - z\,dy$, are negative for all $0 \le t < 2\pi$, hence may be used to define the orientation of $\partial X$ at every point.

**6.6.9** We defined "inward" and "outward" pointing vectors using equations, whereas parametrizations are more natural in this context.

Most of the work is done in example 6.6.10. More specifically, find an $(n-k) \times n$ matrix $A$ such that

$$
\ker A = \operatorname{span}(\vec{\mathbf{v}}_1, \ldots, \vec{\mathbf{v}}_k)
$$

and a linear function $\alpha_i$ such that (equation 6.6.6)

$$
\ker \begin{bmatrix} A \\ \alpha_i \end{bmatrix} = \operatorname{span}(\vec{\mathbf{v}}_1, \ldots, \widehat{\vec{\mathbf{v}}}_i, \ldots, \vec{\mathbf{v}}_k).
$$

Recall that changing the sign of $\alpha_i$ we may assume that $\alpha_i(\vec{\mathbf{v}}_i) > 0$, and that then

$$
P_{\mathbf{x}}(\vec{\mathbf{v}}_1, \ldots, \vec{\mathbf{v}}_k)
$$

is defined in the manifold

$$
M \overset{\text{def}}{=} \{\, \mathbf{y} \in \mathbb{R}^n \mid A\mathbf{x} = A\mathbf{y} \,\}
$$

by the inequalities

$$
\alpha_i(\mathbf{x}) \le \alpha_i(\mathbf{y}) \le \alpha_i(\mathbf{x} + \vec{\mathbf{v}}_i). \tag{1}
$$

In particular, the function

$$
g(\mathbf{y}) = \alpha_i(\mathbf{y}) - \alpha_i(\mathbf{x})
$$

plays the role of $g$ in definition 6.6.15 (which itself refers to definition 6.6.2) at points of $P_{\mathbf{x}}((\vec{\mathbf{v}}_1, \ldots, \widehat{\vec{\mathbf{v}}}_i, \ldots, \vec{\mathbf{v}}_k)$, where the left inequality in (1) is satisfied as an equality. Moreover, we have

$$
[\mathbf{D}g(\mathbf{y})]\vec{\mathbf{v}}_i = \alpha_i(\vec{\mathbf{v}}_i) > 0
$$

so that at such a point, $\vec{\mathbf{v}}_i$ is inward pointing.

At a point of $P_{\mathbf{x}+\vec{\mathbf{v}}_i}((\vec{\mathbf{v}}_1, \ldots, \widehat{\vec{\mathbf{v}}}_i, \ldots, \vec{\mathbf{v}}_k)$, where the right inequality in (1) is satisfied as an equality, an appropriate function $g$ is

$$
g(\mathbf{y}) = \alpha_i(\mathbf{x} + \vec{\mathbf{v}}_i) - \alpha_i(\mathbf{y}).
$$

This time

$$
[\mathbf{D}g(\mathbf{y})]\vec{\mathbf{v}}_i = -\alpha_i(\vec{\mathbf{v}}_i) < 0,
$$

so at such a point $\vec{v}_i$ is outward pointing.

**6.7.1** a.

$$d(x_1x_3\,dx_3 \wedge dx_4) = \Big(D_1(x_1x_3)dx_1 + D_2(x_1x_3)dx_2 + D_3(x_1x_3)dx_3 + D_4(x_1x_3)dx_4\Big) \wedge dx_3 \wedge dx_4$$

$$= (x_3dx_1 + x_1dx_3) \wedge dx_3 \wedge dx_4 = x_3\,dx_1 \wedge dx_3 \wedge dx_4.$$

b. $d(\cos xy\,dx \wedge dy) = (-\sin xy\,dx - \sin xy\,dy) \wedge dx \wedge dy = 0.$

**6.7.3** When computing the exterior derivative, remember that any terms containing $dx_i \wedge dx_i$ are 0.

a. $\quad dW_{\vec{F}_2} = d\left(\dfrac{-y}{x^2+y^2}\,dx + \dfrac{x}{x^2+y^2}\,dy\right) = d\left(\dfrac{-y}{x^2+y^2}dx\right) + d\left(\dfrac{x}{x^2+y^2}dy\right)$

$$= \left(D_1\frac{-y}{x^2+y^2}dx + D_2\frac{-y}{x^2+y^2}dy\right) \wedge dx + \left(D_1\frac{x}{x^2+y^2}dx + D_2\frac{x}{x^2+y^2}dy\right) \wedge dy$$

$$= \frac{y^2-x^2}{(x^2+y^2)^2}\,dy \wedge dx + \frac{(x^2+y^2)-2x^2}{(x^2+y^2)^2}\,dx \wedge dy = 0.$$

b. Set $\vec{F}_3 = \begin{pmatrix} F_1 \\ F_2 \\ F_3 \end{pmatrix}$. Then

$$d\Phi_{\vec{F}_3} = d\Big(F_1\,dy \wedge dz - F_2\,dx \wedge dz + F_3\,dx \wedge dy\Big)$$

$$= d\left(\frac{x}{(x^2+y^2+z^2)^{3/2}}dy \wedge dz - \frac{y}{(x^2+y^2+z^2)^{3/2}}dx \wedge dz + \frac{z}{(x^2+y^2+z^2)^{3/2}}dx \wedge dy\right)$$

$$= \frac{(x^2+y^2+z^2)^{3/2}\cdot 1 - x\frac{3}{2}(x^2+y^2+z^2)^{1/2}\cdot 2x}{(x^2+y^2+z^2)^3}\,dx \wedge dy \wedge dz$$

$$- \frac{(x^2+y^2+z^2)^{3/2} - 3y^2(x^2+y^2+z^2)^{1/2}}{(x^2+y^2+z^2)^3}\,dy \wedge dx \wedge dz$$

$$+ \frac{(x^2+y^2+z^2)^{3/2} - 3z^2(x^2+y^2+z^2)^{1/2}}{(x^2+y^2+z^2)^3}\,dz \wedge dz \wedge dy$$

$$= \frac{3(x^2+y^2+z^2)^{3/2} - 3(x^2+y^2+z^2)^{1/2}(x^2+y^2+z^2)}{(x^2+y^2+z^2)^3} = 0.$$

Solution 6.7.5, part a: If you just integrate everything in sight, this is a long computation. To make it bearable you need to (1) not compute terms that don't need to be computed, since they will disappear in the limit, and (2) take advantage of cancellations early in the game. To interpret the answer, it also helps to know what one expects to find (more on that later).

**6.7.5** a. Computing the exterior derivative from the definition means computing the integral on the right side of

$$d(z^2\,dx \wedge dy)\big(P_{\mathbf{x}}(\vec{v}_1, \vec{v}_2, \vec{v}_3)\big) = \lim_{h \to 0}\frac{1}{h^3}\int_{\partial P_{\mathbf{x}}(h\vec{v}_1, h\vec{v}_2, h\vec{v}_3)} z^2\,dx \wedge dy.$$

Thus we must integrate the 2-form field $z^2\,dx \wedge dy$ over the boundary of the 3-parallelogram spanned by $h\vec{v}_1, h\vec{v}_2, h\vec{v}_3$, i.e., over the six faces of the 3-parallelogram shown in the margin on the next page.

We parametrize those six faces as follows, where $0 \leq s$, $t \leq h$, and

$$\mathbf{x} = \begin{pmatrix} x \\ y \\ z \end{pmatrix};$$ we use proposition 6.6.21 to determine which faces are taken with a plus sign and which are taken with a minus sign:

1. $\gamma \begin{pmatrix} s \\ t \end{pmatrix} = \mathbf{x} + s\vec{\mathbf{v}}_1 + t\vec{\mathbf{v}}_2$, minus. (This is the base of the box shown at left.)

2. $\gamma \begin{pmatrix} s \\ t \end{pmatrix} = \mathbf{x} + h\vec{\mathbf{v}}_3 + s\vec{\mathbf{v}}_1 + t\vec{\mathbf{v}}_2$, plus. (This is the top of the box.)

3. $\gamma \begin{pmatrix} s \\ t \end{pmatrix} = \mathbf{x} + s\vec{\mathbf{v}}_1 + t\vec{\mathbf{v}}_3$, plus. (This is the front of the box.)

4. $\gamma \begin{pmatrix} s \\ t \end{pmatrix} = \mathbf{x} + h\vec{\mathbf{v}}_2 + s\vec{\mathbf{v}}_1 + t\vec{\mathbf{v}}_3$, minus. (This is the back of the box.)

5. $\gamma \begin{pmatrix} s \\ t \end{pmatrix} = \mathbf{x} + s\vec{\mathbf{v}}_2 + t\vec{\mathbf{v}}_3$, minus. (This is the left side of the box.)

6. $\gamma \begin{pmatrix} s \\ t \end{pmatrix} = \mathbf{x} + h\vec{\mathbf{v}}_1 + s\vec{\mathbf{v}}_2 + t\vec{\mathbf{v}}_3$, plus. (This is the right side of the box.)

FIGURE FOR SOLUTION 6.7.5.

The 3-parallelogram spanned by $h\vec{\mathbf{v}}_1, h\vec{\mathbf{v}}_2, h\vec{\mathbf{v}}_3$.

Note that the two first faces are spanned by $\vec{\mathbf{v}}_1$ and $\vec{\mathbf{v}}_2$ (or translates thereof), the base taken with $-$ and the top taken with $+$. The next two faces, with opposite signs, are spanned by $\vec{\mathbf{v}}_1$ and $\vec{\mathbf{v}}_3$, and the two sides, with opposite signs, are spanned by $\vec{\mathbf{v}}_2$ and $\vec{\mathbf{v}}_3$.

To integrate our form field over these parametrized domains we use definition 6.2.1. The computations are tedious, but we do not have to integrate everything in sight. For one thing, common terms that cancel can be ignored; for another, anything that amounts to a term in $h^4$ can be ignored, since $h^4/h^3$ will vanish in the limit as $h \to 0$.

We will compute in detail the integrals over the first two faces. The integral over the first face (the base) is the following, where $v_{1,3}$ denotes the third entry of $\vec{\mathbf{v}}_1$ and $v_{2,3}$ denotes the third entry of $\vec{\mathbf{v}}_2$:

$$-\int_0^h \int_0^h \underbrace{(z + sv_{1,3} + tv_{2,3})^2 dx \wedge dy}_{z^2 dx \wedge dy \text{ eval. at } \gamma \binom{s}{t}} \underbrace{(\vec{\mathbf{v}}_1, \vec{\mathbf{v}}_2)}_{\overrightarrow{D_s\gamma}, \overrightarrow{D_t\gamma}} \, ds \, dt$$

$$= -\int_0^h \int_0^h \Big( \underbrace{z^2}_{\text{term in } h^2} + \underbrace{s^2 v_{1,3}^2 + t^2 v_{2,3}^2 + 2st v_{1,3} v_{2,3}}_{\text{terms in } h^4} + \underbrace{2sz v_{1,3} + 2tz v_{2,3}}_{\text{terms in } h^3} \Big)(v_{1,1}v_{2,2} - v_{1,2}v_{2,1}) \, ds \, dt$$

Note that the integral $\int_0^h \int_0^h z^2 \, ds \, dt$ will give a term in $h^2$, and the next three terms will give a term in $h^4$; for example, $\int_0^h s^2 v_{1,3}^2 \, ds$ gives an $h^3$ and $\int_0^h s^2 v_{1,3}^2 \, dt$ gives an $h$, making $h^4$ in all. These higher degree terms can be disregarded; we will denote them below by $O(h^4)$. This gives the following integral over the first face:

$$-\int_0^h \int_0^h \big(z^2 + 2sz v_{1,3} + 2tz v_{2,3} + O(h^4)\big)(v_{1,1}v_{2,2} - v_{1,2}v_{2,1}) \, ds \, dt$$

$$= -\big(h^2 z^2 + h^3 z v_{1,3} + h^3 z v_{2,3} + O(h^4)\big)(v_{1,1}v_{2,2} - v_{1,2}v_{2,1}).$$

Before computing the integral over the second face (the top), notice that it is exactly like the first face except that it also has the term $h\vec{v}_3$. This face comes with a plus sign, while the first face comes with a minus sign, so when we integrate over the second face, the identical terms will cancel each other. Thus we didn't actually have to compute the integral over the first face at all! In computing the contribution of the first two faces to the integral, we need only concern ourselves with those terms in $(z + hv_{3,3} + sv_{1,3} + tv_{2,3})^2$ that contain $hv_{3,3}$: i.e., $h^2v_{3,1}^2$, $2hsv_{3,3}v_{1,3}$, $2htv_{3,3}v_{2,3}$, and $2zhv_{3,3}$. Integrating the first three would give terms in $h^4$, so the entire contribution to the integral of the first two faces is

$$\int_0^h \int_0^h \left(2zhv_{3,3}+O(h^4)\right)(v_{1,1}v_{2,2}-v_{1,2}v_{2,1})\,ds\,dt = (2zh^3v_{3,3}+O(h^4))(v_{1,1}v_{2,2}-v_{1,2}v_{2,1}).$$

Similarly, the entire contribution of the second pair of faces is

$$\int_0^h \int_0^h \left(2zhv_{2,3}+O(h^4)\right)(v_{1,1}v_{3,2}-v_{3,1}v_{1,2})\,ds\,dt = (2zh^3v_{2,3}+O(h^4))(v_{1,1}v_{3,2}-v_{3,1}v_{1,2}).$$

(The partial derivatives are different, so we have $(v_{1,1}v_{3,2} - v_{3,1}v_{1,2})$, not $(v_{1,1}v_{2,2}-v_{1,2}v_{2,1})$ as before.) And the contribution of the last pair of faces is

$$\int_0^h \int_0^h \left(2zhv_{1,3}+O(h^4)\right)(v_{2,1}v_{3,2}-v_{2,2}v_{3,1})\,ds\,dt = (2zh^3v_{1,3}+O(h^4))(v_{2,1}v_{3,2}-v_{2,2}v_{3,1}).$$

Dividing by $h^3$ and taking the limit as $h \to 0$ gives

$$d(z^2\,dx \wedge dy)\big(P_{\mathbf{x}}(\vec{v}_1,\vec{v}_2,\vec{v}_3)\big) = (2zv_{3,3})(v_{1,1}v_{2,2} - v_{1,2}v_{2,1}) + (2zv_{2,3})(v_{1,1}v_{3,2} - v_{3,1}v_{1,2})$$
$$+ (2zv_{1,3})(v_{2,1}v_{3,2} - v_{2,2}v_{3,1}).$$

Note that in this solution we have reversed our usual rule for subscripts, where the row comes first and column second.

Now it helps to know what one is looking for. Since $d(z^2\,dx \wedge dy)$ is a 3-form on $\mathbb{R}^3$, it is a multiple of the determinant. If you compute

$$\det[\vec{v}_1, \vec{v}_2, \vec{v}_3] = \begin{bmatrix} v_{1,1} & v_{2,1} & v_{3,1} \\ v_{1,2} & v_{2,2} & v_{3,2} \\ v_{1,3} & v_{2,3} & v_{3,3} \end{bmatrix},$$

you will see that

$$d(z^2\,dx \wedge dy)\big(P_{\mathbf{x}}(\vec{v}_1,\vec{v}_2,\vec{v}_3)\big) = 2z\det[\vec{v}_1,\vec{v}_2,\vec{v}_3].$$

b. Using theorem 6.7.2, we have

$$d(z^2\,dx \wedge dy) \overset{\text{part 5}}{=} d\,z^2 \wedge dx \wedge dy$$

$$\overset{\text{part 4}}{=} (D_1z^2\,dx + D_2z^2\,dy + D_3z^2\,dz) \wedge dx \wedge dy$$

$$= 2z\,dz \wedge dx \wedge dy$$

$$\overset{\text{prop. 6.1.22}}{=} 2z\,dx \wedge dy \wedge dz.$$

**6.7.7** a. The four edges of $P_{-\mathbf{e}_2}(h\vec{\mathbf{e}}_2, h\vec{\mathbf{e}}_3)$ are parametrized by

$$1. \quad t \mapsto \begin{pmatrix} 0 \\ -1+h \\ 0 \end{pmatrix} + t \begin{bmatrix} 0 \\ 0 \\ h \end{bmatrix} \qquad 2. \quad t \mapsto \begin{pmatrix} 0 \\ -1 \\ 0 \end{pmatrix} + t \begin{bmatrix} 0 \\ 0 \\ h \end{bmatrix}$$

$$3. \quad t \mapsto \begin{pmatrix} 0 \\ -1 \\ h \end{pmatrix} + t \begin{bmatrix} 0 \\ h \\ 0 \end{bmatrix} \qquad 4. \quad t \mapsto \begin{pmatrix} 0 \\ -1 \\ 0 \end{pmatrix} + t \begin{bmatrix} 0 \\ h \\ 0 \end{bmatrix}.$$

The second and third are taken with a minus sign, and the first and fourth with a plus sign.

Only the first two contribute to the integral around the boundary, since $dx_3$ returns 0 for vectors with no vertical component. They give

$$\int_0^1 (-1+h)^2 h \, dt - \int_0^1 (-1)^2 h \, dt = -2h^2 + h^3.$$

Divide by $h^2$ and take the limit as $h \to 0$, to find $-2$.

b. We find

$$d(x_2^2 dx_3) = 2x_2 \, dx_2 \wedge dx_3,$$

and

$$2x_2 \, dx_2 \wedge dx_3 P_{-\mathbf{e}_2}(\vec{\mathbf{e}}_2, \vec{\mathbf{e}}_3) = -2 \cdot 1 = -2.$$

**6.7.9** First compute $d\omega$:

$$d\omega = d\Big( p(y,z) \, dx + q(x,z) \, dy \Big)$$
$$= D_2 p(y,z) \, dy \wedge dx + D_3 p(y,z) \, dz \wedge dx + D_1 q(x,z) \, dx \wedge dy$$
$$+ D_3 q(x,z) \, dz \wedge dy;$$

comparing this with the formula $d\omega = x \, dy \wedge dz + y \, dx \wedge dz$, we see that

$$x \, dy \wedge dz = D_3 q(x,z) \, dz \wedge dy; \quad \text{i.e.,} \quad D_3 q(x,z) = -x;$$
$$y \, dx \wedge dz = D_3 p(y,z) \, dz \wedge dx; \quad \text{i.e.,} \quad D_3 p(y,z) = -y.$$

In addition,

$$0 = D_1 q(x,z) \, dx \wedge dy - D_2 p(y,z) \, dx \wedge dy; \quad \text{i.e.,} \quad D_1 q(x,z) = D_2 p(y,z).$$

The functions $q(x,z) = -xz$ and $p(y,z) = -yz$ satisfy these constraints, which gives us the 1-forms of the form

$$\omega = -yz \, dx - xz \, dy.$$

**6.7.11** a. This is a restatement of theorem 1.8.1, part 5.

b. Any $k$-form can be written as a sum of $k$-forms of the form

$$a(\mathbf{x}) \, dx_{i_1} \wedge \cdots \wedge dx_{i_k},$$

and any $l$-form can be written as a sum of $l$-forms of the form

$$b(\mathbf{x}) \, dx_{j_1} \wedge \cdots \wedge dx_{j_l}.$$

The result then follows from theorem 6.7.2, part 2 (the exterior derivative of the sum equals the sum of the exterior derivatives) and from proposition 6.1.22 (distributivity of the wedge product). We will work it out in the case of a $k$-form $\varphi = \varphi_1 + \varphi_2$ and an $l$-form $\psi$, where we assume that theorem 6.7.8 is true for $\varphi_1$, $\varphi_2$, and $\psi$.

We have

$$d\big((\varphi_1 + \varphi_2) \wedge \psi\big) = d(\varphi_1 \wedge \psi + \varphi_2 \wedge \psi) = d(\varphi_1 \wedge \psi) + d(\varphi_2 \wedge \psi)$$

$$= \Big(d\varphi_1 \wedge \psi + (-1)^k \varphi_1 \wedge d\psi\Big) + \Big(d\varphi_2 \wedge \psi + (-1)^k \varphi_2 \wedge d\psi\Big)$$

$$= \big((d\varphi_1 + d\varphi_2) \wedge \psi\big) + (-1)^k \big(\varphi_1 \wedge d\psi + \varphi_2 \wedge d\psi\big)$$

$$= d(\varphi_1 + \varphi_2) \wedge \psi + (-1)^k (\varphi_1 + \varphi_2) \wedge d\psi.$$

To move the $b$ next to the $a$ in line [2] we use proposition 6.1.22 (skew commutativity): since $b$ is a 0-form we have

$$dx_{i_1} \wedge \cdots \wedge dx_{i_k} \wedge b$$
$$= (-1)^{0 \cdot k} b \wedge dx_{i_1} \wedge \cdots \wedge dx_{i_k}$$
$$= b \, dx_{i_1} \wedge \cdots \wedge dx_{i_k}.$$

To go to line [2] we also use equation 6.7.8.

To go to line [3] we apply the formula $d(fg) = f \, dg + g \, df$ (theorem 6.7.8, with $k = 0$).

To go to [4] we use the distributivity of the wedge product (proposition 6.1.22).

To go from [4] to [5] we use the commutativity of 0-forms. To go from [5] to [6] we again use skew commutativity, this time moving the 1-form $db$.

c. To simplify notation, in the equation below we write $a(\mathbf{x})$ and $b(\mathbf{x})$ as $a$ and $b$. Recall that we can write the wedge product of a 0-form $f$ and a form $\alpha$ either as $f\alpha$ or as $f \wedge \alpha$.

We now treat the function $ab$ as the function $f$ of theorem 6.7.2, part 5.

$$d\big(\varphi \wedge \psi\big) = d\big(a \, dx_{i_1} \wedge \cdots \wedge dx_{i_k} \wedge b \, dx_{j_1} \wedge \cdots \wedge dx_{j_l}\big) \tag{1}$$

$$= d(ab) \wedge dx_{i_1} \wedge \cdots \wedge dx_{i_k} \wedge dx_{j_1} \wedge \cdots \wedge dx_{j_l} \tag{2}$$

$$= (a \, db + b \, da) \wedge dx_{i_1} \wedge \cdots \wedge dx_{i_k} \wedge dx_{j_1} \wedge \cdots \wedge dx_{j_l} \tag{3}$$

$$= a \, db \wedge dx_{i_1} \wedge \cdots \wedge dx_{i_k} \wedge dx_{j_1} \wedge \cdots \wedge dx_{j_l}$$
$$+ b \underbrace{da \wedge dx_{i_1} \wedge \cdots \wedge dx_{i_k}}_{d\varphi} \wedge dx_{j_1} \wedge \cdots \wedge dx_{j_l} \tag{4}$$

$$= a \, db \wedge dx_{i_1} \wedge \cdots \wedge dx_{i_k} \wedge dx_{j_1} \wedge \cdots \wedge dx_{j_l}$$
$$+ d\varphi \wedge \underbrace{b \wedge dx_{j_1} \wedge \cdots \wedge dx_{j_l}}_{\psi} \tag{5}$$

$$= \underbrace{a \wedge dx_{i_1} \wedge \cdots \wedge dx_{i_k}}_{\varphi} \wedge (-1)^k \underbrace{db \, dx_{j_1} \wedge \cdots \wedge dx_{j_l}}_{d\psi} + d\varphi \wedge \psi \tag{6}$$

$$= d\varphi \wedge \psi + (-1)^k \varphi \wedge d\psi.$$

**6.8.1**

$$\operatorname{grad} f = \begin{bmatrix} 2xy \\ x^2 \\ 1 \end{bmatrix}; \qquad \operatorname{curl} \vec{F} = \begin{bmatrix} D_1 \\ D_2 \\ D_3 \end{bmatrix} \times \begin{bmatrix} -y \\ x \\ xz \end{bmatrix} = \begin{bmatrix} 0 \\ -z \\ 2 \end{bmatrix};$$

$$\operatorname{div} \vec{F} = \begin{bmatrix} D_1 \\ D_2 \\ D_3 \end{bmatrix} \cdot \begin{bmatrix} -y \\ x \\ xz \end{bmatrix} = x.$$

**6.8.3** a. Numbers: $dx \wedge dy(\vec{v}, \vec{w})$, $\vec{u} \cdot (\vec{v} \times \vec{w})$, $\operatorname{grad} f(\mathbf{x}) \cdot \vec{v}$, $\operatorname{div} \vec{F}$

Vectors: $\operatorname{grad} f$, $\operatorname{curl} \vec{F}$

b.
$$\text{grad } f = \vec{\nabla} f$$
$$\text{div } \vec{F} = \vec{\nabla} \cdot \vec{F}$$
$$\text{curl } \vec{F} = \vec{\nabla} \times \vec{F}$$
$$df = W_{\vec{\nabla} f} = W_{\text{grad } f} = D_1 f \, dx_1 + D_2 f \, dx_2 + D_3 f \, dx_3$$
$$dW_{\vec{F}} = \Phi_{\text{curl } \vec{F}} = \Phi_{\vec{\nabla} \times \vec{F}}$$
$$d\Phi_{\vec{F}} = M_{\text{div } \vec{F}} = M_{\vec{\nabla} \cdot \vec{F}}$$

c. Of course more than one right answer is possible; for example, in (i), grad $\vec{F}$ could be changed to curl $\vec{F}$. In (iii), $\Phi_{\vec{F}}(\vec{v}_1, \vec{v}_2, \vec{v}_3)$ is meaningful if it is in $\mathbb{R}^4$.

    i.   grad $f$    ii.  curl $\vec{F}$    iii.  $\Phi_{\vec{F}}(\vec{v}_1, \vec{v}_2)$

    iv.  $W_{\vec{F}}$    v.  unchanged    vi.  unchanged

**6.8.5** a. $\nabla f \begin{pmatrix} x \\ y \end{pmatrix} = \begin{bmatrix} 0 \\ 2y \end{bmatrix}$      b. $\nabla f \begin{pmatrix} x \\ y \end{pmatrix} = \begin{bmatrix} 2x \\ -2y \end{bmatrix}$

     c. $\nabla f \begin{pmatrix} x \\ y \end{pmatrix} = \frac{1}{x^2 + y^2} \begin{bmatrix} 2x \\ 2y \end{bmatrix}$      d. $\nabla f \begin{pmatrix} x \\ y \\ z \end{pmatrix} = \frac{\text{sgn}(x + y + z)}{|x + y + z|} \begin{bmatrix} 1 \\ 1 \\ 1 \end{bmatrix}$

**6.8.7** a. If $\vec{F} = \text{grad } f$, then $D_1 f = F_1$ and $D_2 f = F_2$, so that $D_2 F_1 = D_2 D_1 f$ and $D_1 F_2 = D_1 D_2 f$; these crossed partials are equal since $f$ is of class $C^2$ (theorem 3.3.9).

A less direct way to show this is to say that the exterior derivative of
$$df = W_{\text{grad } f} = W_{\begin{bmatrix} F_1 \\ F_2 \end{bmatrix}} = (F_1 \, dx + F_2 \, dy)$$

must be 0:

$$d(F_1 \, dx + F_2 \, dy) = D_2 F_1 \, dy \wedge dx + D_1 F_2 \, dx \wedge dy = 0, \text{ i.e., } D_1 F_2 - D_2 F_1 = 0.$$

b. If the second derivatives of $f$ are not continuous, then the crossed partials $D_2 D_1 f$ and $D_1 D_2 f$ are not necessarily equal (as you saw in example 3.3.11).

**6.8.9** a. $W_{\begin{bmatrix} x \\ y \\ z \end{bmatrix}} = x \, dx + y \, dy + z \, dz$ and

$$d(x \, dx + y \, dy + z \, dz) = dx \wedge dx + dy \wedge dy + dz \wedge dz = 0.$$

Indeed, a function $f$ exists such that grad $f = \begin{bmatrix} x \\ y \\ z \end{bmatrix}$; it is the function

$$f \begin{pmatrix} x \\ y \\ z \end{pmatrix} = \frac{x^2 + y^2 + z^2}{2}.$$

b. $\Phi_{\begin{bmatrix} x \\ y \\ z \end{bmatrix}} = x\,dy \wedge dz + y\,dz \wedge dx + z\,dx \wedge dy$; the exterior derivative is

$$d(x\,dy \wedge dz + y\,dz \wedge dx + z\,dx \wedge dy)$$
$$= dx \wedge dy \wedge dz + dy \wedge dz \wedge dx + dz \wedge dx \wedge dy$$
$$= 3\,dx \wedge dy \wedge dz.$$

**6.8.11  a.**
$$\operatorname{div} \begin{bmatrix} x^2 y \\ -2yz \\ x^3 y^2 \end{bmatrix} = \begin{bmatrix} D_1 \\ D_2 \\ D_3 \end{bmatrix} \cdot \begin{bmatrix} x^2 y \\ -2yz \\ x^3 y^2 \end{bmatrix} = 2xy - 2z$$

$$\operatorname{curl} \begin{bmatrix} x^2 y \\ -2yz \\ x^3 y^2 \end{bmatrix} = \begin{bmatrix} D_1 \\ D_2 \\ D_3 \end{bmatrix} \times \begin{bmatrix} x^2 y \\ -2yz \\ x^3 y^2 \end{bmatrix} = \begin{bmatrix} 2x^3 y + 2y \\ -3x^2 y^2 \\ -x^2 \end{bmatrix}$$

b.

$$\operatorname{div} \begin{bmatrix} \sin xz \\ \cos yz \\ xyz \end{bmatrix} = \begin{bmatrix} D_1 \\ D_2 \\ D_3 \end{bmatrix} \cdot \begin{bmatrix} \sin xz \\ \cos yz \\ xyz \end{bmatrix} = z\cos xz - z\sin yz + xy$$

$$\operatorname{curl} \begin{bmatrix} \sin xz \\ \cos yz \\ xyz \end{bmatrix} = \begin{bmatrix} D_1 \\ D_2 \\ D_3 \end{bmatrix} \times \begin{bmatrix} \sin xz \\ \cos yz \\ xyz \end{bmatrix} = \begin{bmatrix} xz + y\sin yz \\ x\cos xz - yz \\ 0 \end{bmatrix}.$$

**c. Curl of the vector field in part a**

We will compute $dW_{\vec{F}} = \Phi_{\operatorname{curl}\vec{F}}$, the flux form field of curl $\vec{F}$, and from that we will compute curl $\vec{F}$. Since

$$W_{\vec{F}} = x^2 y\,dx - 2yz\,dy + x^3 y^2\,dz$$

is a 1-form, its exterior derivative is a 2-form. By definition 6.7.1, we have

$$dW_{\vec{F}} = \lim_{h \to 0} \frac{1}{h^2} \int_{\partial(P_{\mathbf{x}}(h\vec{v}_1, h\vec{v}_2))} x^2 y\,dx - 2yz\,dy + x^3 y^2\,dz.$$

If you attempt to compute this using random vectors $\vec{v}_1, \vec{v}_2$, you will regret it.

Rather than computing this with vectors $\vec{v}_1, \vec{v}_2$, remember (theorem 6.1.8) that any 2-form can be written in terms of the elementary 2-forms. So

$$dW_{\vec{F}} = \Phi_{\operatorname{curl}\vec{F}} = a\,dx \wedge dy + b\,dy \wedge dz + c\,dx \wedge dz \qquad (1)$$

for some coefficients $a, b, c$. Theorem 6.1.8 says that to determine the coefficients, we should evaluate $dW_{\vec{F}}$ on the standard basis vectors.

Thus to determine the coefficient of $dx \wedge dy$ we will integrate $W_{\vec{F}}$ over the oriented boundary of the parallelogram spanned by $h\vec{e}_1, h\vec{e}_2$, computing

$$\lim_{h \to 0} \frac{1}{h^2} \int_{\partial P_{\mathbf{x}}(h\vec{e}_1, h\vec{e}_2)} x^2 y\,dx - 2yz\,dy + x^3 y^2\,dz.$$

To do this, we parametrize each edge of the parallelogram shown in the figure at left and give it a plus or minus depending on its orientation:

1. $P_{\mathbf{x}}(h\vec{e}_1)$        parametrized by     $\gamma_1(t) = \mathbf{x} + t\vec{e}_1$

2. $P_{\mathbf{x}+h\vec{\mathbf{e}}_1}(h\vec{\mathbf{e}}_2)$    parametrized by    $\gamma_2(t) = \mathbf{x} + h\vec{\mathbf{e}}_1 + t\vec{\mathbf{e}}_2$

3. $P_{\mathbf{x}+h\vec{\mathbf{e}}_2}(h\vec{\mathbf{e}}_1)$    parametrized by    $\gamma_3(t) = \mathbf{x} + h\vec{\mathbf{e}}_2 + t\vec{\mathbf{e}}_1$

4. $P_{\mathbf{x}}(h\vec{\mathbf{e}}_2)$    parametrized by    $\gamma_4(t) = \mathbf{x} + t\vec{\mathbf{e}}_2$

First, we will determine the orientation of each edge. Using proposition 6.6.21, we have

$$\partial P_{\mathbf{x}}(h\vec{\mathbf{e}}_1, h\vec{\mathbf{e}}_2) = (-1)^0 \left( P_{\mathbf{x}+h\vec{\mathbf{e}}_1}(h\vec{\mathbf{e}}_2) - P_{\mathbf{x}}(h\vec{\mathbf{e}}_2) \right) + (-1)^1 \left( P_{\mathbf{x}+h\vec{\mathbf{e}}_2}(h\vec{\mathbf{e}}_1) - P_{\mathbf{x}}(h\vec{\mathbf{e}}_1) \right),$$

so edges 1 and 2 come with a plus sign, while 3 and 4 get a minus.

We must integrate $x^2y\,dx - 2yz\,dy + x^3y^2\,dz$ over each parametrized edge, but note that for the first edge (and the third), we are only concerned with $x^2y\,dx$, since $dy(\vec{\mathbf{e}}_1) = 0$ and $dz(\vec{\mathbf{e}}_1) = 0$. If you don't see this, recall (equation 6.4.35) how we integrate forms. To integrate

$$x^2y\,dx - 2yz\,dy + x^3y^2\,dz$$

over the first edge, we integrate $x^2y\,dx - 2yz\,dy + x^3y^2\,dz(\overrightarrow{D_t\gamma_1}(t))$ over the first edge parametrized by $\gamma_1$. But $\overrightarrow{D_t\gamma_1}(t) = \vec{\mathbf{e}}_1$.

Thus for edge 1 we compute

$$\frac{1}{h^2} \int_{\partial(P_{\mathbf{x}}(h\vec{\mathbf{e}}_1, h\vec{\mathbf{e}}_2))} x^2y\,dx = \frac{1}{h^2} \int_0^h \overbrace{(x+t)^2 y}^{x^2y \text{ eval. at } \gamma_1(t)} \, dx(\vec{\mathbf{e}}_1)\,dt$$

$$= \frac{1}{h^2} \int_0^h x^2y + 2xyt + t^2y\,dt$$

$$= \frac{1}{h^2} \left( x^2yh + xyh^2 + \cdots \right)$$

FIGURE FOR SOLUTION 6.8.11.

FIGURE FOR SOLUTION 6.8.11.

The parallelogram spanned by $h\vec{\mathbf{e}}_1, h\vec{\mathbf{e}}_2$. In our calculations, we denote by $x, y, z$ the entries of $\mathbf{x}$:

$$\mathbf{x} = \begin{pmatrix} x \\ y \\ x \end{pmatrix}.$$

The $h$ in the denominator might be worrisome – what will happen as $h \to 0$? But we will see that it cancels with a term from another edge. That is the point of having an oriented boundary!

Notice that $t^2y$ gives a term in $h^3$; once we divide by $h^2$ it will be a term in $h$, which will go to 0 as $h \to 0$. So we can ignore it. The terms that count for edge 1 are $\dfrac{x^2y}{h} + xy$, both taken with a +.

For edge 2 we are only concerned with $-2yz\,dy$, since $dx(\vec{\mathbf{e}}_2) = 0$ and $dz(\vec{\mathbf{e}}_2) = 0$. We compute

$$\frac{1}{h^2} \int_0^h -2yz - 2tz\,dt,$$

which gives $\dfrac{-2yz}{h} - z$.

A similar computation for edge 3 gives the terms $\dfrac{x^2y}{h} + xy + x^2$, each term taken with a minus sign; for edge 4 we get $\dfrac{-2yz}{h} - z$, also taken with a minus sign. Thus we have

$$\lim_{h \to 0} \frac{1}{h^2} \int_{\partial(P_{\mathbf{x}}(h\vec{e}_1, h\vec{e}_2))} x^2 y \, dx - 2yz \, dy + x^3 y^2 \, dz$$

$$= \underbrace{\frac{x^2 y}{h} + xy}_{\text{from edge 1}} + \underbrace{\frac{-2yz}{h} - z}_{\text{from edge 2}} - \underbrace{\frac{x^2 y}{h} - xy - x^2}_{\text{from edge 3}} + \underbrace{\frac{2yz}{h} + z}_{\text{from edge 4}}$$

$$= -x^2.$$

We can substitute this for $a$ in equation (1):

$$\Phi_{\text{curl}\,\vec{F}} = -x^2 \, dx \wedge dy + b \, dy \wedge dz + c \, dx \wedge dz.$$

Recall that $\Phi_{\vec{F}} = F_1 \, dy \wedge dz - F_2 \, dx \wedge dz + F_3 \, dx \wedge dy$, so $-x^2$ should be the third entry of curl $\vec{F}$, which is indeed what we got in part a (with considerably less effort!).

For the coefficient of $dx \wedge dz$ we integrate over the boundary of the parallelogram spanned by $h\vec{e}_1, h\vec{e}_3$. To save work, we note that for edges 1 and 3 we are interested only in $x^2 y \, dx$, and for edges 2 and 4 we are only interested in $x^3 y^2 \, dz$. We get

$$\partial\big(P_{\mathbf{x}}(h\vec{e}_1, h\vec{e}_3)\big) = \underbrace{\frac{x^2 y}{h} + xy}_{\text{from edge 1}} + \underbrace{\frac{x^3 y^2}{h} + 3x^2 y^2}_{\text{from edge 2}} - \underbrace{\frac{x^2 y}{h} - xy}_{\text{from edge 3}} - \frac{x^3 y^2}{h} = 3x^2 y^2$$

Since in the formula for $\Phi_{\vec{F}}$ the coefficient of $dx \wedge dz$ is $-F_2$, the second entry of curl $\vec{F}$ should be $-3x^2 y^2$, which is what we got in part a.

A similar computation involving $\vec{e}_2, \vec{e}_3$ gives the coefficient for $dy \wedge dz$, i.e., the first entry of curl $\vec{F}$.

### Div of the vector field in part a

Now let's compute the divergence of the vector field in part a, by computing $d\Phi_{\vec{F}} = M_{\text{div}\,\vec{F}}$. Since

$$\Phi_{\vec{F}} = x^2 y \, dy \wedge dz + 2yz \, dx \wedge dz + x^3 y^2 \, dx \wedge dy$$

is a 2-form, $d\Phi_{\vec{F}}$ is a 3-form and can be written $\alpha \, dx \wedge dy \wedge dz$ for some coefficient $\alpha = \text{div}\,\vec{F}$. We will compute $d\Phi_{\vec{F}}$ by integrating

$$x^2 y \, dy \wedge dz + 2yz \, dx \wedge dz + x^3 y^2 \, dx \wedge dy$$

over the oriented boundary of the parallelogram spanned by $h\vec{e}_1, h\vec{e}_2, h\vec{e}_3$, shown at left.

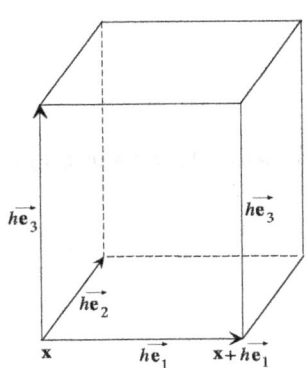

The boundary consists of six faces, which we parametrize as follows, where $0 \le s, t \le h$, and $\mathbf{x} = \begin{pmatrix} x \\ y \\ z \end{pmatrix}$; we use proposition 6.6.21 to determine which faces are taken with a plus sign and which are taken with a minus sign:

1. $\gamma \begin{pmatrix} s \\ t \end{pmatrix} = \mathbf{x} + s\vec{e}_1 + t\vec{e}_2$, minus. (The base of the box shown at left.)

2. $\gamma\begin{pmatrix} s \\ t \end{pmatrix} = \mathbf{x} + h\vec{e}_3 + s\vec{e}_1 + t\vec{e}_2$, plus. (The top of the box.)

3. $\gamma\begin{pmatrix} s \\ t \end{pmatrix} = \mathbf{x} + s\vec{e}_1 + t\vec{e}_3$, plus. (The front of the box.)

4. $\gamma\begin{pmatrix} s \\ t \end{pmatrix} = \mathbf{x} + h\vec{e}_2 + s\vec{e}_1 + t\vec{e}_3$, minus. (The back of the box.)

5. $\gamma\begin{pmatrix} s \\ t \end{pmatrix} = \mathbf{x} + s\vec{e}_2 + t\vec{e}_3$, minus. (The left side of the box.)

6. $\gamma\begin{pmatrix} s \\ t \end{pmatrix} = \mathbf{x} + h\vec{e}_1 + s\vec{e}_2 + t\vec{e}_3$, plus. (The right side of the box.)

For the integral over the first face, we are concerned only with the term $x^3 y^2 \, dx \wedge dy$, since $dy \wedge dz(\vec{e}_1, \vec{e}_2) = 0$ and $dx \wedge dz(\vec{e}_1, \vec{e}_2) = 0$. So we have

$$\frac{1}{h^3} \int_{\partial P_{\mathbf{x}}(h\vec{e}_1, h\vec{e}_2)} x^3 y^2 \, dx \wedge dy = \frac{1}{h^3} \int_0^h \int_0^h \overbrace{\underbrace{(x+s)^3 (y+t)^2 \, dx \wedge dy}_{\overrightarrow{D_s\gamma}, \overrightarrow{D_t\gamma}} (\vec{e}_1, \vec{e}_2)}^{x^3 y^2 dx \wedge dy \text{ eval. at } \gamma\begin{pmatrix} s \\ t \end{pmatrix}} \, ds \, dt$$

After discarding everything that will give terms in $h^4$, dividing by $h^3$, and giving the correct orientation, we are left with the following for the first face:

$$\frac{-x^3 y^2}{h} - \frac{3x^2 y^2}{2} - x^3 y.$$

For the top of the box we get exactly the same terms, with opposite sign. For the third face (the front), we get

$$\frac{2yz}{h} + y.$$

For the fourth face (the back), we get

$$\frac{-2yz}{h} - y - 2z.$$

For the fifth and sixth faces, we are concerned only with $x^2 y \, dy \wedge dz$. The fifth face gives

$$\frac{-x^2 y}{h} - \frac{x^2}{2}.$$

For the sixth face, we get

$$\frac{x^2 y}{h} + 2xy + \frac{x^2}{2}.$$

After cancellations, this leaves $2xy - 2z$. Thus

$$d\Phi_{\vec{F}} = M_{\text{div } \vec{F}} = (2xy - 2z) \, dx \wedge dy \wedge dz.$$

By equation 6.5.13, div $\vec{F} = 2xy - 2z$, which is what we got in part a.

**Curl and div of the vector field in part b**

We do not work out curl and div of the second vector field, since the procedure is the same and you can check your results from part b.

**6.9.1** Theorem 6.8.3 (or table 6.8.1) tells you how to compute exterior derivatives of forms on $\mathbb{R}^3$ using grad, curl and div. Here we have forms on $\mathbb{R}^4$, the vector fields $\vec{\mathbf{E}}$ and $\vec{\mathbf{B}}$ are time dependent, and the derivatives with respect to time have to be taken into account when relevant.

For the term $W_{\vec{\mathbf{E}}} \wedge c\,dt$, the derivatives of $\vec{\mathbf{E}}$ with respect to time do not contribute to the exterior derivative, since they would lead to terms with $dt \wedge dt$. We find

$$dW_{\vec{\mathbf{E}}} \wedge c\,dt = d(E_x\,dx + E_y\,dy + E_z\,dz) \wedge c\,dt$$

$$= (D_y E_x\,dy + D_z E_x\,dz) \wedge dx \wedge c\,dt + (D_x E_y\,dx + D_z E_y\,dz) \wedge dy \wedge c\,dt$$

$$+ (D_x E_z\,dx + D_y E_z\,dy) \wedge dz \wedge c\,dt.$$

Collecting terms, this leads to

$$dW_{\vec{\mathbf{E}}} \wedge c\,dt = (D_y E_z - D_z E_y)\,dy \wedge dz \wedge c\,dt + (D_z E_x - D_x E_z)\,dz \wedge dx \wedge c\,dt$$

$$+ (D_x E_y - D_y E_x)\,dx \wedge dy \wedge c\,dt$$

$$= \Phi_{\vec{\nabla} \times \vec{\mathbf{E}}} \wedge c\,dt.$$

Note that we have simply repeated the computation in part 2 of theorem 6.8.3, with the $c\,dt$ tagging along for the ride.

When we compute $d\Phi_{\vec{\mathbf{B}}}$, the time derivatives of $\vec{\mathbf{B}}$ do contribute to the exterior derivative. We find

$$d\Phi_{\vec{\mathbf{B}}} = d(B_x dy \wedge dz + B_y dz \wedge dx + B_z dx \wedge dy)$$

$$= (D_x B_x dx + D_t B_x dt) \wedge dy \wedge dz + (D_y B_y dy + D_t B_y dt) \wedge dz \wedge dx$$

$$+ (D_z B_z dz + D_t B_z dt) \wedge dx \wedge dy).$$

Separate the terms involving $dt$ from the purely spatial terms, to find

We can move the $dt$'s to the end, writing $dy \wedge dz \wedge dt$ rather than $dt \wedge dy \wedge dz$, since 1-forms and 2-forms commute.

$$d\Phi_{\vec{\mathbf{B}}} = (D_x B_x + D_y B_y + D_z B_z)\,dx \wedge dy \wedge dz$$

$$+ D_t B_x\,dy \wedge dz \wedge dt + D_t B_y\,dz \wedge dx \wedge dt + D_t B_z\,dx \wedge dy \wedge dt$$

$$= M_{\vec{\nabla} \cdot \vec{\mathbf{B}}} + \Phi_{\frac{1}{c} D_t \vec{\mathbf{B}}} \wedge c\,dt.$$

Altogether, this does lead to

$$d\mathbb{F} = \Phi_{\vec{\nabla} \times \vec{\mathbf{E}}} \wedge c\,dt + M_{\vec{\nabla} \cdot \vec{\mathbf{B}}} + \Phi_{\frac{1}{c} D_t \vec{\mathbf{B}}} \wedge c\,dt.$$

When we integrate a vector-valued function in $\mathbb{R}^n$ (or in $\mathbb{R}^3$, as in the case of the Biot-Savart formula), the answer is a vector in $\mathbb{R}^n$ (or in $\mathbb{R}^3$); we compute separately the integral of each entry:

$$\int \begin{bmatrix} a_1 \\ a_2 \\ a_2 \end{bmatrix} = \begin{bmatrix} \int a_1 \\ \int a_2 \\ \int a_2 \end{bmatrix}.$$

**6.9.3** This is one of those unpleasant integrals requiring a couple of changes of variables. First,

$$\begin{bmatrix} I \\ 0 \\ 0 \end{bmatrix} \times \begin{bmatrix} x-s \\ y \\ z \end{bmatrix} = I \begin{bmatrix} 0 \\ -z \\ y \end{bmatrix}.$$

so we are left with showing that

$$\int_{-\infty}^{\infty} \frac{dz}{((x-s)^2 + y^2 + z^2)^{3/2}} = \frac{2}{y^2 + z^2}.$$

To lighten notation, set $y^2 + z^2 = a^2$, and make a first change of variables $x - s = -au$, $ds = a\,du$, leading to

$$\int_{-\infty}^{\infty} \frac{dz}{((x-s)^2 + y^2 + z^2)^{3/2}} = \int_{\infty}^{\infty} \frac{a\,du}{(a^2 u^2 + a^2)^{3/2}} = \frac{1}{a^2} \int_{\infty}^{\infty} \frac{du}{(u^2 + 1)^{3/2}}.$$

Next, set $u = \tan\theta$, $du = d\theta/\cos^2\theta$. This time the region of integration $u \in (-\infty, \infty)$ corresponds to $\theta \in (-\pi/2, \pi/2)$, and the integral becomes

$$\frac{1}{a^2}\int_\infty^\infty \frac{du}{(u^2+1)^{3/2}} = \frac{1}{a^2}\int_{-\pi/2}^{\pi/2}\frac{\cos^3\theta\,d\theta}{\cos^2\theta} = \frac{1}{a^2}\int_{-\pi/2}^{pi/2}\cos\theta\,d\theta = \frac{2}{a^2} = \frac{2}{y^2+z^2}.$$

The length $|\vec{a}\times\vec{b}|$ is the area of the parallelogram spanned by $\vec{a}, \vec{b}$, i.e., $|\vec{a}|\,|\vec{b}|\sin\theta$.

Solution 6.9.5: It seems that this inequality should be an immediate consequence of corollary 1.9.2. But it isn't quite, because that result concerns scalar-valued functions. Moreover, the proof doesn't go through for vector-valued functions, since it is based on theorem 1.9.1, which is false for vector-valued functions. You can apply the proof for the various components of a vector valued function, but then the points **c** you obtain for the components may be different, and conceivably the sum of the squares of the sup's at different points might be bigger than the sup of the sum of squares at one point. The argument given here shows that corollary 1.9.2 is true anyway. The same argument is used in the proof of proposition A5.1.

**6.9.5** Note that for any two vectors $\vec{a}, \vec{b} \in \mathbb{R}^3$ forming an angle $\theta$, we have

$$|\vec{a}\times\vec{b}| = |\vec{a}|\,|\vec{b}|\sin\theta \le |\vec{a}|\,|\vec{b}|.$$

Since

$$\left|\frac{\mathbf{u}}{|\mathbf{u}|^3}\right| = \frac{1}{|\mathbf{u}|^2},$$

we are left with checking that

$$|\vec{j}(\mathbf{x}_1 - \mathbf{u}) - \vec{j}(\mathbf{x}_2 - \mathbf{u})| \le |\mathbf{x}_1 - \mathbf{x}_2|\sup_{\mathbb{R}^3}|[\mathbf{Dj}]|.$$

Set $\mathbf{g}(t) = \vec{j}(\mathbf{x}_1 + t(\mathbf{x}_2 - \mathbf{x}_1))$, so that

$$|\vec{j}(\mathbf{x}_2) - \vec{j}(\mathbf{x}_1)| = |\mathbf{g}(1) - \mathbf{g}(0)|.$$

Now

$$\begin{aligned}
|\mathbf{g}(1) - \mathbf{g}(0)| &= \left|\int_0^1 \mathbf{g}'(t)\,dt\right| \\
&= \left|\int_0^1 [\mathbf{D}\vec{j}(\mathbf{g}(t))](\mathbf{x}_2 - \mathbf{x}_1)\,dt\right| \\
&\le \int_0^1 \left|[\mathbf{D}\vec{j}(\mathbf{g}(t))]\right||\mathbf{x}_2 - \mathbf{x}_1|\,dt \le \sup_{\mathbb{R}^3}|[\mathbf{D}\vec{j}]|\,|\mathbf{x}_2 - \mathbf{x}_1|.
\end{aligned}$$

**6.9.7** Important as this result is, it is a straightforward algebraic calculation. Since $y' = y$ and $z' = z$, we can ignore those terms. Then

$$\begin{aligned}
|x'|^2 - c^2(t')^2 &= \frac{(x-vt)^2}{1 - v^2/c^2} - \frac{(t - vx/c^2)^2}{1 - v^2/c^2} \\
&= \frac{c^2x^2 - 2c^2cvt + c^2v^2t^2 - t^2c^4 + 2c^2tvx - v^2x^2}{c^2 - v^2} \\
&= \frac{c^2(x^2 - t^2c^2) - v^2(x^2 - c^2t^2)}{c^2 - v^2} = x^2 - c^2t^2.
\end{aligned}$$

Solution 6.9.7: A transformation satisfying

$$|\mathbf{x}'|^2 - c^2(t')^2 = |\mathbf{x}|^2 - c^2t^2$$

is called a Lorentz transformation because it preserves the Lorentz pseudolength, analogous to ordinary Euclidean length. Above, the primes denote the new coordinates.

**6.9.9** a. In free space (no charge), one of Maxwell's equations reads

$$\text{div}\,\vec{\mathbf{E}} = g'(x - at) = 0.$$

b. No, unless $g$ is constant. The sign of $\vec{\mathbf{B}}$ (or $\vec{\mathbf{E}}$, take your pick) is wrong. The Faraday 2-form is

$$\mathbb{F} = g(y - ct)\,dx \wedge c\,dt + g(y - ct)\,dx \wedge dy,$$

so

$$d\mathbb{F} = g'(y-ct)\,dy{\wedge}dx{\wedge}c\,dt - g'(y-ct)\,c\,dt{\wedge}dx{\wedge}dy = -2g'(y-ct)\,dx{\wedge}dy{\wedge}c\,dt,$$

and this is 0 only if $g$ is constant.

**6.10.1**  Give $U$ its standard orientation as an open subset of $\mathbb{R}^3$, and $\partial U$ the boundary orientation. Since we have

$$d(z\,dx \wedge dy + y\,dz \wedge dx + x\,dy \wedge dz) = dz \wedge dx \wedge dy + dy \wedge dz \wedge dx + dx \wedge dy \wedge dz$$
$$= 3\,dx \wedge dy \wedge dz,$$

we see that

$$\int_{\partial U} \frac{1}{3}(z\,dx \wedge dy + y\,dz \wedge dx + x\,dy \wedge dz) = \int_U dx \wedge dy \wedge dz,$$

which is the volume of $U$.

**6.10.3**  We will compute the integral in two ways. First we compute it directly. Let $\gamma$ be the inverse of the projection mentioned in the text. If this projection preserves orientation, then so does $\gamma$ and we may use $\gamma$ to perform our integration. Let $U$ be the region we wish to integrate over and set

$$V = \left\{ \begin{pmatrix} x_1 \\ x_2 \\ x_3 \end{pmatrix} \,\middle|\, x_1, x_2, x_3 \geq 0, \text{ and } x_1 + x_2 + x_3 \leq a \right\}.$$

Since $\gamma(V) = U$, definition 6.2.1 tells us that

$$\int_U x_1\,dx_2 \wedge dx_3 \wedge dx_4 = \int_V x_1\,dx_2 \wedge dx_3 \wedge dx_4\left( \vec{D_1\gamma}, \vec{D_2\gamma}, \vec{D_3\gamma} \right) |dx_1\,dx_2\,dx_3|.$$

Since $\gamma \begin{pmatrix} x_1 \\ x_2 \\ x_3 \end{pmatrix} = \begin{pmatrix} x_1 \\ x_2 \\ x_3 \\ a - x_1 - x_2 - x_3 \end{pmatrix}$, with derivative $\begin{bmatrix} 1 & 0 & 0 \\ 0 & 1 & 0 \\ 0 & 0 & 1 \\ -1 & -1 & -1 \end{bmatrix}$,

we have

$$x_1\,dx_2 \wedge dx_3 \wedge dx_4(\vec{D_1\gamma}, \vec{D_2\gamma}, \vec{D_3\gamma}), = x_1 \det \begin{bmatrix} 0 & 1 & 0 \\ 0 & 0 & 1 \\ -1 & -1 & -1 \end{bmatrix} = -1,$$

so we may write our integral as an iterated integral and evaluate:

$$\int_0^a \int_0^{a-x_1} \int_0^{a-x_1-x_2} -x_1\,|dx_3\,dx_2\,dx_1| = -\int_0^a \int_0^{a-x_1} x_1(a - x_1 - x_2)\,dx_2\,dx_1$$
$$= -\int_0^a \frac{x_1(a - x_1)^2}{2}\,dx_1$$
$$= -\frac{1}{2}\left[ \frac{a^2 x_1^2}{2} - \frac{2a x_1^3}{3} + \frac{x_1^4}{4} \right]_0^a = -\frac{a^4}{24}.$$

We can get the same result using the generalized Stokes's theorem. Let $X$ be the region in $\mathbb{R}^4$ bounded by the three-dimensional manifolds given by the equations $x_1 = 0$, $x_2 = 0$, $x_3 = 0$, $x_4 = 0$, and $x_1 + x_2 + x_3 + x_4 = a$. Also let $\partial X_1, \dots, \partial X_5$ be the portions of these manifolds that bound $X$, respectively. We assume all are oriented so that the projection of $\partial X_5$ onto $(x_1, x_2, x_3)$-coordinate space is orientation preserving. We may now apply the generalized Stokes's theorem (theorem 6.10.2), which tells us that

$$\sum_{i=1}^{5} \int_{\partial X_i} x_1 \, dx_2 \wedge dx_3 \wedge dx_4 = \int_X d(x_1 \, dx_2 \wedge dx_3 \wedge dx_4)$$

$$= \int_X dx_1 \wedge dx_2 \wedge dx_3 \wedge dx_4 \,.$$

The integrals over the first four portions, $\partial X_1, \dots, \partial X_4$, contribute nothing to the sum; the form $x_1 \, dx_2 \wedge dx_3 \wedge dx_4$ is uniformly zero over these manifolds. We will spell it out for the first, $\partial X_1$.

Let $\varphi : \mathbb{R}^3 \to \mathbb{R}^4$ take

$$\begin{pmatrix} x_2 \\ x_3 \\ x_4 \end{pmatrix} \mapsto \begin{pmatrix} 0 \\ x_2 \\ x_3 \\ x_4 \end{pmatrix} ;$$

this mapping parametrizes the boundary of $X_1$. The function $x_1$ is $0$ on this manifold, so the form evaluates to zero everywhere:

$$x_1 \, dx_2 \wedge dx_3 \wedge dx_4 (\overrightarrow{D_{x_2}\varphi}, \overrightarrow{D_{x_3}\varphi}, \overrightarrow{D_{x_4}\varphi}) = 0 \det \begin{bmatrix} 1 & 0 & 0 \\ 0 & 1 & 0 \\ 0 & 0 & 1 \end{bmatrix} = 0.$$

Similarly, the boundaries of $X_2, X_3$, and $X_4$ contribute nothing to the integral.

We now have

$$\int_{\partial X_5} x_1 \, dx_2 \wedge dx_3 \wedge dx_4 = \int_X dx_1 \wedge dx_2 \wedge dx_3 \wedge dx_4,$$

and our problem has been reduced to calculating an integral over the region

$$X = \left\{ \begin{pmatrix} x_1 \\ x_2 \\ x_3 \\ x_4 \end{pmatrix} \middle| \; x_1, x_2, x_3, x_4 \geq 0, \; x_1 + x_2 + x_3 + x_4 \leq a \right\}.$$

To compute this integral, we first determine the orientation of $X$. All 4-forms in $\mathbb{R}^4$ are of the form $f(\mathbf{x}) \, dx_1 \wedge dx_2 \wedge dx_3 \wedge dx_4$. In this case it is a good idea to try a constant form. Assume $X$ is oriented by

$$k \; dx_1 \wedge dx_2 \wedge dx_3 \wedge dx_4$$

for some constant $k \neq 0$. Since the vector $\vec{N} = \begin{bmatrix} 1 \\ 1 \\ 1 \\ 1 \end{bmatrix}$ points out of $X$ on

$\partial X_5$, definition 6.6.16 tells us that $\partial X_5$ is oriented by

$$k \; dx_1 \wedge dx_2 \wedge dx_3 \wedge dx_4 \left( \vec{N}, \vec{v}_1, \vec{v}_2, \vec{v}_3 \right).$$

Since we wish the map $\gamma$ to preserve orientation, we require that

$$k\, dx_1 \wedge dx_2 \wedge dx_3 \wedge dx_4 \left(\vec{N}, \overrightarrow{D_{x_1}\gamma}, \overrightarrow{D_{x_2}\gamma}, \overrightarrow{D_{x_3}\gamma}\right) = k\, dx_1 \wedge dx_2 \wedge dx_3 \wedge dx_4 \begin{bmatrix} 1 & 1 & 0 & 0 \\ 1 & 0 & 1 & 0 \\ 1 & 0 & 0 & 1 \\ 1 & -1 & -1 & -1 \end{bmatrix} = -4k$$

be positive. This is satisfied if $k = -1$, so that $-dx_1 \wedge dx_2 \wedge dx_3 \wedge dx_4$ orients $X$.

Now consider the identity parameterization of $X$. The partial derivatives with respect to $x_1, \ldots, x_4$ are $\vec{e}_1, \ldots, \vec{e}_4$. Since $dx_1 \wedge dx_2 \wedge dx_3 \wedge dx_4$ evaluated on these vectors gives 1, the orienting form $-dx_1 \wedge dx_2 \wedge dx_3 \wedge dx_4$ gives $-1$ and the identity parameterization is orientation reversing. So

$$\int_{I(X)} dx_1 \wedge dx_2 \wedge dx_3 \wedge dx_4 = -\int_X |d^4\mathbf{x}|,$$

where $I$ is the identity map on $X$. As an iterated integral this is

$$-\int_0^a \int_0^{a-x_1} \int_0^{a-x_1-x_2} \int_0^{a-x_1-x_2-x_3} dx_4\, dx_3\, dx_2\, dx_1 = -\int_0^a \int_0^{a-x_1} \int_0^{a-x_1-x_2} (a - x_1 - x_2 - x_3)\, dx_3\, dx_2\, dx_1$$

$$= -\int_0^a \int_0^{a-x_1} \frac{(a - x_1 - x_2)^2}{2}\, dx_2\, dx_1$$

$$= -\int_0^a \frac{(a - x_1)^3}{6}\, dx_1 = -\frac{a^4}{24},$$

the same result as before.

**6.10.5** Set $\omega = x\, dy \wedge dz + y\, dz \wedge dx + z\, dx \wedge dy$, and call $A$ the octant of solid ellipsoid given by

$$0 \leq x,\, y,\, z \quad \text{and} \quad \frac{x^2}{a^2} + \frac{y^2}{b^2} + \frac{z^2}{c^2} \leq 1.$$

First, observe that

$$d\omega = 3\, dx \wedge dy \wedge dz.$$

Next, observe that the integral of $x\, dy \wedge dz + y\, dz \wedge dx + z\, dx \wedge dy$ over any subset of the coordinate planes vanishes. For instance, on the $(x,y)$-plane, only the term in $dx \wedge dy$ could contribute, and it doesn't, since $z = 0$ there. So $\int_S \omega = \int_A dx \wedge dy \wedge dz$.

The octant $A$ is the image of the first octant in the unit sphere under the linear transformation $\begin{bmatrix} a & 0 & 0 \\ 0 & b & 0 \\ 0 & 0 & c \end{bmatrix}$, with determinant $abc$. So

$$\text{vol}_3(A) = abc \left(\frac{1}{8}\right)\left(\frac{4\pi}{3}\right) = \frac{abc\,\pi}{6}.$$

Finally, $\displaystyle\int_S \omega = 3\,\text{vol}_3(A) = \frac{abc\,\pi}{2}.$

**6.10.7** It isn't actually easier to do this for a parallelogram than for an arbitrary parametrized surface $S$. Let $\gamma : U \to S$ be an orientation-preserving parametrization; then the mapping

$$\delta : \mathbf{u} \mapsto \frac{\vec{\gamma}(\mathbf{u})}{|\vec{\gamma}(\mathbf{u})|} \quad \text{is a parametrization of } P.$$

So the object is to show that

$$\int_U \frac{1}{|\vec{\gamma}(\mathbf{u})|^3} \det[\vec{\gamma}(\mathbf{u}), \overrightarrow{D_1\gamma}(\mathbf{u}), \overrightarrow{D_2\gamma}(\mathbf{u})]|d^2\mathbf{u}| = \int_U \frac{1}{|\vec{\delta}(\mathbf{u})|^3} \det[\vec{\delta}(\mathbf{u}), \overrightarrow{D_1\delta}(\mathbf{u}), \overrightarrow{D_2\delta}(\mathbf{u})]|d^2\mathbf{u}|.$$

Remembering that

$$|\vec{\gamma}(\mathbf{u})| = \sqrt{\vec{\gamma}(\mathbf{u}) \cdot \vec{\gamma}(\mathbf{u})}$$

we find that

$$\overrightarrow{D_1\delta}(\mathbf{u}) = \frac{1}{|\vec{\gamma}(\mathbf{u})|^3}\left(|\vec{\gamma}(\mathbf{u})|^2\overrightarrow{D_1\gamma}(\mathbf{u}) - (\vec{\gamma}(\mathbf{u})\cdot\overrightarrow{D_1\gamma}(\mathbf{u})\vec{\gamma}(\mathbf{u})\right)$$

and

$$\overrightarrow{D_2\delta}(\mathbf{u}) = \frac{1}{|\vec{\gamma}(\mathbf{u})|^3}\left(|\vec{\gamma}(\mathbf{u})|^2\overrightarrow{D_2\gamma}(\mathbf{u}) - (\vec{\gamma}(\mathbf{u})\cdot\overrightarrow{D_2\gamma}(\mathbf{u})\vec{\gamma}(\mathbf{u})\right).$$

Using $|\vec{\delta}(\mathbf{u})| = 1$, this gives

$$\int_U \frac{1}{|\vec{\delta}(\mathbf{u})|^3}\det\left[\vec{\delta}(\mathbf{u}), \overrightarrow{D_1\delta}(\mathbf{u}), \overrightarrow{D_2\delta}(\mathbf{u})\right]|d^2\mathbf{u}|$$

$$= \int_U \frac{1}{|\vec{\gamma}(\mathbf{u})|^7}\det\left[\vec{\gamma}(\mathbf{u}), \left(|\vec{\gamma}(\mathbf{u})|^2\overrightarrow{D_1\gamma}(\mathbf{u}) - (\vec{\gamma}(\mathbf{u})\cdot\overrightarrow{D_1\gamma}(\mathbf{u}))\vec{\gamma}(\mathbf{u})\right), \left(|\vec{\gamma}(\mathbf{u})|^2\overrightarrow{D_2\gamma}(\mathbf{u}) - (\vec{\gamma}(\mathbf{u})\cdot\overrightarrow{D_2\gamma}(\mathbf{u}))\vec{\gamma}(\mathbf{u})\right)\right]|d^2\mathbf{u}|$$

$$= \int_U \frac{1}{|\vec{\gamma}(\mathbf{u})|^7}\det\left[\vec{\gamma}(\mathbf{u}), |\vec{\gamma}(\mathbf{u})|^2\overrightarrow{D_1\gamma}(\mathbf{u}), |\vec{\gamma}(\mathbf{u})|^2\overrightarrow{D_2\gamma}(\mathbf{u})\right]|d^2\mathbf{u}|$$

$$= \int_U \frac{|\vec{\gamma}(\mathbf{u})|^4}{|\vec{\gamma}(\mathbf{u})|^7}\det\left[\vec{\gamma}(\mathbf{u}), \overrightarrow{D_1\gamma}(\mathbf{u}), \overrightarrow{D_2\gamma}(\mathbf{u})\right]|d^2\mathbf{u}|.$$

In going from the first line to the second, one power of $|\vec{\gamma}(\mathbf{u})|$ factors out of the first column, $\delta(\mathbf{u})$; three powers factor out from the second column, $\overrightarrow{D_1\delta}(\mathbf{u})$; and three more from the third column, giving seven in all. In going from the second line to the third line, since the first column of the matrix is $\vec{\gamma}(\mathbf{u})$, the multiples of $\vec{\gamma}(\mathbf{u})$ in the second and third column do not contribute to the determinant; after canceling them we can factor out a $|\vec{\gamma}(\mathbf{u})|^2$ from each column.

**6.11.1** Since div $\vec{F} = 3$ and (by theorem 6.8.3)

$$d\Phi_{\vec{F}} = M_{\text{div }\vec{F}} = \text{div }\vec{F}\, dx \wedge dy \wedge dz,$$

Stokes's theorem says that

$$\int_S \Phi_{\vec{F}} = \int_X 3\, dx \wedge dy \wedge dz = 3\int_X |dx\, dy\, dz|,$$

where $X$ is the solid torus bounded by $S$. This last integral can be evaluated in many ways: one is to integrate first with respect to $|dx\,dy|$, to find

$$\pi\left((2+\sqrt{1-z^2})^2-(2-\sqrt{1-z^2})^2\right)=4\pi\sqrt{1-z^2},$$

so the final result is

$$12\pi\int_{-1}^{1}\sqrt{1-z^2}\,dz=6\pi^2.$$

**6.11.3** a. In cylindrical coordinates, the integral becomes

$$\mathrm{vol}_3(Z_\alpha)=\int_0^\alpha\left(\int_X r\,|dr\,dz|\right)d\theta=\alpha\int_X x\,|dx\,dz|.$$

One way of saying this is that the volume is the (area of $X$) times the distance that the center of gravity travels. Recall (definition 4.2.1) that the center of gravity of $X$ is the point $\overline{\mathbf{x}}$ whose $x$-coordinate is given by

$$\overline{x}=\frac{\displaystyle\int_X x\,|d^2\mathbf{x}|}{\displaystyle\int_X |d^2\mathbf{x}|},$$

so that

$$\mathrm{vol}_3(Z_\alpha)=\underbrace{\alpha\overline{x}}_{\substack{\text{distance center of}\\\text{gravity has traveled}}}\underbrace{\int_X |d^2\mathbf{x}|}_{\text{area of }X}=\alpha\int_X x\,|dx\,dz|.$$

b. The formula above says that it is $\pi(4\pi)=4\pi^2$.

The doughnut is cut in half vertically, not horizontally like a bagel.

c. Consider the full half-doughnut, bounded by the part of the torus where $y\geq0$, and two circles. Since the vector field is radial, its flow through these two endpoints is 0. Notice that the divergence of our vector field is 3, so the flux of the vector field through the surface is equal to 3 times the volume of the half-doughnut, i.e., $6\pi^2$.

**6.11.5** Denote by $P$ the boundary of the polygon and by $S$ a circle centered at the origin and with radius small enough so that $S$ is inside the polygon (any circle will do, but a small one is convenient). Then the region $U$ between $S$ and the 11-sided polygon is a region bounded by $S$ and the polygon. The curve $P$ carries the boundary orientation of $U$, i.e., it is oriented counterclockwise (see figure 6.6.9). If $S$ is oriented as part of the boundary of $U$, it would be oriented clockwise. Give it the opposite orientation. Green's theorem (Stokes's theorem applied to curves) now says that for any 1-form $\omega$,

$$\int_U d\omega=\int_P \omega-\int_S \omega.$$

We have

$$dW_{\begin{bmatrix}-y/(x^2+y^2)\\ x/(x^2+y^2)\end{bmatrix}}$$

$$= D_2 \frac{-y}{x^2+y^2}\, dy \wedge dx$$

$$+ D_1 \frac{x}{x^2+y^2} dx \wedge dy$$

$$= \frac{-(x^2+y^2)-2y(-y)}{(x^2+y^2)^2} dy \wedge dx$$

$$+ \frac{(x^2+y^2)-2x(x)}{(x^2+y^2)^2} dx \wedge dy$$

$$= \frac{x^2-y^2+y^2-x^2}{(x^2+y^2)^2} dx \wedge dy = 0$$

Now observe the convenient fact that $dW_{\begin{bmatrix}-y/(x^2+y^2)\\ x/(x^2+y^2)\end{bmatrix}} = 0$ (see the margin note). So we have $\int_P \omega = \int_S \omega$:

$$\int_P W_{\begin{bmatrix}-y/(x^2+y^2)\\ x/(x^2+y^2)\end{bmatrix}} = \int_S W_{\begin{bmatrix}-y/(x^2+y^2)\\ x/(x^2+y^2)\end{bmatrix}}$$

$$= \int_S -\frac{y}{x^2+y^2}\, dx + \frac{x}{x^2+y^2}\, dy = 2\pi.$$

**6.11.7** Let $U$ be the region bounded by $C$. Then

$$\int_C W_{\begin{bmatrix}0\\ x\end{bmatrix}} = \int_C x\, dy = \int_U dx \wedge dy,$$

whereas

$$\int_C W_{\begin{bmatrix}y\\ 0\end{bmatrix}} = \int_C y\, dx = \int_U dy \wedge dx;$$

the first returns the area of $U$, and the second returns the negative of the area.

**6.11.9** By Stokes's theorem,

$$\int_{\partial U_R} W_{\vec{F}} = \int_{U_R} \Phi_{\operatorname{curl} \vec{F}}.$$

The curl of $\vec{F}$ is almost constant on a small disc, so

Solution 6.11.9: To express a vector field $\vec{F}$ as its Taylor series, express each component as its Taylor series:

$$\vec{F}(\mathbf{a}+\vec{\mathbf{h}}) = \begin{bmatrix} F_1(\mathbf{a}+\vec{\mathbf{h}}) \\ F_2(\mathbf{a}+\vec{\mathbf{h}}) \\ F_3(\mathbf{a}+\vec{\mathbf{h}}) \end{bmatrix}$$

$$= \begin{bmatrix} F_1(\mathbf{a})+[\mathbf{D}F_1(\mathbf{a})]\vec{\mathbf{h}}+\cdots \\ F_2(\mathbf{a})+[\mathbf{D}F_2(\mathbf{a})]\vec{\mathbf{h}}+\cdots \\ F_3(\mathbf{a})+[\mathbf{D}F_3(\mathbf{a})]\vec{\mathbf{h}}+\cdots \end{bmatrix}$$

$$\int_{U_R} \Phi_{\operatorname{curl}\vec{F}} = \underbrace{\pi R^2}_{\text{area of disc}} \big(\operatorname{curl}\vec{F}(\mathbf{a})\big)\cdot \vec{\mathbf{v}} + \underbrace{o(R^2)}_{\substack{\text{higher-degree terms from}\\ \text{Taylor expansion of } \vec{F} \text{ at } \mathbf{a}.}}$$

Thus

$$\lim_{R\to 0} \frac{1}{R^2} \int_{\partial U_R} W_{\vec{F}} = \pi \big(\operatorname{curl}\vec{F}(\mathbf{a})\big)\cdot \vec{\mathbf{v}}.$$

**6.11.11** You could simply argue that the flux must be 0 because the sphere is symmetrical about the origin, so that the flux through one part is canceled by the flux through another. Being more formal, you can use the divergence theorem to say

$$\int_{\partial\text{ball}} \Phi_{\begin{bmatrix}x^2\\ y^2\\ z^2\end{bmatrix}} = \int_{\text{ball}} d\Phi_{\begin{bmatrix}x^2\\ y^2\\ z^2\end{bmatrix}} = \int_{\text{ball}} (2x+2y+2z)\, dx \wedge dy \wedge dz.$$

Again, at this point you could stop and point out that this integral is 0 by symmetry. But for practice setting up a multiple integrals we can go further:

$$\int_{\text{ball}} (2x+2y+2z)\, dx \wedge dy \wedge dz = \int_{-1}^{1}\left(\int_{-\sqrt{1-z^2}}^{\sqrt{1-z^2}}\left(\int_{-\sqrt{1-y^2-z^2}}^{\sqrt{1-y^2-z^2}} (2x+2y+2z)\, dx\right)dy\right)dz.$$

(Actually computing this integral is unpleasant; take advantage of symmetry whenever you can.)

Where did we use the fact that the surface is oriented by the outward-pointing normal? The surface is the boundary of the ball, which has the standard orientation, so with the boundary orientation, the surface is oriented by the outward-pointing vector (see equation 6.6.23). If the surface had been oriented by the inward-pointing normal, the divergence theorem would say

$$\int_{\partial \text{ball}} \Phi_{\begin{bmatrix} x^2 \\ y^2 \\ z^2 \end{bmatrix}} = -\int_{\text{ball}} d\Phi_{\begin{bmatrix} x^2 \\ y^2 \\ z^2 \end{bmatrix}}.$$

**Solution 6.11.15:** Similar results were sketched in greater generality in exercise 6.3.15

**6.11.13**   This is easiest using Stokes's theorem.   Let $S$ be the disc $x^2 + y^2 \le 1$, $z = 3$, oriented by the downward normal (downward because the circle is oriented clockwise). Then

$$\int_{\partial S} W_{\begin{bmatrix} -3y \\ 3x \\ 1 \end{bmatrix}} = \int_S dW_{\begin{bmatrix} -3y \\ 3x \\ 1 \end{bmatrix}} = \int_S \Phi_{\begin{bmatrix} D_1 \\ D_2 \\ D_3 \end{bmatrix} \times \begin{bmatrix} -3y \\ 3x \\ 1 \end{bmatrix}} = \int_S \Phi_{\begin{bmatrix} 0 \\ 0 \\ 6 \end{bmatrix}} = \int_S 6\, dx \wedge dy.$$

This integral is $-6\pi$: 6 times the area of the circle, with a minus sign because the vector field is pointing upwards, against the downwards orientation of $S$.

**6.11.15** a. Suppose $\vec{w} \in L$ is another vector. Then we can write

$$\vec{w} = (a + bi)(\vec{v}) = a\vec{v} + bi\vec{v} \quad \text{and} \quad i\vec{w} = i(a + bi)(\vec{v}) = -b\vec{v} + a(i\vec{v}).$$

Thus the change of basis matrix from the basis $\vec{w}, i\vec{w}$ to the basis $\vec{v}, i\vec{v}$ is

$$\begin{bmatrix} a & -b \\ b & a \end{bmatrix}$$

whose determinant is $a^2 + b^2 > 0$, so these bases define the same orientation.

b. We will parametrize our domain by the map

$$z \mapsto \begin{pmatrix} z^q \\ -z^p \end{pmatrix}, \quad |z| \le R.$$

Recall that $q$ and $p$ are relatively prime, so that at least one is odd. Assume that $q$ is odd, and note that if we set $z_1 = z^q$ and $z_2 = -z^p$, then points in the image of the mapping satisfy the equation: $(z^q)^p + (-z^p)^q = 0$. (If $q$ and $p$ were both even, this would not be true.)

To show that the mapping parametrizes our domain we also have to show (definition 5.2.3) that it is one to one, of class $C^1$, with locally Lipschitz derivative, and that its derivative is also one to one. Since the coordinate functions $z^q$ and $-z^p$ are polynomials, the mapping is not just $C^1$, it is $C^\infty$; by proposition 2.8.9, its derivative is Lipschitz.

Showing that the mapping is one to one means showing that a given pair $\begin{pmatrix} z^q \\ -z^p \end{pmatrix}$ corresponds to only one $z$. The number $z^q$ has $q$ $q$th roots (proposition 0.7.7). If $\zeta$ is one of them, then the numbers

$$\zeta e^{2\pi i k/q}, \quad k = 0, 1, \ldots, q-1$$

is a list of all of them. If we raise these to the $p$th power, we get

$$\zeta^p e^{2\pi i p k/q}, \; k = 0, 1, \ldots, q-1,$$

and these numbers are all different; indeed if $e^{2\pi i p k/q} = e^{2\pi i p l/q}$, then $e^{2\pi i p(k-l)/q} = 1$, which occurs only if $p(k-l)/q$ is an integer; but $p(k-l)$ is one of the integers $-p(q-1)\ldots, -p, 0, p, \ldots, p(q-1)$, and the only one of these which is divisible by $q$ is 0. Thus our mapping is one to one. It should be clear that the derivative is one to one except at the origin.

Now we will use polar coordinates, writing (as in equation 0.7.10) $z = r(\cos\theta + i\sin\theta)$. In these terms our surface is parametrized by

$$\begin{pmatrix} r \\ \theta \end{pmatrix} \mapsto \begin{pmatrix} r^q \cos q\theta \\ r^q \sin q\theta \\ -r^p \cos p\theta \\ -r^p \sin p\theta \end{pmatrix}.$$

We wish to integrate over the part of $X_{p,q}$ where $|z_1| \le R_1, |z_2| \le R_2$. Thus $|z^q| \le R_1$ and $|-z^p| \le R_2$, which gives $|z| \le R_1^{1/q}, |z| \le R_2^{1/p}$. Set $R = \min(R_1^{1/q}, R_2^{1/p})$. Then the disc of radius $R$ corresponds to the part of $X_{p,q}$ where $|z_1| \le R_1, |z_2| \le R_2$.

This leads to the integral

$$\int_0^{2\pi} \int_0^R (dx_1 \wedge dy_1 + dx_2 \wedge dy_2) \left( \begin{bmatrix} qr^{q-1}\cos q\theta \\ qr^{q-1}\sin q\theta \\ -pr^{p-1}\cos p\theta \\ -pr^{p-1}\sin p\theta \end{bmatrix}, \begin{bmatrix} -qr^q \sin q\theta \\ qr^q \cos q\theta \\ pr^p \sin p\theta \\ -pr^p \cos p\theta \end{bmatrix} \right) dr\, d\theta$$

$$= \int_0^{2\pi} \int_0^R (q^2 r^{2q-1} + p^2 r^{2p-1})\, dr\, d\theta$$

$$= 2\pi \left[ \frac{q^2 r^{2q}}{2q} + \frac{p^2 r^{2p}}{2p} \right]_0^R$$

$$= \pi(qR^{2q} + pR^{2p}).$$

**6.12.1** We will assume that the functions are all defined on either $\mathbb{R}^2$ or $\mathbb{R}^3$, which have no holes. Asking whether a vector field is the gradient of a function is the same as asking whether a function exists whose partial derivatives are the entries of the vector field. We know that $df = W_{\text{grad } f}$, and further that $d\, df = 0$, so it is simply a matter of writing $d\, df = d(W_{\vec{F}})$ and seeing whether this is 0. If not, no such function $f$ exists. If so (assuming the domain has no holes) then such a function does exist.

The vector fields (a)-(e), and (g) are gradients of functions. The vector fields (f), (h), (i), (j), (k), and (l) are not:

a. $W_{\begin{bmatrix} 0 \\ 1 \end{bmatrix}} = dy$,   and   $d(dy) = 0$.

b. $W_{\begin{bmatrix} x \\ 0 \end{bmatrix}} = x\,dx$,   and   $d(x\,dx) = 0$.

c. $W_{\begin{bmatrix} x \\ y \end{bmatrix}} = x\,dx + y\,dy$,   and   $d(x\,dx + y\,dy) = 0$.

d. $W_{\begin{bmatrix} x \\ -y \end{bmatrix}} = x\,dx - y\,dy$,   and   $d(x\,dx - y\,dy) = 0$.

e. $W_{\begin{bmatrix} y \\ x \end{bmatrix}} = y\,dx + x\,dy$   and   $d(y\,dx + x\,dy) = dy \wedge dx + dx \wedge dy = 0$.

f. $W_{\begin{bmatrix} -y \\ x \end{bmatrix}} = -y\,dx + x\,dy$   and   $d(-y\,dx + x\,dy) = 2\,dx \wedge dy$

g. $W_{\begin{bmatrix} y \\ x-y \end{bmatrix}} = y\,dx + (x-y)\,dy$   and

$$d\big(y\,dx + (x-y)\,dy\big) = dy \wedge dx + dx \wedge dy = 0$$

h. $W_{\begin{bmatrix} x-y \\ x+y \end{bmatrix}} = (x-y)\,dx + (x+y)\,dy$   and

$$d\big((x-y)\,dx + (x+y)\,dy\big) = -dy \wedge dx + dx \wedge dy = 2\,dx \wedge dy$$

i. $W_{\begin{bmatrix} x^2-y-1 \\ x-y \end{bmatrix}} = (x^2 - y - 1)\,dx + (x-y)\,dy$   and

$$d\big((x^2-y-1)\,dx + (x-y)\,dy\big) = -dy \wedge dx + dx \wedge dy = 2\,dx \wedge dy$$

**6.12.3**   a. The electric field is the gradient of the potential, i.e.

$$\nabla V \begin{pmatrix} x \\ y \\ z \end{pmatrix} = \frac{1}{x^2 + y^2} \begin{bmatrix} 2x \\ 2y \\ 0 \end{bmatrix}.$$

Parts b and c: The figures at left show the vector fields

b.   $\dfrac{1}{(x-1)^2 + y^2} \begin{bmatrix} 2(-1) \\ 2y \\ 0 \end{bmatrix} + \dfrac{1}{(x+1)^2 + y^2} \begin{bmatrix} 2(+1) \\ 2y \\ 0 \end{bmatrix}$

and

c.   $\dfrac{1}{(x-1)^2 + y^2} \begin{bmatrix} 2(-1) \\ 2y \\ 0 \end{bmatrix} - \dfrac{1}{(x+1)^2 + y^2} \begin{bmatrix} 2(+1) \\ 2y \\ 0 \end{bmatrix}.$

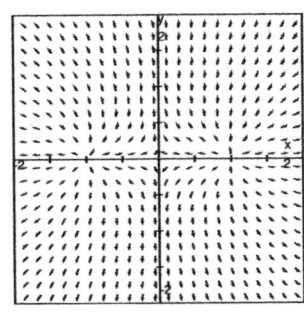

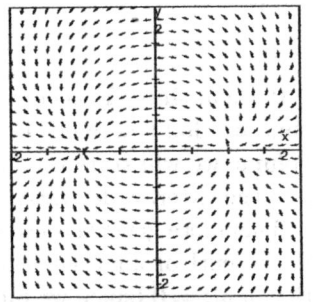

FIGURE FOR SOLUTION 6.12.3.

TOP: The vector field of part b. BOTTOM: The vector field of part c. Since the $z$-coordinate of these vector fields is 0, and since $z$ does not appear in the formulas, it is enough to draw the vector field in the $(x,y)$-plane.

**6.12.5**   a. A necessary but not sufficient condition is that theorem 6.7.7 be satisfied: $d(d\varphi) = 0$. We could either turn $\vec{F}$ into a vector in $\mathbb{R}^3$ by adding a third entry, 0; then we could see whether curl $\vec{F} = \mathbf{0}$. Or we can express $\vec{F}$ as the corresponding 1-form

$$W_{\vec{F}} = F_1\,dx + F_2\,dy = \frac{x\,dx + y\,dy}{x^2 + y^2},$$

and take the exterior derivative:

$$d\frac{x\,dx + y\,dy}{x^2 + y^2} = \frac{2xy - 2xy}{(x^2 + y^2)^2}\,dx \wedge dy = 0.$$

So theorem 6.7.7 is satisfied. However, this isn't enough to tell whether the vector field is a gradient, since theorem 6.12.5 requires the domain to be convex, and our domain is not, since it has a hole at the origin. But in fact

$$\begin{bmatrix} \dfrac{x}{x^2 + y^2} \\ \dfrac{y}{x^2 + y^2} \end{bmatrix} = \vec{\nabla}\left(\frac{1}{2}\ln(x^2 + y^2)\right).$$

b. The divergence of this vector field is 3, so the vector field certainly is not a curl of another vector field; see equation 6.8.13.

**Remark.** Theorem 6.12.5 addresses only the question, when is a vector field the gradient of a function, not the question, when is a vector field the curl of another vector field. However, to get the answer above, we need only work directly from theorem 6.7.7. If $\vec{F} = \text{curl}\,\vec{H}$, then (theorem 6.8.3)

$$dW_{\vec{H}} = \Phi_{\text{curl}\,\vec{H}}$$

and

$$ddW_{\vec{H}} = d(\Phi_{\text{curl}\,\vec{H}}) = d\Phi_{\vec{F}} = M_{\text{div}\,\vec{F}}.$$

So to satisfy $ddW_{\vec{H}} = 0$ we would need div $\vec{F} = 0$.

If we had a vector field $\vec{G}$ such that div $\vec{G} = 0$, then, since $\mathbb{R}^3$ is convex, $\vec{G}$ would be the curl of another vector field, but we don't actually know that with the tools at hand. In a subsequent volume we will state theorem 6.12.5 more generally, asking when is a $k$-form $\varphi$ the exterior derivative of a $(k-1)$-form, i.e., when can $\varphi$ be written $d\psi$. (This more general question includes the question, "when is a vector field the curl of another vector field?") As in the simpler case, the answer is that if $d\varphi = 0$ (since we must have $d\varphi = dd\psi = 0$) and the domain $U$ of $\varphi$ is convex, then $\varphi$ is guaranteed to be the exterior derivative of a $(k-1)$-form; if $d\varphi = 0$, and the domain $U$ of $\varphi$ is not convex, then $\varphi$ may or may not be the exterior derivative of a $(k-1)$-form.    $\triangle$

**6.1** All are numbers except for (c) and (i). The cross product given in (c) is a vector in $\mathbb{R}^3$. In (i), $A^k(\mathbb{R}^k)$ is a 1-dimensional vector space; it is the set of $k$-forms on $\mathbb{R}^k$.

**6.3** If $\varphi$ is a 1-form and $\psi$ is a 3-form, then

$$\varphi \wedge \psi(\vec{v}_1, \vec{v}_2, \vec{v}_3, \vec{v}_4) = \varphi(\vec{v}_1)\psi(\vec{v}_2, \vec{v}_3, \vec{v}_4) - \varphi(\vec{v}_2)\psi(\vec{v}_1, \vec{v}_3, \vec{v}_4)$$
$$+ \varphi(\vec{v}_3)\psi(\vec{v}_1, \vec{v}_2, \vec{v}_4) - \varphi(\vec{v}_4)\psi(\vec{v}_1, \vec{v}_2, \vec{v}_3).$$

**6.5** The tangent vector field $\begin{bmatrix} y \\ 3 - x \end{bmatrix}$ orients the circle clockwise. This corresponds to orientation by $y\, dx + (3 - x)\, dy$, since

$$y\, dx + (3 - x)\, dy \left( \begin{bmatrix} a \\ b \end{bmatrix} \right) = ay + (3 - x)b = \begin{bmatrix} y \\ 3 - x \end{bmatrix} \cdot \begin{bmatrix} a \\ b \end{bmatrix}.$$

How did we find this tangent vector field? The sketch at left suggests one approach.

**6.7**  a. By theorem 3.2.4, the tangent space $T_{\mathbf{x}}S^3$ is the kernel of the derivative at $\mathbf{x}$ of the equation $x_1^2 + x_2^2 + x_3^2 + x_4^2 - 1 = 0$. At the point

$$\mathbf{x} = \begin{pmatrix} 1 \\ 0 \\ 0 \\ 0 \end{pmatrix} \text{ in } S^3, \text{ the derivative is } [2\ 0\ 0\ 0], \text{ so the vectors } \vec{e}_2, \vec{e}_3, \vec{e}_4 \in \mathbb{R}^4$$

are in the tangent space $T_{\mathbf{x}}S^3$. But $dx_1 \wedge dx_2 \wedge dx_3(\vec{e}_2, \vec{e}_3, \vec{e}_4) = 0$. So $dx_1 \wedge dx_2 \wedge dx_3$ is not a nonzero element of $A^3(T_{\mathbf{x}}S^3)$, and it cannot orient $S^3$.

b. The expression

$$(\vec{v}_1, \vec{v}_2, \vec{v}_3) \mapsto \det[\mathbf{x}, \vec{v}_1, \vec{v}_2, \vec{v}_3]$$

is certainly multilinear and alternating as a function of the $\vec{v}_i$, since the determinant is, so $\omega_{\mathbf{x}} \in A^3(T_{\mathbf{x}}S^3)$. The only problem is showing that at no point $\mathbf{x} \in S^3$ it is the zero element of $A^3(T_{\mathbf{x}}S^3)$.

But if $\vec{v}_1, \vec{v}_2, \vec{v}_3$ are linearly independent elements of $T_{\mathbf{x}}S^3$, then $\mathbf{x}$ is orthogonal to $\vec{v}_1, \vec{v}_2,$ and $\vec{v}_3$, so $\mathbf{x}, \vec{v}_1, \vec{v}_2, \vec{v}_3$ are linearly independent elements of $\mathbb{R}^4$, and hence

$$\det[\mathbf{x}, \vec{v}_1, \vec{v}_2, \vec{v}_3] \neq 0.$$

**6.9** We cannot use the chain rule to compute

$$[\mathbf{D}(\gamma_2^{-1} \circ \gamma_1)(\theta)] = [\mathbf{D}\gamma_2^{-1}(\gamma_1(\theta))][\mathbf{D}\gamma_1(\theta)];$$

the problem is that $\gamma_2^{-1}$ is not defined on an open subset of $\mathbb{R}^2$, so it does not have a derivative $[\mathbf{D}\gamma_2^{-1}]$.

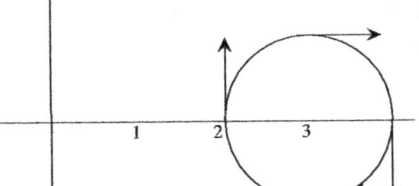

FIGURE FOR SOLUTION 6.5.

At $\begin{pmatrix} 4 \\ 0 \end{pmatrix}$ we attach the vector $\begin{bmatrix} 0 \\ -1 \end{bmatrix}$; at $\begin{pmatrix} 3 \\ -1 \end{pmatrix}$ we attach $\begin{bmatrix} -1 \\ 0 \end{bmatrix}$; at $\begin{pmatrix} 2 \\ 0 \end{pmatrix}$ we attach $\begin{bmatrix} 0 \\ 1 \end{bmatrix}$ and so on. These all correspond to

$$\vec{F}\begin{pmatrix} x \\ y \end{pmatrix} = \begin{bmatrix} y \\ 3 - x \end{bmatrix}.$$

To confirm that this works at all points of the circle, we construct the translated radial vector field $\begin{bmatrix} x - 3 \\ y \end{bmatrix}$; we have

$$\begin{bmatrix} x - 3 \\ y \end{bmatrix} \cdot \begin{bmatrix} y \\ 3 - x \end{bmatrix} = 0,$$

so $\begin{bmatrix} y \\ 3 - x \end{bmatrix}$ is indeed tangent to the circle at every point.

Instead we compute it directly. First we find $\theta_2$ in terms of $\theta_1$ such that

$$\begin{pmatrix} \sin\theta_2 \\ \cos\theta_2 \end{pmatrix} = \begin{pmatrix} \cos\theta_1 \\ \sin\theta_1 \end{pmatrix};$$

this gives $\theta_2 = \pi/2 - \theta_1$. So

$$\theta_1 \underset{\gamma_1}{\longmapsto} \begin{pmatrix} \cos\theta \\ \sin\theta \end{pmatrix} = \begin{pmatrix} \sin(\pi/2 - \theta_1) \\ \cos(\pi/2 - \theta_1) \end{pmatrix} \underset{\gamma_2^{-1}}{\longmapsto} \pi/2 - \theta_1,$$

i.e. $(\gamma_2^{-1} \circ \gamma_1)(\theta_1) = \pi/2 - \theta_1$, which gives

$$\det[\mathbf{D}(\gamma_2^{-1} \circ \gamma_1)(\theta_1)] = \det(-1) = -1.$$

**6.11** a. The 1-form $(x^2 + y^2)\,dz$ corresponds to the vector field

$$\vec{F}\begin{pmatrix} x \\ y \\ z \end{pmatrix} = \begin{bmatrix} 0 \\ 0 \\ x^2 + y^2 \end{bmatrix},$$

sketched at left. This vector field points straight up everywhere. It vanishes on the $z$-axis.

The work of this 1-form is 0 over any path contained entirely in the $(x, y)$-plane or in any plane parallel to the $(x, y)$-plane. It is large (and positive) over any path pointing straight up (in the direction of the positive $z$-axis), except that it is 0 over a path contained in the $z$-axis. The work is large but negative over any path pointing straight down.

If two paths both point straight up and begin at a point in the $(x, y)$-plane, then the work will be larger over the path that starts further from the origin.

A few simple computations confirm these conclusions for sample paths. For instance, take a straight path going from $x = 0$ to $x = 10$ in the horizontal plane at $z = 1$. This path is parametrized by $\vec{\gamma} : t \mapsto \begin{pmatrix} t \\ 0 \\ 1 \end{pmatrix}$, for $0 < x < 10$, so that $\vec{\gamma}'(t) = \begin{bmatrix} 1 \\ 0 \\ 0 \end{bmatrix}$ and $\vec{F}_{\gamma(t)} = \begin{bmatrix} 0 \\ 0 \\ t^2 \end{bmatrix}$. Thus the work of $(x^2 + y^2)\,dx$ over this path is

$$\int_0^{10} \begin{bmatrix} 0 \\ 0 \\ t^2 \end{bmatrix} \cdot \begin{bmatrix} 1 \\ 0 \\ 0 \end{bmatrix} dt = 0.$$

Now parametrize the straight vertical path going from $\begin{pmatrix} 2 \\ 2 \\ 0 \end{pmatrix}$ to $\begin{pmatrix} 2 \\ 2 \\ 1 \end{pmatrix}$ by

$$\vec{\gamma} : t \mapsto \begin{pmatrix} 2 \\ 2 \\ t \end{pmatrix}, \text{ for } 0 < t < 1.$$

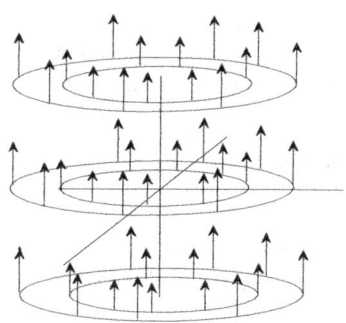

FIGURE FOR SOLUTION 6.11.
The vector field

$$\vec{F}\begin{pmatrix} x \\ y \\ z \end{pmatrix} = \begin{bmatrix} 0 \\ 0 \\ x^2 + y^2 \end{bmatrix},$$

which we saw already in the solution to exercise 1.1.7.

The work is 8, since

$$\int_0^1 \vec{F}\big(\gamma(t)\big) \cdot \vec{\gamma}'(t) = \int_0^1 \begin{bmatrix} 0 \\ 0 \\ 8 \end{bmatrix} \cdot \begin{bmatrix} 0 \\ 0 \\ 1 \end{bmatrix} \, dt = 8.$$

But the work over the vertical path from $\begin{pmatrix} 1/2 \\ 1/2 \\ 0 \end{pmatrix}$ to $\begin{pmatrix} 1/2 \\ 1/2 \\ 1 \end{pmatrix}$ is only $1/2$.

     b. The 1-form $y\,dx - x\,dy - z\,dz$ corresponds to the vector field

$$\vec{F}\begin{pmatrix} x \\ y \\ z \end{pmatrix} = \begin{bmatrix} y \\ -x \\ -z \end{bmatrix},$$

sketched at left and in the solution to exercise 1.1.7. The work of this 1-form will be large over a path spiraling clockwise around the $z$-axis. It will be large on the path going straight down from $\begin{pmatrix} 0 \\ 0 \\ 10 \end{pmatrix}$ to $\begin{pmatrix} 0 \\ 0 \\ 0 \end{pmatrix}$. It will be 0 on any horizonal radial path going straight out from the $z$-axis.

     Again, we can confirm these conclusions with some simple computations. The vertical path from $\begin{pmatrix} 0 \\ 0 \\ 10 \end{pmatrix}$ to $\begin{pmatrix} 0 \\ 0 \\ 0 \end{pmatrix}$ can be parametrized by the map

$\vec{\gamma} : t \mapsto \begin{pmatrix} 0 \\ 0 \\ t \end{pmatrix}$ for $t$ from 10 to 0; the work of $y\,dx - x\,dy - z\,dz$ over this path is

$$\int_{10}^0 \overbrace{\begin{bmatrix} 0 \\ 0 \\ -t \end{bmatrix}}^{\vec{F}(\gamma(t))} \cdot \overbrace{\begin{bmatrix} 0 \\ 0 \\ 1 \end{bmatrix}}^{\vec{\gamma}'(t)} \, dt = \int_{10}^0 -t \, dt = \left[ \frac{-t^2}{2} \right]_{10}^0 = 50.$$

The horizontal radial path from $\begin{pmatrix} 0 \\ 0 \\ 1 \end{pmatrix}$ to $\begin{pmatrix} 1 \\ 1 \\ 1 \end{pmatrix}$ can be parametrized by

$\vec{\gamma} : t \mapsto \begin{pmatrix} t \\ t \\ 1 \end{pmatrix}$, for $0 < t < 1$; the work of $y\,dx - x\,dy - z\,dz$ over this path

is $\int_0^1 \begin{bmatrix} t \\ -t \\ -1 \end{bmatrix} \cdot \begin{bmatrix} 1 \\ 1 \\ 0 \end{bmatrix} \, dt = 0.$

**6.13** The mapping

$$\gamma : \begin{pmatrix} r \\ \theta \end{pmatrix} \mapsto \begin{pmatrix} r\cos\theta \\ r\sin\theta \\ r \end{pmatrix}, 0 \le r \le \sqrt{R}, \ 0 \le \theta \le \frac{\pi}{2}$$

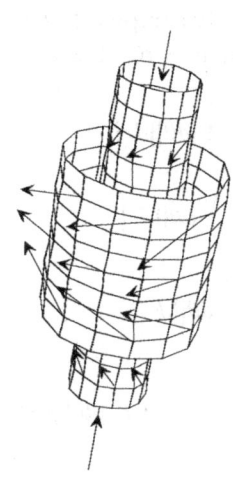

FIGURE FOR SOLUTION 6.11, part b: The vector field

$$\vec{F}\begin{pmatrix} x \\ y \\ z \end{pmatrix} = \begin{bmatrix} y \\ -x \\ -z \end{bmatrix},$$

**Solution 6.13:** If you wish to check orientation using the determinant rather than the cross product, note that the cross product $\overrightarrow{D_1\gamma} \times \overrightarrow{D_2\gamma}$ is normal to the surface. In this case, it points inwards, so we can choose it as our inward-pointing normal. This gives

$$\det\left[\left(\overrightarrow{D_1\gamma} \times \overrightarrow{D_2\gamma}\right), \overrightarrow{D_1\gamma}, \overrightarrow{D_2\gamma}\right]$$
$$= 2r^2 > 0.$$

In higher dimensions, where the cross product is not available, we could use the gradient. The cone is given by

$$f\begin{pmatrix} x \\ y \\ z \end{pmatrix} = x^2 + y^2 - z^2 = 0,$$

so one vector normal to the cone is the gradient of $f$:

$$\vec{\nabla} f \begin{pmatrix} x \\ y \\ z \end{pmatrix} = \begin{bmatrix} 2x \\ 2y \\ -2z \end{bmatrix}.$$

The gradient $\vec{\nabla}$ points in the direction $f$ is increasing fastest, i.e., outwards, so $-\vec{\nabla}$ is an inward-pointing normal.

**Solution 6.15:** In the statement of this exercise, "let $X$ be a subset of $M$" should have been "let $X$ be a subset of $S^3$."

(cylindrical change of coordinates, definition 4.10.9) parametrizes the sector of cone, and

$$\overrightarrow{D_1\gamma} \times \overrightarrow{D_2\gamma} = \begin{bmatrix} -r\cos\theta \\ -r\sin\theta \\ r \end{bmatrix}$$

points inwards, so the parametrization preserves orientation.

Now we can compute the flux; since we have already computed $\overrightarrow{D_1\gamma} \times \overrightarrow{D_2\gamma}$ it is probably easier to use the equivalence $\det[\vec{a}, \vec{b}, \vec{c}] = \vec{a} \cdot (\vec{b} \times \vec{c})$:

$$\int_C \Phi_{\vec{F}} = \int_0^{\sqrt{R}} \int_0^{\pi/2} \det\left[\vec{F}\left(\gamma\begin{pmatrix} r \\ \theta \end{pmatrix}\right), \overrightarrow{D_1\gamma}, \overrightarrow{D_2\gamma}\right] |d\theta||dr|$$

$$= \int_0^{\sqrt{R}} \int_0^{\pi/2} \vec{F}\left(\gamma\begin{pmatrix} r \\ \theta \end{pmatrix}\right) \cdot \left(\overrightarrow{D_1\gamma} \times \overrightarrow{D_2\gamma}\right) d\theta||dr|$$

$$= \int_0^{\sqrt{R}} \int_0^{\pi/2} \begin{bmatrix} r\sin\theta \\ -r \\ r^2\sin\theta \end{bmatrix} \cdot \begin{bmatrix} -r\cos\theta \\ -r\sin\theta \\ r \end{bmatrix} |d\theta||dr|$$

$$= \int_0^{\sqrt{R}} \int_0^{\pi/2} (-r^2\cos\theta\sin\theta + r^2\sin\theta + r^3\sin\theta)|d\theta|\,|dr|$$

$$= \frac{R\sqrt{R}}{6} + \frac{R^2}{4}.$$

**6.15 a.** We denote by $X$ the part of $S^3$ where $x_4 \le 0$. To show that $X$ is a piece-with-boundary of $S^3$ according to definition 6.6.6, we first need to define the smooth boundary (which is the entire boundary), using definition 6.6.2. Let the $U$ in that definition be $\mathbb{R}^4$, and define $\mathbf{f} : U \to \mathbb{R}$ by

$$\mathbf{f}(\mathbf{x}) = x_1^2 + x_2^2 + x_3^2 + x_4^2 - 1,$$

so that $S^3 \cap \mathbb{R}^4$ is defined by the equation $\mathbf{f} = 0$. Clearly the derivative $[\mathbf{Df}(x)] = [2x_i \; 2x_2 \; 2x_3 \; 2x_4]$ is surjective, since it vanishes only at the origin, which is not in $S^3$.

For the $C^1$ function $g$ of definition 6.6.2, we choose $g(\mathbf{x}) = -x_4$, with derivative $[0\;0\;0\;-1]$. At a point $\mathbf{x}$ of the smooth boundary, $g(\mathbf{x}) = 0$, so the derivative

$$\left[\mathbf{D}\begin{pmatrix} \mathbf{f} \\ g \end{pmatrix}(\mathbf{x})\right] = \begin{bmatrix} 2x_1 & 2x_2 & 2x_3 & 2x_4 \\ 0 & 0 & 0 & -1 \end{bmatrix}$$

is surjective: since $x_4 = 0$, at least one of $x_1$, $x_2$, $x_3$ is nonzero.

It follows that the boundary of $X$ consists of the sphere $x_1^2 + x_2^2 + x_3^2 = 1$, which certainly has finite 2-dimensional volume, so $X$ satisfies condition 2 of definition 6.6.6. There is no nonsmooth boundary, so it also satisfies condition 1.

b. At the point $\mathbf{x} = \begin{pmatrix} 1 \\ 0 \\ 0 \\ 0 \end{pmatrix}$, the tangent space $T_{\mathbf{x}}S^3$ is spanned by

$\vec{\mathbf{e}}_2$, $\vec{\mathbf{e}}_3$, $\vec{\mathbf{e}}_4$, and the tangent space $T_{\mathbf{x}}\partial X$ is spanned by $\vec{\mathbf{e}}_2, \vec{\mathbf{e}}_3$. Moreover, $\vec{\mathbf{e}}_4$ is tangent to $S^3$ at $\mathbf{x}$ and outward-pointing from $X$. So $\vec{\mathbf{e}}_2, \vec{\mathbf{e}}_3$ is a direct basis of $T_{\mathbf{x}}\partial X$ if and only if $\vec{\mathbf{e}}_4, \vec{\mathbf{e}}_2, \vec{\mathbf{e}}_3$ is a direct basis of $T_{\mathbf{x}}S^3$, which it is if and only if

$$\omega_{\mathbf{x}}(\vec{\mathbf{e}}_4, \vec{\mathbf{e}}_2, \vec{\mathbf{e}}_3) = \det[\vec{\mathbf{e}}_1, \vec{\mathbf{e}}_4, \ \vec{\mathbf{e}}_2, \ \vec{\mathbf{e}}_3]$$

is positive. This is indeed the case, so $\vec{\mathbf{e}}_2, \vec{\mathbf{e}}_3$ is a direct basis of $T_{\mathbf{x}}\partial X$.

**6.17**   a. The parallelogram $P_{\begin{pmatrix} 1 \\ 2 \\ 3 \end{pmatrix}}(h\vec{\mathbf{e}}_2, h\vec{\mathbf{e}}_3)$ is the left face of the cube in the margin; its boundary consists of $h\vec{\mathbf{e}}_2$, $h\vec{\mathbf{e}}_3$, $\vec{\mathbf{v}}_1$, and $\vec{\mathbf{v}}_2$. We have

$$\mathbf{a} = \begin{pmatrix} 1 \\ 2+h \\ 3 \end{pmatrix}, \quad \mathbf{b} = \begin{pmatrix} 1 \\ 2 \\ 3+h \end{pmatrix}, \quad \mathbf{c} = \begin{pmatrix} 1 \\ 2+h \\ 3+h \end{pmatrix}$$

so $h\vec{\mathbf{e}}_2$ is parametrized by $\gamma_1$ below; $h\vec{\mathbf{e}}_3$ is parametrized by $\gamma_2$; the side $\vec{\mathbf{v}}_1$ is parametrized by $\gamma_4$, and $\vec{\mathbf{v}}_2$ by $\gamma_3$:

$$\gamma_1(t) = \begin{pmatrix} 1 \\ 2+th \\ 3 \end{pmatrix}, \qquad \gamma_2(t) = \begin{pmatrix} 1 \\ 2 \\ 3+th \end{pmatrix},$$

$$\gamma_3(t) = \begin{pmatrix} 1 \\ 2+th \\ 3+h \end{pmatrix}, \qquad \gamma_4(t) = \begin{pmatrix} 1 \\ 2+h \\ 3+th \end{pmatrix},$$

all for $0 \le t \le 1$, and taken with the signs $+$, $-$, $-$, $+$. Using definition 6.2.1, we need to compute, for each parametrization $\gamma_1$, the integral

$$\int_0^1 xyz \, dy \Big( P_{\gamma_i(\mathbf{u})}\big(\gamma_i'(\mathbf{u})\big)\Big)|d^k\mathbf{u}|.$$

These integrals are

$$\int_0^1 1(2+th)(3) \, dy \begin{bmatrix} 0 \\ h \\ 0 \end{bmatrix} dt = \int_0^1 h(6+3th) \, dt = 6h + \frac{3}{2}h^2$$

$$-\int_0^1 1(2)(3+th) \left(dy \begin{bmatrix} 0 \\ 0 \\ h \end{bmatrix}\right) dt = 0$$

$$-\int_0^1 1(2+th)(3+h) \, dy \begin{bmatrix} 0 \\ h \\ 0 \end{bmatrix} dt = -6h - \frac{3}{2}h^2 - 2h^2 - \frac{h^3}{2}$$

$$+\int_0^1 1(2+h)(3+th) \, dy \begin{bmatrix} 0 \\ 0 \\ h \end{bmatrix} dt = 0$$

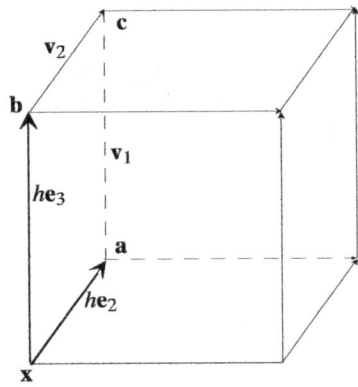

FIGURE FOR SOLUTION 6.17.

Above, $\mathbf{x}$ is the point $\begin{pmatrix} 1 \\ 2 \\ 3 \end{pmatrix}$, and the parallelogram spanned by $h\vec{\mathbf{e}}_2$, $h\vec{\mathbf{e}}_3$ is the left face of the cube.

The boundary orientation of $P_x(h\vec{\mathbf{e}}_2, h\vec{\mathbf{e}}_3)$ is given by proposition 6.6.21. We interpret our sides as parallelograms; then the boundary consists of

$$+h\vec{\mathbf{e}}_2 = P_{\mathbf{x}}(h\vec{\mathbf{e}}_2)$$
$$+\vec{\mathbf{v}}_1 = P_{\mathbf{x}+h\mathbf{e}_2}(h\vec{\mathbf{e}}_3)$$
$$-\vec{\mathbf{v}}_2 = -P_{\mathbf{x}+h\vec{\mathbf{e}}_3}(h\vec{\mathbf{e}}_2)$$
$$-h\vec{\mathbf{e}}_3 = -P_{\mathbf{x}}(h\vec{\mathbf{e}}_3),$$

where we use equation 6.6.24 to determine the signs. This corresponds to walking from $\mathbf{x}$ to $\mathbf{a}$ to $\mathbf{c}$ to $\mathbf{b}$ and back to $\mathbf{x}$.

Our 1-form is $\varphi = xyz\,dy$, which for the first parametrization corresponds to $1(2+th)(3)\,dy$, for the second to $1(2)(3+th)\,dy$, and so on.

Add, divide by $h^2$, and take the limit as $h \to 0$, to find $-2$.

b. We have

$$d(xyz\,dy) = D_1 xyz\,dx \wedge dy + D_2 xyz\,dy \wedge dy + D_3 xyz\,dz \wedge dy$$
$$= yz\,dx \wedge dy + xy\,dz \wedge dy.$$

Evaluated on $\vec{e}_2$, $\vec{e}_3$, only the second term contributes, to give

$$xy\,dz \wedge dy \left( P_{\left(\begin{smallmatrix} 1 \\ 2 \\ 3 \end{smallmatrix}\right)}(\vec{e}_2, \vec{e}_3) \right) = 2\,dz \wedge dy \left( \begin{bmatrix} 0 \\ 1 \\ 0 \end{bmatrix}, \begin{bmatrix} 0 \\ 0 \\ 1 \end{bmatrix} \right) = 2 \det \begin{bmatrix} 0 & 1 \\ 1 & 0 \end{bmatrix} = -2.$$

**6.19**  We have $\mathrm{curl}(\mathrm{grad}\,f) = \vec{0}$ since

$$0 = ddf = dW_{\vec{\nabla}f} = \Phi_{\vec{\nabla} \times \vec{\nabla} f} = \Phi_{\mathrm{curl}(\mathrm{grad}\,f)}.$$

We have $\mathrm{div}\,\mathrm{curl}\,\vec{F} = 0$ since

$$0 = ddW_{\vec{F}} = d\Phi_{\mathrm{curl}\,\vec{F}} = M_{\mathrm{div}\,\mathrm{curl}\,\vec{F}} = (\mathrm{div}\,\mathrm{curl}\,\vec{F})\,dx \wedge dy \wedge dz.$$

**6.21**  a. Write

$$D_1\Big(x(x^2+y^2+z^2)^m\Big) + D_2\Big(y(x^2+y^2+z^2)^m\Big) + D_3\Big(z(x^2+y^2+z^2)^m\Big)$$
$$= m(x^2+y^2+z^2)^{m-1}(2x^2) + (x^2+y^2+z^2)^m + m(x^2+y^2+z^2)^{m-1}(2y^2)$$
$$\qquad + (x^2+y^2+z^2)^m + m(x^2+y^2+z^2)^{m-1}(2z^2) + (x^2+y^2+z^2)^m$$
$$= (x^2+y^2+z^2)^{m-1}\Big(2m((x^2+y^2+z^2)) + 3(x^2+y^2+z^2)\Big)$$
$$= (2m+3)(x^2+y^2+z^2)^m.$$

Solution 6.21: In part a, the exponent $m = -3/2$ corresponds to the vector field $\vec{F} = \vec{r}/r^3$. This gives an inverse square law (not a cube law because the $\vec{r}$ in the numerator has length $r$). This really explains the inverse square laws seen in gravitation and in Gauss's law in electricity, which says that the flux of the electric field through a surface measures the charge in the body bounded by the surface; the flux is not affected by the shape and area of the surface.

The only way for this to vanish is if $m = -3/2$.

b. Here are two possible approaches:

i. To simplify notation, set $\vec{F} = r^{2m}\vec{r}$. Note that $\vec{F}$ is a radial vector field. You should picture the vectors radiating out from the origin and passing through concentric $(n-1)$-dimensional spheres. The length of each vector depends only on the distance from the origin; $|\vec{F}| = r^{2m}|\vec{r}| = r^{2m+1}$, with $r$ being the distance from the origin. If $S_a(0)$ is the $(n-1)$-dimensional sphere of radius $a$ centered at the origin, the integral of $\Phi_{\vec{F}}$ over $S_a(0)$ measures the flow of $\vec{F}$ through that sphere; i.e., the volume of the sphere times the length of the vector field on the sphere:

$$\int_{S_a(0)} \Phi_{\vec{F}} = \underbrace{a^{2m+1}}_{|\vec{F}|} \mathrm{vol}_{n-1}\,S_a(0). \tag{1}$$

For the sphere of radius 1 this gives

$$\int_{S_1(0)} \Phi_{\vec{F}} = \mathrm{vol}_{n-1}\,S_1(0). \tag{2}$$

Further, since $\mathrm{vol}_{n-1} \, S_a(0) = a^{n-1} \, \mathrm{vol}_{n-1} \, S_1(0)$, we have

$$\int_{S_a(0)} \Phi_{\vec{F}} = a^{n-1} a^{2m+1} \, \mathrm{vol}_{n-1} \, S_1(0) \tag{3}$$

We will now use Stokes's theorem to see that if $d\Phi_{\vec{F}} = 0$, the flow of $\vec{F}$ through any sphere centered at the origin is the same. Note that we cannot use Stokes to say that $\int_B d\Phi_{\vec{F}} = \int_S \Phi_{\vec{F}}$, where $B$ is the ball of which $S$ is the boundary; since $d\Phi_{\vec{F}} = 0$, that would give integrals of 0, which contradicts equations (1) and (2). Evidently, $\vec{F}$ is not defined at the origin. But we can set $A$ to be the part of the ball between $S_1(0)$ and $S_a(0)$, which gives

The integral over $S_1(0)$ is taken with a minus sign because the boundary orientation of $S_1(0)$ is the opposite of the boundary orientation of $S_a(0)$.

$$\int_{S_a(0)} \Phi_{\vec{F}} - \int_{S_1(0)} \Phi_{\vec{F}} = \int_A d\Phi_{\vec{F}} = 0, \quad \text{i.e.,} \quad \int_{S_a(0)} \Phi_{\vec{F}} = \int_{S_1(0)} \Phi_{\vec{F}}. \tag{4}$$

Substituting equations (1) and (2) in (4) gives

$$a^{n-1} a^{2m+1} = a^{2m+n} = 1, \quad \text{i.e.,} \quad m = \frac{-n}{2}.$$

ii. We can do the computation as in part a: everything is as above, except that the 3 becomes an $n$. So the condition for the exterior derivative to vanish is $(2m) + n = 0$, i.e., $m = -n/2$.

**6.23** a. The exterior derivative is

$$\left( D_1 \frac{x}{(x^2+y^2+z^2)^{3/2}} + D_2 \frac{y}{(x^2+y^2+z^2)^{3/2}} + D_3 \frac{z}{(x^2+y^2+z^2)^{3/2}} \right) dx \wedge dy \wedge dz$$

$$= \left( \frac{x^2+y^2+z^2-3x^2}{(x^2+y^2+z^2)^{5/2}} + \frac{x^2+y^2+z^2-3y^2}{(x^2+y^2+z^2)^{5/2}} + \frac{x^2+y^2+z^2-3z^2}{(x^2+y^2+z^2)^{5/2}} \right) dx \wedge dy \wedge dz = 0.$$

b. We cannot apply Stokes's theorem to the sphere as the boundary of a ball, since $\varphi$ is not defined at the origin. Instead, parametrize the sphere, for instance by spherical coordinates, with $r = 1$:

$$\gamma\begin{pmatrix} \theta \\ \varphi \end{pmatrix} = \begin{pmatrix} \cos\theta\cos\varphi \\ \sin\theta\cos\varphi \\ \sin\varphi \end{pmatrix}, \quad \text{for} \quad 0 \le \theta < 2\pi, \; \frac{-\pi}{2} \le \varphi \le \frac{\pi}{2},$$

with

$$\vec{D_1}\gamma = \begin{bmatrix} -\sin\theta\cos\varphi \\ \cos\theta\cos\varphi \\ 0 \end{bmatrix} \quad \text{and} \quad \vec{D_1 2}\gamma = \begin{bmatrix} -\cos\theta\sin\varphi \\ -\sin\theta\sin\varphi \\ \cos\varphi \end{bmatrix}.$$

This parametrization preserves orientation: the vector

$$\vec{n} = \begin{bmatrix} x = \cos\theta\cos\varphi \\ y = \sin\theta\cos\varphi \\ z = \sin\varphi \end{bmatrix}$$

is an outward-pointing normal, and $\det[\vec{n}, \vec{D_1}\gamma, \vec{D_2}\gamma] = \cos\varphi$, which is positive for $\varphi$ between $-\pi/2$ and $\pi/2$.

Thus we find (note that the denominator $(x^2 + y^2 + z^2)^{3/2}$ is 1 on the unit sphere)

$$\int_0^{2\pi} \int_{-\pi/2}^{\pi/2} \left(\cos\theta\cos\varphi\, dy \wedge dz + \sin\theta\cos\varphi\, dz \wedge dx + \sin\varphi\, dx \wedge dy\right)\left(\overbrace{\begin{bmatrix} -\sin\theta\cos\varphi \\ \cos\theta\cos\varphi \\ 0 \end{bmatrix}}^{\vec{D_1\gamma}}, \overbrace{\begin{bmatrix} -\cos\theta\sin\varphi \\ -\sin\theta\sin\varphi \\ \cos\varphi \end{bmatrix}}^{\vec{D_2\gamma}}\right) d\varphi\, d\theta$$

$$= \int_0^{2\pi} \int_{-\pi/2}^{\pi/2} \cos\varphi\, d\varphi\, d\theta = 4\pi.$$

c. The unit ball and the cube of side 4 together bound the region between the two, in which $\varphi$ is well defined, so we can apply the divergence theorem, to show that the integral over the boundary of the cube is also $4\pi$.

d. If $\varphi$ is a 2-form and $\varphi = d\psi$, then the integral of $\varphi$ over any oriented closed surface $S$ is 0, since $\partial S = \phi$:

> A compact manifold has empty boundary.

$$\int_S \varphi = \int_S d\psi = \int_{\partial S} \psi = 0.$$

In our case, the integral over the unit sphere is $4\pi$, so this 2-form cannot be written as the exterior derivative of a 1-form.

**6.25**   The divergence theorem says that if $U$ is given the standard orientation,

$$\int_S \Phi_{\begin{bmatrix} x \\ y \\ z \end{bmatrix}} = \int_U \operatorname{div}\begin{bmatrix} x \\ y \\ z \end{bmatrix} dx \wedge dy \wedge dz.$$

Since the divergence above is 3, we find

$$\int_S \Phi_{\begin{bmatrix} x \\ y \\ z \end{bmatrix}} = 3\operatorname{vol}_3(U).$$

**6.27**   Since $\mathbb{R}^3$ has no holes, such a function $f$ exists if $\operatorname{curl}\vec{F} = \vec{0}$, which it does:

> The first two entries are 0 because $F_1$ and $F_2$ are functions of $x$ and $y$ only. The third entry is 0 by assumption.

$$\operatorname{curl}\vec{F} = \begin{bmatrix} D_1 \\ D_2 \\ D_3 \end{bmatrix} \times \begin{bmatrix} F_1(x,y) \\ F_2(x,y) \\ 0 \end{bmatrix} = \begin{bmatrix} -D_3 F_2 \\ D_3 F_1 \\ D_1 F_2 - D_2 F_1 \end{bmatrix} = \begin{bmatrix} 0 \\ 0 \\ 0 \end{bmatrix}.$$

**6.29**   a. Since $d\mathbb{F} = 0$ (see equation 6.9.10), we have

$$0 = \int_U d\mathbb{F} = \int_{\partial U} \mathbb{F}.$$

In this case $\partial U$ is a surface bounding a region in $\mathbb{R}^3$, and since it is in a region at fixed time,

$$\int_{\partial U} W_{\vec{\mathbf{E}}} \wedge c\, dt = 0.$$

The other term is

$$\int_{\partial U} \Phi_{\vec{\mathbf{B}}} = 0;$$

this says that exactly as "much" of $\vec{\mathbf{B}}$ flows in as flows out of $U$, i.e., that $\vec{\mathbf{B}}$ is "incompressible." It is often expressed by saying that there are no *magnetic monopoles*, no sources of the magnetic field.

b. This time, we have

$$4\pi \int_U J = \int_U d\mathbb{M} = \int_{\partial U} \mathbb{M}.$$

Again, obviously

$$\int_{\partial U} W_{\vec{\mathbf{B}}} \wedge c\, dt = 0.$$

The other term is

$$\int_{\partial U} \Phi_{\vec{\mathbf{E}}} = 4\pi \int_U M_\rho.$$

The $4\pi$ in this expression depends on the units; in MKS, the constant is $1/\epsilon_0$.

The flux of the vector field through a surface bounding a region $U$ is $4\pi$ times the total charge in the region. The electric field does have sources (electrons and protons, for instance) and the flux of the electric field through a surface enclosing charge is proportional to this charge.

**6.31** Since $W_{\vec{\mathbf{r}}/r^3} = -d(\frac{1}{r})$, we have $\mathbb{F} = -q \, d\left(\frac{1}{r}\right) \wedge c\, dt$. Thus $\mathbb{F}$ is not defined at the origin, where $1/r$ is not defined. But elsewhere (see equation 6.9.10) we have $d\mathbb{F} = 0$.

In the statement of the problem, we define an $r$-adapted oriented 3-dimensional piece-with boundary $X$, the subset $X_\epsilon \subset X$ and $\partial_{inn} X_\epsilon \subset \partial X_\epsilon$, and finally define "$d\mathbb{F}$" to be the "integrand" such that

$$\int_X d\mathbb{F} \overset{\text{def}}{=} \lim_{\epsilon \to 0} \left( \int_{X_\epsilon} d\mathbb{F} - \int_{\partial_{inn} X_\epsilon} \mathbb{F} \right).$$

In our case, we do have $d\mathbb{F} = 0$ in that sense: indeed, for any such $r$-adapted piece-with-boundary $X$ we have

$$\int_{X_\epsilon} d\mathbb{F} = 0$$

since $d\mathbb{F} = 0$ except at the origin, and

$$\int_{\partial_{inn} X_\epsilon} \mathbb{F} = q \int_{\partial_{inn} X_\epsilon} d\left(\frac{1}{r}\right) \wedge c\, dt = 0,$$

since $1/r = 1/\epsilon$ on $\partial_{inn} X_\epsilon$, in particular is constant, so $d\left(\frac{1}{r}\right)$ vanishes on any vector tangent to $\partial_{inn} X_\epsilon$.

Now we need to find the charges and currents; these arise from

$$d\mathbb{M} = q \, d\Phi_{\frac{\vec{\mathbf{r}}}{r^3}} = 4\pi J.$$

You were asked in exercise 6.10.4 to show that $d\Phi_{\frac{\vec{r}}{r^3}} = 0$ except at the origin, where $\frac{\vec{r}}{r^3}$ is not defined. Let us carry out the computation (using equation 6.5.12 to write the flux as a 2-form in the ordinary way.)

We obtain

$$\Phi_{\frac{\vec{r}}{r^3}} = \frac{x\,dy \wedge dz + y\,dz \wedge dx + z\,dx \wedge dy}{(x^2 + y^2 + z^2)^{3/2}},$$

and a straightforward computation shows that

$$d\frac{x\,dy \wedge dz}{(x^2+y^2+z^2)^{3/2}} = \frac{x^2+y^2+z^2-3x^3}{(x^2+y^2+z^2)^{7/2}}dx \wedge dy \wedge dz$$

$$= \frac{-2x^2+y^2+z^2}{(x^2+y^2+z^2)^{7/2}}\,dx \wedge dy \wedge dz.$$

A similar computation for the other two terms of $\Phi_{\frac{\vec{r}}{r^3}}$ gives

$$d\Phi_{\frac{\vec{r}}{r^3}} = \frac{-2x^2+y^2+z^2+x^2-2y^2+z^2+x^2+y^2-2z^2}{(x^2+y^2+z^2)^{7/2}}\,dx \wedge dy \wedge dz = 0.$$

But this does not tell us about what might be lurking on the line $r = 0$, and indeed, in this case, there is something lurking there. Again we consider an $r$-adapted piece with boundary $X$, and compute

$$\int_X d\mathbb{M} \overset{\text{def}}{=} \lim_{\epsilon \to 0}\left( \int_{X_\epsilon} d\mathbb{M} - \int_{\partial_{inn}X_\epsilon} \mathbb{M} \right).$$

As for $\mathbb{F}$ we have

$$\int_{X_\epsilon} d\mathbb{M} = 0,$$

since $d\mathbb{M} = 0$ except at the origin.

But the other term *does not* vanish. We will assume that $X$ intersects $r = 0$ at a single point, and that at that point the parametrization by $x, y, z$ is orientation preserving. (If there are several such intersection points they all contribute the same amount, with a sign depending on the orientation.) We will see that such an intersection contributes

$$\int_{\partial_{inn}X_\epsilon} \mathbb{M} = 4\pi q.$$

This is obtained by direct computation in spherical coordinates. The map

$$\begin{bmatrix} \varphi \\ \theta \end{bmatrix} \mapsto \epsilon \begin{bmatrix} \cos\varphi\cos\theta \\ \cos\varphi\sin\theta \\ \sin\varphi \\ * \end{bmatrix}, \quad -\frac{\pi}{2} \le \varphi \le \frac{\pi}{2},\ 0 \le \theta \le 2\pi$$

is a parametrization of $\partial_{inn}X_\epsilon$. The star $*$ corresponds to whatever the time-coordinate is: it is some function of $x$, $y$, and $z$, hence of $\varphi$ and $\theta$, which will be irrelevant to the computation, since $\Phi_{\vec{F}/r^3}$ involves no $dt$'s. Moreover, this parametrization is compatible with the boundary orientation of $X_\epsilon$. The computation is a bit lengthy: we find

$$\int_0^{2\pi} \int_{-\pi/2}^{\pi/2} \frac{q}{\epsilon^3} \det \begin{bmatrix} \epsilon \cos\varphi \cos\theta & -\epsilon \sin\varphi \cos\theta & -\epsilon \cos\varphi \sin\theta \\ \epsilon \cos\varphi \sin\theta & -\epsilon \sin\varphi \sin\theta & \epsilon \cos\varphi \cos\theta \\ \epsilon \sin\varphi & \epsilon \cos\varphi & 0 \end{bmatrix} d\varphi \, d\theta = -4\pi q.$$

Thus for any 3-dimensional oriented $r$-adapted piece-with boundary $X$ that intersects $r = 0$ with the correct orientation, we have

$$\int_X d\mathbb{M} = 4\pi q.$$

In other words, the electromagnetic field corresponds to a point charge $q$ at the origin.

# Solutions for the Appendix

**A1.1** Suppose $x$ and $y$ are $k$-close for all $k$ and that $x < y$; moreover we can suppose that $x$ and $y$ are both positive. This means that for all $k$, either $[x]_{-k} = [y]_{-k}$ or $[y]_{-k} = [x]_{-k} + 10^{-k}$.

By "$m$th digit" we mean $m$th digit to the right of the decimal point.

Since $x \neq y$, there must be a first digit in which they differ, say the $m$th. Saying that they are $m$-close means that the $m$th digit of $x$ is one less than the $m$th digit of $y$; in particular, the $m$th digit of $x$ is not a 9. What does saying that they are $(m+1)$-close tell us? Adding $10^{-(m+1)}$ to $[x]_{-(m+1)}$ can only produce a carry in the $m$th position if the $(m+1)$st digit of $x$ is a 9 and the $(m+1)$st digit of $y$ is a 0, and in that case the are $(m+1)$-close. Saying that they are $(m+2)$-close says that the $(m+2)$nd digit of $x$ and $y$ are respectively 9 and 0 as well.

Continuing this way, we see that all the digits of $x$ after the $m$th are 9's, and all the digits of $y$ after the $m$th are 0's; moreover, the $m$th digit of $x$ is one less than the $m$th digit of $y$. That is exactly the condition for $x$ and $y$ to be equal.

**A1.3** a. Let $a$ and $b$ be positive finite decimals. We want to consider $a/b$, as computed by long division, so we may assume (by multiplying both by an appropriate power of 10) that they are both integers. When writing the digits of the quotient after the decimal point, the next digit depends only on the current remainder, which is a number $< b$. Thus the same integer must appear twice, and the digits of the quotient between two such appearances will then be repeated indefinitely.

Suppose we carry out long division to $k$ digits after the decimal point, obtaining a quotient $q_k$ and a remainder $r_k$. These numbers then satisfy

$$a = q_k b + 10^{-k} r_k.$$

By equation A1.3, $b(a/b)$ is

$$\sup_k \inf_{l \geq k} b([a/b]_l) = \sup_k \inf_{l \geq k} b q_l = \sup_k b q_k.$$

A remainder is always smaller than the divisor, so $r_k < b$.

where the inf was dropped since the sequence $b q_k$ is increasing, so the inf is the first element. But we just saw that $a - q_k b = 10^{-k} r_k < 10^{-k} b$. Thus $a$ is the least upper bound of the $q_k b$: it is an upper bound since $10^{-k} r_k \geq 0$, and it is a least upper bound since $10^{-k} r_k$ will be arbitrarily small for $k$ large.

b. Suppose $x \in \mathbb{R}$ satisfies $x > 0$; then $x > 10^{-p}$ for some $p$, since it must have a first nonzero digit. Define

$$\operatorname{inv} x = \inf_{k > p} (1/[x]_k).$$

We need to show that $x \operatorname{inv}(x) = 1$. By equation A1.3, this product of real numbers means

$$x \operatorname{inv}(x) = \sup_k \inf_{l > k} [x]_l [\operatorname{inv}(1/x)]_l = \sup_k [x]_k [\operatorname{inv}(1/x)]_k,$$

242

since again the sequence $[x]_l [\text{inv}(1/x)]_l$ is increasing. So we must show that

$$\sup_k [x]_k [\text{inv}(1/x)]_k = 1.$$

Suppose $x < 10^n$, and choose an integer $p$. Since $\text{inv}(x) = \inf_k 1/[x]_k$, there exists $q$ such that $[\text{inv}(x)]_k - 1/[x]_k \leq 10^{-(p+n)}$ when $k > q$. Then

$$\big|[x]_k [\text{inv}(1/x)]_k - 1\big| = \big|[x]_k [\text{inv}(1/x)]_k - [x]_k (1/[x]_k)\big|$$
$$= [x]_k \big|[\text{inv}(1/x)]_k - (1/[x]_k)\big| \leq x 10^{-(p+n)} \leq 10^{-p},$$

when $k > q$. Thus $\sup_k [x]_k [\text{inv}(1/x)]_k$ is $p$-close to 1 for every $p \in \mathbb{N}$, which means that it is 1.

c. If $x < 0$, define $\text{inv}(x) = -\text{inv}(-x)$. We need to check that for $x < 0$, we have $x \, \text{inv}(x) = 1$, and there is nothing to it:

$$x \, \text{inv}(x) = x\big(-\text{inv}(-x)\big) = \underbrace{(-x)}_{\substack{+ \text{ since} \\ x < 0}} \text{inv} \underbrace{(-x)}_{\substack{+ \text{ since} \\ x < 0}} = 1,$$

We are here assuming the rule of signs that

$$(-a)b = a(-b) = -(ab);$$

how do we know that? Since $(-a)b$ and $a(-b)$ are $\mathbb{D}$-continuous functions, and they agree on the finite decimals, they agree everywhere.

since $x \, \text{inv}(x) = 1$ was proved for $x > 0$ in part b.

**A1.5** This is more fun, after the deadly problems above.

a. Call $a \leq b < c$ the lengths of the sides of original triangle, so that $c$ is the length of the hypotenuse. Let $l_i(\underline{s}) = |\mathbf{x}_i(\underline{s}) - \mathbf{x}_{i+1}(\underline{s})|$. Then

$$\frac{l_i(\underline{s})}{l_{i-1}(\underline{s})} = \frac{a}{c} \text{ or } \frac{b}{c}$$

depending on whether $s_i$ is 0 or 1 and $i$ is even or odd. Indeed, drawing the segment from $\mathbf{x}_i(\underline{s})$ to $\mathbf{x}_{i+1}(\underline{s})$ is dropping the altitude to the hypotenuse of a triangle similar to the original triangle, with one side of length $l_{i-1}$. In the original triangle, the ratio of the height to a side is either $a/c$ or $b/c$.

In particular, the telescoping series

$$\mathbf{x}_0(\underline{s}) + \sum_{i=0}^{\infty} \mathbf{x}_{i+1}(\underline{s}) - \mathbf{x}_i(\underline{s})$$

is convergent, since the corresponding sequence of absolute values $\sum_{i=0}^{\infty} l_i$ is convergent.

b. Suppose the $i$th digit of $\underline{s}$ and $\underline{s}'$ is the first digit that is different. Then $\mathbf{x}_i(\underline{s}) = \mathbf{x}_i(\underline{s}')$, and after that you go either left-right-left-right- ... , or right-left-right-left- .... Both paths lead back to $\mathbf{x}_{i-1}(\underline{s})$.

c. Suppose that $|t_1 - t_2| < 2^{-k}$. Then either

• the first $k$ digits of $t_1$ and $t_2$ coincide; in that case the line from $\mathbf{x}_{k-1}(t_1)$ to $\mathbf{x}_k(t_2)$ is the altitude of a triangle in which both $\gamma(t_1)$ and $\gamma(t_2)$ lie. But the diameter of that triangle is at most $c(b/a)^k$.

or

• there exists $t_3 \in [t_1, t_2]$ that can be written in two ways, and the first $k$ digits of $t_1$ coincide with the first $k$ digits of one way of writing $t_3$, whereas the first $k$ digits of $t_2$ coincide with the first $k$ digits of the other way of writing $t_3$. By the same argument as above,

$$\left|\gamma(t_1) - \gamma(t_2)\right| \leq \left|\gamma(t_1) - \gamma(t_3)\right| + \left|\gamma(t_3) - \gamma(t_2)\right| \leq 2c(b/c)^k.$$

d. Take any point $\mathbf{x} \in T$. Draw the first altitude $[\mathbf{x}_0, \mathbf{x}_1]$, and see if $\mathbf{x}$ is to the left or to the right. Then in the "even/odd" table, look in the second row (since we are writing the first digit of $t$) to see whether the first digit should be 0 or 1. If $\mathbf{x}$ is on the altitude you just drew, then enter either a 0 or a 1 for the first digit. In any case, draw the second altitude $[\mathbf{x}_1, \mathbf{x}_2]$ according to the digit that was either imposed or chosen, and see whether $\mathbf{x}$ is to the left or right of this altitude $t$.

If either is the case, then the first row of the table tells you what the second digit should be. If $\mathbf{x}$ is on this second altitude, the second digit is free. Keep going this, using alternately the top and the bottom row of the table to decide on successive digits. This also creates a sequence of points $\mathbf{x}_1, \mathbf{x}_2, \ldots$ that corresponds to the string of digits chosen. This sequence converges to $\mathbf{x}$, since the successive segments $[\mathbf{x}_i, \mathbf{x}_{i+1}]$ have lengths that decrease at least like a geometric series with ratio $b/c$.

The maximum $k$ such that there exist $t_1 < \cdots < t_k$ with

$$\gamma(t_1) = \gamma(t_2) = \cdots = \gamma(t_k)$$

is 4, and even that happens only when $a$ and $b$ are chosen carefully (for instance, taking $a = b$). Usually, the maximum is 3, achieved at the feet of altitudes that are in the interior of $T$; you get 4 if such a point is a foot of two altitudes, one from each side, as shown in the figure in the margin.

**A2.1** The first step is to set $x = y + \frac{1}{3}$. The equation

$$x^3 - x^2 - x - 2 = 0 \quad \text{becomes} \quad y^3 - \frac{4}{3}y - \frac{65}{27} = 0.$$

The next step is to set

$$y = u - \frac{p}{3u} = u + \frac{4}{9u}.$$

Substituting this in the equation for $y$ leads to

$$\left(u + \frac{4}{9u}\right)^3 - \frac{4}{3}\left(u + \frac{4}{9u}\right) - \frac{65}{27} = u^3 + \frac{4^3}{9^3 u^3} - \frac{65}{27} = 0.$$

Multiply by $u^3$ to find the quadratic equation for $u^3$

$$u^6 - \frac{65}{27}u^3 + \frac{4^3}{27^2} = 0.$$

The discriminant of this quadratic equation is $3969/1729 > 0$, so the equation has exactly one real root, and it is simple. Solve the equation to find

$$u = \left(\frac{65 \pm \sqrt{3969}}{54}\right)^{1/3}.$$

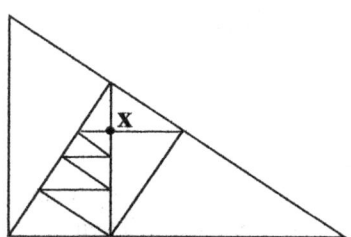

SOLUTION A1.5:    Suppose that the trangle $T$ looks like the triangle above; more particularly, that the two horizontal lines meet exactly at the marked point $\mathbf{x}$. Then there exist four numbers $t_1 < t_2 < t_3 < t_4$ such that

$$\gamma(t_1) = \gamma(t_2) = \gamma(t_3) = \gamma(t_4).$$

It is quite instructive to figure out what they are for this specific example.

This gives two expressions for $x$:

$$\left(\frac{65 + \sqrt{3969}}{54}\right)^{1/3} + \frac{4}{9}\left(\frac{65 + \sqrt{3969}}{54}\right)^{-1/3} + \frac{1}{3} = 2 \quad \text{(surprise!)}$$

and

$$\left(\frac{65 - \sqrt{3969}}{54}\right)^{1/3} + \frac{4}{9}\left(\frac{65 - \sqrt{3969}}{54}\right)^{-1/3} + \frac{1}{3} = 2. \quad \text{(another surprise!)}$$

It certainly isn't obvious that those funny expressions are equal to 2, or to each other, but it is true (try them on your calculator).

This is "explained" by the sentence following equation A2.13 and by exercise A2.2.

The complex roots can be obtained from the complex cube roots:

$$\left(\frac{65 + \sqrt{3969}}{54}\right)^{1/3}\left(\frac{-1 \pm \sqrt{3}}{2}\right) + \frac{4}{9}\left(\frac{65 + \sqrt{3969}}{54}\right)^{-1/3}\left(\frac{-1 \mp \sqrt{3}}{2}\right) + \frac{1}{3} = \frac{-1 + i\sqrt{3}}{2}.$$

We do not claim that this result is obvious, but a calculator will easily confirm it, and it could also be obtained by long division:

$$\frac{x^3 - x^2 - x - 2}{x - 2} = x^2 + x + 1,$$

whose roots are indeed the nonreal cubic roots of 1.

**A2.3** We have

$$\left(y - \frac{a}{3}\right)^3 + a\left(y - \frac{a}{3}\right)^2 + b\left(y - \frac{a}{3}\right) + c = \left(y - \frac{a}{3}\right)\left(y^2 - \frac{2ay}{3} + \frac{a^2}{9}\right) + ay^2 - \frac{2a^2y}{3} + \frac{a^3}{9} + by - \frac{ab}{3} + c$$

$$= y^3 - \frac{2ay^2}{3} + \frac{a^2y}{9} - \frac{ay^2}{3} + \frac{2a^2y}{9} - \frac{a^3}{27} + ay^2 - \frac{2a^2y}{3} + \frac{a^3}{9} + by - \frac{ab}{3} + c$$

$$= y^3 + \frac{a^2y}{9} + \frac{2a^2y}{9} - \frac{a^3}{27} - \frac{2a^2y}{3} + \frac{a^3}{9} + by - \frac{ab}{3} + c$$

$$= y^3 + y\left(b - \frac{a^2}{3}\right) + \frac{2a^3}{27} - \frac{ab}{3} + c.$$

**A2.5** Here the appropriate substitution is $x = u + 7/(3u)$, which leads to

$$\left(u + \frac{7}{3u}\right)^3 - 7\left(u + \frac{7}{3u}\right) + 6 = 0.$$

As above, after multiplying out, simplifying, multiplying through by $u^3$, and setting $v = u^3$, you find the quadratic equation

$$v^2 + 6v + \frac{343}{27} = 0.$$

This time, the roots of the quadratic are complex: $v_1 = -3 + i\frac{10\sqrt{3}}{9}$ and $v_2 = -3 - i\frac{10\sqrt{3}}{9}$. One approach (usually the only one) is to pass to polar coordinates; in this case, you might "just happen to observe" that the cube roots of $-3 + i\frac{10\sqrt{3}}{9}$ are

$$u_1 = 1 + i\frac{2\sqrt{3}}{3} \qquad u_2 = -\frac{3}{2} + i\frac{\sqrt{3}}{6} \quad \text{and} \quad u_3 = \frac{1}{2} - i\frac{5\sqrt{3}}{6}.$$

These lead to

$$u_1 + \frac{7}{3u_1} = 2, \quad u_2 + \frac{7}{3u_2} = -3 \quad \text{and} \quad u_3 + \frac{7}{3u_3} = 1.$$

Indeed, it is easy to check that $(x + 3)(x - 1)(x - 2) = x^3 - 7x + 6$.

**A2.7** a. This is a straightforward computation:

$$\left(x - \frac{a}{4}\right)^4 + a\left(x - \frac{a}{4}\right)^3 + b\left(x - \frac{a}{4}\right)^2 + c\left(x - \frac{a}{4}\right) + d$$

$$
\begin{aligned}
= \quad & x^4 \quad -ax^3 \quad\quad +\tfrac{3}{8}a^2x^2 \quad\quad -\tfrac{1}{16}a^3x \quad\quad +\tfrac{1}{256}a^4 \\
& \quad\quad\quad +ax^3 \quad\quad -\tfrac{3}{4}a^2x^2 \quad\quad +\tfrac{3}{16}a^3x \quad\quad -\tfrac{1}{64}a^4 \\
& \quad\quad\quad\quad\quad\quad\quad +bx^2 \quad\quad -\tfrac{1}{2}abx \quad\quad +\tfrac{1}{16}ba^2 \\
& \quad\quad\quad\quad\quad\quad\quad\quad\quad\quad\quad\quad +cx \quad\quad -\tfrac{1}{4}ac \\
& \quad\quad\quad\quad\quad\quad\quad\quad\quad\quad\quad\quad\quad\quad\quad\quad +d \\
= \quad & x^4 \quad\quad\quad +(-\tfrac{3}{8}a^2 + b)x^2 \quad +(\tfrac{a^3}{8} - \tfrac{ab}{2} + c)x \quad +(-\tfrac{3a^4}{256} + \tfrac{ba^2}{16} - \tfrac{ac}{4} + d)
\end{aligned}
$$

So we see that

$$p = -\frac{3}{8}a^2 + b$$

$$q = \frac{a^3}{8} - \frac{ab}{2} + c$$

$$r = -\frac{3a^4}{256} + \frac{ba^2}{16} - \frac{ac}{4} + d.$$

b.  The equation $x^2 - y + p/2 = 0$ is the definition of $y$, and if you substitute it into the second, you get

$$\left(x^2 + \frac{p}{2}\right)^2 + qx + r - \frac{p^2}{4} = x^4 + px^2 + qx + r = 0.$$

c. When $m = 1$, the curve is a circle. When $m < 0$, it is a hyperbola, when $m > 0$, it is an ellipse.

d. It is exactly the curve $y^2 + qx + r - \frac{p^2}{4} = 0$, which corresponds to $m = \infty$.

e. and f. If the equation $f_m\left(\begin{smallmatrix} x \\ y \end{smallmatrix}\right) = 0$ is a union of two lines, then at the point $\left(\begin{smallmatrix} x \\ y \end{smallmatrix}\right)$ where the lines intersect it cannot represent $x$ implicitly as a function of $y$ or $y$ implicitly as a function of $x$. (We discuss this sort of thing in detail in section 2.10.) It follows that both partial derivatives of

$f_m$ must vanish at that point, so $m$ and $\begin{pmatrix} x \\ y \end{pmatrix}$ satisfy the equations

$$f_m \begin{pmatrix} x \\ y \end{pmatrix} = y^2 + qx + r - \frac{p^2}{4} + m\left(x^2 - y + \frac{p}{2}\right) = 0$$

$$2y - m = 0$$

$$q + 2mx = 0.$$

This gives

$$y = \frac{m}{2} \quad \text{and} \quad x = -\frac{q}{2m},$$

and substituting these values into the first equation, multiplying through by $-4m$, and collecting terms, gives

$$m^3 - 2pm^2 + (p^2 - 4r)m + q^2 = 0.$$

g. The two functions $y^2 + qx + r - \frac{p^2}{4}$ and $x^2 - y + \frac{p}{2}$, when restricted to a diagonal $l_k$, are two quadratic functions that vanish at the same two points. But any quadratic function of $t$ that vanishes at two points $a$ and $b$ is of the form $A(t - a)(t - b)$. In other words, any two such functions are multiples of each other. Thus we have

"The ratios ... must be equal":

If

$$\alpha_1 x^2 + \beta_1 x + \gamma_1$$
$$= A(\alpha_2 x^2 + \beta_2 x + \gamma_2),$$

then

$$\frac{\alpha_1}{\alpha_2} = \frac{\beta_1}{\beta_2} = \frac{\gamma_1}{\gamma_2}.$$

$$\left(k(x - x_1) + y_1\right)^2 + qx + r - \frac{p^2}{4} = A\left(x^2 - k(x - x_1) - y_1 + \frac{p}{2}\right)$$

for some number $A$. It follows that the ratios of the coefficients of $x^2, x$ and constants must be equal, which leads to the equations given in the exercise. Use the first and second, and the first and third, to express $k^3$ as a quadratic polynomial in $k$, and equate these polynomials to get

$$k^2\left(x_1^2 - y_1 + \frac{p}{2}\right) = y_1^2 + qx_1 - \frac{p^2}{4} + r.$$

h. The two parabolas to intersect are $y = x^2 - 2$ and $x = 3 - y^2$, represented in the figure below.

FIGURE FOR SOLUTION A2.7, part h. The parabolas to intersect in order to solve $x^4 - 4x^2 + x + 1 = 0$.

The resolvent cubic is $m^3 + 8m^2 + 12m + 1 = 0$. If we set $m = t - 8/3$ to eliminate the square term, we find the equation

$$t^3 - \frac{28}{3}t + \frac{187}{27} = t^3 + \alpha t + \beta = 0.$$

This is best solved by the trigonometric formula of exercise A2.6, which gives

$$t = \sqrt{-\frac{4\alpha}{3}} \cos\left(\frac{1}{3}\arccos\left(\sqrt{-\frac{3}{4\alpha}}\frac{3\beta}{\alpha}\right)\right).$$

With our values of $\alpha$ and $\beta$, this gives the roots

$$t_1 \approx 2.57816998074169, \quad t_2 \approx -3.37429792821080, \quad t_3 \approx 0.79612794746911.$$

Since $m = t - 8/3$, this gives

$$m_1 \approx -0.08849668592498, \quad m_2 \approx -6.04096459487747, \quad m_3 \approx -1.87053871919756.$$

The corresponding points are

$$\begin{pmatrix} x_1 \\ y_1 \end{pmatrix} \approx \begin{pmatrix} 5.64992908800994 \\ -0.04424834296249 \end{pmatrix}, \quad \begin{pmatrix} x_2 \\ y_2 \end{pmatrix} \approx \begin{pmatrix} 0.08276823877167 \\ -3.02048229743873 \end{pmatrix}, \quad \begin{pmatrix} x_3 \\ y_3 \end{pmatrix} \approx \begin{pmatrix} 0.26730267321838 \\ -0.93526935959878 \end{pmatrix}.$$

If we insert these values into the equation for $k$, we find

$$k_1 \approx 0.29748392549006, \quad k_2 \approx 2.45783738169910, \quad k_3 \approx 1.36767639418013.$$

To find the roots, we must intersect the lines given by the equations $y - y_i = \pm k_i(x - x_i)$, two at a time, of course.

For instance, if we intersect $y - y_1 = -k_1(x - x_1)$ and $y - y_2 = k_2(x - x_2)$, we find the point

$$\begin{pmatrix} X_1 \\ Y_1 \end{pmatrix} \approx \begin{pmatrix} 1.76401492519458 \\ 1.11174865630925 \end{pmatrix}.$$

If we intersect $y - y_1 = -k_1(x - x_1)$ and $y - y_3 = -k_3(x - x_3)$, we find the point

$$\begin{pmatrix} X_2 \\ Y_2 \end{pmatrix} \approx \begin{pmatrix} -2.06149885068464 \\ 2.24977751137410 \end{pmatrix}.$$

If we intersect $y - y_1 = k_1(x - x_1)$ and $y - y_3 = k_3(x - x_3)$, we find the point

$$\begin{pmatrix} X_3 \\ Y_3 \end{pmatrix} \approx \begin{pmatrix} -0.39633853101445 \\ -1.84291576883330 \end{pmatrix}.$$

Finally, if we intersect $y - y_1 = k_1(x - x_1)$ and $y - y_2 = k_2(x - x_2)$, we find the point

$$\begin{pmatrix} X_4 \\ Y_4 \end{pmatrix} \approx \begin{pmatrix} 0.69382245650451 \\ -1.51861039885004 \end{pmatrix}.$$

The points are labeled by the quadrant they are in.

Indeed, the four numbers $X_1$, $X_2$, $X_3$, $X_4$ are (approximations to) the roots of the original polynomial.

Now let us solve the equation $x^4 + 4x^3 + x - 1 = 0$. If we set $x = y - 1$, this becomes $y^4 - 6y^2 + 9y - 5 = 0$, with resolvent cubic $m^3 + 12m^2 + 56m + 81 = 0$.

To solve the resolvent cubic, we need to eliminate the term in $m^2$, by setting $m = t - 4$, giving the equation $t^3 + 8t + 15 = t^3 + Pt + Q = 0$. Then set $t = u - P/(3u)$ and $v = u^3$; as in equation A2.11, this leads to the quadratic equation

$$v^2 + Qv - \frac{P^3}{27} = 0. \tag{1}$$

One root of this equation is

$$v \approx 16.17254074438183,$$

and the three cube roots of $v$ are

$$u_1 \approx 2.52886755571821,$$

$$u_2 \approx -1.26443377785910 + 2.19006354605823\,i,$$

$$u_3 \approx -1.26443377785910 - 2.19006354605823\,i.$$

(If we choose the other root of the equation (1), we would get different values of $u_i$ but the same values of $t_i$; see the text immediately following equation A2.13).

These now give

$$t_1 \approx 1.47437711368740,$$

$$t_2 \approx -0.73718855684370 + 3.10327905690479i,$$

$$t_3 \approx -0.73718855684370 - 3.10327905690479i,$$

and finally, using $m = t - 4$, we get

$$m_1 \approx -2.52562288631260,$$

$$m_2 \approx -4.73718855684370 + 3.10327905690479\,i,$$

$$m_3 \approx -4.73718855684370 - 3.10327905690479\,i.$$

We have now solved the resolvent cubic, and can give the points of intersection of the diagonals of the quadrilateral formed by the roots:

$$\begin{pmatrix} x_1 \\ y_1 \end{pmatrix} \approx \begin{pmatrix} 1.78173868489526 \\ -1.26281144315630 \end{pmatrix},$$

$$\begin{pmatrix} x_2 \\ y_2 \end{pmatrix} \approx \begin{pmatrix} 0.66468621310792 + 0.43542847826296i \\ -2.36859427842185 + 1.55163952845240i \end{pmatrix},$$

$$\begin{pmatrix} x_3 \\ y_3 \end{pmatrix} \approx \begin{pmatrix} 0.66468621310792 - 0.43542847826296i \\ -2.36859427842185 - 1.55163952845240i \end{pmatrix}.$$

We can now compute the slopes of two of the pairs of diagonals, given by

$$k_{1,1} \approx 1.58922084252397,$$

$$k_{1,2} = -k_{1,1},$$

$$k_{2,1} \approx 2.28038824159147 - 0.68042778863371\,i,$$

$$k_{2,2} = -k_{2,1}.$$

We are finally in a position to intersect the pairs of diagonals; the $x$-coordinate of each point of intersection is given by the formula

$$\frac{k_{1,i}\, x_1 - k_{2,j}\, x_2 - y_1 + y_2}{k_{1,i} - k_{2,j}}, \quad \text{for appropriate values of } i \text{ and } j;$$

the four values given by $i = 1, 2$, $j = 1, 2$ give the four roots of the equation

$$y^4 - 6y^2 + 9y - 5 = 0.$$

They are

$$0.79461042126199 + 0.68042778863371i,$$
$$0.79461042126199 - 0.68042778863371i,$$
$$1.48577782032949,$$
$$-3.07499866285346,$$

and finally, the roots of the original equation

$$x^4 + 4x^3 + x - 1 = 0$$

are found from $x = y - 1$ to be

$$-0.20538957873801 + 0.68042778863371i,$$
$$-0.20538957873801 - 0.68042778863371i,$$
$$0.48577782032949,$$
$$-4.07499866285346.$$

In the first printing of the text, there is a typo in the statement of the exercise: In the fifth line, in the denominator, $(\alpha - 1)$ should be $(\alpha + 1)$:

"... is satisfied, with

$$\left|\frac{k}{(\alpha+1)(1-k)}\right|^\alpha \le 1 - k,$$

then ... ."

**A5.1**  First note that

$$\left|\mathbf{f}(\mathbf{u}+\mathbf{h}) - \mathbf{f}(\mathbf{u}) - [\mathbf{Df}(\mathbf{u})]\mathbf{h}\right| = \left|\int_0^1 \Big([\mathbf{Df}(\mathbf{u}+t\mathbf{h})]\mathbf{h} - [\mathbf{Df}(\mathbf{u})]\mathbf{h}\Big)\,dt\right|$$

$$\le \int_0^1 M|t\mathbf{h}|^\alpha\,|\mathbf{h}|\,dt = \frac{M}{\alpha+1}|\mathbf{h}|^{\alpha+1}.$$

To show that $[\mathbf{Df}(\mathbf{a}_1)]$ is invertible, we will use the following lemma:

**Lemma 1.**  *If $P$ and $Q$ are $n \times n$ matrices with $Q$ invertible, and*

$$|I - PQ^{-1}| \le k < 1,$$

*then $P$ is invertible, and*

$$|P^{-1}| \le \frac{1}{1-k}|Q^{-1}|.$$

**Proof.**  Set $X = I - PQ^{-1}$; then $I - X = PQ^{-1}$ is invertible by geometric series. Thus $P^{-1}$ exists, and

$$P^{-1} = Q^{-1}(I + X + X^2 + \cdots),$$

so

$$|P^{-1}| \leq |Q^{-1}| + |Q^{-1}X| + \cdots \leq |Q^{-1}|(1 + |X| + \cdots)$$

$$= \frac{|Q^{-1}|}{1 - |X|} \leq \frac{1}{1-k}|Q^{-1}|. \quad \square$$

The quantity on the third line is $\leq k$ by hypothesis.

To apply this to show that $[\mathbf{Df}(\mathbf{a}_1)]$ is invertible, note that

$$\left| I - [\mathbf{Df}(\mathbf{a}_1)][\mathbf{Df}(\mathbf{a}_0)]^{-1} \right| = \left| \left([\mathbf{Df}(\mathbf{a}_0)] - [\mathbf{Df}(\mathbf{a}_1)]\right)[\mathbf{Df}(\mathbf{a}_0)]^{-1} \right|$$

$$\leq M\,|\mathbf{h}_0|^\alpha \,\left|[\mathbf{Df}(\mathbf{a}_0)]^{-1}\right|$$

$$\leq M\left|[\mathbf{Df}(\mathbf{a}_0)]^{-1}\mathbf{f}(\mathbf{a}_0)\right|^\alpha \,\left|[\mathbf{Df}(\mathbf{a}_0)]^{-1}\right|$$

$$\leq k.$$

Moreover, note that $k$ as defined in the exercise is necessarily strictly less than 1; in fact, it is less than 1/2; see the figure in the margin. So lemma 1 says that $[\mathbf{Df}(\mathbf{a}_1)]^{-1}$ exists and

$$[\mathbf{Df}(\mathbf{a}_1)]^{-1} \leq \frac{1}{1-k}[\mathbf{Df}(\mathbf{a}_0)]^{-1}.$$

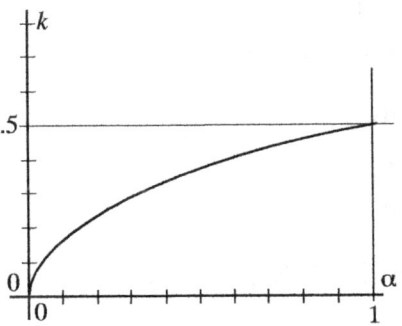

The equation

$$\left(\frac{k}{(\alpha+1)(1-k)}\right)^\alpha = 1 - k$$

represents $k$ implicitly as a function of $\alpha$ for $0 \leq \alpha \leq 1$ (and in fact for all $\alpha \geq 0$). By implicit diferentiation, one can show that this function is increasing; more specifically, it increases from 0 to 1/2 as $\alpha$ increases from 0 to 1.

It follows that we can define

$$\mathbf{h}_1 = -[\mathbf{Df}(\mathbf{a}_1)]^{-1}\mathbf{f}(\mathbf{a}_1) \quad \text{and} \quad \mathbf{a}_2 = \mathbf{a}_1 + \mathbf{h}_1.$$

We need three estimates for the size of $\mathbf{f}(\mathbf{a}_1)$:

$$|\mathbf{f}(\mathbf{a}_1)| = \left|\mathbf{f}(\mathbf{a}_1) - \mathbf{f}(\mathbf{a}_0) - [\mathbf{Df}(\mathbf{a}_0)]\mathbf{h}_0\right|$$

$$\leq \begin{cases} \dfrac{M}{\alpha+1}|\mathbf{h}_0|^{\alpha+1} \\[2mm] \left|\dfrac{M}{\alpha+1}[\mathbf{Df}(\mathbf{a}_0^{-1})]\mathbf{f}(\mathbf{a}_0)\right|^{\alpha+1} \\[2mm] \dfrac{M}{\alpha+1}\left|[\mathbf{Df}(\mathbf{a}_0^{-1})]\mathbf{f}(\mathbf{a}_0)\right|^\alpha|\mathbf{h}_0|, \end{cases}$$

where the three lines on the right after the inequality are all equal.
We find the size of $\mathbf{h}_1$:

$$|\mathbf{h}_1| = \left|[\mathbf{Df}(\mathbf{a}_1)]^{-1}\right|\,|\mathbf{f}(\mathbf{a}_1)|$$

$$\leq \frac{[\mathbf{Df}(\mathbf{a}_0)]^{-1}}{1-k}\frac{M}{\alpha+1}\left|[\mathbf{Df}(\mathbf{a}_0)]^{-1}\mathbf{f}(\mathbf{a}_0)\right|^\alpha\,|\mathbf{h}_0|$$

$$\leq \frac{k}{1-k}\frac{|\mathbf{h}_0|}{\alpha+1}.$$

Next we show that the modified Kantorovich inequality

$$\left|[\mathbf{D}\vec{\mathbf{f}}(\mathbf{a}_0)]^{-1}\right|^{\alpha+1}|\vec{\mathbf{f}}(\mathbf{a}_0)|^\alpha M \leq k$$

holds for $\mathbf{a}_1$:

$$\left|[\mathbf{Df}(\mathbf{a}_1)]^{-1}\right|^{1+\alpha}|\mathbf{f}(\mathbf{a}_1)|^\alpha M \leq \left(\frac{\left|[\mathbf{Df}(\mathbf{a}_0)]^{-1}\right|}{1-k}\right)^{1+\alpha}\left(\frac{M}{\alpha+1}\left|[\mathbf{Df}(\mathbf{a}_0)]^{-1}\mathbf{f}(\mathbf{a}_0)\right|^{\alpha+1}\right)^\alpha M$$

$$\leq \left(\frac{\left|[\mathbf{Df}(\mathbf{a}_0)]^{-1}\right|}{1-k}\right)^{1+\alpha}\left(\frac{k|\mathbf{f}(\mathbf{a}_0)|}{\alpha+1}\right)^\alpha M$$

$$\leq \left(\frac{1}{1-k}\right)^{1+\alpha}\left(\frac{k}{\alpha+1}\right)^\alpha k$$

$$\leq \frac{1}{1-k}\left(\frac{k}{(\alpha+1)(1-k)}\right)^\alpha k \leq \frac{1-k}{1-k}k = k.$$

So we have shown that the method may be iterated; that

$$|\mathbf{h}_{m+1}| \leq \frac{k}{(1-k)(\alpha+1)}|\mathbf{h}_m|,$$

so $\lim \mathbf{a}_j = \mathbf{a}_0 + \sum \mathbf{h}_j$ converges to a point $\mathbf{a}$; and that

$$|\mathbf{f}(\mathbf{a}_{m+1})| \leq \frac{M}{\alpha+1}|\mathbf{h}_m|^{\alpha+1},$$

so $\mathbf{f}(\mathbf{a}) = \mathbf{0}$.

Finally, we show the rate of convergence:

$$1 - [\mathbf{Df}(\mathbf{a}_{m+1})][\mathbf{Df}(\mathbf{a}_0)]^{-1} = \sum_{j=0}^{m}\Big([\mathbf{Df}(\mathbf{a}_j)] - [\mathbf{Df}(\mathbf{a}_{j+1})]\Big)[\mathbf{Df}(\mathbf{a}_0)]^{-1},$$

so

$$\left|1 - [\mathbf{Df}(\mathbf{a}_{m+1})][\mathbf{Df}(\mathbf{a}_0)]^{-1}\right| \leq \sum_{j=0}^{m} M|\mathbf{h}_j|^\alpha\left|[\mathbf{Df}(\mathbf{a}_0)]^{-1}\right| \leq \frac{M|\mathbf{h}_0|^\alpha\left|[\mathbf{Df}(\mathbf{a}_0)]^{-1}\right|}{1 - \left(\frac{k}{(1-k)}\frac{1}{(\alpha+1)}\right)^\alpha}$$

$$= \frac{M\left|[\mathbf{Df}(\mathbf{a}_0)]^{-1}\mathbf{f}(\mathbf{a}_0)\right|^\alpha\left|[\mathbf{Df}(-\mathbf{a}_0)]^{-1}\right|}{1 - \left(\frac{k}{(1-k)(\alpha+1)}\right)^\alpha}$$

$$\leq \frac{k}{1 - \left(\frac{k}{(1-k)(\alpha+1)}\right)^\alpha} < \frac{k}{1-(1-k)} = 1.$$

So, again by lemma 1,

$$\left|[\mathbf{Df}(\mathbf{a}_{m+1})]^{-1}\right| \leq \frac{1}{1 - \dfrac{k}{1 - \left(\dfrac{k}{(1-k)(\alpha+1)}\right)^\alpha}} \leq C\frac{M}{\alpha+1}|\mathbf{h}_m|^{\alpha+1}.$$

**A7.1** We will assume that the function $f : [a,b] \to [c,d]$ is increasing. If it isn't, replace $f$ by $-f$.

1. Pick a number $y \in [c,d]$, and consider the function $f_y : x \mapsto f(x) - y$, which is continuous, since $f$ is. Since

$$f_y(a) = c - y \leq 0 \quad \text{and} \quad f_y(b) = d - y \geq 0,$$

by the intermediate value theorem there must exist a number $x \in [a, b]$ such that $f_y(x) = 0$, i.e., $f(x) = y$.

Moreover, this number $x$ is unique: if $x_1 < x_2$ and both $x_1$ and $x_2$ solve the equation, then $f(x_1) = f(x_2) = y$, but $f(x_1) < f(x_2)$ since $f$ is increasing on $[a, b]$. So set $g(y) \in [a, b]$ to be the unique solution of $f_y(x) = 0$.

Then by definition we have

$$0 = f_y\big(g(y)\big) = f\big(g(y)\big) - y, \quad \text{so} \quad f\big(g(y)\big) = y.$$

This says that $f(g(f(x))) = f(x)$, for any $x \in [a, b]$. But any $y \in [c, d]$ can be written $y = f(x)$, so $f(g(y)) = y$ for any $y \in [c, d]$.

We need to check that $g$ is continuous. Choose $y \in (c, d)$ and $\epsilon > 0$. Consider the sequences $x_n' = g(y - 1/n)$ and $x_n'' = g(y + 1/n)$. These sequences are respectively increasing and decreasing, and both are bounded, so they have limits $x'$ and $x''$. Applying $f$ (which we know to be continuous) gives

$$f(x') = \lim f(x_n') \lim y - 1/n = y \quad \text{and} \quad f(x'') = \lim f(x_n'') \lim y + 1/n = y.$$

Thus $x' = x'' = g(y)$, and there exists $N$ such that $x_N'' - x_N' < \epsilon$. Now set $\delta = 1/N$; if $|z - y| < \delta$, then $g(z) \in [x_N', x_N'']$, and in particular, $|g(z) - g(y)| < \epsilon$. This is the definition of continuity.

2. Choose $y \in [c, d]$, and define two sequences $x_n'$, $x_n'' \in [a, b]$ by the rule $x_0' = a$, $x_0'' = b$, and

$$\begin{cases} x_{n+1}' = \frac{x_n' + x_n''}{2}, \ x_{n+1}'' = x_n'' & \text{if } y \geq f\left(\frac{x_n' + x_n''}{2}\right); \\ x_{n+1}'' = \frac{x_n' + x_n''}{2}, \ x_{n+1}' = x_n' & \text{if } y < f\left(\frac{x_n' + x_n''}{2}\right). \end{cases}$$

Then both sequences converge to $g(y)$. Indeed, the sequence $x_n'$ is nondecreasing and the sequence $x_n''$ is nonincreasing; and both are bounded, so they both converge to something. Moreover, $x_n'' - x_n' = (b - a)/2^n$, so they converge to the same limit. Finally, $f(x_n') \leq y$ and $f(x_n'') \geq y$ for all $n$; passing to the limit, this means that

$$y \geq \lim_{n \to \infty} f(x_n') = \lim_{n \to \infty} f(x_n'') \geq y,$$

so all the inequalities are equalities.

3. Suppose $y = f(x)$, and that $f$ is differentiable at $x$, with $f'(x) \neq 0$. Define the function $k(h)$ by $g(y + h) = x + k(h)$. Since $g$ is continuous, $\lim_{h \to 0} k(h) = 0$. Now

$$\frac{g(y + h) - g(y)}{h} = \frac{x + k(h) - x}{h} = \frac{k(h)}{f(x + k(h)) - f(x)}.$$

Take the limit as $h \to 0$. Since

$$\lim_{h \to 0} \frac{k(h)}{f(x + k(h)) - f(x)} = \frac{1}{f'(x)},$$

we see that

$$\lim_{h \to 0} \frac{g(y + h) - g(y)}{h} = g'(y)$$

also exists, and the limits are equal.

---

Here we choose $y$ to be an interior point; the endpoints $y = c$ and $y = d$ require a slightly different and easier treatment.

**A8.1**  Suppose there were two such continuous functions

$$\mathbf{g}, \mathbf{g}_1 : B_R(\mathbf{b}) \to \mathbb{R}^n \quad \text{with} \quad \mathbf{g}(\mathbf{b}) = \mathbf{g}_1(\mathbf{b}) = \mathbf{a}.$$

Let $R_1$ be the largest number such that $\mathbf{g}$ and $\mathbf{g}_1$ coincide on $B_{R_1}(\mathbf{b})$. If $R_1 = R$, we are done, so we may assume that $R_1 < R$. As shown in the picture in the margin, let $\mathbf{b}_1$ be a point with $|\mathbf{b}_1 - \mathbf{b}| = R_1$ and such that $\mathbf{g}$ and $\mathbf{g}_1$ do not coincide on any neighborhood of $\mathbf{b}_1$; however, we must have $\mathbf{g}(\mathbf{b}_1) = \mathbf{g}_1(\mathbf{b}_1)$ by the continuity of $\mathbf{g}_1$; call $\mathbf{a}_1 = \mathbf{g}(\mathbf{b}_1)$, and $\mathbf{c}_1 = \begin{pmatrix} \mathbf{a}_1 \\ \mathbf{b}_1 \end{pmatrix}$.

Our proof of theorem 2.10.7 (the implicit function theorem) shows that at $\mathbf{b}_1$, the matrix

$$[D_1 F(\mathbf{c}_1), \dots, D_n F(\mathbf{c}_1)]$$

is invertible, hence near $\mathbf{c}_1$, the set

$$\left\{ \begin{pmatrix} \mathbf{x} \\ \mathbf{y} \end{pmatrix} \ \Big| \ F\begin{pmatrix} \mathbf{x} \\ \mathbf{y} \end{pmatrix} = \mathbf{0} \right\}$$

is the graph of a function expressing $\mathbf{x}$ as a function of $\mathbf{y}$. Thus there exist neighborhoods $U_1$ of $\mathbf{a}_1$ and $V_1$ of $\mathbf{b}_1$ such that for every $\mathbf{y} \in V_1$, there is a *unique* $\mathbf{x} \in U_1$ such that $F\begin{pmatrix} \mathbf{x} \\ \mathbf{y} \end{pmatrix} = \mathbf{0}$. So this point $\mathbf{x}$ must be the value $\mathbf{g}(\mathbf{y})$ and the value $\mathbf{g}_1(\mathbf{y})$, meaning that $\mathbf{g}$ and $\mathbf{g}_1$ coincide on some neighborhood of $\mathbf{b}_1$. This contradicts the hypothesis that $R_1 < R$.

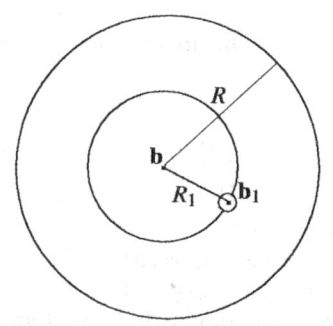

FIGURE FOR SOLUTION A8.1.

Since $R_1$ is the largest number such that $\mathbf{g}$ and $\mathbf{g}_1$ coincide on $B_{R_1}(\mathbf{b})$, and $|\mathbf{b}_1 - \mathbf{b}| = R_1$, we have $\mathbf{g}_1(\mathbf{b}_1) = \mathbf{g}(\mathbf{b}_1)$, but any neighborhood of $\mathbf{b}_1$ includes points in $B_R(\mathbf{b}) - B_{R_1}(\mathbf{b})$, where $\mathbf{g}$ and $\mathbf{g}_1$ do not coincide.

**A9.1**  The final equality in equation A9.6 is justified by the following computation:

$$
\begin{aligned}
\lim_{t \to 0} \frac{g'(c)}{t} &= \lim_{t \to 0} \frac{D_i f(\mathbf{a} + c\vec{\mathbf{e}}_i + t\vec{\mathbf{e}}_j) - D_i f(\mathbf{a} + c\vec{\mathbf{e}}_i)}{t} \\
&= \lim_{t \to 0} \frac{D_i f(\mathbf{a} + c\vec{\mathbf{e}}_i + t\vec{\mathbf{e}}_j) - D_i f(\mathbf{a} + c\vec{\mathbf{e}}_i)}{t} \\
&\quad - \left( \overbrace{\lim_{t \to 0} \frac{D_i f(\mathbf{a} + c\vec{\mathbf{e}}_i + t\vec{\mathbf{e}}_j) - D_i f(\mathbf{a}) - cD_i^2 f(\mathbf{a}) - tD_j D_i f(\mathbf{a})}{t}}^{\text{0 by equation A9.4}} \right) \\
&\quad + \left( \overbrace{\lim_{t \to 0} \frac{D_i f(\mathbf{a} + c\vec{\mathbf{e}}_i) - D_i f(\mathbf{a}) - cD_i^2 f(\mathbf{a})}{t}}^{\text{0 by equation A9.5}} \right) \\
&= \lim_{t \to 0} \frac{D_i f(\mathbf{a}) + cD_i^2 f(\mathbf{a}) + tD_j D_i f(\mathbf{a}) - D_i f(\mathbf{a}) - cD_i^2 f(\mathbf{a})}{t} \\
&= D_j D_i f(\mathbf{a}).
\end{aligned}
\tag{1}
$$

Solution A9.1: This argument would not work if we required only that all second partial derivatives of $f$ exist, instead of requiring that the partials be differentiable at $\mathbf{a}$. If $D_i f$ were a nondifferentiable function with all partial derivatives existing (like the function of example 1.9.3), then we could still write the Jacobian matrix of partial derivatives

$$[D_1 D_i f(\mathbf{a}) \dots D_n D_i f(\mathbf{a})],$$

but we could not substitute it for the derivative $L$ in equation (2).

The most delicate point is that the third line equals 0; this was explained briefly in the textbook. The first line of equation A9.4:

$$0 = \lim_{\sqrt{c^2+t^2} \to 0} \frac{1}{\sqrt{c^2 + t^2}} \Big( D_i f(\mathbf{a} + c\mathbf{e}_i + t\mathbf{e}_j) - D_i f(\mathbf{a}) - [\mathbf{D} D_i f(\mathbf{a})](c\vec{\mathbf{e}}_i + t\vec{\mathbf{e}}_j) \Big)$$

is the equation

$$\lim_{\vec{h} \to 0} \frac{1}{|\vec{h}|} \left( \left( \mathbf{f}(\mathbf{a} + \vec{h}) - \mathbf{f}(\mathbf{a}) \right) - \left( L(\vec{h}) \right) \right) = \mathbf{0} \tag{2}$$

(see theorem 1.7.9), where $\vec{h}$ is replaced by the vector $c\vec{e}_i + t\vec{e}_j$, the function $\mathbf{f}$ is replaced by the function $D_i f$ (differentiable by hypothesis), and $L$ is the derivative of $D_i f$, i.e., the Jacobian matrix $[\mathbf{D} D_i f(\mathbf{a})]$ made of the partial derivatives of $D_i f$. Because the derivative is linear, we can write

$$\begin{aligned}
[\mathbf{D} D_i f(\mathbf{a})](c\vec{e}_i + t\vec{e}_j) &= c[\mathbf{D} D_i f(\mathbf{a})]\vec{e}_i + t[\mathbf{D} D_i f(\mathbf{a})]\vec{e}_j \\
&= c(D_i^2 f)(\mathbf{a}) + t D_j(D_i f)(\mathbf{a}).
\end{aligned} \tag{3}$$

A similar but easier computation justifies the $0$ above the brackets in the fourth line in equation (1).

**A10.1**  It is actually easier to start the induction at $k = 0$: the beginning of the induction is then simply that if $f$ is continuous at $\mathbf{a}$ and $f(\mathbf{a}) = 0$, then

$$\lim_{\vec{h} \to 0} f(\mathbf{a} + \vec{h}) = 0,$$

which is the definition of continuity.

The inductive step as written is just as valid to go from $k = 0$ to $k = 1$ as for any other values; more particularly, in equations A10.4–A10.6, we use the fact that if $f$ is of class $C^1$ with derivatives vanishing at $\mathbf{a}$, then

$$\lim_{\vec{h} \to 0} D_i f(\mathbf{a} + \vec{c}) = 0,$$

which is the definition of continuity for the derivatives.

Why does that imply theorem 1.9.7? There the issue is that if the first partials of $f$ are continuous, then

$$\lim_{\vec{h} \to 0} \frac{1}{|\vec{h}|} \left( f(\mathbf{a} + \vec{h}) - f(\mathbf{a}) - \sum_{i=1}^{n} h_i D_i(f)(\mathbf{a}) \right) = 0.$$

The entire expression $f(\mathbf{a} + \vec{h}) - f(\mathbf{a}) - \sum_{i=1}^{n} h_i D_i(f)(\mathbf{a})$ is a function whose 0th and first partials are continuous and vanish at $\mathbf{a}$, so proposition 3.3.19 applies (and is now proved without using theorem 1.9.7, so the proof isn't circular) and shows that the limit is 0.

**A11.1**  a.  The function $|x|^{3/2}$ is not in $o(|x|^2)$ because it isn't even in $O(|x|^2)$; see part b.

b.  The function $|x|^{3/2}$ is not in $O(|x|^2)$, because

$$\lim_{x \to 0} \frac{|x|^{3/2}}{|x|^2} = \lim_{x \to 0} \frac{1}{|x|^{1/2}} = \infty.$$

c.  The function $|x|^{3/2}$ is in $o(|x|)$, because

$$\lim_{x \to 0} \frac{|x|^{3/2}}{|x|} = \lim_{x \to 0} \frac{1}{|x|1/2} = 0.$$

---

We can replace the $\sqrt{c^2 + t^2}$ in the denominator of the first line of equation A9.4 by $|t|$ in the denominator of the second line because $c$ is between $0$ and $t$, so $\sqrt{c^2 + t^2} < \sqrt{2}|t|$, which goes to $0$ with $t$.

Equation (3): The $i$th column of $[\mathbf{D} D_i f(\mathbf{a})]$ is $[\mathbf{D} D_i f(\mathbf{a})](\vec{e}_i)$, by theorem 1.3.4. Or note simply the definition of the Jacobian matrix, definition 1.7.7.

d. It is also in $O(|x|)$, because it is in $o(|x|)$; see part c.

Recall (equation A11.1) that
$$g \in o(f) \implies g \in O(f).$$

**A11.3** The function $x \log |x|$ is not in $O(x)$; indeed,

$$\lim_{x \to 0} \frac{x \log |x|}{x} = \lim_{x \to 0} \log |x| = -\infty.$$

So in particular, it is not in $o(x)$, or in $O(x^2)$, or in $o(x^2)$ either.

**A12.1** a. This follows immediately from corollary A12.3. We have (proposition 3.4.2)

$$e^x = 1 + x + \frac{x^2}{2!} + \cdots + \frac{x^k}{k!} + \frac{e^c}{(k+1)!} x^{k+1}$$

for some $0 \le c \le x$. Apply this when $x = 1$, so that $e^c < 3$.

b. Multiply the inequality

$$\left| \frac{a}{b} - \left( 1 + 1 + \frac{1}{2!} + \cdots + \frac{1}{k!} \right) \right| \le \frac{3}{(k+1)!}$$

by $k!b$ to get

$$\left| k!a - b \left( k! + k! + \frac{k!}{2!} + \cdots + \frac{k!}{k!} \right) \right| \le \frac{3k!b}{(k+1)!} = \frac{3b}{k+1}.$$

Since $k!a - bm$ is an integer whose absolute value is arbitrarily small, it is 0.

c. As the hint indicates, if you take $k$ to be a prime number $> b$, you have a contradiction: $k$ divides $k!a$ evenly, of course, but neither $b$ nor $m$. Thus $e$ is not rational.

**A12.3** a. Using theorem A12.1, and setting $k = 2$, we get

$$\sin h = h + \frac{1}{2!} \int_0^h (h - t)^2 (-\cos t) \, dt$$

Of course, this also uses equation 3.4.6.

(where $-\cos t$ is the third derivative of $\sin t$). Thus

$$\sin xy = xy + \frac{1}{2!} \int_0^{xy} (xy - t)^2 (-\cos t) \, dt.$$

b. Using part a, the bound is

The first inequality in the third line uses

$$0 \le (x - y)^2 = x^2 - 2xy + y^2,$$

hence $xy \le (x^2 + y^2)/2$. The second inequality uses our assumption that $x^2 + y^2 \le 1/4$.

$$\left| \frac{1}{2!} \int_0^{xy} (xy - t)^2 \cos t \, dt \right| \le \frac{1}{2!} \int_0^{xy} (xy - t)^2 \, dt$$

$$= \frac{1}{2!} \left[ -\frac{(xy - t)^3}{3} \right]_0^{xy} = \frac{1}{6} (xy)^3$$

$$\le \frac{1}{6} \left( \frac{x^2 + y^2}{2} \right)^3 \le \frac{1}{6} \left( \frac{1}{8} \right)^3.$$

**A12.5** a. Clearly $\operatorname{sgn} y$ is differentiable when $y \ne 0$. The chain rule guarantees that $f$ is continuously differentiable unless $y = 0$ or the quantity under the square root is $\leq 0$. But $-x + \sqrt{x^2 + y^2} \geq 0$, and it only vanishes if $y = 0$ and $x \geq 0$.

So we need to show that $f$ is continuously differentiable on the locus where $y = 0$ and $x > 0$. In a neighborhood of a point $\begin{pmatrix} x_0 \\ y_0 \end{pmatrix}$ satisfying $x_0 > 0$, $y_0 = 0$, we can write

$$-x + \sqrt{x^2 + y^2} = -x + x\sqrt{1 + y^2/x^2} = -x + x\left(1 + \frac{1}{2}\frac{y^2}{x^2} + o\left(\frac{y^2}{x^2}\right)\right)$$

$$= \frac{y^2}{2x} + o\left(\frac{y^2}{x}\right),$$

and since $(\operatorname{sgn} y)y^2$ is of class $C^1$, the function is of class $C^1$ on the half-axis $y = 0$, $x > 0$. Combined with what we know from the chain rule, this shows that $f$ is continuously differentiable on the complement of the half-line $y = 0$, $x \leq 0$.

b. We have

$$f(\mathbf{a}) = -\sqrt{\frac{1 + \sqrt{1 + \epsilon^2}}{2}}, \quad f(\mathbf{a} + \vec{\mathbf{h}}) = +\sqrt{\frac{1 + \sqrt{1 + \epsilon^2}}{2}}.$$

So

Here we are dealing with the case $n = 2$, $k = 0$, so theorem A12.5 says that the remainder should be

$$\sum_{I \in \mathcal{I}_2^1} \frac{1}{I!} D_I f(\mathbf{c}) \, \vec{\mathbf{h}}^I$$

$$= [D_1 f(\mathbf{c}) \; D_2 f(\mathbf{c})]\vec{\mathbf{h}}.$$

$$f(\mathbf{a} + \vec{\mathbf{h}}) - f(\mathbf{a}) = 2\sqrt{\frac{1 + \sqrt{1 + \epsilon^2}}{2}},$$

which tends to 2 as $\epsilon \to 0$, and cannot be $[\mathbf{D}f(\mathbf{c})]\mathbf{h}$ for any $\mathbf{c} \in [\mathbf{a}, \mathbf{a} + \vec{\mathbf{h}}]$, i.e., any $\mathbf{c} \in \left[\begin{pmatrix} -1 \\ -\epsilon \end{pmatrix}, \begin{pmatrix} -1 \\ \epsilon \end{pmatrix}\right]$, since $[\mathbf{D}f(\mathbf{c})]$ remains bounded. To see that $[\mathbf{D}f(\mathbf{c})]$ remains bounded, first we compute

$$\left[\mathbf{D}f\begin{pmatrix} x \\ y \end{pmatrix}\right] = \left[\operatorname{sgn} y \frac{-\frac{1}{2} + \frac{x}{\sqrt{x^2+y^2}}}{2\sqrt{\frac{-x+\sqrt{x^2+y^2}}{2}}}, \quad \operatorname{sgn} y \frac{\frac{y}{\sqrt{x^2+y^2}}}{2\sqrt{\frac{-x+\sqrt{x^2+y^2}}{2}}}\right].$$

For $\mathbf{c} = \begin{pmatrix} c_1 \\ c_2 \end{pmatrix} \in [\mathbf{a}, \mathbf{a} + \vec{\mathbf{h}}]$, we have $c_1 = -1$, and the derivative does not blow up because the denominators do not tend to 0 as $c_2 \to 0$:

$$\sqrt{\frac{-(-1) + \sqrt{(-1)^2 + c_2^2}}{2}} \xrightarrow[\text{as } c_2 \to 0]{} \sqrt{\frac{1 + \sqrt{1}}{2}} \neq 0$$

Theorem A12.5 requires that $f$ be of class $C^{k+1}$ on $[\mathbf{a}, \mathbf{a} + \mathbf{h}]$; in our case $k = 0$ and we need $f$ of class $C^1$ on $[\mathbf{a}, \mathbf{a} + \vec{\mathbf{h}}]$, which it isn't: the line of discontinuity of $f$ crosses $[\mathbf{a}, \mathbf{a} + \vec{\mathbf{h}}]$, so we do not have $[\mathbf{a}, \mathbf{a} + \vec{\mathbf{h}}] \subset U$.

**A14.1** By definition 3.8.7, $H = (1/2)(A_{2,0} + A_{0,2})$. Equation 3.8.36 gives us expressions for $A_{2,0}$ and $A_{0,2}$. It is now straightforward to show that $H$ is given by the desired expression. Remember we have set $c = \sqrt{a_1^2 + a_2^2}$, so that $a_1^2 + a_2^2 = c^2$. Thus we have

$$H = \frac{1}{2}(A_{2,0} + A_{0,2})$$

$$= -\frac{1}{2c^2(1+c^2)^{1/2}}\left(a_{2,0}a_2^2 - 2a_{1,1}a_1a_2 + a_{0,2}a_1^2\right) - \frac{1}{2c^2(1+c^2)^{3/2}}\left(a_{2,0}a_1^2 + 2a_{1,1}a_1a_2 + a_{0,2}a_2^2\right)$$

$$= -\frac{1}{2c^2(1+c^2)^{3/2}}\left((1+c^2)\left(a_{2,0}a_2^2 - 2a_{1,1}a_1a_2 + a_{0,2}a_1^2\right) + a_{2,0}a_1^2 + 2a_{1,1}a_1a_2 + a_{0,2}a_2^2\right)$$

$$= -\frac{1}{2c^2(1+c^2)^{3/2}}\left(a_{2,0}(a_2^2 + a_1^2) + a_{0,2}(a_1^2 + a_2^2) + c^2a_{2,0}a_2^2 - c^22a_{1,1}a_1a_2 + c^2a_{0,2}a_1^2\right)$$

$$= -\frac{1}{2(1+c^2)^{3/2}}\left(a_{2,0} + a_{0,2} + a_{2,0}a_2^2 + a_{0,2}a_1^2 - 2a_{1,1}a_1a_2\right)$$

$$= -\frac{1}{2(1+c^2)^{3/2}}\left(a_{2,0}(1 + a_2^2) - 2a_{1,1}a_1a_2 + a_{0,2}(1 + a_1^2)\right).$$

In going to the next-to-last line, the $c^2$ in the $2c^2$ in the denominator cancels the $c^2$ and $a_1^2 + a_2^2$ in the numerator.

**A15.1**   a. Since $0 \le \sin x \le 1$ for $x \in [0, \pi]$, it follows that

$$0 \le \sin^n x \le \sin^{n-1} x \le 1$$

for $x \in [0, \pi]$. So

$$c_n = \int_0^\pi \sin^n x \, dx < \int_0^\pi \sin^{n-1} x \, dx = c_{n-1}.$$

b. We have

$$c_n = \int_0^\pi \sin^n x \, dx = \int_0^\pi \sin x \sin^{n-1} x \, dx$$

$$= \left[-\cos x \sin^{n-1} x\right]_0^\pi + \int_0^\pi \cos x (n-1)\sin^{n-2} x \cos x \, dx.$$

But

$$\left[-\cos x \sin^{n-1} x\right]_0^\pi = 0 \quad \text{for } n \ge 2,$$

so

$$c_n = \int_0^\pi (n-1)\left(\sin^{n-2} x - \sin^n x\right) \, dx.$$

Therefore,

$$c_n = (n-1)(c_{n-2} - c_n), \quad \text{so} \quad nc_n = (n-1)c_{n-2}, \quad \text{i.e.,} \quad c_n = \frac{n-1}{n}c_{n-2}.$$

c.   The formulas $c_0 = \pi$ and $c_1 = 2$ follow immediately from simple integrals. Since

$$(2n-1)(2n-3)\cdots(1) = \frac{(2n)(2n-1)(2n-2)\cdots(1)}{(2n)(2n-2)\cdots(2)}$$

and

$$(2n)(2n-2)\cdots(2) = (2n)\big(2(n-1)\big)\cdots(2\cdot 1) = 2^n\, n!,$$

we have

$$c_{2n} = \frac{2n-1}{2n}\cdot\frac{2n-3}{2n-2}\cdots\frac{1}{2}\pi = \frac{(2n)!}{(2^n n!)^2}\,\pi = \frac{(2n)!}{2^{2n}(n!)^2}\,\pi$$

$$c_{2n+1} = \frac{2n}{2n+1}\cdot\frac{2n-2}{2n-1}\cdots\frac{2}{3}\,2 = \frac{(2^n n!)^2}{(2n+1)!}\,2 = \frac{2^{2n}(n!)^2}{(2n+1)!}\,2.$$

d. Noting that

$$\frac{1+o(1)}{1+o(1)} = 1 + o(1),$$

we get by simple substitution that

$$c_{2n} = \frac{1}{C}\sqrt{\frac{2}{n}}\,\pi\big(1+o(1)\big) \quad\text{and}\quad c_{2n+1} = \frac{C}{\sqrt{2n+1}}\big(1+o(1)\big).$$

So since $c_{2n} > c_{2n+1}$, we have

$$\sqrt{2}\sqrt{\frac{2n+1}{n}}\,\pi\big(1+o(1)\big) \geq C^2\big(1+o(1)\big),$$

but

$$\sqrt{\frac{2n+1}{n}} = \sqrt{2}\big(1+o(1)\big), \quad\text{so}\quad C^2 \leq 2\pi\big(1+o(1)\big).$$

We also have

$$c_{2n} < c_{2n-1}, \quad\text{so}\quad C^2 \geq 2\pi\big(1+o(1)\big).$$

So if there exists $C$ such that $n! = C\sqrt{n}\left(\frac{n}{e}\right)^n\big(1+o(1)\big)$, then $C = \sqrt{2\pi}$.

**A15.3** It is not really the object of this exercise, but let us see by induction that the number of ways of picking $k$ things from $m$, called "$m$ choose $k$," is given by the formula

We have already seen this computation in solution 3.3.5.

$$\binom{m}{k} = \frac{m!}{k!(m-k)!}$$

which you probably saw in high school. One way of seeing this is the relation leading to Pascal's triangle:

$$\binom{m}{k} = \binom{m-1}{k-1} + \binom{m-1}{k},$$

which expresses the fact that to choose $k$ things among $m$, you must either choose $k-1$ things among the first $m-1$, and then choose the last also, or choose $k$ things among the first $m-1$, and then not choose the last.

Suppose that the formula is true for all $m-1$ and all $k$, with the convention that $\binom{m}{k} = 0$ if $k < 0$ or $k > m$. Then the inductive step is

$$\binom{m}{k} = \binom{m-1}{k-1} + \binom{m-1}{k}$$
$$= \frac{(m-1)!}{(k-1)!(m-k)!} + \frac{(m-1)!}{k!(m-k-1)!}$$
$$= \frac{(m-1)!}{(k-1)!(m-k-1)!}\left(\frac{1}{m-k} + \frac{1}{k}\right)$$
$$= \frac{(m-1)!}{(k-1)!(m-k-1)!}\frac{m}{k(m-k)} = \frac{m!}{k!(m-k)!}.$$

Now to our question.

Since the coin is being tossed $2n$ times, there are $2^{2n}$ possible sequences of tosses; saying that the coin is a fair coin is exactly saying that all such sequences have the same probability $1/2^{2n}$.

The number of sequences corresponding to $n+k$ heads is exactly the number of ways of choosing $n+k$ tosses among the $2n$, i.e., $2n$ choose $n+k$:

$$\binom{2n}{n+k} = \frac{(2n)!}{(n+k)!(n-k)!}.$$

We need to sum this over all $k$ such that $n + a\sqrt{n} \le n+k \le n + b\sqrt{n}$.

**A21.1**  a. For any $\mathbf{x}_1, \mathbf{x}_2 \in \mathbb{R}^n$, we have

$$0 \le f(\mathbf{x}_2) = \inf_{\mathbf{y}\in X}|\mathbf{x}_2 - \mathbf{y}| \le \inf_{\mathbf{y}\in X}\left(|\mathbf{x}_2 - \mathbf{x}_1| + |\mathbf{x}_1 - \mathbf{y}|\right) = |\mathbf{x}_2 - \mathbf{x}_1| + f(\mathbf{x}_1).$$

The same statement holds if we exchange $\mathbf{x}_1$ and $\mathbf{x}_2$, so

$$|f(\mathbf{x}_1) - f(\mathbf{x}_2)| \le |\mathbf{x}_1 - \mathbf{x}_2|.$$

This proves that $f$ is continuous.

If $\mathbf{x} \in \overline{X}$, there exists a sequence $\mathbf{y}_i \in X$ with $\mathbf{y}_i \to \mathbf{x}$, so $\inf_{\mathbf{y}\in X}|\mathbf{x} - \mathbf{y}_i| = 0$, hence $f(\mathbf{x}) = 0$. But if $\mathbf{x} \notin \overline{X}$, then there exists $\epsilon > 0$ such that $|\mathbf{x} - \mathbf{y}| > \epsilon$ for all $\mathbf{y} \in X$, hence $f(\mathbf{x}) \ge \epsilon$.

b. There isn't much to show: this function is bounded by 1 and has support in $[0,1]$; it is continuous, (that was the point of including 0 and 1 in $X'$), it satisfies $f \le 0$ everywhere, and $f(x) = 0$ precisely if $x \notin [0,1]$ or if $x$ is in the unpavable set $X'$.

Solution A21.3: The $+$ sign in the exercise, in

$$\left(1 - d(\mathbf{x}, \mathbb{R}^n - X_\epsilon)\right)^+,$$

is defined in definition 4.1.14.

**A21.3**  This has nothing to do with what $\mathbb{R}^n - X_\epsilon$ might be; to lighten notation, set $\mathbb{R}^n - X_\epsilon = Z$. Then by the triangle inequality

$$d(\mathbf{x}, Z) \le |\mathbf{x} - \mathbf{y}| + d(\mathbf{y}, Z)$$
$$d(\mathbf{y}, Z) \le |\mathbf{y} - \mathbf{x}| + d(\mathbf{x}, Z)$$

so

$$d(\mathbf{x}, Z) - d(\mathbf{y}, Z) \le |\mathbf{x} - \mathbf{y}|, \quad d(\mathbf{y}, Z) - d(\mathbf{x}, Z) \le |\mathbf{x} - \mathbf{y}|,$$

and finally $|d(\mathbf{x}, Z) - d(\mathbf{y}, Z)| \leq |\mathbf{x} - \mathbf{y}|$. It then follows immediately that

$$\left|(1 - d(\mathbf{x}, Z)) - (1 - d(\mathbf{y}, Z))\right| = \left|d(\mathbf{y}, Z) - d(\mathbf{x}, Z)\right| \leq |\mathbf{x}, \mathbf{y}|.$$

If $f, g : \mathbb{R}^n \to \mathbb{R}$ satisfy

$$|f(\mathbf{x}) - f(\mathbf{y})| \leq |\mathbf{x} - \mathbf{y}| \quad \text{and} \quad |g(\mathbf{x}) - g(\mathbf{y})| \leq |\mathbf{x} - \mathbf{y}|,$$

then $|\sup(f, g)(\mathbf{x}) - \sup(f, g)(\mathbf{y})\| \leq |\mathbf{x} - \mathbf{y}|$, as you can check by looking at the four cases where each of $\sup(f, g)(\mathbf{x})$ and $\sup(f, g)(\mathbf{y})$ is realized by $f$ or $g$. Applying this to the case $f(\mathbf{z}) = 1 - d(\mathbf{x}, Z)$ and $g(\mathbf{z}) = 0$, we get $h(\mathbf{x}) - h(\mathbf{y})| \leq |\mathbf{x} - \mathbf{y}|$.

Now we will show that $h(\mathbf{x}) = 1$ when $\mathbf{x} \notin X_\epsilon$, and $0 \leq h(\mathbf{x}) < 1$ when $\mathbf{x} \in X_\epsilon$. By definition, $h = 1$ on $Z = \mathbb{R}^n - X_\epsilon$, which is closed. Thus if $\mathbf{z} \notin \mathbb{R}^n - X_\epsilon$, there is a ball $B_\mathbf{z}(r)$ of some radius $0 < r < 1$ around $\mathbf{z}$ such that $B_\mathbf{z}(r) \cap \mathbb{R}^n - X_\epsilon = \phi$, and $h(\mathbf{z}) \leq 1 - r$. Thus $\mathbb{R}^n - X_\epsilon$ is the set of $\mathbf{z} \in \mathbb{R}^n$ such that $h(\mathbf{z}) = 0$. The number $h(\mathbf{z})$ is the supremum of $1 - d(\mathbf{z}, Z)$ and 0, so of course $h(\mathbf{z}) \geq 0$, and since $d(\mathbf{z}, Z) \geq 0$, we have $h(\mathbf{z}) \leq 1$.

**A21.5** The hypotheses of theorem 4.11.21 imply that $\Phi^{-1}$ is bijective and of class $C^1$ with Lipschitz derivative. So the direction " $\implies$ " asserts that if

$$\underbrace{(f \circ \Phi) |\det[\mathbf{D}\Phi]|}_{\text{like } f \text{ in original statement}}$$

is integrable on $U$, then

$$\left(\left((f \circ \Phi) |\det[\mathbf{D}\Phi]|\right) \circ \Phi^{-1}\right) |\det[\mathbf{D}\Phi^{-1}]| = \left((f \circ \Phi) \circ \Phi^{-1}\right) |\det[\mathbf{D}\Phi] \circ \Phi^{-1}| |\det[\mathbf{D}\Phi^{-1}]|$$

is integrable on $V$, and

$$\int_U (f \circ \Phi)(\mathbf{x}) |\det[\mathbf{D}\Phi(\mathbf{x})]| \, |d^n\mathbf{x}| = \int_V (f \circ \Phi \circ \Phi^{-1})(\mathbf{y}) \left|\det\left[\mathbf{D}\Phi\left(\Phi^{-1}(\mathbf{y})\right)\right]\right| |\det[\mathbf{D}\Phi^{-1}(\mathbf{y})]| \, |d^n\mathbf{y}|$$

$$= \int_V f(\mathbf{y}) |\det[\mathbf{D}(\Phi \circ \Phi^{-1})(\mathbf{y})]| \, |d^n\mathbf{y}| = \int_V f(\mathbf{y}) \, |d^n\mathbf{y}|.$$

**A22.1** If you tried to find an unbounded subset of $\mathbb{R}^2$ of length 0 whose projection onto the $x$-axis has positive length, you will have had trouble. What can be found is such a subset whose projection does not have length. For example, the subset of $\mathbb{R}^2$ consisting of points with $x$-coordinate $p/q$ and $y$-coordinate $q$, with $0 \leq p/q \leq 1$, is unbounded, since $q$ can be arbitrarily large; it has length 0, since it consists of isolated points. Its projection onto the $x$-axis is the set of rational numbers in $[0, 1]$, which (as we saw in example 4.4.3) does not have defined volume.

**A23.1** To simplify notation, we will set $\varphi = dx_{i_1} \wedge \cdots \wedge dx_{i_k}$. Thus we want to show that the definition of the wedge product justifies going from the last line of equation A23.6 to the preceding line, i.e.,

$$(df \wedge \varphi)\left(P_0(\vec{\mathbf{v}}_1, \ldots, \vec{\mathbf{v}}_{k+1})\right) = \sum_{i=1}^{k+1} (-1)^{i-1}\left([\mathbf{D}f(0)]\vec{\mathbf{v}}_i\right)\varphi(\vec{\mathbf{v}}_1, \ldots, \widehat{\vec{\mathbf{v}}}_i, \ldots, \vec{\mathbf{v}}_{k+1}).$$

Since $f$ is a 0-form, $df$ is a 1-form, and $\varphi$ is a $k$-form. So definition 6.1.19 of the wedge product says

$$df \wedge \varphi(\vec{\mathbf{v}}_1, \dots, \vec{\mathbf{v}}_{k+1}) = \sum_{\substack{\text{shuffles} \\ \sigma \in \text{Perm}(1,k)}} \text{sgn}(\sigma)\, df(\vec{\mathbf{v}}_{\sigma(1)}) \varphi(\vec{\mathbf{v}}_{\sigma(2)}, \dots, \vec{\mathbf{v}}_{\sigma(k+1)}). \qquad (1)$$

Equation (1): The $\sigma(i)$ indicate that these shuffles are those permutations such that within each group, the vectors come with indices in ascending order.

We called function $f$ in theorem 6.7.2 a 0-form, but recall that by "form" we mean "form field"; we specify constant field when appropriate. So $f$ is a 0-form field. It takes a "parallelogram" formed of zero vectors and anchored at $\mathbf{x}$, and returns a number. Since the anchorage points are in an open subset $U \subset \mathbb{R}^n$, the function $f$ goes from $U$ to $\mathbb{R}$. In the first printing of the textbook, there is an error in definition 6.1.23: "anchored at a point $\mathbf{x} \in \mathbb{R}^n$" should have been "anchored at a point $\mathbf{x} \in U$."

We have $1 + k$ possible $(1,k)$-shuffles. As in the discussion following definition 6.1.19, we use a vertical bar to separate the vector on which the 1-form $df$ is being evaluated from the vectors on which the $k$-form $\varphi$ is being evaluated, and we use a hat to denote an omitted vector:

$$\vec{\mathbf{v}}_1 \big| \widehat{\vec{\mathbf{v}}}_1, \vec{\mathbf{v}}_2, \dots, \vec{\mathbf{v}}_{k+1}$$
$$\vec{\mathbf{v}}_2 \big| \vec{\mathbf{v}}_1, \widehat{\vec{\mathbf{v}}}_2, \dots, \vec{\mathbf{v}}_{k+1}$$
$$\vdots$$
$$\vec{\mathbf{v}}_{k+1} \big| \vec{\mathbf{v}}_1, \vec{\mathbf{v}}_2, \dots, \widehat{\vec{\mathbf{v}}}_{k+1},$$

which we can write

$$\vec{\mathbf{v}}_i \big| \vec{\mathbf{v}}_1, \dots, \widehat{\vec{\mathbf{v}}}_i, \dots, \vec{\mathbf{v}}_{k+1}$$

for $i = 1, \dots, k+1$. When $i$ is odd, the signature of the permutation is $+1$; when $i$ is even, the signature is $-1$. So we can rewrite equation (1) as

$$df \wedge \varphi(\vec{\mathbf{v}}_1, \dots, \vec{\mathbf{v}}_{k+1}) = \sum_{i=1}^{k+1} (-1)^{i-1} df(\vec{\mathbf{v}}_i) \varphi(\vec{\mathbf{v}}_1, \dots, \widehat{\vec{\mathbf{v}}}_i, \dots, \vec{\mathbf{v}}_{k+1}). \qquad (2)$$

Now we can rewrite $df$ as $[\mathbf{D}f]$. Since we want to evaluate $df$ at $\mathbf{0}$, we change the $(\vec{\mathbf{v}}_1, \dots, \vec{\mathbf{v}}_{k+1})$ on the left side of equation (2) to $P_{\mathbf{0}}(\vec{\mathbf{v}}_1, \dots, \vec{\mathbf{v}}_{k+1})$:

$$df \wedge \varphi\big(P_{\mathbf{0}}(\vec{\mathbf{v}}_1, \dots, \vec{\mathbf{v}}_{k+1})\big) = \sum_{i=1}^{k+1} (-1)^{i-1} \big([\mathbf{D}f(\mathbf{0})]\vec{\mathbf{v}}_i\big) \varphi(\vec{\mathbf{v}}_1, \dots, \widehat{\vec{\mathbf{v}}}_i, \dots, \vec{\mathbf{v}}_{k+1}).$$

**A24.1**   a. To show that the pullback $T^*$ is linear if $T$ is a linear transformation, we need to show that for $k$-forms $\varphi$ and $\psi$ and scalar $a$,

$$T^*(\varphi + \psi) = T^*(\varphi) + T^*(\psi);$$
$$T^*(a\varphi) = aT^*\varphi.$$

For the first:

$$T^*(\varphi + \psi)(\vec{\mathbf{v}}_1, \dots, \vec{\mathbf{v}}_k) = (\varphi + \psi)\big(T(\vec{\mathbf{v}}_1), \dots, T(\vec{\mathbf{v}}_k)\big)$$
$$= \varphi\big(T(\vec{\mathbf{v}}_1), \dots, T(\vec{\mathbf{v}}_k)\big) + \psi\big(T(\vec{\mathbf{v}}_1), \dots, T(\vec{\mathbf{v}}_k)\big)$$
$$= T^*\varphi(\vec{\mathbf{v}}_1, \dots, \vec{\mathbf{v}}_k) + T^*\psi(\vec{\mathbf{v}}_1, \dots, \vec{\mathbf{v}}_k).$$

We use definition 6.1.5 of addition of $k$-forms to go from line 1 to line 2.

For the second:

$$T^*\big(a\varphi(\vec{\mathbf{v}}_1, \dots, \vec{\mathbf{v}}_k)\big) = a\varphi\big(T(\vec{\mathbf{v}}_1), \dots, T(\vec{\mathbf{v}}_k)\big) = aT^*\varphi(\vec{\mathbf{v}}_1, \dots, \vec{\mathbf{v}}_k).$$

b. To show that the pullback by a $C^1$ mapping $\mathbf{f}$ is linear, we have

$$\mathbf{f}^*(\varphi + \psi)\Big(P_\mathbf{x}(\vec{\mathbf{v}}_1, \ldots, \vec{\mathbf{v}}_k)\Big) = (\varphi + \psi)\Big(P_{\mathbf{f}(\mathbf{x})}\big([\mathbf{Df}(\mathbf{x})]\vec{\mathbf{v}}_1, \ldots, [\mathbf{Df}(\mathbf{x})]\vec{\mathbf{v}}_k\big)\Big)$$

$$= \varphi\Big(P_{\mathbf{f}(\mathbf{x})}\big([\mathbf{Df}(\mathbf{x})]\vec{\mathbf{v}}_1, \ldots, [\mathbf{Df}(\mathbf{x})]\vec{\mathbf{v}}_k\big)\Big) + \psi\Big(P_{\mathbf{f}(\mathbf{x})}\big([\mathbf{Df}(\mathbf{x})]\vec{\mathbf{v}}_1, \ldots, [\mathbf{Df}(\mathbf{x})]\vec{\mathbf{v}}_k\big)\Big)$$

$$= (\mathbf{f}^*\varphi)\Big(P_\mathbf{x}(\vec{\mathbf{v}}_1, \ldots, \vec{\mathbf{v}}_k)\Big) + (\mathbf{f}^*\psi)\Big(P_\mathbf{x}(\vec{\mathbf{v}}_1, \ldots, \vec{\mathbf{v}}_k)\Big)$$

and

$$\mathbf{f}^*(a\varphi)\Big(P_\mathbf{x}(\vec{\mathbf{v}}_1, \ldots, \vec{\mathbf{v}}_k)\Big) = (a\varphi)\Big(P_{\mathbf{f}(\mathbf{x})}\big([\mathbf{Df}(\mathbf{x})]\vec{\mathbf{v}}_1, \ldots, [\mathbf{Df}(\mathbf{x})]\vec{\mathbf{v}}_k\big)\Big)$$

$$= a\mathbf{f}^*\varphi\Big(P_\mathbf{x}(\vec{\mathbf{v}}_1, \ldots, \vec{\mathbf{v}}_k)\Big).$$

**A25.1** a. Let $A$ be an $n \times k$ matrix, and consider the map $\mathrm{Mat}\,(n, k) \to \mathbb{R}$ given by

$$B \mapsto \det A^\top B.$$

Viewed as a function of the columns of $B$, this mapping is $k$-linear and alternating. It is alternating because exchanging two columns of $B$ exchanges the same two columns of $A^\top B$, and hence changes the sign of $\det A^\top B$. To see that it is multilinear, write

$$B' = \Big[\vec{\mathbf{b}}_1, \ldots, \vec{\mathbf{b}}_{i-1}, \vec{\mathbf{b}}_i', \vec{\mathbf{b}}_{i+1}, \ldots, \vec{\mathbf{b}}_k\Big], \quad B'' = \Big[\vec{\mathbf{b}}_1, \ldots, \vec{\mathbf{b}}_{i-1}, \vec{\mathbf{b}}_i'', \vec{\mathbf{b}}_{i+1}, \ldots, \vec{\mathbf{b}}_k\Big],$$

and

$$B = \Big[\vec{\mathbf{b}}_1, \ldots, \vec{\mathbf{b}}_{i-1}, \alpha\vec{\mathbf{b}}_i' + \beta\vec{\mathbf{b}}_i', \vec{\mathbf{b}}_{i+1}, \ldots, \vec{\mathbf{b}}_k\Big]$$

Then $A^\top B$ is obtained from $A^\top B'$ and $A^\top B''$ by the same process of keeping all the columns except the $i$th, and replacing its $i$th column by

$$\Big(\alpha \cdot \text{the } i\text{th column of } A^\top B'\Big) + \Big(\beta \cdot \text{the } i\text{th column of } A^\top B''\Big),$$

so that $\det A^\top B = \alpha \det A^\top B' + \beta \det A^\top B''$.

b. By part a, the map $B \mapsto \det A^\top B$ defines an element of $\varphi \in A^k(\mathbb{R}^n)$. By theorem 6.1.8, it is of the form

$$\varphi = \sum_{1 \leq i_1 < \cdots < i_k \leq n} c_{i_1, \ldots i_k}\, dx_{i_1} \wedge \cdots \wedge dx_{i_k}$$

for coefficients $c_{i_1, \ldots, i_k}$ that can be evaluated as follows:

$$c_{i_1, \ldots i_k} = \det\left(A^\top\Big[\vec{\mathbf{e}}_{i_1}, \ldots, \vec{\mathbf{e}}_{i_k}\Big]\right)$$

$$= \det\begin{bmatrix} a_{i_1, 1} & \cdots & a_{i_1, k} \\ \vdots & \ddots & \vdots \\ a_{i_k, 1} & \cdots & a_{i_k, k} \end{bmatrix}^\top.$$

c. Thus

$$\varphi(\vec{a}_1, \ldots, \vec{a}_k) = \det A^\top A = \sum_{1 \le i_1 < \cdots < i_k \le n} \det \begin{bmatrix} a_{i_1,1} & \cdots & a_{i_1,k} \\ \vdots & \ddots & \vdots \\ a_{i_k,1} & \cdots & a_{i_k,k} \end{bmatrix}^\top dx_{i_1} \wedge \cdots \wedge dx_{i_k} \begin{bmatrix} \vec{a}_1, \ldots, \vec{a}_k \end{bmatrix}$$

$$= \sum_{1 \le i_1 < \cdots < i_k \le n} \det \begin{bmatrix} a_{i_1,1} & \cdots & a_{i_1,k} \\ \vdots & \ddots & \vdots \\ a_{i_k,1} & \cdots & a_{i_k,k} \end{bmatrix}^\top \begin{bmatrix} a_{i_1,1} & \cdots & a_{i_1,k} \\ \vdots & \ddots & \vdots \\ a_{i_k,1} & \cdots & a_{i_k,k} \end{bmatrix}$$

$$= \sum_{1 \le i_1 < \cdots < i_k \le n} \left( \det \begin{bmatrix} a_{i_1,1} & \cdots & a_{i_1,k} \\ \vdots & \ddots & \vdots \\ a_{i_k,1} & \cdots & a_{i_k,k} \end{bmatrix} \right)^2.$$

Since $A = [\vec{a}_1, \ldots, \vec{a}_k]$, we can write $\varphi : A \mapsto \det A^\top A$ as

$$\varphi(\vec{a}_1, \ldots, \vec{a}_k) = \det A^\top A.$$

To go to line 3, remember (theorem 4.8.7) that the determinant of a matrix equals the determinant of its transpose.

Solution A25.3: The [0] in

$$F^{-1}([0]) \subset \mathbb{R}^n \times \mathrm{Mat}\,(n,n)$$

is the zero matrix among symmetric $n \times n$ matrices.

**A25.3** By definition, the space of rigid motions of $\mathbb{R}^n$ is the space of maps $T : \mathbb{R}^n \to \mathbb{R}^n$ of the form $\mathbf{x} \mapsto A\mathbf{x} + \mathbf{b}$, where $A \in \mathrm{Mat}\,(n,n)$ satisfies $A^\top A = I$. Thus we can think of the space of rigid motions as the subset $F^{-1}([0]) \subset \mathbb{R}^n \times \mathrm{Mat}\,(n,n)$, where $F(\mathbf{b}, A) = A^\top A - I$, and it is important to consider the codomain of $F$ as the space of symmetric $n \times n$ matrices, not the space of all $n \times n$ matrices. Then $F$ is a map from a space of dimension $n^2 + n$ to a space of dimension $n(n+1)/2$. By part d of exercise 3.2.9, we know that the derivative of $F$ is surjective at all the points where $A^\top A = I$. So, by theorem 3.1.10, $F^{-1}([0])$ is a manifold of dimension

$$n + n^2 - \frac{n(n+1)}{2} = \frac{n(n+1)}{2}.$$

**A25.5** This is immediate from Heine-Borel. The manifold $M$ is locally a graph. We have seen many times what this means: for every $\mathbf{x} \in M$, there exist

- subspaces $E_1, E_2 \subset \mathbb{R}^n$ with $E_1$ spanned by $p$ standard basis vectors and $E_2$ spanned by the other $n - p$ standard basis vectors,
- open subsets $U \subset E_1$ and $V \subset E_2$,
- A $C^1$ mapping $\mathbf{g} : U \to V$,

such that $M \cap (U \times V) = \Gamma(\mathbf{g})$. Renumber the variables so that $E_1$ corresponds to the first basis vectors and $E_2$ to the last, and write $\mathbf{x} = \begin{pmatrix} \mathbf{y} \\ \mathbf{z} \end{pmatrix}$. Choose $r(\mathbf{x}) > 0$ such that $B_{r(\mathbf{x})}(\mathbf{y}) \subset U$, and call $Z_{\mathbf{x}}$ the graph of the restriction of the graph of $\mathbf{g}$ to the concentric ball of radius $r/2$. Then the closure of $Z_{\mathbf{x}}$, which is the graph of the restriction of $\mathbf{g}$ to the closed ball of radius $r/2$, is a compact subset of $M$.

The $Z_{\mathbf{y}}$ form an open cover of $Y$, so by Heine-Borel there is a finite subcover: there exist $\mathbf{x}_1, \ldots, \mathbf{x}_m$ such that

$$Y \subset Z_{\mathbf{x}_1} \cup \cdots \cup Z_{\mathbf{x}_m}.$$

If each of the $f(\overline{Z}_{\mathbf{x}_i})$ has $q$-dimensional volume 0, then so does their union, which contains $f(Y)$.

**A25.7** Since $X$ is compact, it is enough to prove the result for a single point $\mathbf{p} \in X$. Clearly we can rotate and translate $M$ so that $\mathbf{p}$ is the origin, and the tangent plane $T_{\mathbf{p}}M$ is $\mathbb{R}^k$. Then in a neighborhood of $\mathbf{p}$, the manifold $M$ is the graph of a $C^1$ map $\mathbf{f} : U \to \mathbb{R}^{n-k}$, where $U$ is a neighborhood of the origin in $\mathbb{R}^k$, with $[\mathbf{Df}(\mathbf{0})] = [0]$. By the compactness of $X$ we may assume that $U$ is the ball of radius $2R$ for some fixed $R$, and that both $\mathbf{f}$ and its derivative depend continuously on $\mathbf{p}$.

Write $F = \begin{pmatrix} F_1 \\ F_2 \end{pmatrix}$ and note that

$$F_2 \begin{pmatrix} \mathbf{x} \\ \mathbf{y} \end{pmatrix} = \mathbf{y} - \mathbf{f}(\mathbf{x})$$

so that $F_2$ is by definition of class $C^1$, with

$$\left[ \mathbf{D}F_2 \begin{pmatrix} \mathbf{0} \\ \mathbf{y} \end{pmatrix} \right] \begin{bmatrix} \mathbf{h} \\ \mathbf{k} \end{bmatrix} = \mathbf{k}.$$

All the trouble is with $F_1$: to show that $F$ is differentiable at the origin with derivative the identity we need to know that

$$\left| \frac{\mathbf{x}}{|\mathbf{x}|} \left| \begin{bmatrix} \mathbf{x} \\ \mathbf{f}(\mathbf{x}) \end{bmatrix} \right| - \mathbf{x} \right| \in o(|\mathbf{x}|).$$

Even after showing this, we will need to show that the derivative of $F$ is continuous.

Computing the length of $\begin{bmatrix} \mathbf{x} \\ \mathbf{f}(\mathbf{x}) \end{bmatrix}$, we find

$$\left| \frac{\mathbf{x}}{|\mathbf{x}|} \left| \begin{bmatrix} \mathbf{x} \\ \mathbf{f}(\mathbf{x}) \end{bmatrix} \right| - \mathbf{x} \right| = |\mathbf{x}| \left( \sqrt{1 + \frac{|\mathbf{f}(\mathbf{x})|^2}{|\mathbf{x}|^2}} - 1 \right).$$

It is not *obvious* that $F_1$ is of class $C^1$. It looks quite a bit like the function in example 1.9.5, the typical function with directional derivatives, but whose derivative is discontinuous: in both cases, there is an $|\mathbf{x}|$ in the denominator.

Since

$$\sqrt{1 + \frac{|\mathbf{f}(\mathbf{x})|^2}{|\mathbf{x}|^2}} \in 1 + o(1), \quad \text{we have} \quad |\mathbf{x}| \left( \sqrt{1 + \frac{|\mathbf{f}(\mathbf{x})|^2}{|\mathbf{x}|^2}} - 1 \right) \in o(|\mathbf{x}|).$$

The computation actually shows more: it shows that $\left[ \mathbf{D}F \begin{pmatrix} \mathbf{0} \\ \mathbf{y} \end{pmatrix} \right] = \mathrm{id}$ for all $\mathbf{y} \in \mathbb{R}^{n-k}$. We still need to show that $F$ is of class $C^1$. Away from $|\mathbf{x}| = 0$, the function $F$ is evidently differentiable, but the derivative is pretty awesome. Write $F = \begin{pmatrix} F_1 \\ F_2 \end{pmatrix}$; then $F_1$ depends only on $\mathbf{x}$, and

$$[\mathbf{D}F_1(\mathbf{x})]\vec{\mathbf{h}} = \vec{\mathbf{h}} \sqrt{\frac{(\mathbf{x} + \mathbf{f}(\mathbf{x})) \cdot (\mathbf{x} + \mathbf{f}(\mathbf{x}))}{\mathbf{x} \cdot \mathbf{x}}}$$

$$+ \frac{\mathbf{x}}{2\sqrt{\frac{(\mathbf{x}+\mathbf{f}(\mathbf{x}))\cdot(\mathbf{x}+\mathbf{f}(\mathbf{x}))}{\mathbf{x}\cdot\mathbf{x}}}} \overbrace{\frac{(\mathbf{x} \cdot \mathbf{x})2(\mathbf{x} + \mathbf{f}(\mathbf{x}))(\vec{\mathbf{h}} + \mathbf{df}(\mathbf{x})(\vec{\mathbf{h}})) - \left((\mathbf{x} + \mathbf{f}(\mathbf{x})) \cdot (\mathbf{x} + \mathbf{f}(\mathbf{x}))\right)2(\mathbf{x} \cdot \vec{\mathbf{h}})}{(\mathbf{x} \cdot \mathbf{x})^2}}.$$

Of course

$$\lim_{\mathbf{x}\to 0}\vec{\mathbf{h}}\sqrt{\frac{(\mathbf{x}+\mathbf{f}(\mathbf{x}))\cdot(\mathbf{x}+\mathbf{f}(\mathbf{x}))}{\mathbf{x}\cdot\mathbf{x}}}=\vec{\mathbf{h}}=\operatorname{id}\vec{\mathbf{h}};$$

we need to show that the second summand has limit 0. Cancel the 2's, notice that $\lim\limits_{\mathbf{x}\to 0}\dfrac{1}{\sqrt{\frac{(\mathbf{x}+\mathbf{f}(\mathbf{x}))\cdot(\mathbf{x}+\mathbf{f}(\mathbf{x}))}{\mathbf{x}\cdot\mathbf{x}}}}=1$, and develop the part of the numerator marked with an overbrace, and which we will now denote by $A$:

$$A \stackrel{\text{def}}{=} (\mathbf{x}\cdot\mathbf{x})(\mathbf{x}\cdot\vec{\mathbf{h}})+(\mathbf{x}\cdot\mathbf{x})\Big(\mathbf{f}(\mathbf{x})\cdot(\vec{\mathbf{h}}+[\mathbf{Df}(\mathbf{x})]\vec{\mathbf{h}})\Big)+(\mathbf{x}\cdot\mathbf{x})(\mathbf{x}\cdot[\mathbf{Df}(\mathbf{x})]\vec{\mathbf{h}})$$

$$-(\mathbf{x}\cdot\mathbf{x})(\mathbf{x}\cdot\vec{\mathbf{h}})-\Big(\mathbf{f}(\mathbf{x})\cdot(\mathbf{x}+\mathbf{f}(\mathbf{x}))\Big)(\mathbf{x}\cdot\vec{\mathbf{h}})-(\mathbf{x}\cdot\mathbf{f}(\mathbf{x}))(\mathbf{x}\cdot\vec{\mathbf{h}}).$$

The terms $(\mathbf{x}\cdot\mathbf{x})(\mathbf{x}\cdot\vec{\mathbf{h}})$ cancel. Since $|\mathbf{f}(\mathbf{x})|\in o(|\mathbf{x}|)$ and $|[\mathbf{Df}(\mathbf{x})]|\in o(1)$, all the other terms are in $o\left(|\mathbf{x}|^3\right)$, and so

$$\frac{A}{(\mathbf{x}\cdot\mathbf{x})^2}\in o\left(\frac{1}{|\mathbf{x}|}\right)\quad\text{and}\quad\frac{\mathbf{x}A}{(\mathbf{x}\cdot\mathbf{x})^2}\in o(1).$$

Since $F_2$ is of class $C^1$, with derivative $\left[\mathbf{D}F_2\binom{0}{0}\right]\begin{bmatrix}\mathbf{h}\\\mathbf{k}\end{bmatrix}=\mathbf{k}$, it follows that $F$ is of class $C^1$. We already showed that it has derivative id at $\binom{0}{0}$, so this completes parts a and b.

c. The point $\binom{\mathbf{x}}{\mathbf{y}}$ is in $(M\cap\partial B_r(0))$ if and only if $\mathbf{y}=\mathbf{f}(\mathbf{x})$ and $|\mathbf{x}|^2+|\mathbf{f}(\mathbf{x})|^2=r^2$, and then

$$F\binom{\mathbf{x}}{\mathbf{y}}=\binom{r\frac{\mathbf{x}}{|\mathbf{x}|}}{0}\in T_0M\cap\partial B_r(0).$$

d. Let $Y_r=M\cap\partial B_r(0)$, and let $\gamma:V\to Y_r$ be a parametrization, where $V$ is an appropriate subset of $\mathbb{R}^{k-1}$. Then

$$\operatorname{vol}_{k-1}Y_r=\int_V\sqrt{\det\Big([\mathbf{D}\gamma(\mathbf{v})]^\top[\mathbf{D}\gamma(\mathbf{v})]\Big)}\,|d^{k-1}\mathbf{v}|,$$

and $F\circ\gamma$ is a parametrization of $F(Y_r)$, so

$$\operatorname{vol}_{k-1}F(Y_r)=\int_V\sqrt{\det\Big([\mathbf{D}(F\circ\gamma)(\mathbf{v})]^\top[\mathbf{D}(F\circ\gamma)(\mathbf{v})]\Big)}\,|d^{k-1}\mathbf{v}|$$

$$=\int_V\sqrt{\det\Big([\mathbf{D}\gamma(\mathbf{v})]^\top\underbrace{[\mathbf{D}F(\gamma(\mathbf{v}))]^\top[\mathbf{D}F(\gamma(\mathbf{v}))]}_{\approx\operatorname{id}}[\mathbf{D}\gamma(\mathbf{v})]\Big)}\,|d^{k-1}\mathbf{v}|.$$

For $r$ sufficiently small, the matrix $[\mathbf{D}F(\gamma(\mathbf{v}))]^\top[\mathbf{D}F(\gamma(\mathbf{v}))]$ is arbitrarily close to the identity: for any $\epsilon>0$, there exists $\rho>0$ such that if $0<r<\rho$ we have

$$\det([\mathbf{D}\gamma(\mathbf{v})]^\top[\mathbf{D}\gamma(\mathbf{v})])\le(1+\epsilon)^2\det\Big([\mathbf{D}\gamma(\mathbf{v})]^\top[\mathbf{D}F(\gamma(\mathbf{v}))]^\top[\mathbf{D}F(\gamma(\mathbf{v}))][\mathbf{D}\gamma(\mathbf{v})]\Big).$$

Thus

$$\operatorname{vol}_{k-1}Y_r\le(1+\epsilon)\operatorname{vol}_{k-1}F(Y_r)\le(1+\epsilon)r^{k-1}\operatorname{vol}_{k-1}S^{k-1}.$$

# INDEX

*A Poets' Agor...*

# WINGS OF THOUGHT

## CONTEMPORARY GREEK AND PHILHELLENE POETRY

EDITED BY
KARINE LENO ANCELLIN AND ANGELA N. LYRAS

*Dedicated to the memory of the Greek poet Katerina Anghelaki-Rooke*
*(1939-2020) whose presence illuminated our inaugural event.*

*In tribute to our founding and perennial mentors,*
*poets Liana Sakelliou and Orfeas Apergis.*

*Αφιερώνεται στην μνήμη της Ελληνίδας ποιήτριας Κατερίνας Αγγελάκη-Ρουκ*
*(1939-2020), η παρουσία της οποίας λάμπρυνε την πρώτη μας εκδήλωση.*

*Ως φόρος τιμής στους ιδρυτικούς και μόνιμους μέντορες μας, την*
*ποιήτρια Λιάνα Σακελλίου και τον ποιητή Ορφέα Απέργη.*

# CONTENTS

## *LULL*

## RISK

## GRAFT

## *VERGE*

## *A Poets' Agora Residency*

**Shamim Azad**

# Πρόλογος

Ανταποκρινόμενη στο αίτημα να κάνω μερικά σχόλια για να προετοιμάσω τους αναγνώστες, τόσο στην Ελλάδα όσο και στο εξωτερικό, θα ήθελα να μοιραστώ τα συναισθήματα και τις σκέψεις μου σχετικά με το ρόλο του ποιητή στην «αγορά» σε συζητήσεις για σύγχρονα κοινωνικά και πολιτικά θέματα. Σήμερα, η «αγορά» υποδηλώνει ένα χώρο με καταστήματα, αλλά την εποχή του Σωκράτη σήμαινε ένα χώρο όπου γίνονταν συζητήσεις για σύγχρονα ζητήματα και ανταλλαγή γενικότερων ιδεών. Πώς ανήκει ένας ποιητής στην Αγορά;

Για τους Έλληνες, όπως κι εγώ, η φράση «Η Αγορά των Ποιητών» προκαλεί ένα μοναδικό, περίπλοκο συναίσθημα που πιθανόν να μην μπορούν να βιώσουν όσοι δεν είναι Έλληνες. Αφενός, η φράση υπενθυμίζει την υπερηφάνεια που μοιραζόμαστε για την κληρονομιά της σημαντικής ποίησης ξεκινώντας από τον Όμηρο. Αυτή η υπερηφάνεια φαίνεται να είναι μέρος του εθνικού μας χαρακτήρα, και έτσι δύσκολα μπορεί να μοιραστεί με ανθρώπους διαφορετικών εθνικοτήτων, φαινομενικά δίνοντας στους ποιητές περισσότερο δικαίωμα να μιλούν για κοινωνικά θέματα. Αφετέρου, υπάρχει μια αίσθηση πίεσης που αισθανόμαστε να δημιουργήσουμε ξανά σπουδαία ποίηση, ενώ ταυτόχρονα κάνουμε τις φωνές μας να ακούγονται στη σύγχρονη αγορά, επειδή τα κοινωνικά ζητήματα είναι τόσο σημαντικά. Στην Ελλάδα, σε σύγκριση με άλλες χώρες όπως η Αμερική, καθώς οι ποιητές και οι συγγραφείς συχνά δεν έχουν πανεπιστημιακές θέσεις, η τέχνη τους δεν περιορίζεται σε μια ακαδημαϊκή δραστηριότητα αλλά ανήκει στην ευρύτερη κοινότητα όπου ζουν. Οι άνθρωποι επιθυμούν να γνωρίζουν τι σκέφτονται οι συγγραφείς και οι φιλόσοφοι.

Όταν οι ποιητές κάνουν τις φωνές τους να ακουστούν, τι κάνουν; Κατά τη γνώμη μου, η δημιουργική τέχνη δίνει τη δυνατότητα στους ποιητές να καταγράψουν τον παλμό της καρδιάς των κοινοτήτων στις οποίες ζουν — να μετρήσουν τον σφυγμό της, αργό ή γρήγορο, σταθερό ή ασταθή, αδύναμο ή δυνατό. Και με αυτόν τον τρόπο οι ποιητές δίνουν φωνή στους συμπολίτες τους που μπορεί να μην έχουν πλήρη συνειδητότητα των συναισθημάτων τους ή δεν έχουν φωνή και πρόσβαση στην Αγορά. Όταν οι άνθρωποι αντιλαμβάνονται τα συναισθήματά τους είναι καλύτερα εξοπλισμένοι για να αποφασίσουν για το μέλλον. Όλες οι φωνές μαζί αντιπροσωπεύουν την κοινωνία καλύτερα από κάθε μία μεμονωμένη φωνή.

Στην Αγορά, εμείς οι ποιητές δεν είμαστε απλώς ουδέτεροι ανταποκριτές των συναισθημάτων των άλλων ανθρώπων. Μιλάμε για σύγχρονα κοινωνικά ζητήματα με βάση τις αξίες που ορίζονται από την ποίηση — ανθρωπιστικές και καλλιτεχνικές. Αυτές οι αξίες χρωματίζουν τις κοινωνικές καταστάσεις με συναισθήματα, με θετικές ή αρνητικές αξιολογήσεις. Ποιες είναι αυτές οι ποιητικές αξίες; Ποια είναι τα κοινωνικά και πολιτικά ζητήματα; Στα ποιήματα που ακολουθούν θα τα ανακαλύψετε και θα τα αφήσετε να σας επηρεάσουν.

Για αυτήν την ευκαιρία να συμμετάσχω στην «Αγορά των Ποιητών», εκ μέρους των συναδέλφων μου ποιητών, του κοινού και των μελλοντικών αναγνωστών, θα ήθελα να εκφράσω την ειλικρινή μου ευγνωμοσύνη στην Καρήν Λένο Ανσελλίν και την Άντζελα Λύρα.

Λιάνα Σακελλίου
Αθήνα, Ιούλιους 2020

# FOREWORD

Asked to write a few comments to prepare readers, both Greek and international, I would like to share my feelings and thoughts about the role of the poet in "the agora": in discussions of contemporary social and political issues. Today, "the agora" suggests a marketplace, but at the time of Socrates it suggested a place where contemporary issues were discussed, and more general ideas were exchanged. How does a poet belong to the "Agora?"

For Greek people, such as myself, the phrase "A Poets' Agora" evokes a unique, complex feeling that non-Greeks may not be able to experience. On the one hand, the phrase brings to mind the pride we share in the heritage of great poetry starting with Homer. This pride seems to be a part of our national character, and thus can hardly be shared with people of different nationalities, seemingly giving poets more rights to speak about social issues. On the other hand, there is a sense of pressure we feel to create great poetry again while also making our voices heard in the contemporary Agora because social issues are so important. In Greece, compared to other countries such as America, since the poets and writers often do not have university positions, their art is not limited to an academic activity but belongs to the greater community where they live. People want to know what writers and philosophers think.

When poets make their voices heard, what are they doing? In my view, creative art qualifies poets to record the heartbeat of the communities in which they live – to take its pulse, slow or fast, steady or unsteady, weak or strong. And in so doing poets give voice to their fellow citizens who might not be fully aware of their own feelings, or do not have voice and access to the Agora. When people are conscious of their feelings, they are better equipped to decide about the future. All the voices together represent society better than any individual voice alone could.

In the Agora, we poets are not just neutral reporters of other people's feelings. We speak about contemporary social issues based on the values prescribed by poetry --- humanistic and artistic ones. These values colour social situations with emotions, with positive or negative evaluations. What are these poetic values? What are the social and political issues? In the poems that will follow you will discover them and will be affected by them.

For this opportunity to participate in *A Poets' Agora,* on behalf of my fellow poets, the audiences, and future readers, I would like to express my heartfelt gratitude to Karine Leno Ancellin and Angela Lyras.

Liana Sakelliou
Athens, July 2020

# Εισαγωγή

Κάθε φθινόπωρο, από το 2015, τρεις ποιητικές φωνές συναθροίζονται στην οικία Κουτζαλέξη, στην Πλάκα. Έτσι ξεκίνησε το *A Poets' Agora* ως χώρος για δίγλωσσες ποιητικές βραδυές, και στην συνέχεια εξελίχθηκε ώστε να περιλαμβάνει Παραχώρηση Διαμονής (Residency) και άλλες συγγραφικές εργασίες. Όταν εγκαινιάσαμε το *A Poets' Agora*, φιλοδοξία μας ήταν να δώσουμε φωνή στις ιδέες και στις μορφές τέχνης που έφθαναν στ'αυτιά μας σαν ουρλιαχτά από διάφορες γωνιές της Αθήνας κατά την δύσκολη περίοδο της οικονομικής κρίσης. Η Άντζελα Λύρα και εγώ εκτιμούμε την ποίηση, και από κοινού αποφασίσαμε να προσφέρουμε ένα φιλόξενο φόρουμ στη τέχνη αυτή.

Εκτός από τον αρχικό δίγλωσσο χαρακτήτρα τους, οι βραδυές εστιάζουν σε ένα θέμα, προσεκτικά επιλεγμένο για να εμπνέει τους ποιητές/ποήτριες και να ανταποκρίνεται στο Zeitgeist της Ελλάδας. Τα διαδοχικά θέματα ήταν: *Muted* το 2015, *Lull* το 2016, *Risk* το 2017, *Graft* το 2018 και *Verge* το 2019. Οι αναγνώσεις εμπλουτίζονται με λογοτεχνικές συζητήσεις, ως μακρινός απόηχος των Γαλλικών λογοτεχνικών σαλονιών. Ένα μικρό ακροατήριο, όχι μεγαλύτερο των πενήντα ατόμων, ανταλλάσει ιδέες με τους ποιητές, και στην συνέχεια μπορεί να έχει τετ-α-τετ συνομιλίες απολαμβάνοντας ποτά και μεζέδες. Οι ενδο-ποιητικές συζητήσεις στο *A Poets' Agora* αντηχούν την ατέρμονα και καλειδοσκοπική τέχνη των λέξεων.

Οι τοίχοι του νεοκλασσικού Κουτζαλέξη όπου πραγματοποιούνται οι εκδηλώσεις, συνεχίζουν έτσι την παράδοση να περιβάλουν τις φωνές των παράνομων αντιπάλων της Απριλιανής Χούντας και τις φωνές άλλων αντιφρονούντων μέσα στα χρόνια.Το λακωνικό αλλά βαθιά συναισθηματικό ποίημα *Άφωνος* της Κατερίνας Αγγελάκη-Ρουκ « *Ἀλλὰ τώρα ἡ ἴδια ἡ ζωή / διατάζει νὰ σωπάσω/ ἀλλοιῶς τὰ λίγα πούχω/ κι αὐτὰ θὲ νὰ τὰ χάσω* » συνοψίζει το κύμα αγανάκτησης και ομορφιάς που βρήκε στέγη στο *A Poets' Agora*.

Η Ελληνική ποίηση έχει μεταμορφωθεί στην διάρκεια των αιώνων από την Ομηρική εποχή, και μολονότι κληρονόμοι αυτής της παράδοσης, οι σύγχρονοι ποιητές που συμμετείχαν στις βραδυές του *A Poets' Agora* αποστασιοποιούν εαυτούς από τον κόσμο του αρχαίου ή κλασικού Ελληνικού πολιτισμού που αβίαστα κάποιοι τους κατατάσσουν. Η παρούσα ανθολογία συγκεντρώνει ετερόκλητες φωνές από τους δημιουργούς της Αθήνας, αλλά και πέραν αυτής. Δεν εκπροσωπούν τίποτε άλλο εκτός από τον εαυτό τους, και δεν υφίσταται μεταξύ τους κανένα συγκεκριμένο ρεύμα ή είδος.

*«Η ποίηση είναι πιό κοντά στη ουσιαστική αλήθεια από ότι η ιστορία»* υποστήριξε ο Πλάτων, δεδομένου ότι καθένας από αυτούς τους ποιητές μεγεθύνει ένα μοναδικό όραμα, ή χαράζει ένα καλειδοσκοπικό αποτύπωμα της φαντασίας του εστιασμένο στο παρόν. Όλοι τους, εκτός της Κατερίνας Αγγελάκη-Ρουκ, βρίσκονται σήμερα εν ζωή. Το φανταστικό που μας προσκαλούν να γευθούμε αποκαλύπτει την αισθητική εκλεκτικών και εγγενών νοημάτων. Μέσα σε πέντε χρόνια, αυτοί οι δεκαπέντε ποιητές επαναλαμβάνουν το σύγχρονο νόημα τής τέχνης τους. Με την παρούσα έκδοση το *A Poets' Agora* φιλοδοξεί να μεταφέρει την αύρα της τέχνης τους πέρα από τον «οίνοπα πόντο» του Αιγαίου.

Καρήν Λένο Ανσελλήν
[*Μετάφραση: Μιχάλης Μοντεσάντος*]

# PREFACE

Since 2015, a trio of poetic voices congregate at the Koutzalexis house for an annual event that occurs in late autumn. Initially, *A Poets' Agora* began as a space for bilingual poetry evenings, and later evolved to include a Residency and various writing projects. When *A Poets' Agora* was launched, our aspiration was to voice the ideas and art forms being lamented around Athens, as the graffiti meme 'βασανίζομαι,' ('vasanizomai' I am tormented) coloured the cement of the city, during the first painful years of the financial crisis. Angela Lyras and I both value the virtuosity of poetry, and together, we concocted an intimate and convivial forum to this art.

Beside their original bilingual character, the evenings are circumstantiated around a theme, carefully designed to be both an inspiration to the poets, and to reflect the zeitgeist of Greece. The uniqueness of *A Poets' Agora's* poetry nights resides in these successive monosyllabic word-themes -*Muted* in 2015, *Lull* in 2016, *Risk* in 2017, *Graft* in 2018 and *Verge* in 2019- that inform our yearly musings, as all the poets compose a poem based on the given theme. These dedicated poems are an exclusive creation for *A Poets' Agora*, and can be found at the beginning of each author's collection. The poetic readings are further punctuated by a literary conversation, later broadcast on *A Poets' Agora* YouTube channel. A small audience of no more than fifty people share ideas with the poets, and are able to engage in one-to-one discussions afterwards, around an offering of drinks and 'meze.' These Agora discussions echo the timeless and multi-fold art of the word.

The walls of the neoclassical Koutzalexis house, wherein these events are held, have thus continued a tradition of welcoming the voices of creative dissent, largely censored during the military Junta of the late 60s. The succinct, yet deeply felt *Mute* poem of Katerina Anghelaki-Rooke "But now, life itself/ orders me to hush/ otherwise the little I have left/ I will lose as well," sums up the surge of indignation and beauty *A Poets' Agora* became home to. Through its artists, the Koutzalexis house has held, diachronically in its rooms, the spirit of its 'polis.' Greek poetry has metamorphosed through the ages, since Homeric times, and even if heirs to that legacy, most of the contemporary poets who take part in A Poets' Agora evenings, dissociate themselves from the ancient or classic realm of Greek culture and identity so easily branded upon them. This anthology brings together eclectic voices, from creative Athenian grounds and beyond, epitomising nothing other than themselves; there is no specific literary movement or genre to represent their wide-ranging individualities.

*"Poetry is nearer to vital truth than history"* suggested Plato, as each one of these poets magnifies a unique vision and etches a kaleidoscopic mind-map honed onto the present times. All, except for Katerina Anghelaki-Rooke, are alive today. In this anthology, collected over five years, these sixteen poets iterate the contemporary expression of their polymorphous art. The visionary worlds they invite us to visit, reveal the aesthetics of free and inherent musings. With this publication, *A Poets' Agora* aspires to take the wonder of their artistry beyond the horizon of the *'wine-eyed'* Aegean Sea.

Karine Leno Ancellin
June 2020

# KOUTZALEXIS HOUSE

### (1801-1868)

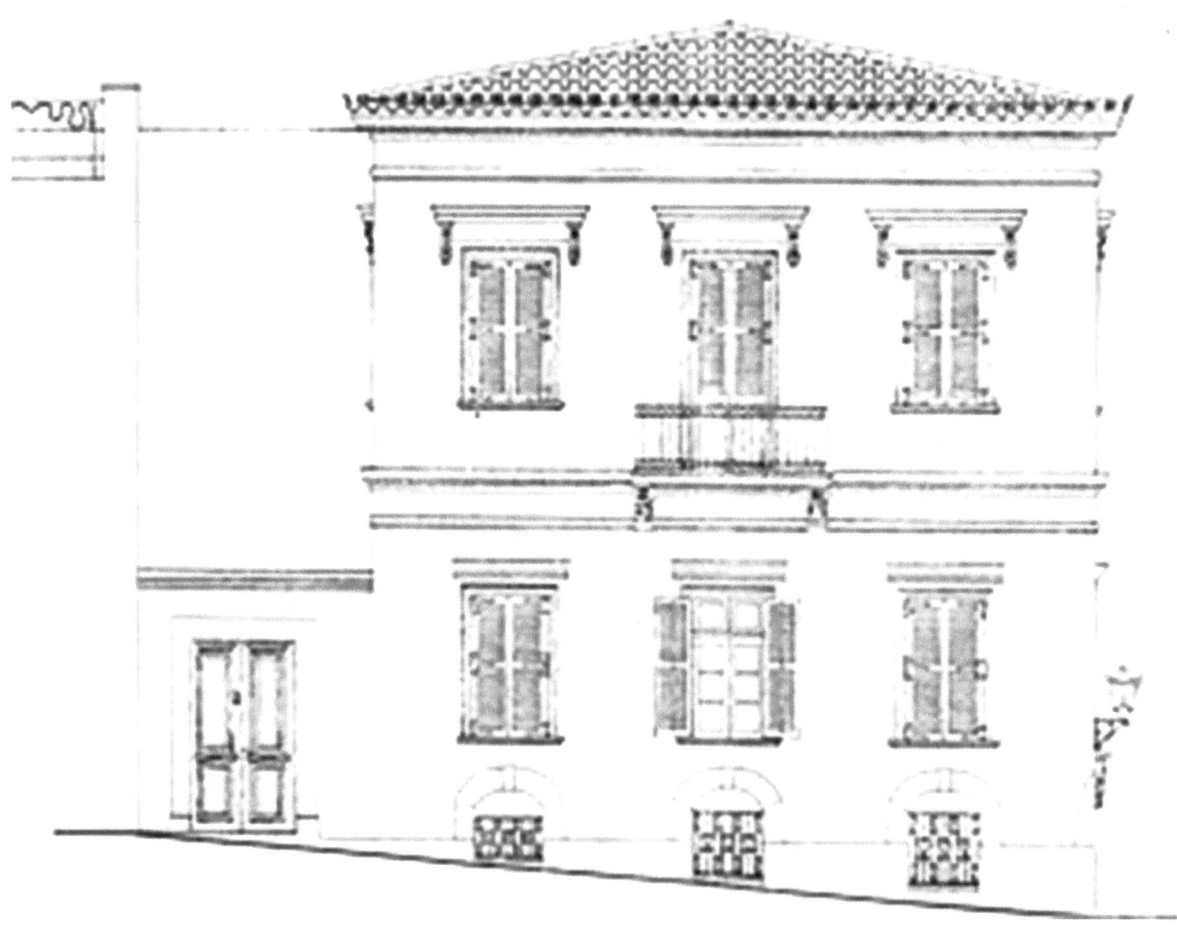

# ΟΙΚΙΑ ΚΟΥΤΖΑΛΕΞΗ

## (ΕΤΗ 1801-1868)

Το νεοκλασικό κτίριο, το οποίο βρίσκεται στην ιστορική γειτονιά της Πλάκας, υψώνεται στην διασταύρωση των οδών Διογένους και Μνησικλέους. Σε μικρή απόσταση από αυτό, συναντάμε τον αρχαιολογικό χώρο της Ρωμαϊκής Αγοράς του Αδριανού και τους Αέρηδες (Ωρολόγιο του Ανδρόνικου Κυρρήστου) που θεωρείται ο πρώτος μετεωρολογικός σταθμός του κόσμου. Κάτω από τα θεμέλια του εν λόγω νεοκλασικού, βρισκόταν το αρχαίο Πάνθεον (Κοινό Ιερό των Θεών), κτίσμα που αποδίδεται στον αυτοκράτορα Αδριανό (130 μ.χ.). Λίγα μέτρα πιο πέρα, εντοπίζεται η μικρή πλατεία του « Πλάτανου » σημείο αναφοράς στην Πλάκα από το 1932, λόγω της ομώνυμης ταβέρνας και του καφενείου του « Ρερέ » που υπήρξε στέκι για πολλούς γνωστούς - ξένους και ντόπιους - συγγραφείς, ποιητές, καλλιτέχνες, καθώς και πολιτικούς της εποχής. Μεταξύ αυτών, σύχναζαν οι Κ. Παλαμάς, Α. Σικελιανός, Π. Καλλιγάς, Π. Βυζάντιος, Κ. Παρθένης, Γ. Σαββάκης, Κ. Βάρναλης, Γ. Σεφέρης, Ο. Ελύτης, Φίλιπ Ζέραρντ, Πάτρικ Λη Φέρμορ, Λώρενς Ντάρελλ και Χένρυ Μίλλερ.

Η τριώροφη οικία διαθέτει ένα αρχικό πυρήνα των οθωμανικών χρόνων. Το 1833 υπήρχε ήδη το σημερινό ημιυπόγειο, ένα λιτό κτίσμα με απλές γραμμές και παραδοσιακά στοιχεία, το οποίο φαίνεται να ήταν τότε ισόγειο, μέχρι που το κτίσμα αγοράστηκε το 1836 από τους αδελφούς Αλέξιο και Ιωάννη Κουτζαλέξη, οι οποίοι στο επόμενο διάστημα, έως το 1868, προσέθεσαν τους επιπλέον ορόφους, προσδίδοντας το επίσημο νεοκλασικό ύφος που διατηρεί μέχρι σήμερα. Το κτίριο στέγασε αρχικά τη Γραμματεία (Υπουργείο) των Ναυτικών (1837), και έκτοτε, στο διάβα του χρόνου, άλλαξε διάφορους ιδιοκτήτες, χρησιμοποιούμενο κυρίως ως κατοικία. Εξαίρεση αποτελούν οι δεκαετίες 1960 και 1970, από τα τέλη του 1965, όταν το ημιυπόγειο φιλοξένησε τη μπουάτ <<Κατακόμβη>> όπου ανάμεσα σε άλλους, ο Γιώργιος Μαρίνος άρχισε την θρυλική μουσικο-θεατρική του καρίερα, ενώ η δημοφιλής τραγουδίστρια, Πόλυ Πάνου, εκτέλεσε του Μίμη Πλέσσα το τραγούδι 'Τι σου 'κανα και πίνεις,' στην ταινία Ολγα, Αγάπη μου, της Φίνος Φιλμ, που γυρίστηκε εκεί το 1967. Περίπου το ίδιο χρονικό διάστημα, στο ανώγειο, εγκαταστάθηκε μπουάτ 'νέου κύματος' - ανακατασκευασμένο άπο τον διάσημο ζωγράφο/σκηνογράφο Γιάννη Τσαρούχη - με το όνομα <<Ταβάνια>> (εμπνευσμένο από τις αξιόλογες οροφογραφίες του 19ου αιώνα, που το κοσμούν), που ξεκίνησε τη λειτουργία του με εμφανίσεις σπουδαίων μουσικών, όπως ο Γιάννης Σπάνος, ο Γιώργος Ζωγράφος, και η Αρλέτα, της οποίας ήταν επίσης κατοικία, για ένα διάστημα.

Οι οροφές υπέστησαν φθορά από το σεισμό το 1999, και οι ταβανογραφίες αποκαταστάθηκαν από την Μαρίζα Αγγελοπούλου και τον Δημήτρη Fresey (Verdigris Architectural Painting) το έτος 2000.

# KOUTZALEXIS HOUSE

## (YEARS 1801-1868)

The neoclassical building is situated in Plaka, the historical quarter of Athens, at the junction of Diogenous and Mnisikleous Streets, and a short walk away from such archaeological sites as the ancient Roman Agora, and the Tower of the Winds monument (The Horologion of Andronikos Kyrrhestes), which is considered to be the world's first meteorological station. Beneath its foundations, once passed the legendary 'Pantheon' of Hadrian (130CE). A few metres away, one discovers the small, yet charming 'Platanos' Square, a landmark in the neighbourhood since 1932, due to the eponymous taverna, as well as the then extant coffee shop of 'Rere', which had become a meeting ground for many well-known foreign and native writers, poets, and fine artists, as well as other intellectuals and politicians of the times; amongst them K.Palamas, A. Sikelianos, P. Kalligas, P. Byzantios, K. Parthenis, G. Savvakis, K. Varnalis, G. Seferis, O. Elytis, Phillip Sherrard, Patrick Leigh Fermor, Lawrence Durrell and Henry Miller.

The three-story home has an original core dating back from the pre-Independence, Ottoman-occupied period. In 1833, the present-day semi-basement level already existed, and was probably the ground floor of an unadorned structure of simple lines and traditional elements of form, until the edifice was sold by the Vrizakis family in 1836 to the brothers Alexios and Ioannis Koutzalexis, who in the following years, through to 1868, added two additional floors, conferring a formal neoclassical style to the structure, which it retains to this day. The building first housed the Naval Ministry in 1837, and has since changed owners several times, serving primarily as a residence. An exception regards the decades of the 1960's and early 1970's, since December of 1965, when the semi-basement hosted the artistic boîte <<Catacomb>> (<<Κατακόμβη>>) where amongst others, George Marinos began his legendary musical and theatrical career, whilst the popular singer Poly Panou performed the Mimis Plessas song 'Τι σου 'κανα και πίνεις,' in the movie *Olga, my Love* (*Όλγα, Αγάπη μου*), which was filmed on location in 1967. Around the same period, the upper floor was reconstructed and decorated by the renowned painter and scenographer, Yiannis Tsarouhis, to become home to the Greek 'New Wave' music boîte <<Ceilings>> (<<Ταβάνια>>), whose name was inspired by the refined 19th-century decorative paintings that grace them.

The ceilings suffered damages during the earthquake of 1999, and underwent a thorough restoration by Mariza Aggelopoulou and Dimitri Fresey (Verdigris Architectural Painting), in the year 2000.

# A Poets' Agora

## MUTED

Katerina Anghelaki - Rooke
Liana Sakelliou Shultz
Orfeas Apergis

*"Sunset with the day also learning from you*
*and steadily maturing*
*there on the balcony of Kallidromiou street,*
*the mulberry trees blossoming again*
*young poets holding their secret poems*
*under their tongues*
*and reciting them for the wind and the street."*

**Katerina Anghelaki-Rooke**

Excerpt from *The Angel Poems* (1998)

# ΑΦΩΝΟΣ

Σίφουνας ἔφερε τόσα δεινά
χάνοντ᾽ ἐλπίδες, σβύνει ἡ ἀστροφεγγιά
κι ἐγώ μένω ἄλλαλη, χωρίς φωνή
χωρίς πνοή
γιά νά ζητήσω
ν᾽ἀποκτήσω
ὅσα ἡ ζωή μοῦ εἶχε ὑποσχεθεῖ.
Ἀλλά τώρα ἡ ἴδια ἡ ζωή
διατάζει νά σωπάσω
ἀλλοιῶς τά λίγα πούχω
κι αὐτά θέ νά τά χάσω.


Κατερίνα Αγγελάκη-Ρουκ

# MUTE

Whirlwind brought so many sufferings
hope disappears, starlight dies
while I remain mute, voiceless
breathless
so I can't ask
for what life has promised me.
But now, life itself
orders me to hush
otherwise the little I have left
I will lose as well.


Katerina Anghelaki-Rooke

# Ὁ ΤΕΛΕΥΤΑΙΟΣ ΕΡΩΤΙΚΟΣ

Κάθε σταγόνα ἀπό σένα
τό τέλος τοῦ ἔρωτα
ὅ,τι ἔχασα στή ζωή
τό ξαναχάνω στά μάτια σου
τό φῶς ἔγινε τρυπάνι
μές τό παράλογο σχῆμα
μιᾶς ἄπιαστης ἐπαφῆς
ὅταν στά πόδια χύνεται
τό κύμα
τό αἰώνιο πέλαγος
τοῦ ζώου - θανάτου.
Δέν ὑπάρχει λόγος νά ζοῦμε
τά κόκαλα θά σπάσουν
σάν καλοκαιρινά καλάμια
κι ὁ μαγνήτης θά γίνει πίσσα
στά λαμπερά νά ρίχνει μαῦρο
Τό σῶμα σου - ἡ Πύλη
ἔκλεισε, ἔφραξε
ἀπ'τόν πολύ πόνο
κι ἐγώ καμένος ἐρημότοπος
τσουρουφλισμένη ἐπιφάνεια
δέν ὑπάρχω
ἐνῶ διπλασιάζονται οἱ μέρες
τοῦ φορτίου
ἡ ἄμμος ξύνει
ὅλο ξύνει τήν πληγή
ἡ καταστροφή εἶναι π'ἀρνιέμαι
νά καταστραφῶ
νά σπάσει τό πιθάρι

νά χυθοῦν τά μέσα
ἔξω
ὅπου συναντιέται ὁλόστητος
ὁ βράχος.
Μές τό δωμάτιο
φάνηκαν ξάφνικα τά ὅρια μου
τά νύχια μου σπασμένα
ἡ σπλήνα μου βαριά τραυματισμένη
κι ὁ ποταμός πού λάτρεψα
σ'ἄλλη κοίτη κυλάει
Ὅλο κι ἀλλοῦ σκάει
τό ρόδι
οἱ χυμοί πού προχωροῦσαν
τή ζωή
ἀλλοῦ λιμνάζουν
ἀπ'τήν ἀφθονία στή στέρηση
πετάω• τρομαγμένη νυχτερίδα
ἡ νεότητα
σκοτάδι τά φτερά της.


Κατερίνα Αγγελάκη-Ρουκ (1975)

# EPITAPH ON LOVE

Every drop from you
is the end of love;
whatever I lost in life
I lose again in your eyes
light becomes a drill
in the absurd shape
of an intangible touch
when at our feet the wave
spreads out
the eternal sea
of the creature-death.
There is no reason to live
the bones will break
like summer reeds
and the loadstone will become tar
casting black on what shines.
Your body – Sublime Gate
shut, blocked up
from so much pain
and me a burned wilderness
a scorched surface
I don't exist
while the days of burden
are doubled,
the sand scraping
scraping at the wound
destruction is what I refuse
to be destroyed
to break the cask

and the inside to pour
out
where the upended rock is met.
In the room
suddenly my limits appeared
my nails broken
my spleen seriously damaged
while the river I had adored
runs in another bed.
The pomegranate bursts
always elsewhere
the saps which advanced life
and I fly from abundance to
privation; youth is a frightened
bat
darkness its wings.

Katerina Anghelaki-Rooke (1975)

# ΜΕΤΑΦΡΑΖΟΝΤΑΣ ΣΕ ΕΡΩΤΑ ΤΗΣ ΖΩΗΣ ΤΟ ΤΕΛΟΣ

Επειδή μέ τή δική μου γλῶσσα
δέν μπορῶ νά σ' αγγίξω
μεταγλωττίζω τό πάθος μου.
Δέν μπορῶ νά σέ μεταλάβω
καί σέ μετουσιώνω,
δέν μπορῶ νά σέ ξεντύσω
ἔτσι σέ ντύνω μ' αλλόφωνη φαντασία.
Στά φτερά σου ἀπό κάτω
δέν μπορῶ νά κουρνιάσω
γι' αὐτό γύρω σου πετάω
καί τοῦ λεξικοῦ σου γυρνάω τίς σελίδες.
Πώς απογυμνώνεσαι θέλω νά μάθω
πώς ξανοίγεσαι
γι' αὐτό μές στίς γραμμές σου
ψάχνω συνήθειες
τά φροῦτα π' ἀγαπάς
μυρουδιές πού προτιμάς
κορίτσια πού ξεφυλλίζεις.
Τά σημάδια σου ποτέ μου δέν θά δῶ γυμνά
ἐργάζομαι λοιπόν σκληρά πάνω στά ἐπίθετά σου
γιά νά τ' ἀπαγγείλω σ' ἀλλόθρησκη λαλιά.
Πάλιωσε ὅμως ἡ δική μου ἱστορία
κανένα ράφι δέν στολίζει ὁ τόμος μου
καί τώρα ἐσένα φαντάζομαι μέ δέρμα σπάνιο
ὁλόδετο σέ ξένη βιβλιοθήκη.
'Επειδή δέν ἔπρεπε ποτέ
ν' ἀφεθῶ στήν ἀσυδοσία τῆς νοσταλγίας
καί νά γράψω αὐτό τό ποίημα
τόν γκρίζο οὐρανό διαβάζω
σέ ἡλιόλουστη μετάφραση.

Κατερίνα Αγγελάκη-Ρουκ (2004)

# TRANSLATING INTO LOVE LIFE'S END

Since I cannot touch you
with my tongue
I translate my passion.
I cannot communicate
so I transubstantiate;
I cannot undress you
so I dress you with the fantasy
of a foreign tongue.
Under your wings
I cannot nestle
so I fly around you
turning the pages of your dictionary.
I want to know how you strip
how you open up
so I look for your habits
in between your lines
for your favourite fruit
your favourite smells
girls you leaf through.
I'll never see your punctuation marks
naked, I work hard on your adjectives
so that I can recite them in the susurrations
of another religion.
But my story has aged
my volume adorns no shelf
and I imagine you now
with a rare gold leather binding
in a foreign library.
Because I should never

have indulged in the luxury of nostalgia
and written this poem
I am reading the gray sky now
in a sun-drenched translation.


Katerina Anghelaki-Rooke (2004)

# ΆΦΩΝΟΙ

Πέθανα στὰ δεκαεννιά μου ἀπὸ τὸν κόκκινο θάνατο – μ᾽ ἔθαψαν στὸ Μοναστήρι.
Πάρτε με ἀπὸ ἐδῶ, γλιτῶστε με, πέφτω στὸν γκρεμό –
ἡ πανώλη.
Ἤμουν Τοῦρκος αἰχμάλωτος τοῦ Βαλκανικοῦ, ἔπαθα ἐξανθηματικὸ τύφο –
μ᾽ ἔθαψαν στὸν Τουρκόδρομο.
Δὲν περιελήφθην στὸν κατάλογο προαγωγῆς ὡς ἀξιωματικὸς τοῦ Ναυτικοῦ –
ἔκανα χαρακίρι.
Ἐπὶ Τουρκοκρατίας μ᾽ ἐκτέλεσαν στὸ Δασκαλειό. Ἦταν ὁ τόπος ἐκτελέσεων –
ἤμουν δασκάλα, εἶπαν ὅλοι.
Ἐκκενῶστε τὰ σπίτια σας, κύριοι. Ἡ φωτιὰ πλησιάζει –

Λιάνα Σακελλίου

# MUTED

Died at nineteen
by the red death—
in the Monastery I was left forever.
Take me from here, save me, I am falling off the cliff—
the plague.
A Turkish hostage of the Balkan War I caught typhoid fever—
they buried me in the Turkish road.
I wasn't nominated for promotion as a Naval Officer—
I committed hara-kiri.
In the Turkish occupation I was executed in Daskaleio; it was the site of executions— I was a teacher,
they said.
Abandon your houses everyone Fire nears—


Liana Sakelliou

# ΥΔΑΤΟΓΡΑΦΙΑ

Κολυμπούσαμε μαζί, αἰῶνες πρίν
σ'ἕναν ὠκεανό κοντὰ στὴν ἔρημο
ἐνῶ ἡ αὔρα σωριαζόταν
στὴ μπλὲ σπηλιὰ τοῦ τίποτα.
Βουτοῦσα σὲ κάτι σημαδεμένο καὶ βαθὺ
καὶ στὸ σκοτάδι του ἔβρισκα ἀνθόζωα
ποὺ κυμαίνονταν σὰν σκελετοὶ
νεογνῶν.
Λύγιζες, ἴσιωνες καὶ πάλι κύρτωνες τὴν πλάτη
σὰν νὰ συμπύκνωνες τὸν πόθο
σὰν νὰ μοῦ ἔδειχνες τὸν τρόμο
κι ὕστερα τὸ θάνατο ποὺ ἔβλεπες τοῦ ἔρωτα.


Λιάνα Σακελλίου

# AQUARELLE

We swam together, centuries ago
in an ocean beside the desert
while the breeze collapsed
into the blue grotto of nothingness.
I dived into something marked and deep
and in its darkness I found anthozoa
that fluctuated like the skeletons
of newborn.
You bent, straightened and arched your back again
as though condensing the desire
as though showing me the fear
and then the death of love that you saw.

Liana Sakelliou
[*translated by David Connolly*]

# ἌΣΚΗΣΗ ΦΛΑΜΑΝΔΟΥ ΔΑΣΚΑΛΟΥ

Κάθεται στὴν πλώρη ἀειθαλής.
Μὲ μία κίνηση τοῦ χεριοῦ
σὰν νὰ τοὺς ζωγραφίζει
μοῦ δείχνει τοὺς ὄγκους τῶν ἀπέναντι βουνῶν.
Θέλει νὰ διασχίσουμε τὴ λίμνη
μέχρι τὰ σύνορα κι ἐκεῖ νὰ τὸν ἀφήσω
σ᾽αὐτὸ ποὺ δὲν ἦταν ποτέ.
Εἶναι σκοτάδι.
Βυθίζω τὶς λεπίδες τῶν κουπιῶν στὴ σκούρα ἔκταση.
Στὴν ἔνταση ἡ βάρκα μυρίζει βλάστηση πυκνή.
Κουκουβάγιες μπαινοβγαίνουν στὰ φυλλώματα
σὲ κάθε ὤθησή της. Οἱ σκαρμοὶ τρίζουν.
Θροΐζω. Δουλεύει στὴ σιωπή.


Λιάνα Σακελλίου

# STUDY BY A FLEMISH MASTER

He sits evergreen in the prow.
With a movement of the hand
as if painting them
he shows me the size of the facing mountains.
He wants us to cross the lake
as far as the borders and for me to leave him
where he never was.
It's dark.
I plunge the oars' blades into the somber expanse.
In the tenseness the boat smells of thick vegetation.
Owls fly in and out of the leafage
at its every propulsion. The rowlocks creak.
I rustle. He works in silence.

Liana Sakelliou
[*translated by David Connolly*]

# Η ΠΥΛΗ

Ὅλη τή νύχτα πλάϊ του
στό ἀτέλειωτο πυροφάνι
ξοδεύω τήν ἄμμο
ξοδεύω τό λάδι
σ' ἕναν ὅρμο ὅπου ἐξαφανίζεται
ἡ Κοιμωμένη.
Δῶσ' μου τό χέρι σου, παιδί μου,
ὁ καιρός πιά εὐνοϊκός.
Ὅμως ἐγώ προσέχω τόν γκιῶνη
καί εἶναι ἀλάνθαστος κί' εἶναι κοφτός
καί δέν σαλεύει.


Λιάνα Σακελλίου

# THE GATE

All night beside him
in the endless glow
of his fishing lamp
I spend the sand
I spend the oil
in a cove where the Sleeping Woman
disappears.
Give me your hand, my child.
The signs for the weather are favorable.
But I set my course by the Scops owl,
which is infallible, blunt
and doesn't budge.


Liana Sakelliou

# Η ΤΑΦΗ

Ό,τι νοιώθω παύει να υπάρχει.
Σκοτώνω με το αίσθημά μου
και ξαναδίνω τη ζωή.
Όλα αυτά που φοβούνται
τη ζωή μου,
για παράδειγμα το νερό πάνω
στο πρόσωπο,
η αποφορά ενός λερωμένου κή-
που,
η μέρα – τρίτη – που οδήγησε
τη νύχτα πιο μακριά από το
τέλος της,
να αυτά
που τρέμουνε
το άγγισμα και
το χωνεμένο βλέμμα
όλα τους
πεθαίνουνε βιαστικά
και χαρούμενα
ότι δεν κατάφερα
να τα νιώσω.

Μετά την ταφή
Θα πεθάνουνε σίγουρα
Όλα αυτά
Που φοβούνται
Το θάνατό μου.
Με την απόσταση πλησιά-
ζω και με το θάνατο ξανα-

δίνω ζωή.
Ο θάνατός μου
η ζωή σας.


Ορφέας Απέργης

# THE BURIAL

What I feel ceases to exit.
I kill with feeling
and give life again.
All those that fear
my life,
for example water
on faces,
the reek of a soiled
garden,
the day – third – that guided
the night farther than
its end,
here they are, those
that shudder
to my touch and
to the accepted gaze
all of them
die in a hurry
and happy
that I didn't manage
to feel them.

After the Burial
They will die for certain
All those
That fear
My death.
By distance I approach
and by death return life.

Matter of my death
and your life.


Orfeas Apergis

# ΤΟ ΤΡΑΠΕΖΙ (απόσπασμα)

Η πόρτα είναι μισάνοιχτη και φαίνονται μέσα της τα νερά του ξύλου
κι από πάνω λίγο φως, αλλά κάπως έντονο,
ίσως από τον ανύπαρχτο κήπο ή από την λυχνία πυρακτώσεως
που όμως δεν φαίνεται καλά.
Δίπλα η κονσόλα με τ' ασημομάχαιρα, τορνευτή, σκαλιγμένη με όντα περίεργα,
επιθυμητικά, ήγουν σάτυροι και νύμφες και μικρά ζωάκια εκπαχύνσεως
με πρησμένα συκώτια, σίγουρα, λιπαρά,
σαν επί θυσία,
έτοιμα,
ακινητούντα,
πιασμένα μέσα στις περιπλοκές του κέδρου.
........
αιωρείται, νομίζεις, το τραπέζι,
μετεωρίζεται μέσα στο άσπρο του, με γύρω τις σκιές του βυσσινιές,
χωρίς να φαίνονται τα πόδια του,
δηλαδή χωρίς στηρίγματα εμφανή, αξιόλογα,
σαν πλάκα ασπρουδερή,
σαν στήλη
δέλτος
χαραγμένη με τους νόμους των φρούτων,
με την νομή της τροφής
και της μαγειρευτής κνίσας.
.........
Πάνω στο τραπέζι,
συνεχίζει,
είναι και το ψωμί το αναπόφευκτο,
.........
Έτσι μεταλαβαίνουμε από τι είμαστε φτιαγμένοι
κι έτσι καθόμαστε,
στα δεξιά της κοπέλας

και συμπληρώνουμε το χαριτωμένο περιβόλι
με τα ρόδια και με τον κόκκο του σιταριού,
ψωμί και αίμα
και η τελική, μοιραία παρένθεση
όπου συνειδητοποιούμε πως
το τραπέζι είναι από ακακία
και πως τα γυαλιά της έχουνε λίγο ραγίσει
κι ακόμα πως δεν στηρίζεται καλά η επιφάνεια,
πως της λείπουνε ποδάρια
πως χρεωκόπησε ο ξυλουργός
και πως όλα τα λεφτά σωθήκανε
και τα δέντρα δεν κόβονται
αλλά γίνονται σαν αγχόνη
και πως πνιγήκαμε παιδιά
μέσα σε δανεισμένη πεποίθηση
και μέσα σε δόλιο εξιλασμό,
ω τι φριχτό κι απέριττο κατασκεύασμα,
ω τι επίσημη ευτυχία,
το τραπέζι το κουτσό
είναι από σανίδες, κούφιο από μέσα,
φαύλον σανιδωτόν,
κοίλον και καταχρυσωμένο,
σαν το ιλαστήριο με τους δύο αγγέλους
και με τα κέρατα στις τέσσερις γωνίες.
Κάτω από το τραπέζι,
– δεν αντέχει ο καιρός αλληγορία –
βλέπεις το μέσα θυσιαστήριο,
τους λέβητες,
τα φτυάρια, τις σχάρες
και τη μεγάλη κρεάγρα με το φρέσκο λίπος,
όλα από χαλκό,
αγοραίο.

Ορφέας Απέργης

# THE TABLE (excerpt)

The door is half-open and you can see in it the waters of the wood
and on it some light, albeit somewhat intense,
perhaps from the bare garden or the incandescent lamp,
that one however cannot make out.
Next to it the drawer with the silverware, round, sculpted with bizarre creatures,
desirous, namely satyrs and nymphs and small animals for slaughter
with swollen livers, assured, unctuous,
as if they were being prepared for sacrifice,
ready,
immotile,
caught inside the intricacies of the cedar wood.
.........
it floats, you would think, the table,
it levitates inside its whiteness, surrounded by its shadows in purple,
its legs never showing,
that is to say without any apparent holdings, worthwhile,
like a white tablet,
like a wax
tablet
engraved with the laws of fruits,
the possession of food
and the burnt offering.
.......
On the table,
it goes on,
it is also the bread that is inescapable,
.......
This is how we commune with what we are made of
and this is how we sit,
on the right of the girl
and fill the lovely orchard

with the pomegranates and the grain of wheat,
bread and blood
and the final, fatal parenthesis
where we realise that
the table is made of acacia
and that that her glasses have slightly cracked
and furthermore the surface is not holding well,
that it's missing legs
that the carpenter has gone under
and that all the money is gone
and the trees are no longer cut
but turn into something like gallows
and that we drowned you guys
inside a borrowed conviction
and inside a devious atonement,
oh what a terrible and austere construct,
oh what an official happiness,
the crippled table
is made of boards, hollow on the inside,
deplorable board,
concave and overly gilded,
like the mercy seat with the two angels
and the horns on four corners.
Underneath the table,
- time cannot withstand allegory -
you can see the inside altar,
the boilers,
the shovels, the grills
and the large fleshhook with the fresh fat,
all made of brass,
vulgar.

Orfeas Apergis

# A Poets' Agora

**Haris Vlavianos    Katerina Iliopoulou    G. F. Zaimis**

*"In the land of shadows*
*naked things wait.*
*Not to jump you or rush you.*
*They have softer ways*
*to break and tatter you.*
*With indiscernible sounds,*
*imperceptible movements*
*they inhabit you.*
*They learn so well*
*you become a passage*
*You will never be able to pass."*

**Katerina Iliopoulou**

Excerpt from 'Penthesileia'
In *Asylum* (2008)
*[translated by Ryan Van Winkle]*

# ΚΥΚΛΑΔΙΚΟ ΕΙΔΥΛΛΙΟ

Χαμήλωσε το βλέμμα.
Όταν η ομορφιά
εισβάλλει με τόση ορμή στη ζωή σου
μπορεί να σε καταστρέψει.
Τα δύο μυρμήγκια
που τώρα τρέχουν βιαστικά
δίπλα στα γυμνά σου πέλματα
θάβουν τα καλοκαιρινά τους όνειρα
βαθιά στο χώμα.
Το φορτίο που κουβαλάνε
δεν πρόκειται να τα συνθλίψει.
Έχουν μετρήσει καλά τις δυνάμεις τους.

Η σκιά σου σβήνει μέσα στη σκιά του δέντρου.
Μαύρο στο μαύρο.
Ενοχή που πρέπει να μείνει στο σκοτάδι
για να συνεχίζει να σε ορίζει.
Το θάμβος όμως αυτών των θραυσμάτων
μπορεί ακόμη να σε κρατήσει.
Δεν έχεις ανάγκη από επίθετα.
Από ανούσιες υπεκφυγές.
Κάθε ερώτηση είναι μια επιθυμία.
Κάθε απάντηση (το ξέρεις πια) μια απώλεια.
Μείνε εκεί που στέκεσαι.
Σε λίγο θα κοπάσει η βροχή.
Τα σύννεφα δεν ρωτούν πού.
Απλώς συνεχίζουν την πορεία τους.

Χάρης Βλαβιανός

# CYCLADIC IDYLL

Lower your gaze.
When beauty
Invades life with such force
It can destroy you.
The two ants
That now run hastily
Near your naked feet
Bury their summer dreams
Deep into the ground.
The weight they carry
Will not crush them.
They have measured their strength well.

Your shadow fades into the shade of a tree.
Black into black.
Guilt that had to remain in the dark
So to keep defining you.
The glow of these fragments, though,
Can still sustain you.
You don't need any adjectives
Any pointless excuses.
Every question is a desire.
Every answer (you know it) is a loss.
Stay where you are.
The storm will lull in a while.
The clouds do not ask where.
They simply continue their journey.

Haris Vlavianos

# ΔΟΞΑΣΙΕΣ ΤΟΥ ΑΥΓΟΥΣΤΟΥ

1. Αν ένας άντρας στα σαράντα του
ζωγραφίζει ακόμη θάλασσες και περιστεριώνες,
αν στη σκέψη του καθρεφτίζεται
ένας ήλιος πιο καθαρός,
πιο σαφής, από τον ήλιο της πραγματικότητας,
αν η λέξη «Αμοργός» δεν είναι απλώς
το προσωπείο μιας φευγαλέας, εφηβικής ανάμνησης,
τότε ανάμεσα στο ποίημα της επιθυμίας
και το ποίημα της ανάγκης
ανασαίνει η αληθινή απώλεια.

2. Οι πρόλογοι εξαντλήθηκαν.
Δεν μπορούν να υποκαθιστούν διαρκώς το θέμα.
Πρέπει ν' αποφασίσει αν μπορεί
να κρατηθεί από αυτή την απόλυτη ιδέα
έστω κι αν έχει πάψει να πιστεύει στη δύναμή της.
Είναι ζήτημα πίστης πλέον.

3. Αλλεπάλληλες μεταμορφώσεις του παραδείσου.
Το μάτι προσπαθεί να ερμηνεύσει το αίνιγμα της ομορφιάς
καθώς στο βάθος ανατέλλει αργά-αργά η Δήλος.
Το καλοκαίρι μοιάζει αιώνιο.
Το ποίημα αρχίζει να επινοεί τον εαυτό του
τη στιγμή που εκείνος στρέφει το πρόσωπό του στο φως.
(Τη στιγμή που η φαντασία απελευθερωμένη
από τη συγκεκριμένη αίσθηση του διάπυρου λευκού
υψώνεται κάθετα στον ουρανό.)

4. Ούτε ένα ιστιοφόρο στον ορίζοντα
να κόβει τον καμβά στα δύο.
Η εικόνα ενός δένδρου
με τα ανεμόδαρτα κλαδιά του να σαρώνουν το χώμα
δεν είναι σήμερα μέρος του σκηνικού.
Όμως η γριά που σέρνεται με τα γόνατα στην ανηφόρα
κρατώντας σφιχτά στο χέρι το εικόνισμά Της
Είναι.

5. Ο άντρας βαδίζει μόνος στην παραλία.
Τον συγκινεί ακόμη ο μελωδικός ψίθυρος των κυμάτων,
ο τρόπος που το νερό νανουρίζει επίμονα τον βράχο.
Η φύση γύρω του
(αρμυρίκια, σάπιες ψαρόβαρκες, κροκάλες)
έχει μια μελαγχολική, ανεπιτήδευτη λαμπρότητα.
Αν ήταν να πεθάνει αυτή τη στιγμή
θα ήθελε να είναι εδώ
στον τόπο αυτό που έχει υπάρξει.
Έστω για λίγο.
Τώρα.


Χάρης Βλαβιανός

# AUGUST MEDITATIONS

1. If a man in his forties
is still drawing seas and dovecotes,
if in his thought is reflected
a sun more transparent,
more lucid than the sun of reality,
if the word "Amorgos" is not just
the mask of a fleeting, adolescent memory,
then between the poem of desire
and the poem of necessity
real loss is throbbing.

2. Prologues have been consumed.
They cannot always substitute the topic.
He must decide whether he can
hold on to this absolute idea
even if he has ceased to believe in its power.

3. Successive metamorphoses of paradise.
The eye tries to interpret the enigma of beauty
while Delos is slowly emerging on the horizon.
Summer feels like an eternity.
The poem begins to invent itself
the moment he turns his face to the light.
(The moment the imagination,
freed from the sensation of the blazing white,
verticality rises in the sky.)

4. Not one sail on the horizon
tearing the canvas apart.
The image of the tree
with its wind-swept boughs scavenging the ground
is not part of the scenery today.
Yet, the old lady, creeping uphill on her knees,
tightly holding her icon, is.

5. The man is walking on the beach alone.
He is still touched by the melodious whisper of the waves,
the way the water is persistently lulling the rock to sleep.
Nature around him
(cedars, rotten fishing boats, shingles)
has a melancholic, unaffected brightness.
If he were to die at this moment,
he would want to be here,
in this place, where he has been before.
Even for a while.
For now.


Haris Vlavianos

# ΤΟ ΠΕΠΛΟ

*"Lift not the painted veil which those who live*
*Call Life…"*
Percy Bysshe Shelley, "Sonnet"
*"Only let down the veil, the veil, the veil."*
Sylvia Plath, "A Birthday Present"

Ὅποιος βρίσκεται μέσα σε λαβύρινθο
                δέν  ἀναζητᾶ τήν  ἀλήθεια·
                        ἀναζητᾶ τήν Ἀριάδνη του.

Ἀλήθεια δέν εἶναι ἡ απογύμνωση
                πού καταστρέφει τό μυστικό·
                        εἶναι ἡ  ἀποκάλυψη πού τό δικαιώνει.

Ἡ ἐμπειρία μας παραμένει δέσμια μιᾶς γνώσης
                πού δεν εἶναι πιά ἐμπειρία μας·
ἡ γνώση μας εὐτελίζεται  ἀπό μιά ἐμπειρία
πού δέν ἔχει  ἀκόμη μετουσιωθεῖ σέ γνώση.

Πρέπει νά ὑφάνουμε νέα ποιητικά σχέδια
                πού θά ὁρίζουν τούς δυνητικούς τύπους μιᾶς  ἀλήθειας
μήτε  αληθινῆς            μήτε ψευδοῦς
μιάς  αλήθειας πού θά εἶναι
                ἀναληθοφανής
                                ἀδιανόητη
                                        ἀσύλληπτη
πού θά μπορεῖ νά μετατρέπει τό σφάλμα (*τό πάθος*)
σέ νῆμα τῆς ζωῆς μας.

Χάρης Βλαβιανός

# THE VEIL

A man locked in a labyrinth
     does not seek truth
         but his Ariadne.

Truth is not an unveiling
     that destroys the secret;
         it is the un-concealment that does it justice.

Our experience remains the captive of knowledge
     that no longer is our experience;
our knowledge is trivialized by an experience
that has not yet become knowledge.

We must leave new poetic paths
that will designate the potential places of a "truth"
neither true       nor false
a truth that will be
         Implausible
            Improbable
              impossible
thereby making error         *(pathos)*

Into the thread of our life.

Haris Vlavianos

# ΝΑΝΟΥΡΙΣΜΑ

Το δωμάτιο αυτό είναι παραγωγός νύχτας
Τα κρεβάτια είναι πηγάδια
Μέσα τους αναπαύεται η αντανάκλασή σου
Εσύ στέκεσαι μπρος στο ανοιχτό παράθυρο
Το παράθυρο βρίσκεται σε τέτοια θέση
Ώστε να πλαισιώνει συμμετρικά το καμπαναριό
Κάθε κτύπος της καμπάνας
Παραμορφώνει τη γεωμετρία του χώρου
Στον τελικό κτύπο το δωμάτιο είναι μια σταγόνα
Το σώμα της αναμονής ιδρώνει.


Κατερίνα Ηλιοπούλου

# LULLABY

This room is a night generator
The beds are wells
Inside them lies your reflection
You stand before the open window
The window is positioned so as
To frame the bell tower symmetrically
Every ring of the bell
Transforms the geometry of the space
In the final ring the room is a water drop
The body of anticipation is sweating


Katerina Iliopoulou

# ΤΑΙΝΑΡΟΝ

Εδώ οι μέρες δεν διαλύονται στον αέρα
Πέφτουν μέσα στο νερό
Σχηματίζοντας μια καταδική τους στιβάδα
Μια επιφάνεια διαχωρισμού.
Ένα γεράκι πετάει πάνω από το σώμα του καλοκαιριού
Βουτάει ξανά και ξανά
Τρέφεται και μεθάει από την πτώση.
Δεν έχει τίποτα εδώ
Μόνο τρελλό αέρα και πέτρες
Και θάλασσα
Μια αναίτια υπόσχεση
Ακονίζει τη λαγνεία μας με τη λάμα του φεγγαριού.
Όταν έφτασα εδώ για πρώτη φορά, στο τοπίο του τέλους,
Ο αέρας έμπαινε στο στόμα μου με τέτοια μανία
Σα να ήμουν ο μοναδικός αποδέκτης του
Μέχρι όλες οι λέξεις μου να εξαφανιστούν.
Κάθε δέντρο υποδέχεται διαφορετικά τον άνεμο
Άλλα υποφέρουν, άλλα πάλι αντιστέκονται
(Έχω συναντήσει μια φοινικιά που γεννούσε τον άνεμο και τον
διένειμε προς κάθε κατεύθυνση)
Άλλα τρέμουν ολόκληρα κι αλλάζουν χρώματα.
Εγώ βέβαια δεν είμαι δέντρο
Κάθισα κάτω και τον φόρεσα παλτό
Έσκυψα το κεφάλι μου και κοίταξα το χώμα
Μέσα από τις ρωγμές του, οι ρίζες του θυμαριού
με τα ιερογλυφικά τους πάσχιζαν να βγουν στο φως.
Τότε οι λέξεις ξαναγύρισαν.

Κατερίνα Ηλιοπούλου

# TAINARON

Here the days do not dissolve in the air
They drop into the water
Forming their very own layer
A surface of separation.
A hawk flies above the body of the summer
It dives again and again
Feeding and getting drunk from falling.
There is nothing here
Only crazy wind and stones
And sea
A random promise
Sharpens our lust with the blade of the moon.
When I arrived for the first time in this landscape of endings
The wind entered my mouth with such fury
As if I were its sole receptacle
Until all my words disappeared.
Every tree receives the wind differently
Some suffer others resist
(I met a palm tree that gave birth to the wind and distributed it
in every direction)
Others shake all over and change colors.
I of course am not a tree
I sat down and wore the wind as a coat
I bent my head and looked at the ground
From its crevices, the roots of thyme
With their hieroglyphics struggled
Then the words came back.

Katerina Iliopoulou
[ by John O'Kane]

# ΤΟ ΤΡΑΓΟΥΔΙ ΤΗΣ ΕΥΡΥΔΙΚΗΣ

Κράτα την υπόσχεσή σου Ορφέα
Κοίταξέ με
Καλλιέργησε με το βλέμμα σου
Τον λειμώνα της περιπλάνησής μου
Σκάψε μου το ταξίδι με το
Στιλέτο των ματιών σου
Ρίξε το δίχτυ σου και
Ανάσυρέ το άδειο
Μάζεψε τις σταγόνες:
Σε καθεμιά
Θα καθρεφτίζεται το πρόσωπό μου.
Εγώ είμαι το σύνορο που συνεχώς υποχωρεί
Ο φύλακας της απόστασης
Και το τραγούδι σου Ορφέα,
Είναι η απόσταση.
Μην αφήνεις τίποτε άθικτο
Καθετί που αγγίζεις
Να μην γίνεται ποτέ δικό σου
Κάθε άγγιγμα τόσο πιο ξένο
Όσο πιο ξένο τόσο πιο αρπακτικό
Κι έτοιμο το άγγιγμα να αντιγυρίσει
Όπως εκείνο ξέρει
Να βάλει μπροστά τη μηχανή της διάλυσης.
Και μ'ένα κράτημα της αναπνοής,
Όλο το θαμποκόκκινο σε παίρνει.
Κράτα το άπνευστο κενό και ύφανέ το.

Κατερίνα Ηλιοπούλου

# THE SONG OF EURYDICE

Keep your promise Orpheus
Look at me
Cultivate with your gaze
The meadow of my wandering
Dig for me the journey with
The stiletto of your eyes
Cast your net and
Draw it up empty
Gather in the drops:
In each one
My face will be mirrored
I am the border which continuously recedes
The guardian of distance
And your song Orpheus
Is distance.
Don't leave anything untouched
Whatever thing you touch
Will never become your own
Every touching all the more foreign
The more foreign all the more gripping
And ready to touch you back
As it alone knows how,
To start up the dissolution machine
And with a holding of your breath
All the blurred red takes you in.
Hold on to the breathless void and weave it.

Katerina Iliopoulou
[translated by John O'Kane]

# LULL

What is the lull that seems

To linger in the in between –

After the why, which then leaves?

G.F. Zaimis

# ΚΟΠΑΖΕΙ

Τι είναι που κοπάζει κι'

αιωρείται μέσα στο ανάμεσα –

μετά το γιατί, που αναχωρεί;

G.F. Zaimis
[μετάφραση: Ορφέα Απέργη]

# PENEL -O- (E)PIC

Sonnet II.

All the while "the wine-dark sea" churned
For you, Penelope, and your days –
Passed slow like time stopped; at bay
As you thumbed clock's page turned
Moments shaped nights, one by one
Wondering, dreaming of him; hope
That he was alive and would soon return home
It was your life-line, this tethered rope
But he had been brought to Calypso's shores.
Pulled from Poseidon's grip, the sea
Captive lover, nymph – someone's whore
Tell me Odysseus, what is your real need?
Is it Ithaca or burning passion, un-bored
Life's lust for adrenaline, don't you see.

G.F. Zaimis

# ΠΗΝΕΛ -Ο- (Ε)ΠΙΚΟ

Σονέτο ΙΙ.

Ενόσω ο «οίνοψ πόντος» άφριζε
Για σένα, Πηνελόπη, κι οι μέρες σου
Αργοκυλούσαν σα να 'χε ο χρόνος σταματήσει· ορμισμένη
Καθώς ξεφύλλιζες του ρολογιού η σελίδα άλλαζε
Τα λεπτά σχημάτιζαν τις νύχτες μία-μία
Αναρωτιόσουν, αυτόν ονειρευόσουν· απ' την ελπίδα κρεμασμένη
Πως ήταν ζωντανός και θα γυρνούσε στην εστία
Στην άκρη κείνου του σωσίβιου σχοινιού δεμένη
Όμως αυτός είχε βρεθεί στις ακτές της Καλυψώς.
Απ' τη λαβή του Ποσειδώνα γλιτωμένος, της θαλάσσιας δίνης
Εραστής δέσμιος, μιας νύμφης – της πόρνης κάποιου
Πες μου, Οδυσσέα, ποια η ανάγκη σου στ' αλήθεια;
Είναι η Ιθάκη ή το διάπυρο το πάθος, της α-πληκτικής
Ζωής ο πόθος –μη δεν το βλέπεις;– αδρεναλίνης.


G.F. Zaimis
[μετάφραση: Σεσίλ Ιγγλέση Μαργέλλου]

# PSYCHE'S CUP

Come beloved one, drink from the river Lethe
Hypnotic waters that sparkle, dizzy
To forget the parted memory; of the twin soul
Not the other half, but rather your whole.
She is obliged, the river goddess to him,
Hades of whom she flows like fluid emotion
As sleep; into the netherworld of oblivion –
Imbibe her essence and allow the dark to dim
In complete forgetfulness; made numb
Drink from the daughter of strife –
And heal your woes clean; from earthly life
Your reunion comes with the chaliced cup
Tonic to relieve the heart's wounds; soothe
Come child drink; embrace divine truth.


G. F. Zaimis

# ΚΥΠΕΛΛΟ ΨΥΧΗΣ

Έλα αγαπημένε, πιες απ' τον ποταμό της Λήθης
Νερά υπνογόνα που σπιθίζουν, ζαλισμένα
Για να ξεχάσεις τη φευγάτη μνήμη· της δίδυμης ψυχής
Όχι τ' άλλο μισό, το όλο σου μάλλον απαρχής.
Οφείλει η θεά του ποταμού να υπηρετήσει
Τον Άδη απ' όπου χύνεται σαν ρέουσα συγκίνηση
Σαν ύπνος· στον κάτω κόσμο της λησμοσύνης –
Ρούφηξε την ουσία της, άσε το σκότος της να σβήσει
Σε πλήρη αμνημοσύνη· ναρκωμένος
Πιες απ' την κόρη της έριδας νερό –
Τις λύπες γιάτρεψε· της επί γης ζωής σου
Το ξανασμίξιμό σας, δες, το κύπελλο χαρίζει
Ανακουφίζει κάθε πόνο της καρδιάς – τονωτικό
Έλα παιδί να πιεις· τη θεία αλήθεια ασπάσου.

G. F. Zaimis
[μετάφραση: Σεσίλ Ιγγλέση Μαργέλλου]

# A Poets' Agora

## 'RISK'

"We will have become ourselves yet no one will recognize us:
the future will carve itself in stone in its attempt to steal our form
on the day we return from the cities in the centre of the world.
In this boundless space we will pick spots where there will be no
reflection– sticker ads will seal off the fractures where the cities join
with one another; we will have become ourselves yet no one will
recognize us."

**Theodoros Chiotis**

Excerpt from *'Quasar (Future Biology)'*
Original English version written in early 2011,
and published in issue 24 of *Otoliths* (February 2012).

# ΑΥΤΟΣ ΔΕΝ ΕΙΝΑΙ ΕΝΑΣ ΔΕΡΒΙΣΗΣ

## Πλατεία Ταξίμ, Κωνσταντινούπολη, 2013

Πράσινο σκοτεινό βελούδο τού κυπαρισσιού
ρελιασμένο με μαύρο φαρδύ καραβόπανο
απλωμένο φουσκώνει καθώς περιστρέφεται
με τα χέρια ανοιχτά
σε ουρανό και σε γη
μες στο πλήθος που όρθιο
χτυπά παλαμάκια
πάνω στην ασφαλτοστρωμένη
πληγή της πλατείας
Μα στο πρόσωπο –
Μα το πρόσωπο
δεν είναι αγίου
ούτε ανθρώπου
Το πρόσωπο είναι μουτσούνα
τερατόμορφης πάπιας
Με καπάκια στα μάτια
με το ράμφος ραμμένο
βουλωμένο κι αυτό
–για τα δακρυγόνα–
Άσπρο καουτσούκ νοσοκομείου
–δίχως βλέμμα ή ανάσα–
πάνω από το κυπαρισσένιο
φύλλωμα που κυματίζει
Πάψτε
Πάψτε
να χτυπάτε τα χέρια
Αρπάξτε τον

Σώστε τον
Μες στο άσπρο σκάφανδρο
δεν θα του φτάσει ο αέρας
Θα σωριαστεί στο χώμα
και δεν θα φυτρώσει


Παναγιώτης Ιωαννίδης

# THIS IS NOT A DERVISH

## Taksim Square, Istanbul, 2013

Dark cypress-green velvet
with wide black canvas hem
billows unfurled as he turns and turns
arms open
to the sky and earth
in the standing crowd
clapping its hands
on the bitumen-laid
wound of the square
But on his face –
But the face
is neither saint's
nor man's
The face is a mask:
a monstrous duck
Eyes capped
the beak sewed up
sealed tight
–against the tear-gas–
White hospital rubber
–gazeless, breathless–
over the swelling
cypress foliage
Stop
Stop
your clapping
Seize him

Save him
In his white diving mask
the air will soon be spent
He'll sink to the ground
and won't spring up


Panayotis Ioannidis

# ΠΟΥ ΧΑΘΗΚΕ

Είδα έναν φαύνο
μες στον υπόγειο –
τα μάτια του ερεθισμένα
απ' το ηλεκτρικό φως
έβλεπαν σ' άλλα δάση
Τα πόδια του σφιγμένα μέσα
σ' ορειβατικά μποττάκια
κάρφωναν τις πλάκες
Το στέρνο τα πλευρά του
διαγράφονταν με βία
σε κάθε ανάσα –
αλαφιασμένο ζώο
Ξένος τόπος ήταν
– κι η πέτρα στο λαιμό του
άχρηστο φυλαχτό
Τέντωνε το κεφάλι κοίταζε
γύρω με δυσκολία
Όμως τα βλέφαρά του
βάσταγαν ακόμη
τη βαθιά σκιά

Παναγιώτης Ιωαννίδης

# LOST

I saw a faun
down in the Underground –
his eyes, irritated
by the electric lights
looked out to other forests
His feet, wedged inside
climbing boots
nailed down the slabs
His chest his ribs
were violently redrawn
with every breath –
panicking beast
This was an alien land
– and that stone round his neck
a useless amulet
He craned his head, he strained
to look around
His eyelids though
still held
the deepest shade

Panayotis Ioannidis
[*Translated by the author and Stefanos Bacigal*]

# ΑΠΟΚΑΘΗΛΩΣΗ

Εκ των υστέρων όλα εξηγούνται
Γιατί όλοι αποκοιμήθηκαν
γιατί ο φίλος δείλιασε
γιατί θα είναι πάντα ένας άγνωστος
αυτός που τρυφερά θα εκτελεί
την τελευταία χειρονομία
Το σώμα κατεβαίνει
βαρύ απόκοσμο
Τσακίζουνε τα χέρια για να το αποθέσουν
στο άστρωτο σεντόνι
*

Αντικρίζοντας έχανα το πρόσωπο
Η εικόνα παύει – η δύναμή της σκορπίζεται
Από τώρα κι ύστερα θραύσματα φαντασίας
μόνο
θα δοκιμάζουν να την ανασυνθέσουν
– μάταια με υλικά που θα απέχουν πάντα απ'
το να είναι
ακριβώς τα κατάλληλα
Η εικόνα σπάζει – όμως επιτέλους
το σώμα θα αναπαυθεί
το σώμα που ένιωσε πολύ αργά τι είχε
αναλάβει
για τι προοριζόταν σιωπηλά από την αρχή
Αφού αυτά έχουν ξανασυμβεί
Είναι απλώς μια επανάληψη
γνωστών κινήσεων αναπόφευκτης
Αλληλοδιαδοχής

Παναγιώτης Ιωαννίδης

# THE DEPOSITION

In retrospect it all gets explained
Everyone falling asleep
the friend acting the coward
the way it's always a stranger
who tenderly performs
the final gesture
The body's taken down,
heavy, other-worldly
They snap the arms to lay it out
on the pristine sheet
*

I'm looking but I can't see the face
The image stops working – its
strength disintegrates
Now only fragments of imagination
will try to recompose it
– in vain using materials that are far
from appropriate
The image breaks – but at last
the body can rest
the body that felt only too late
what it had undertaken
for what it was always silently
destined
since all this has happened before
It's just a repetition
of familiar movements in a fatal sequence

Panayotis Ioannidis
[translated by Clare Durey]

# ΜΑΧΗ Η ΦΥΓΗ

τσουκνίδα, αγιάγκαθο, άγρια ρόδα
κράτα κοντά
το πέρασμα στενεύει
το νερό δεν πίνεται
τα παπούτσια είναι μαγεμένα
και λιώνουν γρηγορότερα
τρέχεις με την ντουλάπα στην πλάτη
σε σταματάνε να σε ρωτήσουν
αν θες το κόκκινο ή το μπλε χάπι
πνιγηρή νυχτερινή ανατομία
λαθομανή σκιάχτρα
αποδιοπομπαία τρομοκρατία
σκοτεινά σημάδια
στους καρπούς των χεριών
και στα πλευρά
χάλκινο νόμισμα
κάτω απ' τη γλώσσα
κόκκινη προστασία απέναντι
στη μαύρη μαγεία
το καπνιστό κρέας του πολιτισμού
στριφογυρίζει στο έντερο
ρευστή καταλυτική πυρόλυση
ατμοπυρόλυση
καταλυτική αναμόρφωση
τεντώνεται η ραχοκοκαλιά
αδρεναλίνη
πόθοι και ανάγκες
αστερίσκοι βουλιαγμένοι
σε πετροχημικά

Κυόκο Κισίντα

# FIGHT OR FLIGHT

nettle, thistle, wild rose
keep by your side
the passage narrows
undrinkable water
bewitched pair of shoes
soles that wear out fast
you run
a closet on your back
do you want the red pill
or the blue one, they ask
stifling nocturnal anatomy
errorist scarecrows
terrorist scapegoats
obscure marks on
your wrists and ribs
a copper coin
underneath the tongue
red protection
against black magic
civilisation's smoked meat
twirls in the gut
fluid catalytic cracking
steam cracking
catalytic reforming
stretching of the spine
adrenaline
desires and demands
asterisks drowned
in petrochemicals

Kyoko Kishida

# ΜΠΑΣΟ-ΜΠΑΣΟ

τόσες συχνότητες
στο μονοπάτι σου
καμιά τους μπάσα

—

σοφές ατραποί
σε κάθε τους άκρια
ηχούν οι τίγρεις

—

μπάσα ξεσπάσματα
της νύχτας το βάθος
υποδαυλίζει

—

μαύρο ξυράφι
το ξόδι του πάνθηρα
κρύβει στο στόμα


Κυόκο Κισίντα

# BASHO-BASS

so many frequencies
all along your passageway
none of them bass

—

a few wise walkways
on every side of them
tigers echoing

—

bass overflows in
to the deepness of the night
it foments, inflames

—

a black razorblade
the panther's funeral
hides in the mouth


Kyoko Kishida

# ΛΩΤΟΦΑΓΟΙ

Αυτή η σιωπηλή κατανόηση
δε χωρά στις ντουζίνες χρόνια που μου
δόθηκαν.
Φόνευσε τη.
Δηλητηρίασέ τη σταδιακά
όπως μόλυνε κι αυτή εμάς
που πάντα την αποστρεφόμασταν.
Ποιός φύτεψε χαλίκια μέσα στα ζαχαρωτά;
Ποιός έδεσε αμόνια στα φτερά των γλάρων;
Γιατί χρησιμοποιούμε ακόμα το επίρρημα
<<στωικά>>;
Σεβασμό;
Σε ποιους;
Στους λωτοφάγους;


Κυόκο Κισίντα

# THE LOTUS EATERS

This unspoken understanding
doesn't fit into the dozens of years I have left.
Murder it.
Poison it piecemeal
just as it contaminated us-
we who had always turned away from here.
Who planted pebbles in the candy?
Who tied anvils to the seagulls' wings?
Why do we still use the adverb 'stoically'?
Respect?
For whom?
The lotus eaters?

Kyoko Kishida
*[Translated by Rachel Hadas]*

# ΑΦΟΡΙΣΜΟΣ
## (Η Προδοσία των Εικόνων 5)

Ο Κάφκα γράφει στους Αφορισμούς:
      *"Από τον πραγματικό αντίπαλο, κουράγιο δίχως τέλος ρέει μέσα σου".*
Αλλά τι
      είναι αυτό από το οποίο
προσπαθείς να δραπετεύσεις
      σε
αυτό τον σκοτεινό δρόμο;
Όλος αυτός ο
      παραπανίσιος χρόνος
θα σου
      αφαιρεθεί.
Τα ερείπια παραμένουν
στο ίδιο σημείο.

Η μηχανή των
      συναλλαγών μας
    αφήνει πίσω της
Ανταλλακτικά
        &
στραβά
Δόντια.

Μια αυτοπροσωπογραφία δεν είναι αναγκαίο να είναι ολοκληρωμένη για να αποκαλύψει το πρόσωπο: μόνο οι θεατές χρειάζεται να φαίνονται ολόκληροι ακόμη κι αν είναι από μέσα διάτρητοι.

                    Σαν τον Jude,
κανείς συνεχίζει να
επιθυμεί να εξαφανιστεί
        από τον κόσμο &
κανείς λανθασμένα μπερδεύει την πόλη
        με το άτομο.

Θοδωρής Χιώτης

# APHORISM
## (The Betrayal of Images V)

Kafka writes in the Aphorisms:
> *"From the true opponent, a limitless courage flows into you"*.

But what
       is it that
you seek to escape
       in
this dim road?
All of this
       additional time
will be taken
       from you.
The ruins linger
in the same spot.

The machine of our
       interactions
    leaves behind
Spare
       parts &
crooked
    Teeth.

A self-portrait does not have to be complete to show the face: only viewers need to appear whole even though inside they might be full of holes.

                Like Jude,

one continues to
wish oneself

              out of the world &

one mistakes the city
       for the person.

Theodoros Chiotis

# '21'

*"Ένα νησί δεν παύει να είναι ερημωμένο επειδή κατοικείται"*
(Gilles Deleuze)

Κάθε χάρτης
είναι η τεκμηρίωση ενός
Εξ
      ορ
   Κ
        ισμού:
φυτεύουμε πλοκάμια
στη γη των ματιών μας
και        μαθαίνουμε στους εαυτούς μας να
βλέπουμε      τα σύνορα
        από απόσταση ασφαλείας.

Το σώμα χαράζει μια διαδρομή που αρχικά φαίνεται άγνωστη, στη συνέχεια γνώριμη και μετά από καιρό απειλητική. Πρώτα γίνεται η εγγραφή του σώματος στον χώρο και μετά η μνήμη του χώρου αναδιπλώνεται και εγγράφεται σε κάθε σημείο του σώματος.

      Μασουλάμε τη ρητίνη που εκπέμπει το φως των φαναριών της πόλης και
φτύνουμε χάρτες:

περνάμε από τη μια νησίδα
στην άλλη μέχρι να
φύγουμε από την
πόλη και
τα πρόσωπά μας
αποκτούν ράμφη που ξηλώνουν
τις ραφές των πεζοδρομίων.

Το μεγάφωνο

στη μέση της πλατείας βραχυ

          Κυκλώνει:

  "Αισθήματά

      Μόδες        ηδονής

ΣΠΛΕΝΤΙΤ».
Το σώμα επιστρέφει εκεί
όπου παλιά
α
να
τρι
χι
α
Ζε

τα κόκκαλα τρίβονται
η ακτινογραφία αποκαλύπτει
ότι οι ώμοι έχουν γίνει
      χωράφια -

ποιος έβαλε την κόκκινη σαύρα
      που τυλίγεται
      γύρω από το μπράτσο
            του δέντρου
            που έχει γύρει
            προς τη θάλασσα

                    (αν εδώ συμπληρώναμε
                    ότι το δέντρο γέρνει
                    προς την θάλασσα
                    χωρίς ποτέ να την αγγίζει,
                    θα βλέπαμε
                    πόσο πολύ
                    αναζητούμε
                    το κλισέ - είναι και αυτό
                    μια επιστροφή)

      ο χρόνος
            ακρόλιθο της
Αμνηστίας.

Θοδωρής Χιώτης

Σημείωση: οι στίχοι 28-29 προέρχονται από το μυθιστόρημα Έξι νύχτες στην Ακρόπολη του Γιώργου Σεφέρη
(Αθήνα: Ερμής, 1998, σ. 17-18).

# '21'

"An island doesn't stop being deserted simply because it is inhabited."
(Gilles Deleuze)

Every map
is the evidence of an
Ex
     or
  C
       ism:
we sow tentacles
in the soil of our eyes
and        we teach ourselves to
observe      the borders
        from a safety distance.

The body traces a route that initially seems unfamiliar, then familiar and finally menacing. First, the body is inscribed on space and then the memory of space unfolds inscribing itself everywhere on the body.

      We chew on the resin emitted by the city stoplights and spit out
maps:

we jump from one islet
to another until we
leave the
city
and our faces
grow beaks that unpick
the seams of the pavements.

The loudspeaker
in the middle of the square short

                    Circuits:

  "Sentiments

        Fashions        of desire

SPLENDID".

The body returns to the point where
it used to
qu
i
ver

the bones are ground to dust
the x-ray reveals
the shoulders have become
        flatlands -

who was it that placed the crimson lizard
        wrapping itself
        around the arm
                of the tree
                which stoops
                towards the sea

                                                (if we noted here
                                            that the tree stoops
                                                towards the sea
                                    without ever touching it,
                                                we would see
                                                how earnestly
                                                we are looking
                            for a cliche - and that is in itself
                                                    a return)

        Time
        a spandrel
to amnesty.


Theodoros Chiotis

Note: the lines 28-29 come from the novel 'Six Nights at the Acropolis' by George Seferis
(Athens: Ermis, 1998, p. 17-18)

# ΑΤΡΑΚΤΟΣ (παραλλαγή)

Το σώμα                    αρχιπέλαγος

                            [ὀφθάλμοις δὲ μέλαις νύκτος ἄωρος . .]

νομίζουμε        ότι κοιμόμαστε

αλλά στα μάτια μας

         παραμονεύουν

οι ώρες

         η διχαλωτή γλώσσα      που φτιάχνει

    μια καμπύλη       που διεκδικεί

                  [τ]η[ν]        αναπνοή

    το σώμα ένας κύκλος

που                 γίνεται

μισοφέγγαρο και μετά

        χάν

            ε

     ται στο

          νερό.

                                  Νομίζουμε ότι κοιμόμαστε

                                    κρατάμε

         στα χέρια μας

         τα φύλλα και

         τις σαύρες

        που μαζεύονται στο κρεβάτι

         & τρέχουν πάνω κάτω

        στην σπονδυλική στήλη που

        μοιραζόμαστε με το αγόρι

      που τραβάει φωτογραφίες συνέχεια

         μπροστά

        από τις κάμερες του

     κλειστου κυκλωματος παρακολούθησης.

λες:

"είναι          ώρες τώρα που είμαι εδώ πάνω και δεν

το παίρνω απόφαση να πεθάνω",

τα δάκτυλα      ενωμένα με μεμβράνη

φτερά νυχτερίδας:
οι αρτηρίες πάλλονται
οι λέξεις εντοπίζουν πλέγματα.

Νομίζουμε ότι κοιμόμα στε

πέτρες που τις λιώνει
το αίμα που κολλάει
στα πέλματα.

Δεν γνωρίζουμε τα ονόματα
τα χέρια βυθίζονται στους χάρτες
οι οπές του προσώπου κα λύπτονται
από χώμα
το στόμα ανοιγοκλείνει
είμαστε σίγουροι ότι είμαστε άλλοι μα
όλοι ονειρευόμαστε
ότι είμαστε ο ίδιος
άνθρωπος: μοιραζόμαστε το ίδιο δέρμα:
ίσως έτσι να κάνουμε αυτό που θέλουμε
                              [με το σώμα του άλλου]:
μετά θα εξαφανίσουμε
το κουφάρι.

Νομίζουμε ότι κοιμόμαστε

κολλάμε τη μύτη μας πά νω στην οθόνη
που παίζει σε λούπα
ένα βιντεο από το κλειστό κύκλωμα τηλεόρασης: το ίδιο
πρόσωπο μπλε
σκούρο μπλε σχεδόν
μαύρο.

Θοδωρής Χιώτης
Σημείωση: το παράθεμα της Σαπφούς στο στίχο 2 προέρχεται από τον τόμο Greek Lyric, Volume I: Sappho and Alcaeus (Loeb Classical Library No. 142). Επιμ. David A. Campbell (Harvard University Press: Cambridge, MA, 1982).

# SPINDLE (variation)

The body                          an archipelago
                                  [and on the eyes

black sleep of night]
        we think              we are asleep
but behind our eyes
it is the hours that
        are lurking
        the forked tongue     draws
        a curve       competing
                              for     [a] breath
        the body a circle
which                   transforms into a
halfmoon before
                its im
                        mer
                sion in water.

                                        We think we are asleep
                                                we clutch
                                        in              our hands
                                                the leaves and
                                                the      lizards
                                        which gather on the bed
                                        running up and down
                                                the spine we
                                        share   with    the boy
                                        taking   pictures incessantly
                                                in front of
                                                the CCTV
                                                cameras.

you say:
"I have been up here for        hours      and
I have yet to make the decision to die",
the space between the fingers              is now a fleshy cobweb

            wings of a        bat:
the arteries are throbbing
                the words locate grids.


        We think we are asleep
        rocks eroded by
the blood sticking to
        the soles of the feet.


We do not know the names
        the hands sink inside the maps
        the holes on the face are covered
by soil
the mouth opens and then shuts


we proclaim we are unlike others but
when we dream,
        we are the same
person: we are sharing the same skin:
perhaps we might even be able to do what it is we want
                                    [using some one else's body]:
afterwards we shall simply toss
the carcass aside.


We think we are asleep


we stick our nose on the screen
playing on a loop a video
        from the CCTV: it is the same
face blue
        dark blue nearly
black.




Theodoros Chiotis
Note: The offset quotation in lines 2-3 is from Anne Carson,
If Not, Winter: Fragments of Sappho. Vintage Books: New York, 2002.

# A Poets' Agora

## 'GRAFT'

*"When I ask my mother about Saigon, she remembers the rhythmic swishing of straw brooms along the streets. I was eight when we left.*
*Perhaps anyone who says they will tell you a whole story is already in the midst of a fabrication. Maybe there are no whole stories unless we shape them as such, adding and subtracting accordingly to reflect another fabrication."*

**Adrianne Kalfopoulou**

Excerpt from *'Ruin: Essays in Exilic Living'*
Red Hen Press, 2014.

"Μην κάνετε" είπε "γάμους κι αρραβώνες,
γιορτές και γεννητούρια πια. Ήρθαν οι έσχατοι χρόνοι.
Πλυθείτε και ζωστείτε και προσμένετε
την μια και μεγάλη Γιορτή."
Μετά έφυγε κάπου προς τα πάνω.

Τότε όλοι παράτησαν μέσα στα παλαιά τείχη της πόλης
ονόματα και διακριτικά.
Έγινε ένας ψηλός σωρός
από χρυσές κανάτες, περικνημίδες, σκήπτρα και ζυγούς.
Βγήκανε στα λιβάδια σαν κοπάδι.

Τα άστρα γύρισαν πάνω απ' το κεφάλι τους
κάμποσους γύρους. Αυτοί τα είδαν "Οι ουρανοί"
είπαν "περιελίχθηκαν σα ρούχο κι άλλαξαν. Είμαστε νέο τάγμα."

Ο άνεμος σκόρπισε κάθε σοφία.

Ευθύγραμμα πορεύτηκαν κινώντας στο διάβα τους όρη.
Τα διέταζαν και τα 'ριχναν στη θάλασσα.
"Άνω και Κάτω" είπανε γελώντας "έχουν από καιρό καταργηθεί."

Ύστερα κάποιοι είδαν και πάλι την Ισχύ σε όνειρο.
Ήτανε κάτι κάθετο μπηγμένο στο λιβάδι.
Μ' ένα βραχίονα έδειχνε Ψηλά
και με τον άλλον Χαμηλά. Λες κι είχε σημασία.

Από το νέο τάγμα των Αρνιών
βγήκαν και πάλι παλιές φύτρες.
Σιωπηρά κινήθηκαν στα άνω λιβάδια της υπεροχής.
Έκλωσαν το μαλλί και το 'παν αργυρόλευκο.
Το γάλα τους ονόμασαν χιόνι απ' τις απάτητες κορφές.
(Τα όρη επέστρεψαν οριστικά στη θέση τους.)
Εξέθρεψαν, εξέθρεψαν το νέο-παλιό μόσχευμα
το 'βγαλαν με ωδίνες μέσα από τις λαγόνες τους, ένα λεπτοφυές αναρριχητικό
και το 'παν Genius.
Τότε οι λειμώνες φύτρωσαν μαρμάρινα μνημεία.
Πάνω τους κούρνιασαν όσοι είχανε το μπόλι
και κουμαντάρανε από ψηλά το ποίμνιο.
Μα αυτό συνέχισε να περπατά ηδονικά
ανάμεσα σε παπαρούνες κι άγρια μέντα,
Περιέλίχθη
σαν άσπρο κύμα γύρω απ' τα μνημεία .
Αφήνοντάς τα πίσω του,
κέρδισε πάλι τον ευθύγραμμό του δρόμο.


Λένια Ζαφειροπούλου

"Do not", he said, "make nuptials and betrothals,
feasts and birthings anymore. The end of time is nigh.
Go wash and girdle up and wait
for the one great Feast."
Then he was gone, upwards somewhere.

Then they all abandoned, inside the old city walls,
names and insignia.
A tall heap was made,
out of golden goblets and greaves, sceptres and scales.
Out they came unto the fields, like a herd of cattle.

The stars revolved above their heads
many times round. They saw them. "The skies",
they said, "are twisted like a garment and are changed.
We belong to a new order."

The wind dispersed all wisdom.

They marched away in a straight line, moving some mountains as they passed.
They would command and throw them into the sea.
"Above and Below", they chortled, "have long been made defunct."

Then, some saw Power in their dream again.
'Twas something perpendicular thrust deep into the meadow.
With one arm it pointed up High
and with the other down Below. As if it mattered.

Out of the new order of the Lambs
sprang the old budding family roots again.
In silence they moved towards the upper meadows of supremacy.
They spun their wool and called it silvery-white.
Their milk, they called it snow from peaks untrodden.
(The mountains did return to their original places, for good.)
They nurtured, they nurtured the new-old graft,
they pushed it out through their loins by the pangs of their labour,
a finely-grown creeping plant,
and called it Genius.

Then out of the meadows sprang some monuments of marble.
Upon them came to nestle those that did possess the seedling graft
and they would handle the flock from up high.
But it carried on its sensual walk
amongst the poppies and the wild mint.
It twisted
like a white wave round the monuments.
Leaving them behind,
it gained its linear course again.


Lenia Safiropoulou
[*translated by Orfeas Apergis*]

Όταν πρωτοανέβηκες στον άμβωνα της δυστυχίας,
άρχισες να κηρύττεις ασυγκράτητος,
μπροστά στα δακρυσμένα μάτια των θεατών.
Ελλείψει μαυροπίνακα
σημείωσες στο σώμα σου:
"Το Σθένος" είπες
"δεν είναι δα και τόσο άχρηστο νόμισμα, όλο και κάτι κερδίζεις μ' αυτό."
Τι κέρδισες λοιπόν πέρα απ' την ατελείωτη ορθοστασία
πάνω στον ίδιο ετούτο άμβωνα;
(Στο ένα γόνατο έπεφτες σπάνια και μοναχά
για να ελέγξεις τα κορδόνια στα αθλητικά παπούτσια σου.)
Έκανες και το λάθος στα τελευταία σου γενέθλια,
να εγείρεις πρόποση στους θεατές με κώνιο αντί σαμπάνιας.
Τα πόδια σου ήδη δε νιώθουν την κούραση, σε λίγο ούτε η μέση
Τα μάτια όλων έχουν στεγνώσει από καιρό.
Γλιστράνε τα βιτρώ στο δάπεδο, στους τοίχους σκαρφαλώνουν, πέφτει νύχτα .
Αν δεν τ' αποφασίσεις να γκρεμιστείς από 'κει πάνω,
θα γίνεις μόνιμο ανθρωπόμορφο
στολίδι του Καθεδρικού.
Και τα αγάλματα δεν έχουνε φωνή.


Λένια Ζαφειροπούλου

When you first climbed upon the altar of unhappiness,
you started preaching unrestrained,
before the tearing eyes of the spectators.
Lacking a blackboard
you jotted on your body:
"Fortitude", you said,
"is no mean currency; there's always something to be gained from it."
So what exactly did you gain, other than the endless standing
upon this very altar?
(You'd fall on one knee only rarely, only
to check the shoelaces of your sneakers weren't undone.)
Plus, you made the mistake, at your last birthday,
to propose a toast to the audience with hemlock instead of champagne.
Your feet already cannot feel any fatigue, in a short while your back won't either.
All eyes have dried up, since long ago.
The stained glass slips onto the floor, crawls up the walls, the night is falling.
If you don't make up your mind to come tumbling down from all the way up there,
you shall become some permanent human-like
ornamental fixture in the Cathedral.
And statues do not have a voice.


Lenia Safiropoulou
[*translated by Orfeas Apergis*]

Πότε άρχισε αυτή η φαντασίωση;

Πως θα βουτήξεις κάθετα
Στην ανοιχτή πληγή του χρόνου, χωρίς να κομματιάσεις το κρανίο σου,
Μέσα από βιολογική βρωμιά, σήψη και τη χυδαία ανθοφορία,
Θα φτιάξεις σήραγγα
Και θα περάσεις από τα σήματα ασυρμάτου, όλο το δρόμο πέρα ως πέρα,
Τους ποταμούς του αίματος, την κληρονομική πληροφορία ,
Χωρίς να πέσουνε ποτέ μπροστά σου οι μπάρες…
Κι η σήραγγά σου τάχα θα σε βγάλει ξανά στην επιφάνεια
Πίσω στην εποχή πριν απ' το ρήγμα , πριν απ' τη διαρροή αισθημάτων, πριν —
Και πώς θα ήταν δυνατόν να αναδυθείς από ένα έδαφος χωρίς κρατήρες; —

Και πως θα δεις το δέρμα άθικτο του χρόνου
Την παγερή αδιαφορία των βρεφών.
Σαν άμμο της ερήμου
Σαν στέππα απάτητη

ή σαν τη Γη πριν όλως δι' όλου πράξει.

Πριν πιάσει ν' αναδεύεται στο στρώμα της τις νύχτες
Ο μόνος άρρωστος του Σύμπαντος
Υπό τον κλινικό λαμπτήρα της Σελήνης.


Λένια Ζαφειροπούλου

Whenever did this fantasizing start?

How will you dive a vertical dive
into the open wound of time, without splintering your skull,
Through biological stench, rot and vulgar flowerings
Will make a tunnel
And will come through the wireless, all the way to the end,
Crossing the rivers of blood, the hereditary information,
Without toll-bars blocking your way…
And will your tunnel lead you, perchance, back to the surface
Back to the time before the breach, before the leaking feelings, before —
And how could you possibly surface out of some craterless ground? —

And how will you look upon the untouched skin of time
the icy indifference of infants.
Like the desert sand
Like the untrodden steppe

                                                    or like the Earth before it ever acted.


Before it started writhing on its mattress at night
The sole sick man of the Universe
Under the clinical light-bulb of the Moon.



Lenia Safiropoulou
[*translated by Orfeas Apergis*]

# ΟΜΟΙΟΣΤΑΣΗ

Στην επιφάνεια του καφέ, η Χαλκιδική
Ο περσικός στόλος συντρίβεται
Ο Άθωνας υψώνεται
– Μια δάφνη θροΐζει

Είναι ακόμα νωρίς

Στον καλό τόπο
οι ψαράδες σταματούν
και η άγκυρα
δαγκώνει το βυθό
Αυτοί ψαρεύουν
και τους βλέπουν τα βουνά
– Ο καφές θα μπει στο θερμός –
Κανείς δεν βλέπει
τον κόπο του βυθού
Τότε η άγκυρα
σπάει μια πέτρα


Χρήστος Σιορίκης

# HOMEOSTASIS

On the coffee's surface, Halkidiki
The Persian fleet is vanquished
Athos rises
– A bay tree rustles

It's early yet

In the good spot
the fishermen stop
The anchor
bites the seabed
And they fish
and are seen by the mountains
– The coffee will be put into the thermos –
Nobody notices
the effort of the seabed
Then the anchor
breaks a stone


Christos Siorikis
[*translated by Chris Sakellaridis*]

# CALLED BACK

Πριν επιστρέψω:
ν' αγγίξω
τις κουμαριές
εκεί που πυκνώνουν, στο σύνορο
να θυμάμαι τα χρώματα
και πώς πιάνουν
τα χέρια δυο λέξεις
ενώ κόβεις
τα κούμαρα
και νερό
δεν ζητάς
ο αέρας δροσίζει
κι ακόμη
δεν έχει
κρυώσει τη γη


Χρήστος Σιορίκης

# CALLED BACK

Before I go back:
let me touch
the arbutus shrubs
where they grow thicker, at the border
so I may remember the colours
and how hands
hold two words
as you pluck
the arbutus fruit
and do not ask
for water
the air growing cooler
hasn't yet
chilled the earth


Christos Siorikis
[*translated by Panayotis Ioannidis*]

# ΓΕΡΜΑΝΙΚΗ ΣΩΤΗΡΙΑ

Με βρήκε στο ποτάμι
που ξέπλενα το πόδι μου
από το δάγκωμα του λύκου
Δε μου 'βγαινε φωνή
όμως με είδε
Κι εγώ τον είδα
μα δεν ήξερα
πώς είναι η βοήθεια
και η πληγή
Και φώναξα
hallo! hallo!
Και επειδή
δεν έβλεπε το πόδι μου
από εκεί που ήταν
έκλαψα
και φώναξα
Rot! Rot! *

* «κόκκινο» στα γερμανικά

Χρήστος Σιορίκης

# GERMAN SALVATION

He found me at the river
rinsing my leg –
from the wolf's bite
No voice would come out of me
but still he saw me
I saw him too
yet didn't know
how to say help
or wound
And so I shouted
hallo! hallo!
And since
he couldn't see my leg
from where he stood
in tears
I shouted
Rot! Rot! *

* "red" in German

Christos Siorikis
[*translated by Panayotis Ioannidis*]

# SENTENCE

Order a beginning
Unbidden, sometimes bundled
A gift of bark
Compost —
Compose your particular
Possibility — citizen without
Passport, unless bartered —
The tongue
A grammar bridge
Shelter despite the storm-torn body of it
Call it love
Call it graft
This transplant of the unlikely
Care given
A meal
You used to count on
This is space
In the midst of rot
Burgeoning the lineage
Each seedling
Where
The line breaks
New tissue

Adrianne Kalfopoulou

# ΑΠΟΦΑΣΗ

Όρισε μια αρχή
Αυτόκλητη, ίσως τυλιγμένη
Το κομμάτι ενός φλοιού σαν δώρο
Λίπασμα-
Σύνθεσε το μοναδικό σου
ενδεχόμενο- πολίτης χωρίς
διαβατήριο, πλην της διαπραγμάτευσης-
Η γλώσσα
Μια γραμματική γέφυρα
Καταφύγιο παρά το καραβοτσακισμένο
σώμα της
Πες το αγάπη
Πες το μπόλι
Αυτό το μόσχευμα του απίθανου
Φροντίδα που δίνεται
Ένα γεύμα
Στο οποίο βασιζόσουν
Αυτό είναι χώρος
Καταμεσής της σήψης
Βλασταίνει η κληρονομιά
Κάθε φιντάνι
Εκεί
Που σπάει ο στίχος
Καινούργιο δέρμα


Αδριάννα Καλφοπούλου
[translated by Katerina Iliopoulou]

# FALL GRAPES

We didn't know the acrid scent of trodden grapes
stewing in their ferment. We mistook the flushed skins

for sweet juices, bit the thick-fleshed fruit,
learned these clusters were meant for the barrels,

ready to be mashed: pulp and stems, seeds, stray leaves
churned to sift the liquid out of bitterness, what we did

in love without admitting the skin we licked
along each other's necks was mustos, the taste of

what our bodies could not change, what others turn into wine.


Adrianne Kalfopoulou

# ΦΘΙΝΟΠΩΡΙΝΑ ΣΤΑΦΥΛΙΑ

Δεν διακρίναμε την στυφή ευωδία του πατημένου σταφυλιού
που σιγοβράζει στη ζύμωσή του. Παρεξηγήσαμε τη ξαναμμένη φλούδα

για τους γλυκούς χυμούς, δαγκώναμε του φρούτου την πυκνή σάρκα,
μαθαίνοντας πως οι χυμοί αυτοί είχαν γραφτεί για τα βαρέλια,

έτοιμοι να συνθλιβούν: πολτός και τσαμπιά, κουκούτσια, φύλλα ορφανά
αγκαλιασμένα σφιχτά να σουρώσουν τη πικράδα, αυτό που κάναμε

ερωτευμένοι δίχως την ομολογία πως το δέρμα που γλύψαμε
ο ένας στου άλλου το λαιμό ήταν ο μούστος, η γεύση

που τα κορμιά μας δεν μπόρεσαν ν αλλάξουν, και άλλοι το κάνουνε κρασί.


Αδριάννα Καλφοπούλου
[*translated by Korina Gougouli*]

# THE HISTORY OF TOO MUCH

There is too much here, the sapphire, the thistle,
the oregano blooms in June, everything extravagant –
the rich peat of what decays, the ruins that don't decay,
these especially are too much, the temples and statues
in their stark marble glow, that simplicity which is not simple at all.
This sheen of time, the wear of wars, the famine years
of Occupation, lucent as the columns standing stoic, Doric –
their weight has whittled the people: the weight of that antiquity,
of those stones, the grandeur and pride – too much
in this moment, this present crushed by the evidence,
the result of living with beheaded gods, and maimed still
beautiful torsos, the muscled limbs in chipped robes.
They plague our dreams, what was once achieved is now
incomplete, these pieces of the golden age aging
in the midst of traffic, too much, the yelling and honking,
the protests in the middle of everything – people are impatient;
how can anyone be patient, overwhelmed as they are.
Even the oregano's thick perfume, the sapphire sea, remind people
of extravagant loves and sacrifice, while here, now,
ghosts live on as gods and their impossibility.


Adrianne Kalfopoulou

# Η ΙΣΤΟΡΙΑ ΤΗΣ ΥΠΕΡΒΟΛΗΣ

Όλα είναι πάρα πολλά εδώ, το ζαφείρι του νερού, τα αγκάθια,
η ρίγανη ανθίζει τον Ιούνιο, όλα υπερβολικά-
η πλούσια τύρφη της παρακμής, τα ερείπια που δεν παρακμάζουν,
αυτά ειδικά είναι πάρα πολλά, οι ναοί και τα αγάλματα
μες στην άψογη μαρμάρινη λάμψη τους, αυτή η απλότητα που δεν είναι απλή.
Η στιλπνότητα του χρόνου, η φθορά των πολέμων, τα χρόνια της πείνας
στην Κατοχή, λαμπερά σαν τις κολόνες που στέκουν στωικές, Δωρικές-
το βάρος τους έχει λυγίσει τους ανθρώπους, το βάρος της αρχαιότητας,
της πέτρας, το μεγαλείο και η περηφάνεια – υπερβολικά
αυτή τη στιγμή, μέσα στο συντετριμμένο από τα τεκμήρια παρόν,
το αποτέλεσμα του να ζεις με αποκεφαλισμένους θεούς, και σακατεμένους
ωραίους κορμούς, μυώδη μέλη μέσα σε σπασμένους χιτώνες.
Κατατρύχουν τα όνειρά μας, εκείνο που κάποτε κατορθώθηκε είναι τώρα
λειψό, αυτά τα θραύσματα της χρυσής εποχής γερνούν
μέσα στην κίνηση των δρόμων, όλα τα υπερβολικά, οι φωνές και οι κόρνες,
οι διαδηλώσεις – οι άνθρωποι ανυπομονούν
πως γίνεται ενώ κατακλύζεσαι από όλα αυτά να κάνεις υπομονή.
Ακόμα και η πυκνή οσμή της ρίγανης, η ζαφειρένια θάλασσα, μας θυμίζουν
ακραίους έρωτες, και θυσίες, ενώ τώρα, εδώ,
τα φαντάσματα συνεχίζουν να ζουν αδιέξοδα, όπως οι θεοί.


Αδριάννα Καλφοπούλου
[translated by Katerina Iliopoulou]

# A Poets' Agora

# 'VERGE'

"(The nothing rising underfoot). Then later
The high-dive at the pool, the tree-house perch,
Ferris wheels, balconies, cliffs, a penthouse view,
The merest thought of airplanes. You can call
It a fear of heights, a horror of the deep;
But it isn't the unfathomable fall
That makes me giddy, makes my stomach lurch,
It's that the ledge itself invents the leap."

**A. E. Stallings**

Excerpt from *Fear of Happiness*, Poetry magazine, Vol.195, No.6, March 2010, Chicago

# SANCTUARY OF ARTEMIS AT BRAURON
## (A Prayer for Daughters) For Myrto and Atalanta

*τάς τε κόρας, Λιμνᾶτι, κόρᾳ κόρα, ὡς ἐπιεικές*

Day-trip from Athens, on a day
Too fine for anyone to stay

Within the walls—and so we splurge
On time and gas, to river's verge,

Where columns, reconstructed, stand
On tiptoes on the boggy land:

Cracked capitals—the nests of sparrows,
That, newly fledged, shoot forth like arrows

Into the blue—hold mostly pure
Empyrean entablature;

The stoa sometimes seems to stretch
Both upwards, and below the vetch,

In pools by wispy cirrus troubled
In which their fluted drums are doubled.

No one stands guard, or catalogues
The visitors, but belching frogs,

And from tall reeds we hear the words,
Though untranslatable, of birds.

The girls, our daughters, on the verge
Of growing up, up hillsides surge

In search of—climbing, pinecones, flowers,
Footholds, skinned knees, superpowers?

They're near the age when other girls
Dedicated severed curls,

A tambourine or headband, ball
And brazen mirror, favorite doll,

And sometimes even jeweled rings,
Leaving behind their childish things

For Artemis, who never crosses
Age's sill of gains and losses.

This realm of Artemis, this pool,
This temple, is just vestibule:

No mortal stays forever there,
But passes through, as "little bear."

It is a simile that rubs
Both ways—see how they climb like cubs,

And how, on two feet, nothing wild
So much resembles a girl child!

Here in this sanctuary, here
Dropped in this spring that still springs clear,

Archaeologists have found
Bronze mirrors, toys, and jewelry drowned,

Where now, our muddy-footed daughters
Poke with sticks the tad-poled waters

Here Mneso offers, may you bless,
The votive of this frog-green dress.

The moment that seems, like the spring,
From stillness sprung, is on the wing,

As dragonflies—which are instead
Of jaded green, carnelian red

And hang like ornaments that stopped
Mid-air the instant they were dropped

Into the pool—seem in no hurry,
And yet their beating wings are blurry

With all the work of staying still,
As water weeps and flows downhill.

The two friends want of course to stay
A little longer. Call it play,

This state of being, outside time,
When it is not yet work to climb.

Goddess of girlhood, hear my prayer
For her, and my own, little bear:

Lady of wilderness, grant that she
May dwell here long and happily

Before she leaves these hills for good
And crosses into womanhood,

(That busy city, where we go
With fretful list and task in tow),

Leave in her something else, unnamed,
Untrammeled, liminal, untamed.


A.E. Stallings

# ΙΕΡΟΝ ΤΗΣ ΑΡΤΕΜΙΔΟΣ ΕΝ ΒΡΑΥΡΩΝΙ
## (Προσευχή για Κόρες) Για τη Μυρτώ και την Αταλάντη

*τάς τε κόρας, Λιμνᾶτι, κόρα κόρα, ὡς ἐπιεικές[1]*

Απ' την Αθήνα ημερήσια εκδρομή –
Πώς κανείς να μην εκμεταλλευτεί

Τέτοιον καιρό; – οπότε εμείς, οι εντός των τειχών, σκορπάμε
Χρόνο και καύσιμα μέχρι να πάμε

Στου ποταμού το χείλος, όπου κίονες αναστυλωμένοι
Ακροπατούν πάνω στη γη που είναι απ' το έλος ποτισμένη:

Σπασμένα κιονόκρανα –φωλιές για το σπουργίτι,
Που, νεοσσό, πετάγεται σα βέλος απ' το σπίτι

Ως μες στο κυανό στερέωμα– απάνω τους στηρίζεται όλος,
Σαν ένα αέτωμα, ο καθαρός παραδεισένιος θόλος·

Μοιάζει ν' απλώνεται η στοά εν τω αρχαίω οίκω
άνω και κάτω απ' το λαθούρι και το βίκο,

Καθώς οι ραβδωτές κολώνες καθρεφτίζονται διπλά
Μες στα στεκάμενα νερά, που ενοχλούνται από τα ελικόσχημα φυτά.

---

[1] Ανωνύμου: στίχος από το επίγραμμα 280 του βιβλίου VI της Παλατινής Ανθολογίας. Το πλήρες επίγραμμα, σε πρόχειρη πεζή δική μου μετάφραση, έχει ως εξής (με πλάγιους χαρακτήρες επισημαίνω τα μέρη του στίχου που χρησιμοποιεί η Stallings ως προμετωπίδα του ποιήματός της): «Η Τιμαρέτη, κόρη του Τιμάρετου, πριν απ' τον γάμο της, αφιέρωσε σ' εσένα, θεά Άρτεμη της Λίμνης [Σ.τ.Μ. εννοεί την Βραυρωνία Αρτέμιδα], τα τύμπανά της και την αγαπημένη της μπάλα και το διχτάκι που της κράταγε τα μαλλιά, και τις κούκλες της με τα φορέματά τους, δώρο από κόρη παρθένο σε παρθένο, ως αρμόζει [Σ.τ.Μ. εννοεί από την μικρή κοπέλα στην παρθένο Άρτεμη]. Όμως τώρα, κόρη της Λητώς, καλύπτοντάς την με το χέρι σου, κράτησε αυτή αγνή με αγνότητα».

Κανείς δεν τα φρουρεί αυτά, ούτε εξετάζει
Τους επισκέπτες, παρά μόνο το βατράχι που κοάζει,

Και μέσ' απ' τα καλάμια ακούγονται λογιών-λογιών
Κουβέντες αμετάφραστες πουλιών.

Τα δυο κορίτσια, οι κόρες μας, στο χείλος της ενήλικης ζωής,
Στις παρειές των λόφων σκαρφαλώνουν μεθ' ορμής,

Εις άγραν – αναρρίχησης, κουκουναριών, ανθών,
Και πατημάτων και γονάτων πληγωμένων και δυνάμεως θεών;

Είναι κοντά στην ηλικία μ' εκείνα τα κορίτσια
Που αφήναν πίσω τους τα παιδικά καπρίτσια,

Για τη θεά Αρτέμιδα: ένα τουμπελέκι
Ή μια ταινία για τα μαλλιά, ή όποιο άλλο τσουμπλέκι,

Μπάλες, διαμαντικά, καθρέφτες ή και κούκλες,
Μαζί με τις κομμένες παιδικές τους μπούκλες,

Όλα αφιερωμένα στην Αρτέμιδα, που δε διαβαίνει
Ποτέ της ηλικίας το κατώφλι, ούτε κέρδη και ζημιές προσμένει.

Ετούτη η επικράτεια της Άρτεμης, ετούτα τα νερά
Κι ετούτος ο ναός δεν είναι άλλο τίποτα παρά

Προθάλαμος: κανείς θνητός δε μένει εδώ για πάντα αυθυπάρκτως,
Παρά διασχίζει και περνά σαν αρκουδάκι – «άρκτος»[2]

---

[2]    Πάπυρος: «Στον ναό αυτόν της Αρτέμιδος (Βραυρώνιον) […] έμεναν κάποια κοριτσάκια ηλικίας 5 με 10 ετών, προσφέροντας υπηρεσίες στη θεά. Ήταν εκεί για ένα ορισμένο χρονικό διάστημα. Ονομάζονταν τα κορίτσια αυτά, χαρακτηριστικά, «άρκτοι», δηλαδή αρκούδες. Ο λόγος που έμεναν εκεί ήταν ότι τα είχαν υποσχεθεί οι μανάδες τους πριν γεννηθούνε στην Άρτεμη, για να βγουν γερά. Ως εκ τούτου εκείνη ήταν προστάτιδα των λεχώνων και των επιτόκων. Όταν τα κορίτσια αυτά μεγάλωναν πιστευόταν ότι είχαν τη βοήθεια της θεάς σαν γίνονταν σύζυγοι». [Σ.τ.Μ., με κάποιες τροποποιήσεις]

Είναι μια παρομοίωση που λειτουργεί κι από τις δυό –
Δες το παιδί πώς σκαρφαλώνει σαν αρκούδι μοναχό,

Και πώς, όταν ορθώνεται στα δυό,
Κανένα ζώο άγριο δεν μοιάζει τόσο με ανθρώπινο μικρό!

Εδώ σ' αυτό το ιερό εδώ,
Μες στο νερό που απ' την πηγή πετιέται απάνω καθαρό,

Οι αρχαιολόγοι έχουν βρει
Καθρέφτες και παιχνίδια και κοσμήματα που 'χουν πνιγεί,

Εδώ που οι κόρες μας με πόδια λασπωμένα
Σκαλίζουν με τα ξύλα τους νερά βατραχωμένα,

Εδώ η Μνησώ προσφέρει, για να ευλογηθεί,
Το ανάθημα ρούχου βατραχοπράσινου που έχει φορεθεί.

Εδώ η στιγμή, που μοιάζει να ορίζεται
Απ' την πηγή, μέσα στην άκρα ηρεμία μετεωρίζεται,

Καθώς οι λιμπελλούλες –που αντί
Για πράσινο του αχάτη είναι καρνεόλιο ροδακί

Και αιωρούνται σαν στολίδια που έχουν σταματήσει
Εν τω μέσω του αέρα, τη στιγμή που τα 'χεις ρίξει

Μες στα νερά– δεν μοιάζουν να ζορίζονται
Παρ' ότι τα φτερούλια τους στριφογυρίζονται

Μ' εντατικό ρυθμό που αχνά τα ορίζει,
Καθώς το νερό τρέχει αργά-αργά σα να δακρύζει.

Οι δύο φίλες θέλουν βέβαια να μείνουν
Λίγο ακόμα, πιο πολύ να παρατείνουν

Πες το παιχνίδι, αυτόν τον τρόπο να υπάρχεις εκτός χρόνου,
Όταν ακόμα σκαρφαλώνεις άνευ πόνου.

Θεά της κοριτσίστικης ζωής άκου την προσευχή μου
Για εκείνη, τη μικρούλα άρκτο, τη δική μου:

Θεά της άγριας φύσης δώσε της να κατοικήσει
Εδώ χρόνο πολύ και να ευτυχήσει
Προτού ετούτες τις πλαγιές ν' αφήσει,
Της γυναικείας φύσης το κατώφλι να διασχίσει,

(Αυτής της πόλης όπου όλες πάμε
Κι απ' τις πολλές δουλειές βαρυγκωμάμε),

Άφησε μέσα της κάτι μικρό που να μην ονομάζεται,
Κάτι ανεμπόδιστο και οριακό – να μη δαμάζεται.


Αλίσια Ε. Στόλινγκς
*[μετάφραση και σημειώσεις στα ελληνικά: Ορφέας Απέργης]*

# SUNSET, WINGS

Crows descry the sky,
desecrate the cyanic,
scrying and crying.
Swallows, I swear, not
swifts; but swift--swoop, swivel--whose
scissored silhouettes,
belated, become
a quibble of pipistrelles,
tippling acrobats.
Who haunts the hill? Lo,
one-note woe: oh well, twilight
throws in the towel.


A.E. Stallings

# ΔΥΣΗ, ΦΤΕΡΑ

Κόρακες κόβουν τον ουρανό,
ξεσκίζουν το κυανό,
κρώζοντας, κλαίγοντας.
Χελιδόνια, λέω, δες, όχι
σταχτάρες· γοργά –στρίβουν, χυμούν–
περιγράμματα ψαλιδιστά,
γίνονται τελικά
ένσταση νυχτερίδων,
ακροβάτες μπεκρήδες.
Ποιός στοιχειώνει τη χώρα; Να,
νότα μία βογγά: πάει, η δύση
παραδίδεται.


Αλίσια Ε. Στόλινγκς
[μετάφραση: Παναγιώτης Ιωαννίδης]

# ADVICE TO PSYCHE

## Hem Audax et Temeraria Lucerna

Resist the ugly sisters and their wisdom.
They cannot hold a candle to your flame.
Keep your word so there will be no story;
A happy ending ends it just the same.

Be blind, uncurious. Resist the urge
To look upon Desire in the light—
Not lest it turn out monstrous, and dark—
But singed, and beautiful, and winged for flight.


A.E. Stallings

# ΣΥΜΒΟΥΛΕΣ ΣΤΗΝ ΨΥΧΗ

## Hem Audax et Temeraria Lucerna

Στις αδελφές τις δύσμορφες και στην σοφία τους αντιστάσου.
Δεν φτάνουν ούτε το μικρό σου δαχτυλάκι.
Κράτα τον λόγο σου, ιστορία να μην υπάρξει.
Το χάππυ εντ, είναι τέλος κι αυτό, όπως και να 'χει.

Τυφλή, με δίχως περιέργεια, μη μπεις στον πειρασμό
Να κοιτάξεις τον Πόθο μες στο φως—
Όχι μήπως αποδειχθεί πως είναι τερατώδης, σκοτεινός—
Αλλά τσουρουφλισμένος, ωραίος, και φτερωτός.


Αλίσια Ε. Στόλινγκς
[μετάφραση: Παναγιώτης Ιωαννίδης]

# Η ΑΤΛΑΝΤΙΚΗ ΚΑΤΑΣΤΑΣΗ ΜΟΥ

*Θαλασσοψιττακέ, του είπα,*
ποντοπόρε, πήγαινες ως την αρκτική, καημένε μου, κι αν υπερωκεάνιος, το
μόνο σου αυγό διεκδικούσες. Αυλαία ο βράχος, πληγωμένη – και να 'θελε
να γίνει κιμωλία, και να 'βγαζε απάνω στο σκαρφάλωμα πνιχτή βραχνή
φωνή. Θα διασχίζεις τα κύματα, θα φτάνεις στον ουρανό του τράγου και
στην αβάπτιστη φωλιά του τοκετού.

Γιατί κι εγώ περπάτησα ως εκεί και είδα μιαν ακρόπολη στο χείλος.
Και, σοβαρά, δεν είχα τι να κάνω, ανέσυρα της μιας προϊστορίας τη
φοβερή ανάπαυλα, ηλιόλουστος στο μέσο της ανάσας. Το πέλμα μου σαν
νόημα πετά, μέχρι τα δυτικά, κι αποτυπώνεται ασθμαίνον κι ιδρωμένο.

*Το γένος της χαράδρας, του ψιθύρισα, να περπατάς καμαρωτό κι*
*ύστερα να βυθίζεσαι στη στάλα των αφρών. Φωλιάζεις σε σχισμές, μικρή*
*μου καλογραία, και σκάβεις τα λαγούμια σου σ' ολισθηρές πλαγιές. Θα*
*επιστρέφεις, θα ψάχνεις το ζευγάρωμα του χρόνου, θα ίπτασαι σε κύκλους*
*και την άνοιξη στην ξηροτέρα γη και την πατροπαράδοτη.*

Όδευα προς τη θάλασσα κι εγώ, αφήνοντας την κούρνια μου –
μεταίχμιο προτού ν' αποδημήσω. Ήταν που μόλις είχα πτερωθεί και δεν θα
ξαναγύριζα για χρόνια, καλοθρεμμένος και χορτάτος νεοσσός. Μια νύχτα
βγήκα, φάνηκα, περπάτησα γι' αρχή, όλο και πιο ταχέως. Έτρεξα, μα δεν
ήθελα και ν' απογειωθώ. Έφτασα στο νερό και, ως να ξημερώσει, έλαμνα
δίχως κόπο και διάλειμμα. Μακριά από την ακτή, δεν θα 'ψα χνα το είδος
μου για να συναθροιστώ. Να λείπω ήθελα.

*Λουσμένος απ' το χώμα και κυρίαρχος της πόρτας σου φρουρός,*
*του υπενθυμίζω, ενώ εκείνη ξεραίνει το χορτάρι μες στις σήραγγες, τη*
*λάσπη σκόνη θρύμματα να κάνει. Αχ, το ατλαντικό μου απολίθωμα, πόσο*
*νερό χρειάζεσαι κρυψώνα;*

Κι εγώ, πώς έτσι κάμφθηκα; Πώς από νέος ήμουν πιο θαμπός,
τα πιο χλωμά μπαλώματα φορούσα; Ο ρυπαρός ευάλωτος, εισπνέοντας
κινδύνους – και πού να βρω σαράντα ψάρια την ημέρα; Κι αν

ερωτοτροπούσα με το ράμφος μου, κι αν πάνω στο νερό αποκοιμιόμουν, κι αν γνώριζα πως θα γεννήσουμε αυτό που θα ζυγίζει ακόμη πιο πολύ κι από το βάρος μας, κι αν έπρεπε βδομάδες να περάσουν, ώσπου να γίνει η όλη ζέστη μας ρωγμή, και πάλι θα θυσίαζα την πτήση, να 'μαι κολυμβητής. Και πάλι θα συμβίβαζα το πτέρωμα, να 'μαι πελαγικός, να 'ναι τα δάχτυλά μου σε μεμβράνες.

*Θαλασσοψιττακέ, του είπα,*
*ποντοπόρε, εσύ πάντα το γνώριζες ότι δεν έχει τέρμα , πως όλο περαιτέρω*
*θα διαβαίνουμε. Απ' όσα διαμείφθηκαν εδώ κι απ' όσα υπαινίχθηκε ο*
*ορίζοντας, τι άλλο διδαχθήκαμε; Θέα στο άκρον άωτον του κόσμου κι ό,τι*
*τυχόν ξανάμαθα, μιλώντας στα γκρεμνά.*


Γιάννης Δούκας

# MY ATLANTIC STATE

I said to him, *Sea-Parrot,*
*navigator, up to the arctic you were going, poor you, transoceanic though*
*you were, your only egg is what you claimed. The rock, a wounded curtain*
*– and would it want to be a chalk and make a croaky muffled sound upon*
*the climbing. You'll cross the waves and you will reach the heavens of the*
*goat and the unchristened nest of childbirth.*

Because I walked up there myself and, right on the verge, I saw an
acropolis. And, seriously, I had nothing to do, I just produced the terrible
respite of a prehistory, sunshiny in the middle of a breath. My foot flies like
a meaning to the west and gets imprinted gasping and all-sweaty.

*The genus of the gorge, I whispered, in pride you walk and then*
*you sink into the drop of foam. You lair in slits, my little nun, and dig your*
*holes on slippery slopes. You will be coming back and looking f or the mating*
*of the year, you'll fly in circles in the springtime on dried and ancestral*
*Land.*

And I was headed to the sea myself, leaving behind my roost – the
verge before migration. Because I had grown feathers and I would not
come back for years, well-fed, full-grown already. And I went out one night,
I showed myself, I walked in the beginning, faster, faster. I ran, but I would
not take off. I reached the water and I paddled until dawn, without fatigue
or lull. And off the coast, I wouldn't look for congregation with my kind. I
wanted to be absent.

*Bathed by the soil and the guarding master of your door, I remind*
*him, while she is drying the grass inside the tunnels, she makes dust out of*
*mud and into crumbles. Oh, my Atlantic fossil, how much water do you*
*need to hide?*

And me, how was I bent like that? How had I been the dimmest
from so young, and did I wear the palest patches? Squalid fragile, inhaling
dangers – and where to find my forty fish a day? And if I flirted with my

beak and if I fell asleep upon the waters and if I knew that we'd give birth
to something weighing more than us and if we had to spend whole weeks
before our warmth became a crack, still I would sacrifice my flight, to be
a swimmer. Still, I would compromise the plumage, to be pelagic and have
my fingers in membranes.

    I said to him, *Sea-Parrot,*
*navigator, you always knew that there's no ending, that we'll forever keep*
*on going forward. From everything that has been said, from what the horizon*
*insinuated, what else have we been taught? The view to the world's far*
*end, and what I happened to relearn by talking to the cliffs.*

Yiannis Doukas

# Η ΣΑΠΦΩ ΝΟΙΚΟΚΥΡΑ

Πολιορκώντας από μέσα ένα τριάρι
Πρωτοβουλία για παράθυρο θα πάρει
Και με τις γλάστρες στη βεράντα της θα ζει
Το μεσημέρι πάντα θα την ξεσηκώνει
Θα νιώθει κάπως ξεχασμένη, κάπως μόνη
Ώσπου στο τέλος θα καλμάρει κάθε αυγή

Θα δώσει χώρο στο παράλογο
Κι όταν ανοίξει τον διάλογο
Με τ' άπειρο
Πάνω σε πάπυρο
Θ' αφήσει τ' όνομά της
Κληρονομιά της

Βροχή, θα σκέφτεται, και πολυκατοικία
Και θα βυθίζεται σε μια μελαγχολία
Σαββατοκύριακο, τσιγάρο και καφές
Σε μια παρένθεση που θα χωράει τόσα
Όσα δεν ήξερε ποτέ να πει η γλώσσα
Κι όσα δεν μπόρεσαν να κρύψουν οι βαφές


Γιάννης Δούκας

# SAPPHO THE HOUSEWIFE

She besieges her apartment from within
Aiming at a window with a view
Living among the potted plants on the veranda
She gets wound up around noon
Feeling forgotten and forlorn
Until at sunrise she calms down

She is conversing with infinity
Giving space to the absurd
And leaves her name behind
On parchments of a shopping list

She sinks into a weekend depression
Considering the rain and condominiums
With coffee and cigarettes and anything
That language cannot say or hair dye will hide
Looks to the meaning in the bracket between
Two items on the parchments of a shopping list


Yiannis Doukas
*[Translated by the poet himself, with help from Sam Buchan-Watts]*

# ΕΠΙΤΑΦΙΟΣ

*...γιατί τ' αγάλματα δεν είναι πια συντρίμμια,*
*είμαστε εμείς.*
ΓΙΩΡΓΟΣ ΣΕΦΕΡΗΣ

Μετά δε ταύτα σβήσαμε τα φώτα,
έγειρες πάνω μου απαλά και είπες:
«θα ζήσουμε μπαλώνοντας τις τρύπες
της ιστορίας· τίποτα όπως πρώτα

δεν θα μπορέσει πια να ξαναγίνει».
Τουλάχιστον, εκείνο τ' «όπως πρώτα»,
καθώς το καταργούν τα γεγονότα,
κι εμείς να τ' αρνηθούμε. Τι θα μείνει

σ' αυτήν τη γη, σε χρόνο ενεστώτα
απ' όλο το αναμάσημα που λίγη
ανάσα έχει ακόμη και μας πνίγει,
την ώρα που του στρέφουμε τα νώτα;

Στεφάνους καταθέτουμε και κλαίμε,
μα είμαστε ό,τι θάβουμε, ό,τι καίμε.

Γιάννης Δούκας

# EPITAPH

*...because the statues are no longer debris,*
*we are.*
GEORGE SEFERIS

We turned off the lights after all of this,
you gently leaned on me and said:
"We shall live stitching with a thread
the holes of history; whatever was before this

can never be restored".
At least, that "before",
as events negate it forevermore,
we should reject that too. What is to be stored

on this earth, in the present
out of all this rumination
whose remaining breath leads us to our obliteration
at the same time we turn our backs to what is
Everpresent?

We lay wreathes and weep,
but we are what we burn, we are what we bury deep.


Yiannis Doukas
[*translated by Theodoros Chiotis*]

Από τότε που άρχισε να με απασχολεί η υπόθεσή του, βάζω το ραδιόφωνο να παίζει σε μέτρια ένταση για να καταφέρω να κοιμηθώ. Ξυπνώ συνήθως στο άκουσμα μιας είδησης. Μου είναι δύσκολο να ανακαλέσω αν ήταν πρωί ή μεσημέρι, πάντως σίγουρα δεν επρόκειτο για απογευματινό δελτίο, γιατί η εκφωνήτρια, που τη φωνή της άκουγα για πρώτη φορά, ανακοίνωνε πως, πριν λίγη ώρα, κάτω από τον δυνατό ήλιο, σ' ένα από

τα φωτεινότερα σημεία της πόλης, τρεις ερευνητές, που απέφυγαν να δώσουν περισσότερα στοιχεία για τη φύση και την πορεία της έρευνάς τους, ντυμένοι με πολυχρωματικά κουστούμια και μακιγιαρισμένοι με κηλίδες μελάνης ρόρσαχ, αναπαρέστησαν, με μόνο μέσο τη σύνθεση και τις δυνατότητες των σωμάτων τους, ένα μέρος των αφηγήσεων για την περίπτωση του Ιβάν Ισμαήλοβιτς, όπως παρουσιάστηκε στον ημερήσιο τύπο. Πετάχτηκα από το κρεβάτι με την βεβαιότητα πως ο σκοπός τους

δεν ήταν διαφορετικός από τον δικό μου –να καταλάβουν, δηλαδή, τι συμβαίνει. Εκτίμησα πως θα γινόταν ατύχημα αν έπαιρνα το ποδήλατό μου για να φτάσω γρηγορότερα στο σημείο της αναπαράστασης, και έτσι το έκοψα με τα πόδια. Πολύ σύντομα , χρειάστηκε να γονατίσω στο πεζοδρόμιο κλείνοντας τα μάτια. Αδύνατο να υπολογίσω πόσος χρόνος πέρασε –βρισκόμουν από ώρες, πριν ακόμη ξυπνήσω, σε εκείνο που είναι το χείλος. Ευτυχώς, κανείς δεν προθυμοποιήθηκε να με τραβήξει από το έδαφος - η δοκιμασμένη στάση παραπλανά πως πρόκειται για προσευχή.

Όταν σηκώθηκα και έφτασα, σέρνοντας βιαστικά τα πόδια μου, στο σημείο

της αναπαράστασης, δεν συνάντησα κανέναν, εκτός από τον ραδιοφωνικό ανταποκριτή. Χωρίς να ρωτήσω, με ενημέρωσε πως η περφόρμανς είχε τελειώσει, και πως ελάχιστοι άνθρωποι που περνοδιάβαιναν, παρακολούθησαν, κατά σύμπτωση, μικρά μέρη της. Ούτε και ο ίδιος πρόλαβε να συναντήσει τους τρεις ερευνητές, που δήλωσα ν, πάντως, πως θα συνεχίσουν την μελέτη του φαινομένου που ακούει στο όνομα Ιβάν

Ισμαήλοβιτς. Ο ανταποκριτής ήταν δυσαρεστημένος με το γεγονός πως κανένας από τους ανθρώπους που παρευρέθηκαν δεν δέχτηκε να μοιραστεί τις εντυπώσεις του. Ήμουν έτοιμη να τον ρωτήσω πώς πληροφορήθηκ ε την ώρα και το σημείο, αλλά καλύτερα να σιωπώ. Από τότε που με απασχόλησε για πρώτη φορά η ιστορία του, με κατατρέχει μια από τις εντονότερες έγνοιες που θυμάμαι: Είναι κακή ιδέα να μαθαίνω περισσότερα από όσα ο Ιβάν Ισμαήλοβιτς θα ήθελε να γνωρίζω, γιατί στέκεται στην άκρη, κι αν φτάσει ως εκείνον πως γνωρίζω, κι α ν αυτό τον οδηγήσει να αισθανθεί ορατός ενώ παλεύει να ακροπατήσει, θα σκεφτεί πως υπάρχει μόνο ένας τρόπος να διαφύγει: πέφτοντας στο κενό.


Παυλίνα Μάρβιν

Since his story started to trouble me, I put the radio on, at moderate volume in order to be able to fall asleep. I usually wake up to the sound of some news item. I have difficulty recalling if it was morning or noontime, but it was certainly not the afternoon news bulletin, because the newscaster, whose voice I heard for the first time, announced that, some minutes ago, under the bright sun, at one of the most shining parts of the city, three investigators, who were evasive about the nature and the progress of their work, dressed in multi-colored costumes and made up with Rorschach ink spots, reconstructed, by means only of the arrangement and the potentialities of their bodies, some of the narratives concerning the case of Ivan Ismailovic, as they were reported in the daily press. I jumped out of bed being certain that their aim was no different from mine; to understand what is happening. I reckoned that there would be an accident if I threw myself on my bicycle in order to reach the spot of the performance as soon as possible, so I started striding across the streets. Soon enough, I had to kneel on the ground with my eyes shut. It is impossible to estimate how much time elapsed; I was, for quite a few hours, even before waking up, on the brink. Fortunately, nobody volunteered to lift me up on my feet; this tried and tested posture is mistaken for praying. When I got up and reached the scene of the performance, I met no one except for the radio reporter. Although I didn't ask, he informed me that the performance was over, and that very few passers-by watched by chance some part of it. Not even he had the opportunity to meet the three investigators, who in any case stated that they would go on studying the phenomenon under the name of Ivan Ismailovic. The radio reporter was displeased with the fact that none of the bystanders were willing to share his views. I was about to ask him how he was informed about the time and the place, but I felt it was better to keep my mouth shut. Since his story started to trouble me, I am haunted by one of the most devastating worries I ever remember; it is quite a bad idea to know more than what Ivan Ismailovic wants me to know, because he is tiptoeing on the edge, and if he heard that I know, if this made him feel visible, while he was struggling to balance on his toes, he might think that there is only one way out; to fall into the void.


Pavlina Marvin
*[Translated by Anastasia Lambropoulou]*
*[Edited by Dimitris Gkioulos]*

Η πορεία είναι κάτι σχετικό. Συντελείται χωρίς να προϋποθέτει απαραιτήτως ποδαρόδρομο. Πορεία σημαίνει πως ένας φύλακας άνοιξε την πύλη προς το μέρος χωρίς τη συγκατά θεσή μου. Ή πως ο χειμώνας επιβλήθηκε αυτοβούλως για να εξυπηρετήσει μια συνθήκη ομιχλώδους παγετού με φωτεινό σκοπό. Πληροφορούμαι με ευχαρίστησή μου, πως ο Ιβάν Ισμαήλοβιτς είναι ακατάλληλα ντυμένος για τις περιστάσεις. Έτσι ακριβώς: για να κατορθώσει την πορεία πρέπει να ξεγελάσει τον εαυτό του πως περπατάει αλλού: εκεί που βρισκόταν προηγουμένως –στην έρημο.

Παυλίνα Μάρβιν

The route is something relative. It occurs without necessarily requiring the march. Route means that a guard has opened the gate to the place without my permission. Or that winter was imposed by its own volition in order to serve a condition of misty frost with a luminous target. I am pleased to hear that Ivan Ismailovic is improperly dressed for the occasion. In that exact manner: in order to accomplish the route, he must fool himself to believe that he is walking elsewhere: at the place where he previously was –in the desert.

Pavlina Marvin
[Translated by Christiana Mygdali]

# ΟΥΚΡΑΝΙΚΗ ΙΣΤΟΡΙΑ

Με τον Βάλτερ γνωριστήκαμε τυχαία και έκτοτε κρατήσαμε επικοινωνία, που με τον καιρό έπαιρνε να πυκνώνει. Εκείνος, έμενε με την οικογένειά του σ' ένα όμορφο αγρόκτημα της Ζαπορίζιας, κάπου στον 15ο αιώνα, κι εγώ στο Κίεβο, σ' ένα σκοτεινό δια μέρισμα του 15ου ορόφου, το 1984. Για μεγάλο διάστημα είχαμε μια μάλλον πετυχημένη σχέση από απόσταση. Επικοινωνούσαμε κυρίως με πουλιά και με μπουκάλια, καμιά φορά κατάφερνε να βγάλει τη φωνή του από το στόμιο του νεροχύτη μου κι εγώ μετά βίας στρίμωχνα τα μουρμουρητά μου στην καμινάδα του. Όποτε ευκαιρούσαμε, δίναμε ο ένας στον άλλον εξαιρετικές συμβουλές, τόσο εύστοχες, που λίγη σημασία είχε ότι ποτέ δεν έφταναν ολόκληρες ή στην ώρα τους. Δεν κοιμόμουν χωρίς τη σκέψη του. Δεν ξυπνούσε χωρίς τη δική μου. Τα απογεύματα πίναμε σχεδόν μαζί —εγώ, ελάχιστο λικέρ Σαρλόττα, κι εκείνος ουίσκι σκέτης βύνης, όχι κάτω από τέσσερις μερίδες. Τον βοηθούσα να μη φοβάται το μέλλον και εκείνος μου μιλούσε για το παρόν λεπτομερώς, προσφέροντάς μου την μοναδική ικανότητα να αναπαριστώ με ακρίβεια τα παρελθόντα. Ο Βάλτερ, το ακριβό μυστικό μου. Κι αν στην αρχή δεν είχαν σημασία τα φιλιά που δεν παραδίδονταν, όπως πάντοτε, σιγά-σιγά, οι αποστολές ξεκίνησαν να μπερδεύονται. Ενώ διατύπωνα σαφή ερώτηση για τα φέουδα, η απάντηση που έπαιρνα αφορούσε στα υλικά της χωριάτικης σούπας. Ενώ η κουκουβάγια συνέχισε να έρχεται τα βράδια, τα πόδια της ήταν άδεια, έμοιαζε πλέον παχιά και δυσκίνητη.
Η κορυφαία παρανόηση δεν άργησε. Μπουκάλι στη μπανιέρα μου, και το μήνυμα με κάρβουνο: «Δε μ'αγαπάς». Μπουκάλι στο σακί με το αλεύρι, και το μήνυμα με φτηνό μολύβι για τα μάτια: «Δε μ'αγαπάς». Ο ουρανός κι η θάλασσα, κατάλαβαν πρώτοι, καθώς αρμόζει, και έπαψαν να πηγαινοφέρνουν τα μηνύματα. Έγινα εξαίρετη ιστορικός. Ξέχασα, χωρίς χαρά, αλλά και χωρίς λύπη. Σε πρόσφατες ανασκαφές, με κάλεσαν για την αποκωδικοποίηση κάποιου μεσαιωνικού χειρογράφου. Μου είπαν πως επρόκειτο για το ημερολόγιο ενός μυλωνά. Οι τελευταίες φράσεις: «Δεν το κατάλαβα ποτέ, Γκαλίνα. Εμείς, είχαμε αλλάξει δαχτυλίδια και σε λίγο θα παντρευόμαστον.» Στον υπεύθυνο αρχαιολόγο, έγραψα βιαστικά πως βρήκα μόνον υπολογισμούς παρακολούθησης εσόδων. Άνευ σημασίας.

Παυλίνα Μάρβιν

# UKRAINIAN STORY

I met Valter by chance but we stayed in touch, which kept growing with time. He lived with his family in a beautiful farm in Zaporizhia, sometime in the 15th century, and I in Kiev, in a dark flat on the 15th floor, in 1984. For a long time, we had a rather successful long-distance relationship. We communicated mainly through birds and bottles; sometimes, he would manage to squeeze his voice through my kitchen sink's pipes, and I would almost push my whispers through his chimney. Whenever we had time, we gave each other marvelous advice, so appropriate it hardly mattered that it never arrived intact or on time. I never slept without thinking of him. He never woke without thinking of me. In the evenings, we would almost drink together – I, a drop of Charlotte liqueur, and he, a single malt whisky, no fewer than four shots. I helped him to not fear the future, and he told me about the present in full detail, offering me the unique ability to accurately represent the past. Valter, my precious secret. And if, at first, the never received kisses didn't matter, as always happens, little by little, the dispatches began to get messed up. I phrased a precise question concerning feudalism, and got an answer related to the ingredients of a rustic soup. Though the owl still came at night, its claws were bare, it now seemed fat and less agile. The misunderstanding to end all misunderstandings wasn't late in coming. Bottle in my bathtub, message in charcoal: "You don't love me". Bottle in the flour sack, message in cheap eye pencil: "You don't love me". The sky and the sea were first to understand, and stopped transferring messages backwards and forwards. I became an exceptional historian. I forgot, with no joy, but also with no grief. Last week, I was invited to decode a medieval manuscript. They told me it was a miller's diary. Its last phrases were: "I never understood, Galina. We had exchanged engagement rings and would be married soon". To the chief archaeologist, I wrote hastily that I had only found calculations about credit accounts. Unimportant.

Pavlina Marvin
[*Translated by Panayotis Ioannidis*]

# A POETS' AGORA RESIDENCY

A POETS' AGORA RESIDENCY

In May 2019, *A Poets' Agora* initiated a two-week Residency program at the historical Koutzalexis house. The selected applicant, British Bengali poet Shamim Azad, had the opportunity to meet the Greek authors affiliated with our poetry circle over two literary dinners and a few informal gatherings. From there sprang ideas, projects and, most of all, concrete realisations, such as an experience with a migrant organisation and the translation of Greek poets into Bengali. Shamim Azad introduced the Greek writers to the London Literary Festival 'BSKL Boi-lit 2019: Identity quest of Bengalis in Britain.' For this festival she published a translation of Adrianne Kalfopoulou's poem, *'Let Yourself Forgive'*, and she read the poetry of Orfeas Apergis at the British Bangladeshi Poetry Collective.

Charcoal sketch of the 'Koutzalexis House' at the foothills of the Acropolis
by Ioanna Trachana, 2020

# CONSENT

I wake up in the middle of an Athenian humid night
while the sun and her flowers were asleep.
The notes of a Greek Ballad were fading
from under an ancient Sycamore tree.
As day's consequences gradually grip people's mind at night-
I wake from sleep to a dreadful dream
I wake to an era that I only read in history
I wake with the raging roar of fascists raping Greece
I hear the whispers of poets' past passed
from this neoclassical building of 1837
I hear their silent sighs coming out of layers of anxiety.
I hear people fighting back fearlessly
I hear their noises of resistance outside and
cheers at an ultimate victory.
As I open eyes in my bed,
they meet two huge and heavy rectangular grey shutters
they don't shut back, rather open gasping at me.
I tread tenderly on the wooden floor
I place my weight softly on their footmarks
to keep from erasing the inscriptions, the music
I barely burden them with my responsibility to reincarnation.
And yet, they give me a positive push from below,
from the dark basement
they give me affirmation to tell their stories,
rebuild their untold words for restoration!

Shamim AZAD

Plaka, Athens, 11 May 2019

# BIOGRAPHIES OF THE POETS

**Katerina Anghelaki-Rooke** was born in 1939 and died in January of 2020. She was the godchild of Nikos Kazantzakis, who was her father's close friend. She studied foreign languages at the University of Athens, Nice and Geneva, graduating in 1962. She was a Fulbright Visiting Lecturer in 1980-81, teaching courses on Modern Greek poetry and on Nikos Kazantzakis at Harvard, Utah and San Francisco State universities. Anghelaki-Rooke was awarded the Greek State Poetry Prize in 1985 and in 2000 the Ouranis Poetry Prize of the Academy of Athens. Among her translations into Greek are works by Samuel Beckett, Andrei Voznesensky, Edward Albee, Dylan Thomas, Seamus Heaney, Pushkin, and a volume of 20 contemporary American poets. She had published more than twelve collections of poems. She has been translated into more than ten languages.

Born in Athens, Greece, **Liana Sakelliou** studied English at the University of Athens (B.A.), Edinburgh (Grad Diploma), Essex (M.A.), and The Pennsylvania State University (Ph.D.). She is Professor in English and Creative Writing at the Department of English Language and Literature at The National and Kapodistrian University of Athens. Her 18 books with poems, scholarly articles, essays, and translations have been widely published in Greece, France and the U.S.A. For her academic and creative-writing activities she received two Fulbright awards, and grants from Princeton University, University of Coimbra, University of Sussex (West Dean), and The British Council. Her poems have appeared in several international anthologies, as well as in many journals and magazines. She is a member of various professional societies including The Hellenic Authors' Society.

**Orfeas Apergis** (Athens, 1974) is a poet, essayist and performer. He has worked variously as a hospital doctor, army doctor, real estate agent, political consultant, classical singer, but mainly as a school teacher in Athens. He is a member of the editorial board of frmk [φρμκ] a biannual journal on poetry poetics and visual arts; and he has been publishing in other Greek literary journals and newspapers since 2006. His collected poems appeared in 2011, under the title "Y". He has translated Byron, Brontë, Browning, MacNeice, Larkin, Hill, Muldoon, Williams, Bishop and Duncan, among others. In 2013, he was poet-in-residence at King's College, London, and visiting poet at the University of Barcelona. His work has been included in the English anthology *Futures: Poetry of the Greek Crisis* (2015), the Spanish anthology *La Busqueda del Sur [In Search of the South]* (2016), and the German anthology *Dichtung mit Biss [Poetry* *with a Bite]* (2018). His work has also been translated into Swedish and French. His latest volume of poetry, entitled *I glossa tous [Their tongue]*, came out in early 2019.

**Haris Vlavianos** was born in Rome in 1957. He studied Economics and Philosophy at the University of Bristol; and History and Political Theory at Oxford University. He has published ten collections of poetry, including *Vacation in Reality* (2009), which won the prestigious "Diavazo" Poetry Prize, and was short-listed, along with his later *Sonnets of Despair* (2011) for the National Poetry Prize. He has published collections of thoughts and aphorisms on poetry and poetics, and a book of essays entitled *Does Poetry Matter? Thoughts on a Useless Art* (2009). *History of Western Philosophy in 100 Haiku* (Daedalus Press) was published in 2014 and translated into English by Peter Mackridge. Vlavianos has translated the works of poets such as Ezra Pound, e.e.cummings, Wallace Stevens, Michael Longley, William Blake, Fernando Pessoa, and Anne Carson. His books have been translated into many languages. He is editor of the influential journal *"Poetics"*, as well as poetry editor at the *'Patakis'* publishing house, and is currently a Professor of History and Politics at the American College of Greece. In 2014 he published his first novel, *Blood into Water*. His latest work, *Hitler's Secret Diary* (Patakis 2016), is a study of Hitler's personality through the fictional 'recreation' of the lost diary he kept while in prison in 1924.

**Katerina Iliopoulou**, is a poet, artist and translator, who lives and works in Athens. She is the author of four poetry books: *Mister T.* (2007; first prize for a new author by the literary journal "Diavazo"), *Asylum* (2008), *Book of the Soil* (2011), and *Every place only once, and completely* (2015); a book of poetry and photography, *Gestus,* together with visual artist Yiannis Isidorou (FRMK Editions, Athens 2014); a book of short stories, *It isn't yet* (Melani Editions, Athens 2019); she is also one of seven authors of *A Conversation on Poetry Now* (FRMK Editions, Athens 2018), a collective book of essays on poetics. Her translations in book form include extensive selections from the work of Sylvia Plath and Walt Whitman, while she has also published translations of Mina Loy, Robert Hass and Ted Hughes, among others. Her own poetry has been translated and published in literary reviews, journals and anthologies in several languages, and she has participated in numerous international poetry festivals (including the "Poetry Parnassus" held in London in its 'olympic' year 2012). *Mister T.* and *The Book of the Soil* have been published in French, while the former has also been published in Turkish and Italian. She is the editor in chief of "FRMK", a biannual journal on poetry, poetics and the visual arts, and co-editor of the bilingual web platform greekpoetrynow.com.

**Ginger F. Zaimis** is an American polymath, writer/poet/editor/literary translator, lecturer and adviser as well as a Southerner, New Yorker Emeritus and Philhellene who has read arts & architectural history, economics and the classics. She specializes in integrating perennial archetypes through contemporary modernisms and comparative literature which re-unite the humanities and sciences as one. Her grammatology is imbued with six monographs including *Therapy with Antigone and the Trilogy Verses*, *Prometheus Rebound and Other Mythologies* and her forthcoming collection entitled, *Is and Not*. Her writing and research address themes of architecture, philosophy, progressed mythologies and poetry as well as quantum physics, bioenergetics and consciousness. Ginger is the Literary & Arts Chair (for Greece) of the International Friends of Bibliotheca Alexandrina, the Library of Alexandria. She is the architect of two new poetic forms, the *Portico* and *Triptych*, and a literary translator from Ancient Greek to English verse of the Stoic philosopher, Cleanthes' *Hymn to Zeus*, Aristophanes' Frogs, as well as portions of: *the Socratic Dialogues, Hippocratic Corpus*, the Stoic Marcus Aurelius' *Meditations* (a work in progress) and other selections of mythos and logos. Her work has been presented at international centres for literature, architecture & art, biennials, world libraries, museums, academic institutions as well as the Research Centre for Greek Philosophy at The Athens Academy, and her poems have been translated into Modern Greek, Bengali as well as French and Arabic (forthcoming).

**Panayotis Ioannidis** was born in 1967 in Athens, where he now lives. His poems have been published in three books (*The lifesaver*, 2008; *Uncovered*, 2013; *Poland*, 2016 [shortlisted for the "Anagnostis" Prize] – all by Kastaniotis Editions); various journals (in Greek, and also in English, Polish, Swedish, and Turkish); and included in two English-language anthologies: *Futures: Poetry of the Greek Crisis* (Penned in the Margins, 2015) and *Austerity Measures:The New Greek Poetry* (Penguin, 2016; New York Review Books, 2017). He has translated several English-language poets, including S. Heaney, R. Creeley, T. Gunn, E. Bishop, R. Duncan, A. Motion, and D. Harsent. He is poetry editor for the Greek monthly *'The Books' Journal'*; on the editorial board of the biannual journal for poetry, theory and the visual arts, frmk [φρμκ]; and curator of the monthly poetry readings, *"Words (can) do it"*. He teaches poetry as creative writing to children (on the Cavafy Archive's educational programme) and adults (in the Athens British Council Poetry Group).

**Kyoko Kishida** (pen name) is an Athens-based poet, translator and editor, born in 1983. As a founding co-editor of the poetry magazine *Teflon*, which publishes cutting edge literature from Greece and the world, she has introduced the works of Audre Lorde, Nanni Balestrini, Yosano Akiko, Keston Sutherland, Nicole Brossard, and many other contemporary and experimental poets to a Greek readership. Her most recent translations include the young American poets Hala Alyan, Max Ritvo, Wendy Trevino and Ari Banias. She also writes on topics as varied as the poetics of hip hop and the poetry of African-American lesbian poets (Pat Parker, Cheryl Clarke etc.). Her translations include Valerie Solana's *Scum Manifesto* and Mike Davis' *Dead Cities*. In 2014 her translation of Debbie Drechsler's graphic novel *Daddy's Girl* was nominated for Comicdom's best translated comic award. She has organized and participated in various interdisciplinary performances, including *"Poetry Is Just Words in the Wrong Order"*, that won the 'Soundout! New Ways of Presenting Literature' Award (Berlin, 2014). She has collaborated with the poet Jazra Khaleed for the poetry short film *Gone is Syria, Gone* (2016), that was selected for the 'Internationale Kurzfilmtage Winterthur,' the 'Kasseler Dokfest' and 'L'Alternativa.' Her poems were recently included in the anthology *Austerity Measures: The New Greek Poetry* (Penguin, 2016; New York Review Books, 2017).

**Theodoros Chiotis** is a poet and translator. He writes poetry in Greek, English and in programming languages; he is currently researching the collaborative use of neural networks in writing poetry. His publications include *Screen* (in collaboration with photographer Nicholas Ventourakis - Paper Tigers Books, 2017) and *limit.less: towards an assembly of the sick* (Litmus, 2017). He is the editor and translator of the anthology *Futures: Poetry of the Greek Crisis* (Penned in the Margins, 2015). In 2017 he was awarded the 'Dot Award for New Media Writing' for his *Mutualised Archives* project by the the Institute for the Future of the Book/Bournemouth University, while in 2018 his poem *"Interference"* received a High Commendation at the 'Forward Prizes for Poetry.' He has presented his work in literary festivals in Greece, Great Britain, Germany, Belgium, Poland, United States, while his work has been published in journals and anthologies in Greece, Great Britain, Germany, New Zealand, Australia, Estonia and Croatia. In 2018 he was commissioned by the 'Bayerische Staatsoper' to write poems for the book accompanying the retrospective of the work of Royal Ballet resident choreographer Wayne McGregor. He is a member of the editorial board of the Greek literary journal [φρμκ] and the British literary journal *Hotel*. He has presented his visual poetry in publications and exhibitions in Greece and abroad. As Coordinator of Scholarly Research and Digital Development of the Cavafy Archive (Onassis Foundation), he spearheaded the complete digitisation of the Alexandrian poet's physical archive and its uploading on the web as a free-to-access archive. He has designed more than 60 educational workshops revolving around poetry for secondary and tertiary education for both the public and the private sector in Greece and abroad. He has translated contemporary British and US poets into Greek. He has also translated Aristophanes into English for the Athens & Epidaurus Festival. He lives and works in Athens.

**Lenia Safiropoulou** is a classical singer, poet and translator. She studied in Stuttgart and London with scholarships by the Onassis and Callas Foundations and by the Royal Opera House Covent Garden. She performs with orchestra, chamber and early music ensembles, and pianists in Greece and abroad. Her first personal album *Sunless Loves* was released in 2017 by the British label *First Hand Records*. She has published three poetry books, *Paternoster Square* (Polis edition, Anagnostis Prize, Athens 2012), *It is hard to stumble on stones* (Patakis edition, 'Athens Woman of the Year' award 2016) and *Hall of the Lost Steps-26 bareheaded sobs* (Polis edition, Laskaridis foundation award 2018). Her poems have been translated into German and included in the anthologies: *Kleine Tiere zum Schlachten* (parasitenpresse, Köln 2017) and *Dichtung mit Biss* (Romiosini, Berlin 2018). She has translated Goethe, Heine, Pushkin and Kafka into Greek, as well as the complete Sonnets by William Shakespeare (Gutenberg, 2016). Lenia is a producer at the Greek classical Radio ERT3 and a Professor of singing at the Athens Conservatoire. www.leniasafiropoulou.gr

**Christos Siorikis** (1989) grew up in Agrinio and currently lives in Athens. He studied in the Department of Primary Education at the University of Athens and then specialised in literature didactics. He teaches Spanish to children and adults. He has organised creative literature activities in collaboration with the Library of the Cervantes Institute and uses poetry and other literary texts in his teaching of the Spanish language. He studies the works of Greek writer Zacharias Papantoniou and has edited the section of the September 2016 issue of literary journal *Νέα Εστία* dedicated to the aforementioned writer. He is a member of the group, *Ομάδα Άστυ*, which organises walks in Athens that aim in the public's contact with the city's various locations and history, often through literature. He writes poems and translates Spanish literature. He has translated amongst others: Julio Cortazar's book, *The Bear's Discourse*, for its illustrated version (Papyros Publishing, 2015). His poems have been published in journals and anthologies and have been translated in German and Hebrew. His book titled *Η Πρώτη Φορά* (*The FirstTime*) has been published by Αντίποδες Publications (2018).

**Adrianne Kalfopoulou** is the author of three poetry collections, two books of essays, and several chapbooks. *A History of Too Much* (Red Hen Press, 2018) is her most recent publication. Poems, essays, blog posts, and assemblages have appeared in online and print publications such as *The Harvard Review* online, *Hotel Amerika, Slag Glass City, Superstition Review* and elsewhere. A collection of poems *Xeni, Xenos, Xenitia* (Melani Press 2013) was translated into Greek in collaboration with the poet Katerina Iliopoulou. Her poems have been anthologized in *Futures: Poetry of the Greek Crisis* (Penned in the Margins, 2015), and *Borderlands and Crossroads: Writing the Motherland* (Demeter Press 2016). She teaches American literature and creative writing at Deree College, and is the McGee professor of Creative Writing at Davidson College for 2020-2021.

**A.E. Stallings** studied classics at the University of Georgia (in Athens, Georgia), and Oxford University. She has resided in Athens, Greece since 1999. She published four collections of poetry: *Archaic Smile* (University of Evansville Press), recipient of the Richard Wilbur Award, *Hapax* (TriQuarterly Books), *Olives* (TriQuarterly Books), a finalist for the National Book Critics Circle Award, and *Like* (Farrar, Straus & Giroux), a finalist for the Pulitzer Prize. Among her other publications are a verse translation of Lucretius' philosophical epic, *The Nature of Things*, and Hesiod's 8th Century B.C. Almanac, *Works and Days*, shortlisted for the Runciman Award, with Penguin classics. An illustrated and annotated translation of the pseudo-Homeric poem, *The Battle of the Frogs and the Mice*, is forthcoming from Paul Dry Books. Stallings has received a translation grant from the National Endowment of the Arts, the Willis Barnstone Translation Prize, the 2008 Poets' Prize, and the Benjamin H. Danks Award from the American Academy of Arts and Letters. A member of the American Academy of Arts and Sciences, she was a 2011 Guggenheim fellow and a 2011 MacArthur fellow. She lives with the journalist John Psaropoulos and their two children, Jason and Atalanta.

**Yiannis Doukas** was born in Athens in 1981. He studied philology at the National and Kapodistrian University and Digital Humanities at King's College London. For the past few years, he lived in Galway, Ireland, where he was working towards a Ph.D. on intertextuality in late antique epic poetry and methods for its digital representation. He has published the books: *The World as I Came and Found it* (short stories, Kedros, 2001), *Inner Borders* (poetry, Polis, 2011, *Diavazo* journal 'Debut Poetry Collection' Award) and *The Stendhal Syndrome* (poetry, Polis, 2013, 'G. Athanas' Award of the Academy of Athens). Some of his poems have been included in anthologies and translated into English, French, German, Serbian, Dutch and Polish. Two poems from *Inner Borders* have been set to music by Nikos Platyrachos in the album *Ta Astega*, (2015). He has written the lyrics for two songs by Thanos Mikroutsikos, included in the album *Stin Omihli ton Kairon* (2017). He translates from English. He worked, along with Haris Vlavianos, on the translation of '*erotic poems*' by e. e. cummings (Patakis, 2014, bilingual edition). He has published book reviews in Greek newspapers and literary magazines.

**Pavlina Marvin** was born in Athens in 1987, but grew up in Hermoupolis on the island of Syros. She studied history at the National and Kapodistrian University of Athens. She is writing her PhD on Greek national book policy. She was a co-publisher and co-editor of Teflon poetry magazine (2008-2011). She studied poetry at the biennial workshop of the Takis Sinopoulos Foundation (2007-2009). Her first book, *"Histories from all around my world"*, was published by Kichli Publications (2017) and was awarded the 'Yannis Varveris' prize by the Hellenic Authors Association. As a writer and performer, she has been invited to participate in a range of interdisciplinary arts projects and festivals, in Greece and abroad. Parts of her work have been translated into English, French, German, Italian, Spanish, Bengali and Serbo-Croatian.

**Shamim Azad** poet in Residence is originally from Sylhet, Bangladesh. She is a British Bengali bilingual author, poet and storyteller in the UK. After her career in teaching, journalism and television in her home country, Azad came to London in 1990 to further develop her literary vocation. She has published more than 35 books including novels, short stories and collections of poems in English and Bengali. Her poems and translations have appeared in various international magazines including *The New Yorker*. She has also performed at the Edinburgh Fringe Festival. She is the founder and chair of Bishwo Shahitto Kendro (World Literature Centre) London, and a trustee of Rich Mix Centre. Shamim Azad received the Year of the Artist Award, London Arts in 2000, the UK Civic Award in 2004, and the Bangla Academy Syed Waliullah Award in 2016.

# ΒΙΟΓΡΑΦΙΚΑ ΤΩΝ ΠΟΙΗΤΩΝ

Η **Κατερίνα Αγγελάκη-Ρουκ** γεννήθηκε στην Αθήνα το 1939 και απεβίωσε τον Ιανουάριο του 2020. Ήταν βαπτισιμιά του Νίκου Καζαντζάκη που ήταν στενός φίλος του πατέρα της. Σπούδασε στην Αθήνα, στη νότια Γαλλία και αποφοίτησε στη Γενεύη με το δίπλωμα Μεταφραστών και Διερμηνέων (ελληνικά, αγγλικά, γαλλικά, ρωσικά). Πρωτοδημοσίευσε το 1956 στην *Καινούργια Εποχή*. Άρθρα της για την ποίηση και τη μετάφραση της ποίησης έχουν δημοσιευτεί σε περιοδικά και εφημερίδες ανά τον κόσμο. Ποιήματά της έχουν μεταφραστεί σε περισσότερες από δέκα γλώσσες και βρίσκονται σε παγκόσμιες ανθολογίες. Το 1984 της απονεμήθηκε το Κρατικό Βραβείο Ποίησης, και το 2000 το Βραβείο Ουράνη της Ακαδημίας Αθηνών. Το μεταφραστικό της έργο επικεντρώνεται στην ποίηση. Είχε εκδώσει περίπου 20 ποιητικές συλλογές.

Η **Λιάνα Σακελλίου** γεννήθηκε στην Αθήνα και σπούδασε Αγγλική Φιλολογία στο Πανεπιστήμιο Αθηνών. Συνέχισε μεταπτυχιακές σπουδές στη Βρετανία και την Αμερική. Έχουν κυκλοφορήσει 18 βιβλία της με ποιήματα, δοκίμια, άρθρα και μεταφράσεις στην Ελλάδα, τη Γαλλία και τις ΗΠΑ. Είναι καθηγήτρια δημιουργικής γραφής και αμερικανικής λογοτεχνίας στο Τμήμα Αγγλικής Γλώσσας και Φιλολογίας του Πανεπιστημίου Αθηνών. Για τις επιστημονικές και συγγραφικές της δραστηριότητες έλαβε υποτροφίες από το Ίδρυμα Φουλμπράιτ, από το Τμήμα Ελληνικών Σπουδών στο Πανεπιστήμιο του Πρίνστον, από το Πανεπιστήμιο της Κοίμπρα, από το Πανεπιστήμιο του Σάσσεξ (Γουέστ Ντιν) και από το Βρετανικό Συμβούλιο. Ποιήματά της έχουν μεταφραστεί σε οκτώ γλώσσες και εμφανιστεί σε περιοδικά και σε εφημερίδες καθώς και σε έξι διεθνείς ανθολογίες. Είναι μέλος πολλών επιστημονικών εταιρειών καθώς και της Εταιρείας Ελλήνων Συγγραφέων.

Ο **Ορφέας Απέργης** (Αθήνα, 1974) είναι ποιητής, δοκιμιογράφος, μεταφραστής και performer. Ποιήματα, μεταφράσεις και δοκίμιά του έχουν δημοσιευτεί μεταξύ άλλων στα περιοδικά *Ποίηση, Ποιητική, Νέα Εστία* και *Books' Journal* από το 2006 και εξής. Από το 2013 συνεργάζεται κυρίως με το λογοτεχνικό περιοδικό *Φάρμακο*, του οποίου είναι συντάκτης. Κρατάει τη στήλη πολιτισμικής κριτικής «Τεχνολογίες» στα *Νέα* και τη στήλη «Ut poesis» στο *Books' Journal*. Πιο πρόσφατα βιβλία του «Η γλώσσα τους» (ποιήματα, εκδόσεις Νεφέλη, 2019) και «Ποίηση Τώρα» (συλλογικός τόμος με δοκίμια, εκδόσεις φρμκ, 2018). Συγκεντρωτική έκδοση των ποιημάτων του, με τίτλο «Υ», κυκλοφόρησε το 2011 από τις εκδόσεις Πατάκη. Ποιήματά του έχουν μεταφραστεί στα Αγγλικά, Γαλλικά, Γερμανικά, Ισπανικά και Σουηδικά, και περιλαμβάνονται στις πρόσφατες ανθολογίες *Futures: Poetry of the Greek Crisis* [Μέλλοντα: Ποίηση της ελληνικής κρίσης, 2015, στα Αγγλικά], *La Busqueda del Sur* [Σε αναζήτηση του Νότου, 2016, στα Ισπανικά], *Kleine Tiere zum Schlachten* [Μικρά ζώα επί σφαγή, 2018, στα Γερμανικά] και *Dichtung mit Biss* [Ποίηση με πείσμα, 2018, επίσης στα Γερμανικά]. Έχει μεταφράσει, μεταξύ άλλων, Byron, Emily Brontë, Browning, MacNeice, Larkin, Hill, Muldoon, Williams, Bishop και Duncan. Το 2018 έγραψε το θεατρικό *(Βρικόλακες)*, εμπνευσμένο από τον Ίψεν, που ανέβηκε στο Φεστιβάλ Αθηνών-Επιδαύρου σε σκηνοθεσία Μιχάλη Κωνσταντάτου, και απέδωσε στα Αγγλικά τη διασκευή των *Ορνίθων* που ανέβασε ο Νίκος Καραθάνος στη Νέα Υόρκη (St. Ann's Warehouse). Το 2020-21 θα κυκλοφορήσουν η ποιητική συλλογή του με τίτλο «Σαν χρέος», από τις εκδόσεις Νεφέλη και, σε μετάφρασή του, *Η Αποκάλυψη του Ιωάννη*, από τις εκδόσεις Πατάκη και το *Night Boat to Tangier* [Νυχτερινό πλοίο για Ταγγέρη] του Kevin Barry, από τις εκδόσεις Gutenberg.

Ο **Χάρης Βλαβιανός** γεννήθηκε στη Ρώμη το 1957. Σπούδασε Οικονομικά και Φιλοσοφία στο Πανεπιστήμιο του Μπρίστολ και Ιστορία και Πολιτική Θεωρία στο Πανεπιστήμιο Οξφόρδης. Έχει εκδώσει δέκα ποιητικές συλ-λογές, με πιο πρόσφατες τις *Διακοπές στην πραγματικότητα* (2009) [Βραβείο του περιοδικού *Διαβάζω* (2010)] και *Σονέτα της συμφοράς* (2011), καθώς και τρεις συλλογές δοκιμίων. Το βιβλίο του, *Η ιστορία της δυτικής φιλοσοφίας σε 100 χαϊκού: από τους Προσωκρατικούς έως τον Ντεριντά*, μεταφρασμένο στα αγγλικά από τον Peter Mackridge, κυκλοφόρησε το 2014 (εκδόσεις Dedalus Press). Το 2015 κυκλοφόρησε το *Γιατί γράφω ποίηση* (εκδόσεις Άγρα). Έχει μεταφράσει σημαντικά έργα κορυφαίων Αμερικανών και Ευρωπαίων ποιητών: Walt Whitman, Ezra Pound, John Ashbery, William Blake, Zbigniew Herbert, Fernando Pessoa, e.e. cummings, Michael Longley, Wallace Stevens και Anne Carson. Ποιήματά του έχουν μεταφραστεί σε πολλές ευρωπαϊκές γλώσσες και συλλογές του έχουν εκδοθεί στην Αγγλία, τη Γαλλία, τη Γερμανία, τη Σουηδία, την Ολλανδία, την Ιρλανδία και την Ισπανία. Διευθύνει το περιοδικό *Ποιητική*. Διδάσκει Ιστορία και Ιστορία των ιδεών στο Αμερικάνικο Κολέγιο Ελλάδος και είναι υπεύθυνος εκδόσεων στις εκδόσεις «Πατάκη». Το 2014 εξέδωσε το πρώτο του μυθιστόρημα με τίτλο, *Το αίμα νερό*. Το τελευταίο έργο του, *Το κρυφό ημερολόγιο του Χίτλερ* (εκδόσεις «Πατάκη» 2016), συνδυάζει την μυθοπλασία με την ιστορική τεκμηρίωση σε μιά μελέτη που σκιαγραφεί την προσωπικότητα του Χίτλερ, με έμπνευση το χαμένο ημερολόγιο που κρατούσε στη φυλακή το 1924.

Η **Κατερίνα Ηλιοπούλου** δημοσιεύει ποιήματα και κείμενα για την ποίηση από το 2001. Έχει εκδώσει τέσσερα βιβλία ποίησης από τις εκδόσεις Μελάνι: *Ο κύριος Ταυ, Άσυλο, Το βιβλίο του χώματος, Μια φορά κάθε τοπίο και ολότελα* (2015), καθώς και μια συλλογή διηγημάτων (*Δεν είναι ακόμα*, Μελάνι 2019), ένα βιβλίο με ποίηση και φωτογραφία σε συνεργασία με τον Γιάννη Ισιδώρου (*Gestus*, 2014, εκδ. ΦΡΜΚ) και ένα βιβλίο με δοκίμια για την ποίηση *Μια συζήτηση για την ποίηση τώρα* (συλλογικό, 2018, εκδ. ΦΡΜΚ). Έχει συμμετάσχει σε πολλά διεθνή φεστιβάλ ποίησης, ενώ ποιήματά της έχουν μεταφραστεί σε πολλές γλώσσες και έχουν συμπεριληφθεί σε περιοδικά και ανθολογίες. Το πρώτο της βιβλίο, *Ο κύριος Ταυ* (βραβείο πρωτοεμφανιζόμενου συγγραφέα περιοδικού *Διαβάζω*), έχει εκδοθεί στα Γαλλικά, τα Τουρκικά και τα Ιταλικά, ενώ το *Βιβλίο του χώματος*, έχει εκδοθεί στα γαλλικά (ed. Desmos). Από το 2008 επιμελείται την δίγλωσση ιστοσελίδα για την σύγχρονη ελληνική ποίηση, *greekpoetrynow*, ενώ από το 2013 διευθύνει το εξαμηνιαίο περιοδικό ΦΡΜΚ, για την ποίηση, την ποιητική και τα εικαστικά. Η μεταφραστική της εργασία αφορά το έργο σύγχρονων ποιητών όπως ο Ρόμπερτ Χας, η Μίνα Λόυ, ο Τεντ Χιουζ κ.ά. Σε συνεργασία με την Ελένη Ηλιοπούλου έχει μεταφράσει δύο βιβλία με ποίηση της Σύλβια Πλαθ (*Άριελ*, Μελάνι 2012 και *Σύλβια Πλαθ, Ποιήματα*, Κέδρος 2003), καθώς και μια εκτεταμένη ανθολογία με ποίηση του Walt Whitman (*Φύλλα Χλόης*, Κέδρος 2019).

Η **Ginger F. Zaimis** είναι αμερικανίδα ποιήτρια, δοκιμιογράφος και πολυμαθής, που ειδικεύεται στις αρχιτεκτονικές μορφές. Κατέχει την έδρα Τεχνών και Λογοτεχνίας του Διεθνή Συλλόγου Φίλων της Βιβλιοθήκης της Αλεξάνδρειας. Η γραμματολογία της συνενώνει τις διασταυρώσεις των σύγχρονων μοντερνισμών, τη συγκριτική λογοτεχνία και τη μυθολογία ώστε να συνδέσει τον διεπιστημονικό διάλογο με τη γλώσσα, την ιστορία και τη φιλοσοφία, ενώ παράλληλα συνδέει τις τέχνες και τις επιστήμες. Το 2014 της απονεμήθηκε το βραβείο PEN AMERICA's Joyce Osterweil για την καλύτερη πρώτη ποιητική συλλογή. Έχει γράψει δύο ποιητικές συλλογές, *Prometheus Rebound and Other Mythology* και *Excavated Athens to Alexandria* και *The Portico Convention* και *Triptych*. Έχει ασχοληθεί με την αρχιτεκτονική των ποιητικών μορφών. Έχει γράψει - σε συνεργασία με τον Ευάγγελο Μουτσόπουλο, Μέλος της Ακαδημίας Αθηνών, το έργο *Philosophy and Poetry*. Η τελευταία της συλλογή με τίτλο *Therapy with Antigone and the Trilogy Verses* (Spuyten Duyvil Press, Νέα Υόρκη) κυκλοφόρησε το 2017.

Ο **Παναγιώτης Ιωαννίδης** γεννήθηκε το 1967 στην Αθήνα όπου και ζει. Τα ποιήματά του έχουν δημοσιευθεί σε τρία βιβλία (*Το σωσίβιο*, 2008 - *Ακάλυπτος*, 2013 - *Πολωνία*, 2016 [υποψήφιο για το βραβείο του «Αναγνώστη»] – όλα από τις εκδ. Καστανιώτη)· στα ελληνικά και –μεταφρασμένα στα αγγλικά, πολωνικά, σουηδικά, και τουρκικά– σε ξένα περιοδικά. Έχουν επίσης συμπεριληφθεί σε δύο αγγλόγλωσσες ανθολογίες: *Futures: Poetry of the Greek Crisis* (Penned in the Margins, 2015) και *Austerity Measures: The New Greek Poetry* (Penguin, 2016 - New York Review Books, 2017). Έχει μεταφράσει διάφορους αγγλόφωνους ποιητές, όπως τους: Σ. Χήνυ, Ρ. Κρήλυ, Τ. Γκανν, Ε. Μπίσοπ, Ρ. Ντάνκαν, Ά. Μόσιον, και Ντ. Χάρσεντ. Είναι υπεύθυνος για την ποίηση στο μηνιαίο περιοδικό *The Books' Journal* – μέλος της συντακτικής ομάδας του εξαμηνιαίου περιοδικού για την ποίηση, τα εικαστικά και την θεωρία, «ΦΡΜΚ»· και επιμελητής των μηνιαίων ποιητικών εκδηλώσεων «Με τα λόγια (γίνεται)». Διδάσκει ποίηση ως δημιουργική γραφή σε παιδιά (στα εκπαιδευτικά προγράμματα του Αρχείου Καβάφη) και ενήλικες (στην Ομάδα Ποίησης του Βρετανικού Συμβουλίου).

Η **Κυόκο Κισίντα** (λογοτεχνικό ψευδώνυμο) είναι ποιήτρια, μεταφράστρια και επιμελήτρια. Ως ιδρύτρια και αρχισυντάκτρια του ποιητικού περιοδικού *Τεφλόν* έχει μεταφράσει και παρουσιάσει στο ελληνόφωνο κοινό το έργο σημαντικών σύγχρονων ποιητών και ποιητριών από όλο τον κόσμο, όπως ο Nanni Balestrini, η Nicole Brossard, ο Keston Sutherland, η Audre Lorde και η Τσιμάκο Τάντα και έχει γράψει κριτικά κείμενα για θέματα όπως η ποίηση των Αφροαμερικανίδων λεσβιών ποιητριών (Pat Parker, Cheryl Clark, κ.ά) και η ποιητική του χιπ χοπ. Οι πιο πρόσφατες μεταφράσεις της συμπεριλαμβάνουν νέες φωνές της αμερικανικής ποίησης (Hala Alyan, Max Ritvo, Ari Banias κ.ά). Από τα βιβλία που έχει μεταφράσει ξεχωρίζουν το κόμικ *Το κορίτσι του Μπαμπά* της Debbie Drechsler, το οποίο προτάθηκε για το βραβείο καλύτερης μεταφρασμένης έκδοσης του Comicdom Con, και *Οι νεκρές πόλεις* του Mike Davis. Από το 2009 έχει διοργανώσει μια σειρά από ποιητικές περφόρμανς στην Αθήνα. Το 2014 συμμετείχε στην περφόρμανς *Poetry Is Just Words in the Wrong Order*, η οποία κέρδισε το βραβείο Soundout! New Ways of Presenting Literature, στο Βερολίνο. Ποιήματά της δημοσιεύτηκαν πρόσφατα στην ανθολογία *Austerity Measures: The New Greek Poetry*. Ζει και εργάζεται στην Αθήνα.

Ο **Θοδωρής Χιώτης** είναι ποιητής και μεταφραστής. Γράφει στα ελληνικά, στα αγγλικά και χρησιμοποιώντας γλώσσες προγραμματισμού και σε συνεργασία με νευρωνικά δίκτυα. Έχει εκδώσει τις συλλογές *Screen* (σε συνεργασία με τον φωτογράφο Νικόλα Βεντουράκη - Paper Tigers Books, 2017) και *limit.less: towards an assembly of the sick* (Litmus, 2017). Είναι ο επιμελητής και μεταφραστής της ανθολογίας *Futures: Poetry of the Greek Crisis* (Penned in the Margins, 2015 - η ελληνική εκδοχή της ανθολογίας θα κυκλοφορήσει το 2021). Έχει λάβει το Dot Award for New Media Writing για το έργο *Mutualised Archives* από το Institute for the Future of the Book/Πανεπιστήμιο του Bournemouth, το 2017, και Υψηλό Έπαινο για το ποίημα *"Interference"* από τα Forward Prizes for Poetry, το 2018. Έχει παρουσιάσει τη δουλειά του σε φεστιβάλ σε Ελλάδα, Αγγλία, ΗΠΑ, Γερμανία, Βέλγιο, και Πολωνία, ενώ η δουλειά του έχει δημοσιευτεί σε περιοδικά και ανθολογίες σε Ελλάδα, Αγγλία, Γερμανία, Νέα Ζηλανδία, ΗΠΑ, Αυστραλία, Εσθονία και Κροατία. Το 2018 του ζητήθηκε από την Κρατική Όπερα της Βαυαρίας να γράψει ποιήματα για τη συνοδευτική έκδοση της ρετροσπεκτίβας του έργου του φιλοξενούμενου χορογράφου (resident choreographer) του Βασιλικού Μπαλέτου της Αγγλίας, Wayne McGregor. Είναι μέλος των συντακτικών επιτροπών των περιοδικών *[φρμκ]* και *Hotel*. Έχει παρουσιάσει οπτικά ποιήματα σε εκδόσεις και εκθέσεις στην Ελλάδα και το εξωτερικό. Ως Συντονιστής Επιστημονικής

Έρευνας και Ψηφιακής Ανάπτυξης του Αρχείου Καβάφη (Ίδρυμα Ωνάση), ήταν υπεύθυνος για την πλήρη ψηφιοποίηση και ανάρτηση με ελεύθερη πρόσβαση του αρχείου του Αλεξανδρινού ποιητή στο διαδίκτυο. Έχει σχεδιάσει πάνω από 60 εκπαιδευτικά ποιητικά εργαστήρια για τη δευτεροβάθμια και τριτοβάθμια εκπαίδευση για δημόσιους και ιδιωτικούς φορείς σε Ελλάδα και εξωτερικό. Έχει μεταφράσει σύγχρονους βρετανούς και αμερικανούς ποιητές στα ελληνικά. Έχει επίσης μεταφράσει Αριστοφάνη στα αγγλικά για λογαριασμό του Ελληνικού Φεστιβάλ. Ζει και εργάζεται στην Αθήνα.

Η **Λένια Ζαφειροπούλου** είναι λυρική τραγουδίστρια, ποιήτρια και μεταφράστρια. Σπούδασε τραγούδι, πιάνο και Lied στη Στουτγάρδη και όπερα στο Λονδίνο (Guildhall School of Music, National Opera Studio) με υποτροφίες των Ιδρυμάτων Μ. Κάλλας και Α. Ωνάση καθώς και του Royal Opera House Covent Garden. Εμφανίζεται σε συναυλίες και παραστάσεις στην Ελλάδα και στο εξωτερικό. Κυκλοφορεί ο προσωπικός της δίσκος *Sunless Loves* από την βρετανική εταιρία *First Hand Records*. Ποιητικά βιβλία: *Paternoster Square* (Πόλις, βραβείο περιοδικού Αναγνώστη), *Σκληρό να σκοντάφτεις σε πέτρες* (Πατάκης, βραβείο Women of the Year), *Αίθουσα των χαμένων βημάτων* (Πόλις, βραβείο Ιδρύματος Αικ. Λασκαρίδη). Μεταφράσεις: *Όταν ο Νους σου βράζει κι η Καρδιά* (Πατάκης), *Σαίξπηρ: Τα Σονέτα* (Gutenberg) *Πούσκιν: Τσάρος Σαλτάν* (Πατάκης), *Θραύσματα από τον Κάφκα* (Νεφέλη) Η Λένια είναι παραγωγός του Τρίτου Προγράμματος της ΕΡΤ και καθηγήτρια τραγουδιού στο Ωδείο Αθηνών. (www.leniasafiropoulou.gr)

Ο **Χρήστος Σιορίκης** (1989) μεγάλωσε στο Αγρίνιο και ζει στην Αθήνα. Σπούδασε στο Παιδαγωγικό Τμήμα Δημοτικής Εκπαίδευσης του Πανεπιστημίου Αθηνών και εξειδικεύτηκε στη διδακτική της λογοτεχνίας. Εργάζεται ως δάσκαλος ισπανικών σε παιδιά και ενήλικες. Έχει επιμεληθεί δημιουργικές δραστηριότητες στη βιβλιοθήκη του Ινστιτούτου Θερβάντες της Αθήνας, ενώ χρησιμοποιεί την πεζογραφία και την ποίηση για τη διδασκαλία της ισπανικής γλώσσας. Μελετά το έργο του Ζαχαρία Παπαντωνίου και έχει επιμεληθεί το αφιέρωμα της *Νέας Εστίας* στο συγγραφέα το 2016. Συμμετέχει στην Ομάδα Άστυ, η οποία διοργανώνει περιπάτους στην Αθήνα με στόχο τη γνωριμία με το τοπίο και την ιστορία της πόλης μέσα και από τη λογοτεχνία. Γράφει ποιήματα και μεταφράζει από τα ισπανικά. Έχει μεταφράσει, μεταξύ άλλων, το βιβλίο του Χούλιο Κορτάσαρ *Ο Λόγος της Αρκούδας*, για την εικονογραφημένη έκδοσή του (Εκδόσεις Πάπυρος, 2015). Ποιήματά του έχουν δημοσιευτεί σε περιοδικά και ανθολογίες και έχουν μεταφραστεί στα γερμανικά και στα εβραϊκά. Το βιβλίο του *Η Πρώτη Φορά* κυκλοφορεί από τις εκδόσεις Αντίποδες (2018) και ήταν υποψήφιο για το Βραβείο του περιοδικού *Αναγνώστη* για πρωτοεμφανιζόμενο ποιητή.

Η **Αντριάνα Καλφοπούλου** έχει συγγράψει τρείς ποιητικές συλλογές, δύο συλλογές δοκιμίων και αρκετά δημοσιεύματα. Το *History of Too Much* (Red Hen Press, 2018) είναι η πιό πρόσφατη της δημοσίευση. Ποιήματά της, δοκίμια, και συλλογές έχουν εμφανιστεί σε διαδικτιακές και έντυπες εκδόσεις όπως το *The Harvard Review* online, *Hotel Amerika, Slag Glass City, Superstition Review* μεταξύ άλλων. Η συλλογή ποιημάτων της *Xeni, Xenos, Xenitia* (Melani Press 2013) έχει μεταφραστεί στα Ελληνικά υπο την επιμέλεια της ποιήτριας Κατερίνας Ηλιοπούλου. Ποιήματά της έχουν συμπεριληφθεί σε ανθολογίες όπως *Futures: Poetry of the Greek Crisis* (Penned in the Margins, 2015) και *Borderlands and Crossroads: Writing the Motherland* (Demeter Press, 2016). Διδάσκει Αμερικανική Λογοτεχνία και Δημιουργική Γραφή στο Αμερικανικό Κολλέγιο Ελλάδας- Deree College και θα κατέχει τη θέση της καθηγήτριας Δημιουργικής Γραφής στο Davidson College για την περίοδο 2020-2021.

Η **Αλίσια Ε. Στόλινγκς** σπούδασε κλασική φιλολογία στο Πανεπιστήμιο της Τζόρτζια και στο Πανεπιστήμιο της Οξφόρδης. Ζει στην Αθήνα από το 1999. Έχει εκδώσει τέσσερις ποιητικές συλλογές: *Archaic Smile* (University of Evansville Press), που τιμήθηκε με το Βραβείο Richard Wilbur, *Hapax* (TriQuarterly Books), *Olives* (TriQuarterly Books), που βρέθηκε στη μικρή λίστα για το Εθνικό Βραβείο του Κύκλου Κριτικών Βιβλίου και *Like* (Farrar, Straus & Giroux), το οποίο ήταν υποψήφιο στη μικρή λίστα για το Βραβείο Pulitzer. Έχει δημοσιεύσει στους Penguin Classics μια μετάφραση σε στίχους του φιλοσοφικού έπους του Λουκρήτιου *De Rerum Natura* και το *Έργα και Ημέραι* του Ησιόδου, που ήταν υποψήφιο για το Βραβείο Runciman. Μια εικονογραφημένη και σχολιασμένη μετάφρασή της του ψευδοομηρικού έργου *Βατραχομυομαχία* θα κυκλοφορήσει προσεχώς από τις εκδόσεις Paul Dry books. Η Στόλινγκς έχει λάβει υποτροφία μετάφρασης από το National Endowment of the Arts, έχει τιμηθεί με το Willis Barnstone Translation Prize, το Βραβείο Ποιητών το 2008 και το Βραβείο Benjamin H. Danks Award από την Αμερικανική Ακαδημία Τεχνών και Επιστημών. Μέλος της Αμερικανικής Ακαδημίας Τεχνών και Επιστημών, ήταν το 2011 υπότροφος των Ιδρυμάτων Guggenheim και MacArthur. Ζεί με τον δημοσιογράφο Γιάννη Ψαρόπουλο και τα δύο παιδιά τους, Ιάσονα και Αταλάντη.

Ο **Γιάννης Δούκας** γεννήθηκε στην Αθήνα το 1981. Σπούδασε φιλολογία στο Εθνικό και Καποδιστριακό Πανεπιστήμιο και Digital Humanities στο King's College του Λονδίνου. Τα τελευταία χρόνια έζησε στο Γκόλγουεϊ της Ιρλανδίας, όπου και εκπονούσε τη διδακτορική του διατριβή με θέμα τη διακειμενικότητα στην επική ποίηση της Ύστερης Αρχαιότητας και τις μεθόδους ψηφιακής της αναπαράστασης. Έχει γράψει τα βιβλία: *Ο κόσμος όπως ήρθα και τον βρήκα* (σύντομα πεζά, Κέδρος, 2001), *Στα μέσα σύνορα* (ποιήματα, Πόλις, 2011, Βραβείο 'Πρωτοεμφανιζόμενου Ποιητή' του Περιοδικού *Διαβάζω*) και *Το Σύνδρομο Σταντάλ* (ποιήματα, Πόλις, 2013, Βραβείο 'Γ. Αθάνα' της Ακαδημίας Αθηνών). Ποιήματά του έχουν ανθολογηθεί και μεταφραστεί στα αγγλικά, τα γαλλικά, τα γερμανικά, τα σερβικά, τα ολλανδικά και τα πολωνικά. Δυο ποιήματα από το βιβλίο *Στα μέσα σύνορα* («Δυο δρόμοι» και «Στην εποχή του κάτι σαν») μελοποιήθηκαν από τον Νίκο Πλατύραχο στον δίσκο του *Τα Άστεγα* (2015). Δυο τραγούδια σε στίχους του («Στην ίδια πόλη υπό βροχήν» και «Μπιλιάρδο») έχουν συμπεριληφθεί *Στην ομίχλη των καιρών*, δίσκο του Θάνου Μικρούτσικου (2017). Μεταφράζει από τα αγγλικά. Έχει συνυπογράψει, με τον Χάρη Βλαβιανό, τη μετάφραση των 'ερωτικών ποιημάτων' του e.e. cummings με τον τίτλο *λοιπόν ας φιληθούμε* (Πατάκης, 2014). Κείμενά του έχουν δημοσιευτεί σε ηλεκτρονικά και έντυπα μέσα, κυρίως στο *Διαβάζω* και την *Ποιητική*, στης οποίας τη συντακτική ομάδα μετέχει.

Η **Παυλίνα Μάρβιν** γεννήθηκε στην Αθήνα το 1987, αλλά μεγάλωσε στην Ερμούπολη της Σύρου. Σπούδασε Ιστορία στο Εθνικό και Καποδιστριακό Πανεπιστήμιο Αθηνών. Εκπονεί διδακτορική διατριβή με θέμα την εθνική πολιτική για το βιβλίο. Υπήρξε συνεκδότρια του περιοδικού *Τεφλόν*. Μελέτησε ποίηση στο διετές εργαστήριο του Ιδρύματος Τάκη Σινόπουλου (2007-2009). Το πρώτο της βιβλίο, *Ιστορίες απ' όλον τον κόσμο μου*, κυκλοφόρησε από τις Εκδόσεις Κίχλη (2017) και τιμήθηκε με το βραβείο πρωτοεμφανιζόμενου ποιητή της Εταιρείας Συγγραφέων «Γιάννης Βαρβέρης». Ως συγγραφέας και performer, έχει συμμετάσχει σε πολλά διακαλλιτεχνικά γεγονότα και φεστιβάλ, στην Ελλάδα και το εξωτερικό. Μέρος της δουλειάς της έχει μεταφραστεί στην αγγλική, γαλλική, γερμανική, ιταλική, ισπανική, ινδική (Bengali) και σερβοκροάτικη γλώσσα.

# SELECTED BIBLIOGRAPHY

## MUTED

**Photo by Derek Wagon, Mnisikleous street, Plaka (1961)**

**Poems by Katerina Anghelaki-Rooke**

✦ *Ἄφωνος – Mute*

✦ *Ὁ Τελευταῖος Ἐρωτικός – Epitaph On Love* (1975)
  From the collection: *The Body is the Victory and the Defeat of Dreams*
✦ *Μεταφράζοντας σὲ ἔρωτα τῆς ζωῆς τό τέλος – Translating into Love Life's End* (2004)
  From the collection: *Translating into Love Life's End*
  [All translations by the author]

**Poems by Liana Sakelliou**

✦ *Ἄφωνοι – Muted*
  [translated by the author]

✦ *'Ὑδατογραφία' – 'Aquarelle'*
✦ *Ἄσκηση φλαμανδοῦ δασκάλου – Study by a Flemish Master*
  Ἀπό τήν συλλογή: *Πάρε με σάν φωτογραφία*
  [translated by David Connolly]

**Poems by Orfeas Apergis**

✦ *Η ταφή – The burial*
✦ *Το τραπέζι - The Table*
  From the collection 'Y'
  [All translations by the author]

# LULL

**Illustration: Angela Lyras, *Love Returned* (2014), colour photograph.**

## Poems by Haris Vlavianos

- ✦ *Κυκλαδικό Ειδύλλιο – Cycladic Idyll*
- ✦ *Δοξασίες του Αυγούστου – August Meditations*
- ✦ *Το Πέπλο – The Veil*
  From the Collection: *Affirmation - Selected Poems* 1986 – 2006 (Dedalus Press, 2007)
  [All translations by the author]

## Poems by Katerina Iliopoulou

- ✦ *Νανούρισμα – Lullaby*
  [translated by the author]

- ✦ *Ταίναρον – Tainaron*
- ✦ Το τραγούδι της Ευρυδίκης – The song of Eurydice
  Από την συλλογή: *Το βιβλίο του χώματος* (Μελάνι, 2011)
  [Translated by John O'Kane] *The Book of the Soil* (Melani, 2011)

## Poems by Ginger F. Zaimis

- ✦ *Lull – Κοπάζει*
  [μετάφραση: Ορφέα Απέργη]

- ✦ *Penel -o- pic – Πηνελο (ε) πικό*
  *Psyche's cup – Κύπελλο Ψυχής*
  From the collection: *Philosophy and Poetry* (2014)
  (The Library of Alexandria - International Friends of Bibliotheca Alexandrina, Greece)

# RISK

**Illustration: Laline Pierrakos, *Blue is for Victory* (2017) collage on paper.**

## Poems by Panayotis Ioannidis

✦ *Αυτός δεν είναι ένας δερβίσης – This is not a Dervish*
[translated by the author]

✦ *Που Χάθηκε - Lost* from the book *The Lifesaver*
[Translated by Panayotis Ioannidis and Stefanos Bacigal] Kastaniotis Editions, (2008)

✦ *Αποκαθήλωση – The Deposition* from the book *Uncovered*
[Translated by Clare Durey - Kastaniotis Editions, 2013]

## Poems by Kyoko Kishada

✦ *Μάχη η Φυγή – Fight or Flight*
[translated by the author]

✦ *Basho Bass* Haiku [translated by the author]
(Published in *Teflon*, issue no 3, summer 2010)

✦ *Λωτοφάγοι- The Lotus Eaters* From the collection *Austerity Measures,* edited by Karen Van Dyck
[Translated by Rachel Hadas] -Published by *The New Greek Poetry* Penguin, 2016; New York Review
of Books, (2017)

## Poems by Theodoros Chiotis

✦ *Αφορισμός – Aphorism*
✦ *'21'* First published in *NICE! Is return possible?*
(ed. Yiannis Grigoriadis & Yiannis Isidorou. *Lo and behold!* Athens, 2016)
✦ *'Ατρακτος (Παραλλαγή) - Spindle (variation)* First published in the *Hypnos* issue
(Onassis Cultural Centre: Athens, 2016)
[All translations by the author]

# GRAFT

**Illustration: Eliza Jackson, *Untitled* (2004), watercolour.**

**Poems by Lenia Safiropoulou**

All poems are without title.

[All translations by Orfeas Apergis]

**Poems by Christos Siorikis**

✦ *Ομοιόσταση' – Homeostasis*
   [translated by Chris Sakellaridis]

✦ *Called Back* published in *The Books' Journal* (vol. 72, 2016)
✦ *Γερμανική Σωτήρια - German Salvation* published in Farmako (vol. 9, 2017)
   [translated by Panayotis Ioannidis]

**Poems by Adrianne Kalfopoulou**

✦ *Sentence – Απόφαση*
   [translated by Katerina Iliopoulou]

✦ *Fall Grapes - Φθινοπωρινά Σταφύλια,*
   from the collection *Passion Maps* published by Red Hen Press (2009) [translated by Korina Gougouli]
✦ *The History of Too Much - Η Ιστορία της Υπερβολής*
   from her latest collection A History of Too Much
   Published by Red Hen Press (2018) [translated by Katerina Iliopoulou]

# VERGE

**Illustration: Apolonia Sokol, *Abdy* (2016), oil on linen. Courtesy of *THE PILL* Gallery.**

## Poems by A. E. Stallings

✦ *Sanctuary of Artemis at Brauron (a prayer for daughters)* – Ιερόν της Αρτέμιδος εν Βραβρώνι (προσευχή για κόρες)
[translated by Orfeas Apergis]

✦ *Sunset, Wings* - Δύση, Φτερά Poem 11, first published in *LIKE*, 2019
(Finalist for The Pulitzer Prize for Poetry) [translated by Panayotis Ioannidis]

✦ 'Συμβουλές στην Ψυχή/Hem audax et temeraria lucerna' - *Advice to Psyche/Hem audax et temeraria lucerna* - Συμβουλές στην Ψυχή/Hem audax et temeraria lucerna
Poem 39, first published in *OLIVES*, 2012, [translated by Panayotis Ioannidis]

## Poems by Yiannis Doukas

✦ *Η Ατλαντική Κατάστασή μου* - *My Atlantic Slate*
[Translated from Greek by the poet himself]

✦ *Η Σαπφώ Νοικοκυρά* - *Sappho The Housewife* [Translated from Greek by the poet himself, with help from Sam Buchan-Watts.]
Πρώτη δημοσίευση: Στα μέσα σύνορα (Πόλις, 2011), σελ. 33.
First published in English in *Five Dials 27* (September 2013)

✦ *Επιτάφιος* – *Epitaph* Πρώτη δημοσίευση: Το Σύνδρομο Σταντάλ (Πόλις, 2013), σελ. 61.
[translated by Theodoros Chiotis]
First published in *Futures: Poetry of the Greek Crisis* (Penned in the Margins, 2015)

## Poems by Pavlina Marvin

✦ *'(24)'*
[translated by Anastasia Lambropoulou, Edited by Dimitris Gkioulos]

- ✦ *'(4)'*
  [translated by Christiana Mygdali]
  First published in Rosa Luxemburg Institute's Anthology *ξύπνησα σε μια χώρα* (2019)
  From *Turn off the lighthouses for Ivan Ismailovic/ Σβήστε τους φάρους για τον Ιβάν Ισμαήλοβιτς* (unpublished)
- ✦ *Ουκρανική Ιστορία - Ukrainian Story* [translated by Panayotis Ioannidis]
  From *Ιστορίες απ' όλον τον κόσμο μου,* Kichli Publications, Athens, (2017)
  First published in Greek in *Εντευκτήριο* Magazine (Thessaloniki)
  First published in English by *Und*, Athens.

Historical sources for Koutzalexis House: Leonidas Kallivretakis, www.eie.gr
Photos by Panos Kokkinias of details from the decorative ceilings at the Koutzalexis House, cover, p. xv and p.137.

# ACKNOWLEDGEMENTS

We extend our sincere gratitude to the people who have inspired and nurtured this endeavour, amongst them our three mentors mentioned at the incipit of this work: Katerina Anghelaki-Rooke, Liana Sakelliou and Orfeas Apergis, as well as the additional thirteen poets who took part in this first five-year phase of our poetic evolution.

Very many thanks to all those who helped *A Poets' Agora* in so many different ways leading to this publication; Mihalis Georgiou and Gdesign who have given our yearly edition its amazing aesthetic, as well as Mihalis Montesantos for his generous and relentless support in editing and translating many of our publications, and our special gratitude to all others who encouraged and assisted in the creation of this unique and diverse anthology.

Lastly, we personally extend our wholehearted appreciation to all the less visible people who have collaborated at the actual happening, helping the setting and actualizing of the event with their artworks, seating, film and sound engineering, online streaming, cooking and all the small details that have made these evenings so special.

Karine Leno Ancellin and Angela Lyras

*A Poets' Agora* is a registered non-profit association promoting
Greek, and Greek-inspired poetry. Each year, in late autumn,
we host a convivial bilingual event destined for the wider
literary community in Athens. The readings take the form
of an open conversation around the poems of three
contemporary authors. The intra-poetic discussions of
*A Poets' Agora* echo the timeless
and polymorphous art of the word.

*A Poets' Agora* conducts a Residency program during the spring of each year.

*A Poets' Agora* is an initiative of Karine Leno Ancellin and Angela N. Lyras

Contact: apoetsagora@gmail.com

www.apoetsagora.com

YouTube channel: A Poets Agora
Facebook page: A Poets Agora

To order additional copies of this book, contact:
Xlibris
844-714-8691
www.Xlibris.com
Orders@Xlibris.com

CPSIA information can be obtained
at www.ICGtesting.com
Printed in the USA
BVHW022113270921
617630BV00002B/14